QA 280 G1F

Probability and Mathematical Statistics
LEHMANN • Testing Statistical Hypoth...
LEHMANN • Theory of Point Estimation
MATTHES, KERSTAN, and MECKE • Infinitely Divisible Point Processes
MUIRHEAD • Aspects of Multivariate Statistical Theory
OLIVER and SMITH • Influence Diagrams, Belief Nets and Decision Analysis
PRESS • Bayesian Statistics: Principles, Models, and Applications

BECK and ARNOLD • Parameter Estimation in Engineering and Science
BELSLEY, KUH, and WELSCH • Regression Diagnostics: Identifying Influential Data and Sources of Collinearity

*Now available i...   ...assics Library

(cardplate continues in back of book)

Nonlinear Multivariate
Analysis

Nonlinear Multivariate Analysis

Albert Gifi
University of Leiden
The Netherlands

JOHN WILEY & SONS
Chichester • New York • Brisbane • Toronto • Singapore

Copyright © 1990 by John Wiley & Sons Ltd.
Baffins Lane, Chichester
West Sussex PO19 1UD, England

Reprinted with corrections August 1991

Other Wiley Editorial Offices

John Wiley & Sons, Inc., 605 Third Avenue,
New York, NY 10158-0012, USA

Jacaranda Wiley Ltd, G.P.O. Box 859, Brisbane,
Queensland 4001, Australia

John Wiley & Sons (Canada) Ltd, 22 Worcester Road,
Rexdale, Ontario M9W 1L1, Canada

John Wiley & Sons (SEA) Pte Ltd, 37 Jalan Pemimpin   05-04,
Block B, Union Industrial Building, Singapore 2057

Library of Congress Cataloguing-in-Publication Data:

Gifi, Albert
  Nonlinear multivariate analysis / Albert Gifi.
    p. cm.
  Includes bibliographical references.
  ISBN 0 471 92620 5
  1. Multivariate analysis.   2. Nonlinear theories.   I. Title.
QA278.G535   1990
519.5'35—dc20   89-24944
                                                    CIP

British Library Cataloguing in Publication Data:

Gifi, Albert
  Nonlinea multivariate analysis.
  1. Multivariate analysis. Applications of computer
  systems. Programming
  519.5'35'028851

  ISBN 0 471 92620 5

Printed in Great Britain by Antony Rowe Ltd, Chippenham, Wiltshire

FOREWORD

This book by Gifi is a nonstandard treatment of some of the more important parts of multivariate data analysis. A heavy emphasis has been placed on plotting observations and/or variables in low-dimensional Euclidean space. This clearly shows its psychometric origins, because of the obvious relationships with factor analysis and component analysis. Much less emphasis than usual is paid to regression and analysis of variance, and the various multivariate generalizations of the linear model.

There are other ways in which the book is nonstandard. It is organized, as many other multivariate data analysis books are, around various key techniques, such as principal component analysis, canonical analysis, and so on, but in this book these techniques are identified, much more than is usually done, with computer programs and the type of output they reproduce. A technique is defined, more or less, as a tool into which you feed data of a particular type and format, and which then reproduces output of a particular type and format. This very instrumental interpretation of data analysis is contrasted, in many places, with the model-oriented approach of classical statistics. Models do have a place in Gifi's philosophy: not as tentative approximations to the truth, but as devices to sharpen and standardize the data analytic techniques. At the same time the inferential methods of statistics are explicitly introduced as one of various ways in which the stability of data analysis tools can be investigated.

Of course we have to realize that there is, according to many statisticians, one appropriate way to analyse data: first formulate a model on the basis of prior knowledge, then compute the likelihood of the data given the model, and then estimate and test the appropriateness of the model by maximizing the likelihood function. There are some variations on this scheme, depending on what exactly this prior knowledge includes, but these variations are minor compared to the prescription seemingly advocated by Gifi. First choose a technique (implemented in a computer program) on the basis of the format of your data, then apply this technique, and study the output. The prior knowledge is not mentioned explicitly, but more importantly it is not isolated in any one particular time and place in the data analysis process. The classical recipe suggests that prior knowledge (theory, prejudice, experience, expectations) enters in the formulation of the model, but then it should not be used at all during the other stages of the process. These are supposed to be mechanical, almost as applying the gigantic computer program MAXLIK, which handles data of any type, or the even more gigantic program TBAYES, which takes data and prior information as input and transforms this to posterior information. In the book by Gifi many other places where prior information, and expertise, play a role in data analysis are indicated. Not all thought and creativity of the investigator, and the history of his science, is concentrated in the mysterious phase in which the model is formulated.

The techniques actually discussed in the book have a very strong emphasis on geometry, in the sense that they make pictures of data. As we said, this places the techniques rather firmly in the tradition of factor analysis, or perhaps even more clearly in the tradition of multidimensional scaling. Another relationship with multidimensional scaling is that variables are transformed optimally, using monotone transformations or splines, in order to improve the fit of the low-dimensional representation. This is different from traditional factor analysis, in which the fit is improved by computing additional factors and then rotating them to simple structures. Gifi incorporates the additional factors in the nonlinear transformations, and keeps the pictures simple and preferably in two or at most three dimensions. The pictures are, of course, descriptive devices. In most cases they have to represent an enormous amount of information, and they can approximate that by showing the dominant relations in the data, its most coarse features. This may be useless, or even offensive, to people who want specific answers to specific questions. Again the psychometric (or social science) origins of Gifi's system become apparent here, because in the large scale survey type of investigations in which multivariate methods are likely to be used the questions are usually not very specific.

Gifi was one of the first investigators to use the bootstrap and jackknife to study the stability of multivariate representations. The reasons for this are clear: the amount of prior knowledge one needs to use these techniques is minimal, especially if one uses the perturbation interpretation discussed in Chapter 12 of the book. If we use the idea that a statistical technique is a tool, that can be applied in many different situations, although it is not necessarily optimal for any one specific situation, then this fits in rather nicely with the ideas behind bootstrapping, jackknifing, cross validating, and using randomization tests. The tools have their own quality control devices built in, as it were, and because of that they become more useful and more informative at relatively little extra cost.

The application of a computer program, or a set of related computer programs, as a statistical toolbox around which various applications can be organized, is much more common now than it was during the Gifi project. We now have books built around GLIM, BMDP, the DataDesk, SPSS, and so on. The emphasis on computations to replace assumptions, and the use of resampling and Monte Carlo in that context, is very important in modern statistics. Optimal transformations occur with great regularity in the statistical journals these days, in combination with semi-parametric models, splines, smoothing, and Box–Cox families. Correspondence analysis (sometimes disguised as ACE) can now be discussed among consenting statisticians, without causing frothing at the mouth and the throwing of stones. It was not always like this! The Gifi project started as a one-person effort in 1968. It became a more-than-one-person effort in 1974 and a quite-a-lot-of-persons effort in 1978. It was generously supported by ZWO, the Netherlands Organization for the Advancement of Pure Research, but was frowned upon by official statisticians, and led to the types of semi-religious debates that typically occur in the foundations of statistics area. That this resistance was eventually largely overcome was not in the least due to internal developments in the discipline of statistics itself.

There are a few persons I would like to thank personally, on behalf of Albert Gifi, of course: in the first place John Van de Geer, who taught us the geometric approach to multivariate analysis that pervades the book, who provided us with a Department to work in, and with the freedom we needed; secondly, Doug Carroll and Joe Kruskal at Bell Laboratories in Murray Hill, who invited several of us to come over there and who provided access to many resources, tools and organizations; thirdly, Forrest Young and Yoshio Takane, who provided invaluable assistance at a very critical point in the development of these techniques; fourthly, Chris Haveman, of SPSS, who remembered from his days at the Computing Center of the University of Leiden that people there were doing useful things in multivariate data analysis; and finally, Richard Gill, formerly of the Center for Mathematics and Informatics in Amsterdam, who provided valuable support and connections to the world of academic statistics. There are innumerable other persons who provided various services at various points in time, and I would like to thank them collectively as well.

As explained elsewhere in the book, Gifi is not quite dead yet. Many of the topics mentioned in the book, but not worked out in detail there, have been taken up in dissertations and research monographs. The basic approach has been extended to path models and dynamic systems, and variations based on multinormal likelihood theory are also available. Links to multidimensional scaling have been strengthened, as have the links to classical multivariate statistical theory. However, the book, this book, is the only comprehensive, programmatic, encompassing statement of the general Gifi philosophy, and the way in which it is implemented on the digital computer.

Los Angeles, 22 April, 1989   Jan de Leeuw

PREFACE

'It's difficult to understand why statisticians commonly limit their
inquiries to Averages, and do not revel in more comprehensive
views. Their souls seem as dull to the charm of variety as that of
the native of one of our own flat English counties, whose
retrospect of Switzerland was that, if its mountains could be
thrown into its lakes, two nuisances would be got rid of at once.
An average is but a solitary fact, whereas if a single other fact be
added to it, an entire Normal Scheme, which nearly corresponds
to the observed one, starts potentially into existence.
'Some people hate the very name of statistics, but I find them full
of beauty and interest. Whenever they are not brutalised, but
delicately handled by the higher methods, and are warily
interpreted, their power of dealing with complicated phenomena is
extraordinary. They are the only tools by which an opening can be
cut through the formidable thicket of difficulties that bare the path
of those who pursue the Science of man.'

Francis Galton (1889), *Natural Inheritance,* London: Macmillan
(from D.W. Forrest (1974), *Francis Galton: The Life and Work of
a Victorian Genius,* New York: Taplinger Publishing Co.)

This volume is a revised version of a mimeographed text that was brought into
limited circulation in June 1981. At that time it served as reading material for
the post-doctoral course *Nonlinear Multivariate Analysis,* prepared by the
Department of Data Theory and attended by approximately fifty participants
from Belgium and The Netherlands. A previous Dutch version (Albert Gifi,
Niet-lineaire Multivariate Analyse, Leiden, 1980) served a similar purpose for
a course held in March 1980. Since then a large number of other courses and

seminars, both at home and abroad, have been built around the 'Gifi system' (a name coined by Jan de Leeuw in 1984 when the endeavour showed signs of losing momentum), and major parts of the software have been developed into reliable and portable procedures. In the course of time the 1981 version of the text saw five completely unrevised impressions.

Who is Albert Gifi? From the preface of the 1981 text we quote: 'The text is the joint product of the members of the Department of Data Theory of the Faculty of Social Sciences, University of Leiden. Albert Gifi is their nom de plume. The portrait, however, of Albert Gifi shown here is that of the real Albert Gifi to whose memory this book is dedicated, as a far too late recompense for his loyalty and devotion, during so many years, to the Cause he served.'

Since the composition of a department changes over time it is proper to lift the veil of anonymity a little bit. The members of the Gifi team who contributed to the original texts and programs are Bert Bettonvil, Eeke van der Burg, John van de Geer, Willem Heiser, Jan de Leeuw, Jacqueline Meulman, Jan van Rijckevorsel, and Ineke Stoop. Assistance in organizing the Gifi courses was provided by Steef de Bie, Judy Knip, and Dré Nierop. Editorial assistance in an earlier enterprise to revise the 1981 text was provided by Peter van der Heijden and Adriaan Meester. Throughout the years Peter Neufeglise has supported our group by lending his patient technical assistance to many problems on numerous computer systems; he has also been responsible for software maintenance. Renée Verdegaal contributed to the development of OVERALS, only a rough outline of which was documented in the 1981 book. Peter Verboon contributed to the present volume by redoing numerous graphics, and Ivo van der Lans assisted in proof-reading. Gerda van den Berg and Patrick Groenen took care of the final stage of converting the software into SPSS procedures.

In March 1987 the present editorial team started with the final revision. First we merely aimed at two things: (a) correcting typing errors and (b) getting the distribution of sections and paragraphs more even (while reading we could perhaps just as well make some notes for an index). However, soon more desiderata popped up, like having a more uniform level of presentation and getting the notation consistent across chapters. We also thought of making fresh drawings of the figures, checking all analyses with updated programs, and so on. These ambitions almost led to a dead end, until we seized the opportunity – starting in December 1987 – to stay a couple of times in the cultural centre 'De Pauwhof' in Wassenaar, where we found the right atmosphere to fully concentrate on the job. We decided to put Chapter 4 (old) after Chapter 8 (old) for reasons of balance, and we adjusted all notation and cross-referencing as well as we could. We wrote a number of short introductions and added extra explanations where the original text really was obscure. An important decision was made to refrain from inserting new insights and/or references to new work in the body of the text. In our opinion this would have resulted in an unbalanced mixture of the original version with a possible start of a completely new book. We chose to add a section called 'Epilogue' to each chapter; the epilogues give some guidance to new developments, especially when it is inspired by – or relevant to – the Gifi system, without trying to be exhaustive. When the sign '➻' appears in the text comments on the current topic may be found in the last section of the chapter.

After eight years of informal circulation Gifi's work has now reached the present form. We are in debt to students, colleagues, and friends for many helpful corrections and suggestions during this time. Especially the detailed remarks by Shizuhiko Nishisato, Jos ten Berge, Charley Lewis, and Ivo Molenaar were very welcome. A final word of thanks is due to Susañña Verdel, who was involved in rounds and rounds of word processing, and who finished the job with great accuracy and care.

The Editors

De Pauwhof,
Wassenaar, The Netherlands
9 May, 1989

Willem Heiser
Jacqueline Meulman
Gerda van den Berg

CONTENTS

CHAPTER 1
CONVENTIONS AND CONTROVERSIES
IN MULTIVARIATE ANALYSIS

In this introductory chapter we shall try to give a definition of multivariate analysis (MVA), taking into account the definitions given by others. We do this by giving a brief content analysis of the prefaces and introductory chapters of some of the more popular books on MVA. It turns out that there is considerable agreement over the definition, but the main reason for this is that the definition which is most frequently used is not very specific. In our content analysis of the various books we repeatedly find a number of problems or controversies that seem interesting enough to be analysed in more detail. This is done in the second part of the chapter.

After the discussion of controversies a number of data analytic principles is stated which indicate our position in these matters and which form the basis of this book. Essentially, we adopt the rather pragmatic point of view that there is no real opposition between a model-driven and a technique-driven approach to data analysis; models can be used as *gauges* for a technique, and techniques can be accompanied by *stability studies* under specific models. Next we state our position on a number of recurrent problems that are specific for MVA:

(a)   Should we always be guided by the multinormal distribution, or not?
(b)   How do we summarize the strength of bivariate association among multiple categorical variables?
(c)   How do we deal with the empty cell problem in large multidimensional contingency tables?
(d)   Is it possible to infer causal relationships in the absence of experimental manipulation or control?

This discussion leads to a definition of MVA that emphasizes the *asymmetric role* of variables and objects (units of analysis, cases) and treats population aspects and sample aspects on an equal footing.

There are two ways to distinguish specific forms of MVA that are characteristic of this book. The first one is a distinction into linear, monotone, and nonlinear MVA; the second one is a distinction into bivariate, multivariate, and joint bivariate MVA. After explaining these distinctions the chapter concludes by outlining the major technical ingredients on which the book relies heavily. These include join and meet problems, join and meet loss functions, and join and meet techniques, as well as the use of optimal scaling and alternating least squares.

1.1 CONTENT ANALYSIS OF MVA BOOKS

1.1.1 Roy (1957)

Roy does not pretend to give a complete coverage of MVA. In most of the chapters he analyses multinormally distributed variables, and more specifically he derives bounds for confidence intervals for certain classes of parametric functions. He mentions factor analysis, classification, and variance component analysis as his most important omissions. It is especially important for our purposes that Roy's chapter 15 is about nonparametric generalizations of MVA, by which he means techniques very similar to the modern loglinear models for multidimensional contingency tables:

> Despite all the mathematical elegance and comparative simplicity of 'normal variate' analysis of variance and multivariate analysis, one cannot help feeling that the nonparametric approach (whether of this variety or of other varieties) is far more realistic and physically meaningful, and is likely, in the future, to supplant, to a large extent, the existing techniques of 'normal variate' analysis of variance and multivariate analysis, including those discussed in the first fourteen chapters of this monograph (Roy, 1957, p. viii).

Although Roy does not give an explicit definition, it seems that he interprets MVA as a class of techniques that can be used to test a restricted number of hypotheses about the relationships between correlated variables.

1.1.2 Kendall (1957, 1975)

Kendall does give a definition: 'We may thus define multivariate analysis as the branch of statistical analysis which is concerned with the relationships of sets of dependent variables' (Kendall, 1957, p. 6). He also gives a number of examples of typical MVA problems, and he says that such a list of problems is usually more informative than a simple definition. In his book he makes the important distinction between the analysis of *dependence* and the analysis of *interdependence*. In the analysis of dependence we investigate if and how a group or set of variables depends on another group. The first set consists of the so-called *dependent* variables, the second group of the *independent* variables. There is a certain amount of *asymmetry* in the analysis, or, to put it differently, the direction of causal influence is from the independent to the dependent variables. In the analysis of interdependence the sets of variables are treated symmetrically; there is no distinction between dependent and independent variables. The main example of interdependent analysis is principal components analysis, the most familiar example of dependent analysis is multiple regression.

In the modernized version of Kendall's book, published in 1975, the introduction is considerably longer. The definition of MVA has not changed, and is repeated with slightly different phrasing on page 1. Kendall then proceeds to list the most important purposes of MVA techniques:

(a) Structural simplification
(b) Classification
(c) Grouping of variables
(d) Analysis of dependence
(e) Analysis of interdependence
(f) Construction and testing of hypotheses

He also mentions a number of important problems in the further development of MVA techniques:

(a) In many practical situations we cannot make the usual statistical assumptions. There is often no random sample from a population; either the population is not defined or the sample is not random. 'It is a mistake to try and force the treatment of such data into a classical probabilistic mould, even though some subjective judgment in treatment and interpretation may be involved in the analysis of the results' (Kendall, 1975, p. 4).
(b) Although it is practically impossible to apply MVA without the use of a computer, there are certain dangers in the uncritical use of 'canned' computer programs.
(c) Even if we have a random sample it is often impossible to assume multivariate normality: 'Most theoretical work in multivariate statistics is based on the assumption that the parent population is multinormal, and its robustness under departures from normality is very often difficult to determine with any exactitude' (Kendall, 1975, p. 4).

(d)   Pictures and graphs are very important in MVA. It is, however, very difficult to make satisfactory graphic representations of data in more than two dimensions.

(e)   There is no single, natural definition of order between points in multi-dimensional space. As a consequence there is no satisfactory nonparametric MVA, comparable to univariate nonparametric statistics.

We remember from Kendall's discussion that in MVA we deal with observations on a number of interdependent variables. However, the variables are not necessarily stochastic, not necessarily continuously measurable, and the observations are not necessarily independent and identically distributed. Finally, it is remarkable that Kendall, like Roy, has a last chapter on the analysis of categorized or discrete multivariate data. There is no such chapter in the other books we discuss. If we compare the two editions of Kendall's book we find, certainly in the introductory chapter, a shift from multivariate statistical analysis and multinormal analysis to a more general description of MVA.

1.1.3 Anderson (1958)

'In this book we shall concern ourselves with statistical analysis of data that consist of sets of measurements on a number of individuals or objects. ... The mathematical model on which analysis is based is a multivariate normal distribution or a combination of multivariate normal distributions' (Anderson, 1958, p. 1). This is certainly clear enough. It is interesting to study Anderson's reasons for this apparent loss of generality:

(a)   The multinormal distribution often turns out to be a good description of the distribution of multivariate data arising in practice.

(b)   The multinormal distribution follows from the multidimensional central limit theorem, and is consequently a good approximation if the observations can be interpreted as the sum of a large number of independent small effects.

(c)   Distribution theory based on the assumption of multinormal parent populations is mathematically relatively simple, and consequently many interesting results have been derived in the last 75 years.

Anderson interprets MVA in most of his book as a generalization of familiar normal theory inferential statistics to multinormal situations. As a consequence, typical multivariate techniques such as principal components analysis and canonical analysis do not get much attention in his book. ⇥

1.1.4 Cooley and Lohnes (1962, 1971)

The first version of the book demonstrates how the usual MVA techniques must be implemented on a computer. This means, of course, that it is now completely out of date. The definition of MVA is the same as Kendall's. The second version, published in 1971, has an interesting preface. Somewhere between 1962 and 1971 Cooley and Lohnes found their identity: they are now multivariate data analysts. This explains the change in the title of the book. The reason for the change is, of course, Tukey. 'His gift to us was our professional identity. ... Tukey made our interest in multivariate heuristics rather than multivariate inference sound respectable' (Cooley and Lohnes, 1971, p. v). This is a remarkable statement. We must remember to investigate why social scientists studying MVA did not have a professional identity before 1962, and we must also find out if the only effect of Tukey's paper (1962) and subsequent efforts was that some shady practices now *sound* respectable.

Cooley and Lohnes use Kendall's definition in their 1971 book; they also use the distinction between analysis of dependence and interdependence throughout the book. There is much popularization, application, and computation in the book.

1.1.5 Morrison (1967, 1976)

Morrison's definition is more or less standard: 'Multivariate statistical analysis is concerned with data collected on several dimensions of the same individual' (Morrison, 1967, p. vii). Morrison imposes more restrictions than Kendall, however, because he assumes explicitly that the individuals are a random sample from an infinite population, and he assumes multivariate normality almost everywhere. These are possibly the reasons why Morrison, like Anderson, uses the term 'multivariate *statistical* analysis'. The major difference from Anderson is that Morrison has an extensive treatment of principal components analysis, canonical analysis, and factor analysis, and that principal components analysis in particular is described as a descriptive data-reduction technique. Morrison's book also has an extensive introduction to matrix algebra. The changes in the 1976 edition are not very interesting for our purposes.

1.1.6 Van de Geer (1967, 1971)

On the cover of the Dutch version of Van de Geer's book we read: 'Multivariate analysis, being the art of describing the relationship between several variables by using mathematical techniques'. In Cooley and Lohnes (1962) we saw emphasis on computer programs, no statistics, and only a little algebra. In Van de Geer's book there is also no statistics, but matrix algebra takes up almost half the book. MVA only comes at the end: 'In these last chapters multivariate analysis is discussed as a data-reduction technique only. In other words the book does not discuss statistical questions in the sense of inferential statistics' (Van de Geer, 1967, p. 12). The computer is mentioned as an important source of inspiration; the time we do not have to spend any more behind the adding machine can now be used to gain insight into what we are really doing. It is clear that the insight that Van de Geer is trying to teach is mainly geometrical; wherever possible he uses figures. Bringing insight is truly bringing *in sight*.

In the much enlarged English edition of 1971 the starting points are stated even more clearly: 'In my opinion, statistical theory is a substantially more advanced subject than is required to understand what multivariate techniques really do with data' (Van de Geer, 1971, p. ix). Van de Geer mentions the main advantages of his approach to MVA: insight instead of a mere copying of computer output, emphasis on the similarities of the various forms of MVA, and the ensuing 'cross-fertilization' between the various social sciences that use some form of MVA. Important additions in the English version are path analysis and structural equations, discussed as two new variations on the same theme.

According to Van de Geer, MVA is linear analysis of data matrices, and its most important purposes are data-reduction and geometric representations. MVA is applied linear algebra, or, which amounts to the same thing, applied linear geometry. ➤ The emphasis on pictures (also pictures in the form of graphs with arrows) is also to be found, to a lesser degree, in Cooley and Lohnes (1971) and Kendall (1975). Van de Geer does not seem to have trouble with his professional identity; in fact his geometrical approach fits into the psychometric tradition starting with multiple factor analysis and culminating with the book of Coombs (1964).

1.1.7 Dempster (1969)

Dempster's book has a fine introductory chapter. We find the following description there:

> The purpose of this book is to describe certain methods of analysis of statistical data arising from multivariate samples. A basic aim of such data analysis is to reduce large arrays of numbers to provide meaningful and reasonably complete summaries of whatever information resides in sample aggregates. Another aim is to draw inferences from sample aggregates to larger population aggregates from which the samples are drawn; that is, to understand how certain information about a sample provides uncertain information about a population (Dempster, 1969, p. 3).

Thus data analysis and statistics are distinguished from the start. They are also treated separately in the book, starting with data analysis:

> While most books on statistical theory start out with sampling theory and attempt to make methods of data analysis follow, the attitude in this book is that the methods of data analysis are carried out largely because of the intrinsic appeal of the sample quantities computed. Such, at least, were the historical origins of the methods described here. Moreover, when viewed as producing descriptive or summary statistics, the methods have value even when assumptions like randomness of samples and normality of populations are quite unwarranted (Dempster, 1969, p. 3).

The separate treatment gives the impression that the two approaches have nothing in common. In his introduction Dempster states that statistics can be used to show that some data analysis techniques have attractive properties or to show how data analysis techniques can be improved. In the book itself we do not find any examples of this type of interaction.

Before he starts his treatment of MVA, Dempster, like Van de Geer, gives an extensive introduction to the theory of matrices and finite-dimensional linear spaces. The introduction tries to avoid the use of coordinates, and is consequently very geometrical. The computational translation of the linear operations is discussed separately. Dempster also gives a list of three important omissions:

(a)   Categorical data
(b)   Specialized techniques such as factor analysis and structural equations
(c)   Cluster analysis and related subjects

As a consequence of these omissions Dempster's data analysis is limited to the discussion of various computational aspects of the linear model. He mentions principal components analysis and canonical analysis and there is an example of the explorative use of canonical analysis, but these typical multivariate techniques do not get much attention.

1.1.8 Tatsuoka (1971)

There are two different definitions of MVA on the first page of this book: 'Multivariate statistical analysis, or *multivariate analysis* for short, is that branch of statistics which is devoted to the study of multivariate (or multi-dimensional) distributions and samples from those distributions' (Tatsuoka, 1971, p. 1). Tatsuoka says that this is the definition used by the statistician and that it is not very useful because it sounds tautological. We may add that it also sounds rather imperialistic to define multivariate analysis as a mere abbreviation of multivariate statistical analysis. The second definition is the one used by the data analyst:

> In applied contexts, particularly in educational and psychological research, multivariate analysis is concerned with a group (or several groups) of individuals, each of whom possesses values or scores on two or more variables such as tests or other measures. We are interested in studying the interrelations among these variables, in looking for possible group differences in terms of these variables, and in drawing inferences relevant to these variables concerning the populations from which the sample groups were chosen (Tatsuoka, 1971, p. 1).

The book contains much matrix algebra and a fair amount of statistics. There is considerably less geometry than in Van de Geer or Dempster; the approach of the linear model is the same as in Anderson or Morrison. Familiar univariate techniques are generalized as far as possible: 'Pointing out the analogy between a given multivariate technique and the corresponding univariate method is one of the principal didactic strategies used throughout this book' (Tatsuoka, 1971, p. 3). The main disadvantage of this approach, as we have seen before, is that the multivariateness of the data is treated as a sort of nuisance and that it is difficult to fit typical multivariate techniques without univariate analogues into the framework. Principal components analysis consequently does not get much attention but canonical analysis is discussed fairly extensively, although the treatment is statistical, as in Anderson, and based on the multinormal model.

1.1.9 Harris (1975)

There is a certain pattern in our analyses so far. In the more recent books the data analytic aspects of MVA get more attention and the limitations of inferential multinormal analysis are stated more clearly. The book by Harris is in some respects a reaction, a sort of inferential backlash. There is an interesting first chapter, defending the inferential approach. Harris even has the courage to defend the null hypothesis and the associated significance test, while this illustrious duo is considered to be dead and buried in most data analytic circles. We do not discuss his arguments in detail here, but we shall come back to some of them further on in this chapter.

Harris' reasoning roughly goes as follows. Statistical methods are a form of quality control for scientific production. They are necessary because it has been demonstrated many times that the opinion of the investigator is not a sufficient basis for believing in the generalizability of his results. We need formal methods to study generalizability. Formal methods can be classified as descriptive or inferential. Inferential methods are designed specifically to prevent conclusions being drawn that are not generalizable, conclusions that are only a consequence of particularities of the specific sample or the specific technique:

> As will become obvious in the remaining sections of this chapter, the present Primer attempts in part to plug a 'loophole' in the current social control exercised over researchers' tendency to read too much from their data. It also attempts to add a collection of rather powerful techniques to the descriptive tools available to behavioral researchers. Van de Geer (1971) has in fact written a textbook on multivariate statistics which deliberately omits any mention of their inferential applications (Harris, 1975, p. 4–5).

What are the tools that Harris uses for this plumbing job? In the first place he chooses the now-familiar starting point that MVA techniques generalize the existing univariate techniques. They often do this by replacing groups of variables by single variables which are linear combinations of the variables in the group. The coefficients of the linear combinations are chosen to optimize some intuitively or statistically attractive criterion. Data analysts are interested primarily in the optimal coefficients, they call them 'loadings' (of the observed variables on the constructed variables), while statisticians are interested primarily in the maximum value of the criterion. However, it is clear that the two approaches complement each other; one way of computing the maximum is by explicitly computing the maximizer. Harris consequently emphasizes optimality criteria with a relatively simple statistical interpretation, which leads naturally to optimum coefficients. Thus he prefers the union–intersection approach of Roy (1957) to the more usual likelihood ratio method, because this last method usually does not give a unique set of coefficients and consequently frustrates the data analyst. Matrix algebra does not get much attention in Harris' book, because it is only a technique for efficient optimization, and obviously there is almost no geometry in the book. Multivariateness is a nuisance, even more so than in Tatsuoka's book; multivariate data must be made univariate as soon as possible. Harris' attempt to integrate multivariate statistical analysis and multivariate data analysis is interesting, but doomed to fail. We have already seen that data analysts are not primarily interested in coefficients, but in pictures. A single set of coefficients does not give very interesting pictures.

1.1.10 Dagnelie (1975)

It is nice to read the usual definition in French for a change, but if we translate it back into English it sounds a bit disappointing; 'In a general sense analysis of several variables or multidimensional analysis or multivariable analysis or multivariate analysis can be interpreted as a set of statistical techniques which have the purpose of studying the relationships that exist between several dependent or interdependent variables' (Dagnelie, 1975, p. 11). Dagnelie gives this definition after comparing the definitions given by Anderson, Cooley and Lohnes, Kendall, Morrison, and Press. This is somewhat disappointing, his book is oriented heavily to the Anglo-American literature, it does not introduce anything new, and it does not show a typical French approach to MVA. Dagnelie discusses the usual MVA techniques, using the classification criterion that for each technique the variables are partitioned into sets containing one or more variables. There is not much algebra, not much geometry, a fair amount of cookbookery statistics, and there are some nice examples from ecology: 'The user obviously must have a sufficiently precise idea what the general principles and the conditions for application of these methods are, but he must primarily concentrate his attention on the interpretation of the results he has obtained' (Dagnelie, 1975, p. 18).

1.1.11 Green and Carroll (1976)

This is a different type of book. At first sight it does not even belong in our list, because its explicit purpose is to add some useful tools to the mathematical toolbox of the researcher. On second sight, however, the book wants considerably more, and it can be interpreted as fitting in the trend we have discovered already in Van de Geer. MVA is applied linear algebra and linear geometry. There is a nice introductory chapter, in which the ideas of the authors are explained. We give some representative quotations: 'Completion of this book should provide both a technical base for tackling most application-oriented multivariate texts and, more importantly, a geometric perspective for aiding one's intuitive grasp of multivariate methods' (Green and Carroll, 1976, p. xii), 'In function, as well as in structure, multivariate techniques form a unified set of procedures that can be organized around relatively few prototypical problems' (p. 1), 'The heart of any multivariate analysis consists of the data matrix, or in some cases, matrices. The data matrix is a rectangular array of numerical entities whose informational content is to be summarized and portrayed in some way' (p. 3), 'To a large extent, the study of multivariate techniques is the study of linear transformations' (p. 14).

The message is clear. If these mathematical tools are understood, then one can recognize MVA techniques in their various disguises, and one can construct or choose the appropriate MVA technique that one needs in any practical situation. There are no statistics in the book. In the preface we get the impression that Green and Carroll see their book as an introduction to Tatsuoka, or Harris, or Morrison, but it is quite possible that some readers of the book come to the conclusion that they do not need any additional education. This is why we have interpreted the book as a book about MVA, and not only about the mathematical toolbox. Of course, Green and Carroll may not agree with our point of view. ⇥

1.1.12 Cailliez and Pagès (1976)

In France data analysis is very popular, mainly because of the work and the influence of J.P. Benzécri. The book by Cailliez and Pagès is about this French form of data analysis, which differs in some important respects from the Anglo-American form. French linear algebra, for example, is considerably more modern and abstract, and uses coordinates and matrices to a far lesser extent. This is in the tradition of modern French mathematicians such as Bourbaki, Dieudonné, Cartan, and Choquet. The book by Cailliez and Pagès has a very extensive introduction to this type of linear algebra, and has very extensive chapters on regression, principal components analysis, canonical analysis, and on Benzécri's version of metric multidimensional scaling. There is also a chapter on correspondence analysis, a technique for nonlinear principal components analysis developed by Benzécri, which is also one of the main techniques discussed in this book. There is a useful introductory chapter on sets, relations, and functions, and a useful final chapter on classification and clustering, but by far the largest part of the book is on applied linear algebra: 'We have made a definite choice in our presentation of data analytic techniques: we constantly use the notion of a projector, of an M-symmetric application, of quantification, and above all the "duality diagram", which is an efficient and comprehensive notational device to present all aspects of the techniques of linear algebra' (Cailliez and Pagès, 1976, p. vii). The book has an interesting preface by G. Morlat, which discusses the differences between data analysis and classical statistics in considerable detail. We shall come back to that discussion later in the chapter.

1.1.13 Giri (1977)

Giri's starting point is the invariance of testing problems under groups of transformations, the book can be interpreted as an application of aspects of decision theory in multivariate situations. The emphasis is on hypothesis testing, not on estimation. Everything is normally distributed. There is some material on matrix algebra and transformation groups in the book, but very little material on principal components analysis, factor analysis, and canonical analysis.

1.1.14 Gnanadesikan (1977)

The definition of MVA is the usual one, but the approach used in the book is quite unusual. Gnanadesikan is a data analyst:

> Much of the theoretical work in multivariate analysis has dealt with formal inference procedures, and with the associated statistical distribution theory, developed as extensions of and by analogy with quite specific univariate methods, such as tests of hypotheses concerning location and/or dispersion matrices. The resulting methods have often turned out to be of very limited value for multivariate data analysis (Gnanadesikan, 1977, p. 2).

Gnanadesikan emphasizes graphical representations, but of a slightly different nature than the ones we have met earlier. The book uses probability plots and related graphics, there is less linear algebra than in Cailliez and Pagès and less matrix algebra than in Van de Geer, and there is considerable attention for goodness-of-fit (of the multinormal distribution) and for outlier detection.

Gnanadesikan lists the most important purposes of MVA techniques:

(a)   Reduction of dimensionality
(b)   Study of dependence
(c)   Multidimensional classification
(d)   Investigation of statistical methods
(e)   Data reduction and clarification

He also mentions the most important problems, in a list that resembles the one given by Kendall:

(a)   It is difficult to find out what the client wants to know exactly.

(b)   If we have decided on going multivariate, it is difficult to determine the number of variables.

(c)   Even with the present generation of computers there is a number of MVA techniques that can only be applied to data with a relatively small number of variables and/or individuals.

(d)   Multidimensional pictures and graphs can easily become complicated and confusing.

(e)   There is no natural order in multidimensional space.

1.1.15 Kshirsagar (1978)

This is a thick book, and also a very good book, but nevertheless we can be brief. Multinormal statisticians have been very active since Anderson. A lot of computing has been done and many expansions have more terms now than in 1958. New test statistics have been thought out and new techniques have been developed to derive complicated distributions more elegantly. Kshirsagar's book is the most complete summary of these multinormal results so far. There are no examples, there is no factor analysis, and principal components analysis and canonical analysis are treated multinormally: 'The theory of multivariate analysis developed so far almost invariably assumes that the joint distribution of the random variables is a multivariate normal distribution' (Kshirsagar, 1978, p. 1). If one *defines* multivariate analysis as multinormal analysis this is a trivial conclusion; if one defines multivariate analysis as multivariate statistical analysis it is incorrect because of the existence of multivariate multinomial analysis, already discussed by Kendall and Roy. Kshirsagar mentions as most important omissions factor analysis, multiple time series, and categorical data. He thinks that regression analysis is the most important statistical technique, which is a typical multinormal point of view. On the other hand he does mention dimension reduction as an important purpose of MVA:

> The aim of the statistician undertaking multivariate analysis is to reduce the number of variables by employing suitable linear transformations and to choose a very limited number of the resulting linear combinations in some optimal manner, disregarding the remaining linear combinations in the hope that they do not contain much significant information. The statistician thus reduces the dimensionality of the problem (Kshirsagar, 1978, p. 2).

As in the book of Harris this point of view dictates the preferred technique. Harris, a one-dimensional multivariate analyst, uses the union–intersection approach and the largest root criterion. Kshirsagar uses the Bartlett partitioning of the multinormal likelihood ratio statistic.

1.1.16 Thorndike (1978)

Thorndike considers himself to be a data analyst, like Cooley and Lohnes. In his data analytical work he often meets the situation where his clients have a fair comprehension of analysis of variance techniques, but only a little understanding of techniques based on correlation coefficients. It is not difficult to guess that Thorndike's clients have been mis(in)formed by the ancient rituals usually known as 'statistics for psychologists'. Thorndike proposes his book as an antidote: 'The approach is largely intuitive and geometric rather than mathematical ...' (Thorndike, 1978, p. vi). This is a somewhat unfortunate distinction. The next quotation is less controversial:

> The organization of the book reflects a conviction that understanding can be most readily developed by showing the essential unity and orderly progression of concepts in multivariate statistics. ... The geometric interpretation of the correlation coefficient as the angle between two vectors in a people space is readily generalizable to multiple and canonical correlation (Thorndike, 1978, p. vii).

There is very little mathematics in the book, as promised: 'The work of other authors, notably Quinn McNemar, Harry Harman, and Bill Cooley and Paul Lohnes, is frequently cited for the reader who wants or needs the mathematical foundations of the topics discussed' (Thorndike, 1978, p. vii). Incredibly, this is a real quotation. Thorndike's book could very well become *the* book on MVA for the *fifties*. It is remarkable that the reader who needs the mathematical foundation must go a very long way. Thorndike refers him to Cooley and Lohnes, who refer him to Tatsuoka, who refers him to Morrison, who refers him to Anderson. In the meantime the unfortunate seeker has gone back twenty years in time, and has finally arrived at a book of unquestionable quality, but completely about multinormal analysis. It would perhaps be better to refer directly to a book about linear algebra.

Thorndike also uses the Bartlett–Kendall distinction between analysis of interdependence and dependence. He calls this internal and external factor analysis, which makes factor analysis a form of internal factor analysis. There is also some useful information in the book, for example about interpretation of canonical analysis results.

1.2 CORRESPONDENCE ANALYSIS OF TABLES OF CONTENT

A somewhat more objective analysis of the contents of the MVA books we have discussed is also possible. In Table 1.1 we have indicated the number of pages in the books for each of the following seven subjects:

MATH: *Mathematics other than statistics, i.e. linear algebra, matrices, transformation groups, sets, relations*

CORR: *Correlation and regression, including path analysis, linear structural and functional equations*

FACT: *Factor analysis and principal components analysis*

CANO: *Canonical correlation analysis*

DISC: *Discriminant analysis, classification, cluster analysis*

STAT: *Statistics, including distributional theory, hypotheses testing, and estimation; also statistical analysis of categorical data*

MANO: *MANOVA, and the general multivariate linear model*

Table 1.1 Number of pages of MVA books devoted to several subjects

		MATH	CORR	FACT	CANO	DISC	STAT	MANO
1	Roy (1957)	31	0	0	0	0	164	11
2	Kendall (1957)	0	16	54	18	27	13	14
3	Kendall (1975)	0	40	32	10	42	60	0
4	Anderson (1958)	19	0	35	19	28	163	52
5	Cooley and Lohnes (1962)	14	7	35	22	17	0	56
6	Cooley and Lohnes (1971)	20	69	72	33	55	0	32
7	Morrison (1967)	74	0	86	14	0	84	48
8	Morrison (1976)	78	0	80	5	17	105	60
9	Van de Geer (1967)	74	19	33	12	26	0	0
10	Van de Geer (1971)	80	68	67	15	29	0	0
11	Dempster (1969)	108	48	4	10	46	108	0
12	Tatsuoka (1971)	109	13	5	17	39	32	46
13	Harris (1975)	16	35	69	24	0	26	41
14	Dagnelie (1975)	26	86	60	6	48	48	28
15	Green and Carroll (1976)	290	10	6	0	8	0	2
16	Cailliez and Pagès (1976)	184	48	82	42	134	0	0
17	Giri (1977)	29	0	0	0	41	211	32
18	Gnanadesikan (1977)	0	19	56	0	39	75	0
19	Kshirsagar (1978)	0	22	45	42	60	230	59
20	Thorndike (1978)	30	128	90	28	48	0	0

Of course it is easy to criticize this classification. There are some subjects in the same category that are marginally related (such as discriminant analysis and cluster analysis). Moreover canonical analysis in Anderson or Kshirsagar is quite different from canonical analysis in Van de Geer or Cailliez and Pagès. We have chosen this classification because we wanted to be able to classify most of the material, while we do not want too many zero entries in the data matrix. Of course our selection of MVA books is also not exactly a random sample. We have included some of the books that are used most, some personal favourites, some rather extreme books, and some others that happened to be in the library at the time. The entries in Table 1.1 are probably

Table 1.2 Projections of books, projections of subjects, and singular values from correspondence analysis of Table 1.1

1	ROY	−1.11	−0.61	−0.34
2	KEN 1	0.07	0.70	0.25
3	KEN 2	−0.21	0.46	−0.49
4	ANDE	−0.78	−0.11	0.16
5	COL 1	0.03	0.41	1.05
6	COL 2	0.36	0.70	0.09
7	MOR 1	−0.16	−0.16	0.46
8	MOR 2	−0.25	−0.20	0.39
9	GEE 1	0.73	−0.19	−0.05
10	GEE 2	0.68	0.24	−0.18
11	DEMP	0.03	−0.37	−0.44
12	TATS	0.27	−0.45	0.28
13	HARR	0.02	0.51	0.52
14	DAGN	0.12	0.48	−0.19
15	GREC	1.08	−1.33	0.03
16	CAPA	0.65	−0.07	−0.13
17	GIRI	−0.98	−0.40	−0.25
18	GNAN	−0.40	0.33	−0.34
19	KSHI	−0.75	0.08	−0.01
20	THOR	0.57	0.81	−0.35
A	MATH	1.14	−1.53	0.07
B	CORR	0.80	1.36	−1.25
C	FACT	0.29	0.98	0.65
D	CANO	0.26	0.87	0.84
E	DISC	0.26	0.48	−0.79
F	STAT	−1.56	−0.50	−0.61
G	MANO	−0.67	0.21	2.51
	Singular value	0.60	0.51	0.34

also not very reliable, because we have added number of pages of various chapters, and chapters are usually not completely about one single subject. Somebody else, using the same categories, would certainly arrive at different page counts. Not too different, we hope.

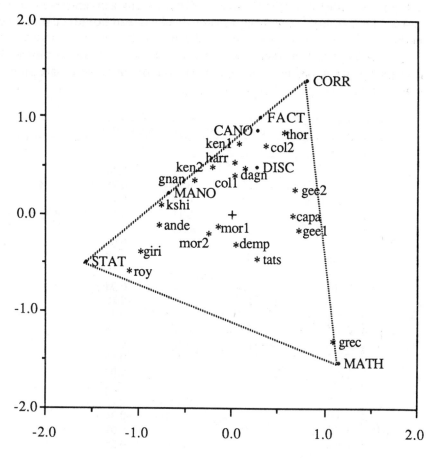

Figure 1.1 Correspondence analysis of MVA books (complete).

We applied correspondence analysis to Table 1.1. We shall not discuss the precise nature of this technique here as this will be done in Chapter 8 of this book. ➤ For now it is sufficient to know that correspondence analysis is a form of multidimensional scaling, which represents row and column points in a low-dimensional joint space. Two books are close together in the space (or in the picture) if they have similar contents; two subjects or topics are close if they occur in the same books in the same degree; and a book is close to a subject if the book pays relatively much attention to the subject. In correspondence analysis as we have used it each book point is a weighted mean of subject points, with the weights given by the entries of Table 1.1. Thus Table 1.1 tells us that ROY = [(31×MATH) + (164×STAT) + (11×MANO)] / (31 + 164 + 11). It follows that the books are all in the convex hull of the subject points.

The projections of the books and the subjects on the three most important dimensions are given in Table 1.2. Correspondence analysis is a singular value–singular vector technique, in which the importance of a dimension is

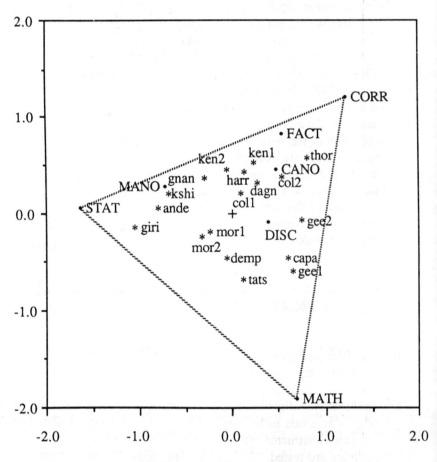

Figure 1.2 Correspondence analysis of MVA books (reduced).

defined as the value of the corresponding singular value. The three largest singular values are given in the last row of Table 1.2, the three remaining singular values are 0.21, 0.17 and 0.10. The projections on the first two dimensions are also represented in Figure 1.1. In this figure we have drawn a triangle with corners CORR, STAT, and MATH. The clearest, most pure representatives of these three subjects are, respectively, THOR, ROY, and GREC. Most of the books are on the line between CORR and STAT; these are the most classical books which vary, mainly, in degree of difficulty. Typical data analysis books are near the line between CORR and MATH, at least if we define data analysis in this context as applied linear algebra. Graphical data analysis as in GNAN fits better into the classical scale CORR–STAT. In the interior of the triangle we find four books which pay a lot of attention to both statistics and linear algebra. The book COL1 is not in the 'correct' position (i.e. not where we expect it – we expect it closer to COL2 and THOR); the third dimension in Table 1.2 shows that this happens because COL1 gives much more attention to MANO than we expect on the basis of its 'technical level'. Of course, COL1 is a manual for computing MVA solutions and its treatment of MANO is not at all technical. We have also repeated the analysis without the 'extremists' GREC and ROY. The projections on the first two dimensions are shown in Figure 1.2.

The position of the remaining books stays roughly the same, but the first two singular values are now considerably smaller (0.56 and 0.38). The third singular value is 0.33, about the same as before, and the third dimension continues to contrast COL1 with the rest. It is clear that correspondence analysis shows us the same dimensions as our content analysis, only in a more compact and comprehensive form. Of course this is due partly to our selection of the books.

1.3 A SHORT SUMMARY AND SOME PROBLEMS

Some important points emerge from our discussion of the MVA books. In the first place there seems to be a certain difference, or even conflict, between the statistical and the data analytical approach to MVA. The statistical approach starts with a statistical model, usually based on the multinormal distribution. The model is assumed to be true, and within the model certain parametric hypotheses are constructed. The remaining free parameters are estimated and the hypotheses are tested. The data analytic approach does not start with a model, but looks for transformations and combinations of the variables with the explicit purpose of representing the data in a simple and comprehensive, and usually graphical, way. It is certainly necessary to discuss the role and the meaning of the statistical model with the corresponding tests and estimators. In particular we want to know more about the importance of the multinormal distribution for MVA.

A second problem is the importance of categorical data or, in other terms, the role of nominal and ordinal variables in MVA. The only books that pay at least some attention to categorical variables are the ones by Roy and Kendall; in the other books they only occur as codings in the context of MANOVA and discriminant analysis. In this case they are often called 'dummy variables'. In general they are interpreted as nonstochastic and are part of the design matrix, and thus independent variables. Actually they are not even variables; they are only used to write down parametric multinormal models in compact matrix notation. A possible exception to this rule is the notion of quantification in the book by Cailliez and Pagès.

In our list of MVA books we have restricted ourselves to general, often introductory, books with a considerable amount of overlap. There is also a number of books that deal especially with the analysis of multivariate categorical data. The most important ones are Haberman (1974), Bishop, Fienberg, and Holland (1975), Gokhale and Kullback (1978), Fienberg (1977), and Goodman (1978). The content of these books is generally quite different from those of the books we have discussed. The models are formulated in the tradition of classical statistics, but the emphasis is on discrete distributions and on asymptotic methods. Linear algebra and geometry are not important at all. It is clear that these books constitute a completely different tradition, which is a bit strange because both forms of MVA start directly from the work of Fisher and Pearson. In classical general handbooks of statistics, such as those of Fisher, Yule and Kendall, Cramér, Kendall and Stuart, Wilks, and Rao, the two forms of MVA are both treated, but they are not or almost never related to each other. There is a gap between the discrete and the continuous approach.

Of course there is also an exception to this rule – the books by Kullback (1959) and Lancaster (1969). Kullback discusses statistical procedures based on the information–theoretic measures of divergence, which are distance measures between parametric distributions, defined in exactly the same way for discrete and continuous distributions. To some extent the same thing is true for other works in theoretical statistics that rely heavily on general exponential models of which both the multinormal and the multinomial are special cases. Lancaster's book is partly about the decomposition of multivariate probability distributions by using orthogonal functions on the marginals, and this technique can be applied to both discrete and continuous distributions too. Another problem we shall have to discuss in this chapter is the relationship between discrete and continuous MVA.

There are also some books that are, at least in some respects, similar to our book. We mention them here because they also do not fit into the list we used earlier in the chapter. Volume II of the treatise by Benzécri (1973) deals exclusively with theory and applications of correspondence analysis. There is also a very interesting recent book by Nishisato (1980) which deals with the applications of 'dual scaling' to various forms of categorical data. The amount of overlap with our book is considerable as far as results are concerned, mainly with our Chapters 3 and 8. Our general approach to MVA, however, is quite different from that of Nishisato.

1.4 DATA ANALYSIS AND STATISTICS

1.4.1 Tukey's definition of data analysis

We start by discussing the ideas of John Tukey, who was the first to present data analysis as an independent discipline, distinct from statistics:

> All in all, I have come to feel that my central interest is in *data analysis*, which I take to include, among other things: procedures of analyzing data, techniques for interpreting the results of such procedures, ways of planning the gathering of data to make its analysis easier, more precise or more accurate, and all the machinery and results of (mathematical) statistics which apply to analyzing data.
> Large parts of data analysis are inferential in the sample-to-population sense, but these are only parts, not the whole. Large parts of data analysis are incisive, laying bare indications which we could not perceive by simple and direct examination of the raw data, but these too are only parts, not the whole. Some parts of data analysis, as the term is here stretched beyond its philology, are allocation, in the sense that they guide us in the distribution of effort and other valuable considerations in observation, experimentation, and analysis. Data analysis is a larger and more varied field than inference, or incisive procedures, or allocation (Tukey, 1962, p. 2).

Tukey's reasons for not wanting to be called a statistician any more are clear from these quotations. In the first place, data analysis is more general than inferential statistics and, in the second place, there are parts of (mathematical) statistics that are outside data analysis:

> To the extent that pieces of *mathematical statistics* fail to contribute, or are not intended to contribute, even by a long and tortuous chain, to the practice

of data analysis, they must be judged as pieces of *pure* mathematics, and criticized according to its purest standards. Individual parts of mathematical statistics must look for their justification toward either data analysis or pure mathematics. Works which obey neither master, and there are those who deny the rule of both for their own work, cannot fail to be transient, to be doomed to sink out of sight (Tukey, 1962, p. 3).

Tukey is obviously not very happy with the historical development of statistics:

What is needed is progress, and the unlocking of certain rigidities (ossifications?) which tend to characterize statistics today. Whether we look back over this century, or look into our crystal ball, there is but one natural chain of growth in dealing with a specific problem of data analysis, viz:

(a1')   recognition of problem;
(a1")   one technique used;
(a2)   competing techniques used;
(a3)   rough comparisons of efficacy;
(a4)   comparison in terms of a precise (and thereby inadequate) criterion;
(a5')   optimization in terms of a precise, and similarly inadequate criterion;
(a5")   comparison in terms of several criteria.
         (Number of primes does not indicate relative order.)

If we are to be effective in introducing novelty, we must head two main commandments in connection with new problems:

(A)   Praise and use work which reaches stage (a3), or only stage (a2), or even stage (a1").
(B)   Urge the extension of work from each stage to the next, with special emphasis on the earlier stages.

One of the signs of the lassitude of the present cycle of data analysis is the emphasis of many statisticians upon certain of the later stages to the exclusion of the earlier ones. Some, indeed, seem to equate stage (a5') to statistics – an attitude which if widely adopted is guaranteed to produce a dried-up, encysted field with little chance of real growth (Tukey, 1962, p. 7).

The point of view that (a5') can be identified with statistics is discussed in somewhat more detail:

The view that 'statistics is optimization' is perhaps but a reflection of the view that 'data analysis should not *appear to* be a matter of judgment'. Here 'appear to' is in italics because many who hold this view would like to suppress these words, even though, when pressed, they would agree that the optimum *does* depend on the assumptions and criteria, whose selection may, perhaps, even be admitted to involve judgment (1962, p. 9).
   In data analysis we must look to a very heavy emphasis on judgment. At least three different sorts or sources of judgments are likely to be involved in almost every instance:

(a1) judgment based upon the experience of the particular field of subject matter from which the data come,
(a2) judgment based upon a broad experience with how particular techniques of data analysis have worked out in a variety of fields of application,
(a3) judgment based upon abstract results about the properties of particular techniques, whether obtained by mathematical proofs or empirical sampling (Tukey, 1962, p. 9).

The most important maxim for data analysis to heed, and one which many statisticians seem to have shunned, is this: 'Far better an approximate answer to the *right* question, which is often vague, than an *exact* answer to the wrong question, which can always be made precise.' Data analysis must progress by approximate answers, at best, since its knowledge of what the problem really is will at best be approximate (Tukey, 1962, pp. 13–14).

Tukey's paper has been very influential, for several reasons. In the first place, as we have seen in Section 1.1.4, it gave a number of people their professional identity:

It was a stroke of genius to realize that to render 'a deed without a name' respectable, you should name it (or perhaps I should say rename it), and we are all grateful for the name 'Data Analysis'. This important part of our subject can now be studied without apology or shame, and courses on it are taught and may be attended by consenting adults (Box, 1979, p. 3).

In the second place Tukey's paper has pointed to some really controversial problems in the development of mathematical statistics. His conclusions have been repeated and stated even more forcefully by Wolfowitz (1969):

Except perhaps for a few of the deepest theorems, and perhaps not even these, most of the theorems of statistics would not survive in mathematics if the subject of statistics itself were to die out. In order to survive the subject must be more responsive to the needs of application (p. 748).
What we must guard against is the development of a theory which, on the one hand, bears little or no relation to the actual problems of statistics, and which, on the other hand, when viewed as pure mathematics, is not interesting per se nor likely to survive (p. 749).

It is also clear that the same problems are the reason why the *Annals of Mathematical Statistics* have been replaced by two journals: the *Annals of Probability,* and the *Annals of Statistics.*

On the other hand there has always been a considerable difference between 'English' and 'American' statisticians. Wolfowitz (1969, p. 745) severely criticizes Barnard, because he has questioned the value of rigorous proofs in one of his papers. Kiefer (1964) is highly critical of the monumental treatise of Kendall and Stuart: 'My main criticism of the book is that it has a very high density of errors in statements and proofs. Another negative aspect is the exclusion, in a work of this encyclopaedic nature, of much of the content and almost all of the spirit of modern mathematical statistics' (Kiefer, 1964, p. 1371). Kendall, on the other hand, views the situation quite differently:

> The early statisticians of the present century were competent at mathematics, but they were not great creative mathematicians. Karl Pearson was trained in mathematics, but Edgeworth was a classical scholar and Yule an engineer by training. Fisher, who *was* a creative mathematician, criticized his predecessors for the clumsiness of their style; but even he wrote in the tradition of English mathematics, which does not care much about extreme generalization or extreme rigor as long as it gets the right answer to its problems. The consequence was that, with a few exceptions, theoretical statistics in the forties could be understood by anybody with moderate mathematical attainment, say at the first year undergraduate level. I deeply regret to say that the situation has changed so much for the worse that the journals devoted to mathematical statistics are now completely unreadable. Most statisticians deplore the fact, but there is not very much they can do about it (Kendall, 1972, p. 205).

As a consequence, the English statisticians, who do not identify with the Annals and with the emphasis on extreme generality, are not in favour of replacing 'statistics' by 'data analysis':

> Whereas the contents of Tukey's remarks is always worth pondering, some of his terminology is hard to take. He seems to identify 'statistics' with the grotesque phenomenon generally known as 'mathematical statistics' (Anscombe, 1967, p. 3).
>
> The elevation of Data Analysis to its proper place as a subject meriting serious study makes me as happy as I would be if some neglected but important activity of the carpenter, such as the use of the saw or the chisel, had at last received proper recognition and study. But my enthusiasm for the naming of Data Analysis does not extend to the renaming of Statisticians as 'Data Analysts', any more than I should be happy to hear a carpenter described as a sawyer or a chiseler (Box, 1979, p. 3).

It is clear that Box, in this quotation, identifies data analysis with what Tukey has called *exploratory* data analysis, which corresponds with the incisive aspects of general data analysis of his 1962 paper.

In order to prevent possible misunderstandings we also state our own position on these issues. We agree with Anscombe and others that data analysis in the broad sense of Tukey's 1962 paper is identical with statistics, and that it is consequently unnecessary to rename all statisticians. We agree with Tukey and Wolfowitz that mathematical statistics is interesting to the extent that it has practical applications or consequences, but we also think that even very abstract and general work *can* have these practical consequences, although it will sometimes be difficult to trace the exact path of influence that a paper, or class of papers, has. To be more precise, it is very difficult for us to point out any paper in the *Annals* that has surely not influenced the practice of data analysis. We agree with Kiefer and Wolfowitz that if a proof is given then it should be rigorous. Many of the controversies over 'what Fisher meant', both in statistics and genetics, are simply due to the fact that Fisher skipped steps and did not state restrictions or assumptions. We shall come back to the role of optimization in a later section.

1.4.2 Benzécri's definition of data analysis

The proper translation of 'analyse des données' seems to be data analysis. However, the principles of French data analysis, explained by Benzécri (1973), are quite different from those of the Tukey school. We translate them first, and then give some brief comments.

> *Principle 1:* Statistics is not the same thing as probability theory. Under the name 'mathematical statistics' several writers (who, I tell you this in French, seldom write in our language ...) have built a pompous discipline, which abounds in hypotheses that are never satisfied in practice. We cannot expect a solution of our typological problems from these authors.

> *Principle 2:* The model must follow the data, and not the other way around. This is another error in the application of mathematics to the human sciences: the abundance of models, which are built *a priori* and then confronted with the data by what one calls a 'test'. Often the 'test' is used to justify a model in which the number of parameters to be fitted is larger than the number of data points. And often it is used, on the contrary, to strongly reject as invalid the most judicial remarks of the experimenter. But what we need is a rigorous method to extract structure, starting from the data.

> *Principle 3:* It is convenient to treat simultaneously information on as many dimensions as possible. As a consequence the problem of the validity of a 'test' – which, we admit, is sometimes troublesome – is not that important any more. Nobody knows if the inequality $0.5 \neq 0.7$ must be interpreted in these practical cases as a certain empirical result, or as if it is only a result of chance. But finding that in a space of two dimensions fifty points are approximately arranged on a circle is certainly a discovery (at least if the computing method does not deceive us!).

> *Principle 4:* For the analysis of complex facts and notably for the analysis of social facts we cannot do without the computer. This principle is obviously true ... but what would our Gaulish fathers have thought about it fifteen years ago?

> *Principle 5:* Using a computer means that all techniques designed before the advent of automatic computing must be abandoned. I say techniques, not science: the geometric and algebraic principles of our programmes were known to Laplace, 150 years ago. But Laplace was also the author of a treatise on celestial mechanics, that has just been republished for usage by space

technicians. ... And that treatise did not suffice for Napoleon to conquer the moon!

(The translation is as literal as possible; the principles are on pages 3, 6, 9, 12, and 15 of Benzécri, 1973.)

The problem with these principles in general is that Benzécri seems to imply that large, unstructured multivariate data sets are the only types that are possible. Principles 3, 4, and 5 only seem relevant in that context. Another problem, with principles 2 and 3, is that if they are applied consistently they lead to the type of blind empiricism that has terrorized psychometrics for a long time. If we continue to add empirical variables, without the guidance of any theory or control whatsoever, then we end up with a lot of explained variance but with no possibilities for prediction. We do need theory or, if you prefer, a model, because we need sensible rules on how to add variables. If the structure is there, it will become more clear when we add variables from the same domain. It will disappear when we add enough variables from other domains. In the context intended by Benzécri, we agree with principles 2 and 3 and their corollaries. We shall come back both to tests and to the empiricism–rationalism problem in a later section of this chapter.

Principle 1 is quite important. The important part is stated more explicitly in Benzécri (1973):

> The mathematical foundations of statistical analysis are more algebraic and geometric (and if many dimensions are involved geometric ideas merge with algebraic calculations) then probabilistic; it is better to speak of the average and the principal axes, etc. ... which are defined in terms of a finite number of actual data points, than to speak of the expected value, etc. ... defined in a potentially infinite universe. But probabilistic concepts and ideas can suggest algebraic operations and sometimes can be used to evaluate their usefulness (p. 6).

Compare this with the Kendall quotation on the 'classical probabilistic mould' in Section 1.1.2 and also with:

> A common feature of problems of this kind is that the stochastic model so central to classical statistical analysis is either absent altogether or is playing a very subordinate role. This, in my view, is inherent, and nothing would be more misguided than an attempt to force such problems into an ill-fitting classical statistical mould (Sibson, 1972, p. 311).

It is also interesting that Rao's excellent *Linear Statistical Inference and Its Applications* of 1965 has a 50-page introductory chapter on the algebra of vectors and matrices *and* a 50-page introductory chapter on probability theory. Our content analysis of the books on MVA in Section 1.2 also shows the importance of distinguishing MATH and STAT.

1.4.3 Robust statistics

The important review papers by Huber (1972), Hampel (1973), and Bickel (1976) clearly show that the robust statisticians are trying to make a step in the direction of data analysis:

> Often in statistics one is using a parametric model, such as the common model of normally distributed errors, or that of exponentially distributed observations. Classical (parametric) statistics derives results under the assumption that these models are strictly true. However, apart from some simple discrete models perhaps, such models are never exactly true (Hampel, 1973, pp. 87–88).

This is similar, although much more moderate, to Benzécri's principle 1. It is also similar to: 'The hallmark of good science is that it uses models and "theory" but never believes them' (Wilk, quoted by Tukey, 1962, p. 7).

> We should also remember that we never know the exact distribution of ordinary data; and even if we did, or as far as we do, there remain serious questions about how to handle the excess knowledge of details. After all, a statistical model has to be simple (where 'simple', of course has a relative meaning, depending on state and standards of the subject matter field); Ockham's razor is an essential tool for the progress of science (Hampel, 1973, p. 89).

Thus robust statistics defines more general 'supermodels' and finds estimates that are as good as possible for all models in the supermodel. A very important tool is the study of measures of robustness of estimators, developed mainly by Hampel:

> If we look briefly to some related fields, we can recognize parallel developments there. Robustness may be viewed as a set of stability requirements, analogous to stability of ordinary differential equations, for example. Only quite recently, numerical analysts have become dissatisfied with such possibilities as obtaining a negative variance with the familiar formula $\{\sum x_i^2 - (\sum x_i)^2 / n\} / (n-1)$ and a correctly operating computer; so they started to investigate numerical stability. And in parts of prediction and control theory, engineers resorted to 'sub-optimal solutions' after finding the 'optimal' solutions offered to them inappropriate in practice (Hampel, 1973, p. 90).

We must remember from the quotations of Hampel that robust statistics does not trust the usual parametric models any more (although it trusts the supermodels) and that the emphasis on optimization is replaced partly by an emphasis on stability.

Bickel remarks that the history of statistics has been marked by factional striving between various schools involved with the so-called foundations:

> A conflict which I find more significant and which has serious consequences in the field rather than just in academia is that between what I would like to call the optimists and the pessimists. The optimist confronted with a problem uses his intuition, physical knowledge, perhaps his prior opinions to construct a mathematical model for the data. Once constructed the model is unchangeable and analysis proceeds using methods optimal for that model. The pessimist, lately called a data analyst, uses models only provisionally and is always ready to change his view of the structure of the data in the light of the values that he sees. Purists of either persuasion are fortunately rare.
>
> I view robustness ideas as attempts to bridge the gap between these points of view (Bickel, 1976, p. 145).

We think that the use of 'optimists' and 'pessimists' is somewhat unfortunate. In many social science situations optimism is not enough, mathematical models in these situations can only be constructed by *idealists*. It is also not true that purists of either persuasion are rare, there are many fanatical optimists in mathematical statistics.

There is another interesting problem we mention in this context. The optimists optimize the methods given the model. It has been pointed out by Tukey, who quotes Mallows and Barnard, that it is equally rational to optimize models given the methods (Tukey, 1962, p. 10). The Mallows quotation is: 'But it seems to me to be no more reprehensible to start with an intuitively attractive design and then to search for optimality criteria which it satisfies, than to follow the approach of the present paper, starting from (if I may call it so) an intuitively attractive criterion, and then to search for designs which satisfy it.' The paper Mallows mentions is on optimal designs by Kiefer; the discussion seems to indicate that English statisticians are natural pessimists.

Box (1979) discusses robustness in a wider perspective:

> Some of us have had a preoccupation with optimal or best procedures. But the best, of course, is not necessarily very good. For instance, to bring in the aspect of everyday life, if I ever *had* to decide between cutting my throat with a razor blade or with a rusty nail, I suppose I would choose the razor blade. But although not strictly relevant to the problem as posed, one question that might cross my mind would be, 'Have I considered all my options?' A principle that is being given more attention these days is that of 'robustification'. Here one doesn't attempt to guarantee that things will be optimal over some tractable, but perhaps very narrow set of circumstances. Instead one tries to ensure that they will be fairly good over a wide range of possibilities *likely to happen in practice*. Look at the human hand, for example. I doubt if there is any single thing that it does that could not be done better by some special instrument, but it is very good at doing a very large number of things that come up in facing the world as it actually is. Another way to say this is that there is really nothing wrong with optimization per se, but that we ought to try to optimize over *that distribution of circumstances which the world really presents to us*. The mistake is choosing the best over too narrow a set of alternatives,

suboptimization. It is sometimes argued that by doing simplified exercises we
can at least obtain useful pointers. However, I feel that such pointers are very
likely to indicate the *wrong* direction, as might be true in the case of the razor
blade and the rusty nail (Box, 1979, p. 1).

Again we see the shifting emphasis from optimization to stability (robustness). ➡

1.4.4 Exploration and confirmation

These two terms are becoming very popular these days, but again they are
used by different people with various different meanings. We have *exploratory* factor analysis and *exploratory* multidimensional scaling; we also have
confirmatory factor analysis and *confirmatory* multidimensional scaling. The
difference between the two in this context is that the exploratory models are
the most general factor analysis and scaling methods; the confirmatory models
impose restrictions on the parameters and representations, presumably by incorporating prior knowledge. On the other hand, the point of view can also be
defended that both factor analysis and multidimensional scaling are inherently
exploratory, because they impose relatively little structure and deal with a large
number of variables, and because constraint equations can also be used in a
tentative (exploratory) way. Finally, confirmatory can be used as meaning the
same thing as inferential, which makes, for example, all Jöreskog's factor
analysis programs (Jöreskog and Sörbom, 1979) and Ramsay's multidimensional scaling programs (Ramsay, 1978) confirmatory, as are also the ones
that are called exploratory in the first meaning we have discussed. ➡

In order to explain in which sense we use the words, we start with some
recent 'definitions' by Tukey:

> If we need a short suggestion of what exploratory data analysis is, I would
> suggest that
>
> 1. It is an attitude, AND
> 2. A flexibility, AND
> 3. Some graph paper (or transparencies, or both).
>
> No catalogue of techniques can convey a willingness to look for what can be
> seen, whether or not anticipated. Yet this is at the heart of exploratory data
> analysis. The graph paper – and transparencies – are there, not as a technique,
> but rather as a recognition that the picture-examining eye is the best finder we
> have of the wholly unanticipated (Tukey, 1980, p. 24).

Tukey disagrees, for example, with Parzen, who proposed in a recent paper to
identify exploratory data analysis with confirmatory *nonparametric* statistical

data analysis (and to identify confirmatory data analysis with confirmatory *parametric* statistical data analysis). Tukey objects strongly to bringing the two under a common denominator: 'Replacing chicken by tuna in chicken salad will not give fish wings or train chickens to swim' (Tukey, in the discussion of Parzen, 1979, p. 122). There are other interesting contributions in the discussion of Parzen's paper, and in a sense they have much in common. They tend to emphasize the point of view that scientific discovery is a process of 'conjectures and refutations', and that exploratory data analysis is there to provide us with the conjectures, while confirmatory data analysis is there to provide us with the refutations. Statistics should cover both activities, should provide conjecturing techniques and refuting techniques, and should also emphasize the fact that refutation of a theory or model leads to new conjectures in the form of modifications of the model. Thus the process is circular and not linear:

> It is widely recognized that the advancement of learning does not proceed by conjecture alone, nor by observation alone, but by iteration involving both. Certainly, scientific investigation proceeds by such iteration. Examination of empirical data inspires a tentative explanation which, when further exposed to reality, may lead to its modification. This modified explanation is again put into jeopardy by further exposure to reality, and so on, in a continued alternation between induction and deduction. I am continually surprised that statisticians, even good ones, still seem to ignore this iterative aspect of investigation and talk as if the movement from an initial (perhaps ill-posed) question, to design, to data collection, to analysis of the data, to the answer were a one-shot affair (Box, 1979, p. 2).

It seems to us that the words 'exploratory' and 'confirmatory' should be used in this sense of conjectures and refutations. Tukey puts this in a historical perspective:

> Once upon a time, statisticians only explored. Then they learned to confirm exactly – to confirm a few things exactly, each under very specific circumstances. As they emphasized exact confirmation, their techniques inevitably became less flexible. The connection of the most used techniques with past insights was weakened. Anything to which a confirmatory procedure was not explicitly attached was decried as 'mere descriptive statistics', no matter how much we had learned from it. Today, the flexibility of (approximate) confirmation by the jackknife makes it relatively easy to ask, for almost any clearly specified exploration, 'How far is it confirmed?'. *Today exploratory and confirmatory can – and should – proceed side by side* (Tukey, 1977, p. vii).

We also want to emphasize that this 'side by side' often has the effect of blurring the distinction between exploratory and confirmatory. This is perfectly all right. Many techniques have both exploratory and confirmatory aspects, although it is perhaps true that in general exploratory techniques use graph paper and confirmatory techniques use tests of hypotheses, confidence regions, and the like. Most techniques, however, can be used both to conjecture and to refute, and perhaps all good techniques must have both aspects. Because many techniques in data analysis are not based on any explicit probabilistic assumptions it is also clear that at least in these cases refutation must be possible without using the techniques of inferential statistics (and in fact where would the natural sciences be if this were not true). The words 'conjectures' and 'refutations', of course, are taken from the work of Karl Popper:

> The way in which knowledge progresses, and especially our scientific knowledge, is by unjustified (and unjustifiable) anticipations, by guesses, by tentative solutions of our problems, by *conjectures*. These conjectures are controlled by criticism; that is, by attempted *refutations*, which include severely critical tests. They may survive these tests; but they can never be positively justified: they can neither be established as certainly true nor even as 'probable' (in the sense of the probability calculus) (Popper, 1963, p. vii).

There are two other points of view with which we do not agree. The first one is explained most clearly by Morlat, in his preface to Cailliez and Pagès (1976). According to him, data analysis is the modern form of 'descriptive statistics', and much more powerful forms of description are now possible because of the computer. Tukey emphatically disagrees:

> Some have suggested that 'exploratory data analysis' is just 'descriptive statistics' brought somewhat up to date. Much effort, much intelligence and understanding has been devoted in recent years to convince us that 'the map is not the region'. Perhaps an equal effort, at least among statisticians, is needed to persuade us of the equally true statement 'the usual bundle of techniques is not a field of intellectual activity'! (Tukey, 1980, p. 24).

On the other hand, Morlat's interpretation of data analysis is understandable, because of the claims of Benzécri and his school, which constitute the other point of view with which we disagree. Morlat summarizes it as follows:

> And the description proves to be so effectual that some do not hesitate to conclude that the essential parts of statistics, or even all of statistics, must belong to data analysis. Classical mathematical statistics, according to them, consists only of a number of somewhat arbitrary mind-games, whose only purpose is to furnish the rooms of statisticians who, at the time, did not have computers to solve more realistic problems (Morlat, preface to Cailliez and Pagès, 1976, p. iii).

Both points of view are unnecessarily extremist and polemic, and they are a serious threat to the arrival of the 'whole statistician', desired by Box (1979), by Tukey (1980), and also by us.

1.4.5 Inference

We have used the word 'inferential' a number of times, and it is consequently desirable to specify what we mean by it. If we use the word we assume that the data are a random sample (in some sense of the word) from a (finite or infinite) population. We have *certain* knowledge about the sample, and we want to use this to obtain *uncertain* knowledge about the population. The problem is, of course, what we mean by uncertain knowledge and how we must obtain it. There has been and still is a lot of discussion on these 'foundations of statistics', much of it very polemic and much of it not very enlightening. The key issues are philosophical and have been discussed (as 'the problem of induction') in philosophy for a very long time.

We do not want to become involved in this debate. In general we agree that scientists should behave 'rationally' or 'coherently'. It would be highly irresponsible not to agree. On the other hand, we also think that 'irrational' and 'incoherent' behaviour has advanced scientific knowledge on many occasions in the past. We also take the point of view that current definitions and systems of rational statistical behaviour are 'persuasive' in the technical sense of Stevenson (1938) and Black (1949). And we remain unconvinced. Both decision theory and subjective Bayesianism are very beautiful structures and heroic attempts to codify scientific behaviour in the face of uncertainty. We think that their practical relevance *as prescriptive systems* is limited, although they have had and will continue to have considerable influence on the practice of data analysis. Something like Carnap's principle of tolerance is clearly called for in the debate: it is not our business to set up prohibitions, but to arrive at conventions. We obviously do not agree with statements such as: 'In statistics, the rules are those of probability. A data analyst, not obeying these rules, will be incoherent and often do stupid things' (Lindley, in the discussion of Parzen, 1979, p. 127). In most practical situations that we are aware of, a decision theorist or subjective Bayesian who insists on being completely coherent will be forced to do nothing at all, and even doing nothing at all will not necessarily be coherent according to his own criteria. Of course, we do not intend to build a system that *requires* people to systematically do stupid things. On the other hand, we find it difficult to agree with systems that try to define stupidity in a scientific context and that, in the higher level of analysis in which choice of system is included in the process, must then be classified as stupid according to their own criteria.

1.5 DATA ANALYTIC PRINCIPLES OF THIS BOOK

1.5.1 Model and technique

The procedure usually adopted in statistics is the following one. We start with a question or problem and then build a probabilistic model which we assume to have generated the data. The question is then imbedded in the model as an additional specification of the model, and the assumptions made so far also provide us with an 'optimal' technique for answering the question. Thus, given a question and an optimality criterion, the model automatically provides us with a technique. This approach tends to suggest the 'linear' model of science: question → data → model → technique → answer. We have seen that people like Tukey, Box, Barnard, and Mallows have not been particularly happy with this linear model and have suggested building a feedback loop that returns from 'answer' to 'model', i.e. that modifies the model by considering the results, or even building a feedback loop from 'answer' to 'data', which modifies the data by considering the results, for example by deleting offending observations. Classical statistics has very little to say about this feedback; in fact it usually advises us to start the whole process from scratch, with possibly the same question but certainly with new data. We have also seen an important variation of this linear process that is much less common. Given the question and the optimality criterion the technique can also be used to provide us with an optimal model. Superficially the two variations have much in common: they both try to establish a one-to-one correspondence between techniques and models. In practice, however, starting with the technique invites us to consider many different optimality criteria, and indeed there are many techniques that are not optimal for any model. Although the alternative approach seems equally linear and a simple 'dual' method for approaching the same one-to-one correspondence, it tends to break the chain and to introduce feedback, although perhaps not explicitly.

Robust statistics can be considered as a very interesting attempt to relax the theoretical one-to-one correspondence between models and methods. The performance of a technique, for example computing the mean or median, is studied for a class of models. In addition, properties of a class of models that one is interested in are used to derive a technique that has good performance for all models in the class. This two-way approach and the study of classes of techniques and models are important liberalizations, but the current formalizations of robust statistics tend to proceed along classical lines. The model is replaced by a supermodel, which is often a parametricized family of simple models, and the optimality criteria change accordingly. Instead of looking for a procedure that is optimal in a simple model, we look for a procedure that minimizes the worst possible performance in any simple model in the supermodel. This may be more realistic, but it certainly is not essentially different from the classical linear approach. Consequently much practical work in robust statistics is in the spirit of data analysis, while much of the theoretical work is in the spirit of mathematical statistics.

In this book we adopt the point of view that, given some of the most common MVA questions, it is possible to start either from the model or from the technique. As we have seen in Section 1.1 classical multivariate statistical analysis starts from the model, generally using the multivariate normal distribution. Categorical or discrete MVA starts from a multinomial or Poisson model, which may often be more realistic but then proceeds along the same lines. A conventional optimality criterion such as asymptotic variance or asymptotic covering probability is chosen in both cases, and optimal procedures are derived or at least approximated. In many cases, however, the choice of the model is not at all obvious, choice of a conventional model is impossible, and computing optimum procedures is not feasible. In order to do something reasonable in these cases we start from the other end, with a class of techniques designed to answer the MVA questions, and postpone the choice of model and of optimality criterion.

We do not think that this is the only appropriate starting point in MVA or even that it is the best starting point in MVA, but we do think that it is the most useful starting point in exploratory multivariate situations if we want to say something about the relationship between models and techniques. Thus the technique is taken as *a priori* given in our theoretical work, in much the same way as the model is in multivariate statistical analysis, and we study its properties by applying it to a number of interesting models of various kinds. The results then validate or invalidate the technique in the situation under consideration, and the results can be used in a feedback process that modifies the technique. This explains at the same time the choice of techniques in this book: they have been validated in this sense in a number of interesting situations. We now discuss in more detail some of our major tools in this validating process.

1.5.2 Gauging

What do we mean by gauging of a technique? We construct a model, with known properties, apply the technique to the model, and see if and how the technique recovers or represents the known properties. There are many types of gauges, and we mention some of the important ones.

(a) *Probabilistic gauges*. The technique can be applied to a parametric family of probability distributions. It is interesting to see how the parameters are represented by the technique. The major probabilistic gauge in MVA is, of course, the multinormal distribution, but other interesting examples are the Rasch model (Rasch, 1960) for binary multivariate data and the Markov chain for time series.

(b) *Statistical gauges*. Instead of studying the population, as in (a), we now study the sample. It is interesting to compute theoretically what aspects of the model are 'estimated', and how well. Statistical gauges can also be compared with the corresponding probabilistic gauges to assess the effect of sampling.

(c) *Monte Carlo gauges*. As in (b), but now we do not derive formulas, we only do computation. There are many examples in psychometrics, of which the most familiar one is perhaps computation of the null distribution of Kruskal's stress (Kruskal, 1964a, 1964b).

(d) *Algebraic gauges*. The data can also be generated by an algebraic model without probabilistic structure. Again we investigate what aspects of the model are represented and how well. In numerical analysis the Hilbert matrix is a familiar algebraic gauge; in psychometrics there are many scaling models formulated purely in algebraic terms such as the Guttman scale and the conjunctive and disjunctive models of Coombs (1964). Sometimes, as in the study of the Spearman model in factor analysis (Spearman, 1927), it is convenient to separate the algebraic and the probabilistic aspects when constructing gauges.

(e) *Empirical gauges*. If we have data with well-established properties, for example measurements on a physical process that is theoretically well understood, then we can use the technique to find out if it gives results that are in accordance with the theory. Interesting examples are in Wilson (1926), Wilson and Worcester (1939), and Stigler (1977).

We emphasize that the use of statistical gauges can sometimes lead to the result that there is a model for which the technique is in some sense optimal. Consequently, it is possible to study at least some of the questions that interest statisticians in this gauging framework. We also agree with data analysts such as Benzécri that algebraic gauges are sometimes at least as important as probabilistic ones, especially in exploratory MVA. This is also in the tradition of psychometric scaling theory.

1.5.3 Stability

Even more important than gauging is the analysis of the stability of a technique. In general stability means that a small and unimportant change in data, model, or technique should lead to a small and unimportant change in the results. Both 'small' and 'unimportant' can be defined in various ways, and consequently there are many types of stability. We list the most important ones.

(a)   *Replication stability*. If we replicate the experiment under the same conditions and apply the technique to the new data, then the results should not change dramatically. This requirement is, of course, fundamental for all scientific investigation, and it depends both on the properties of the technique and on the quality of the experiment (we use 'experiment' in the widest possible sense). In the social sciences independent replications are often impossible. Consequently replication stability is often not investigated directly, but the stability question is imbedded in a statistical model. If the data are a simple random sample, then the model will tell you what will happen 'on the average' if we replicate the experiment a large number of times. This is one of the reasons why statistics is needed in the social sciences.

> You asked me to speak of the statistical methods of treating data. I wish you had not. It is a mean subject. Those of you who have read the biography of the great Lord Rayleigh by his son will recall his statement that he even doubts the utility of averaging values to obtain a mean, though he admits that this is carrying disbelief rather far. We find very little statistical analysis in experimental physics or chemistry to-day, a smaller relative amount, I think, than was found a generation ago; and even in astronomy, for which the method of least squares was developed by Gauss and in which it was universally applied in the past, there is a strong tendency to short-cut formal statistical processes. It is now to the biologist and the economist that you must go for complicated statistical analysis. Why this state of affairs? May it perhaps lie in a contrast of the experimental and observational methods, in a difference of degree of attainable control? Shall we say that when the control is good, when we are working in a field in which control is easy or when we are sufficiently astute or fortunate to design experiments so that those consequences in which we are interested are independent of the other variations, then we have no need of statistics and can go along with Lord Rayleigh? Shall we admit that statistics belongs rather in the field of observation and serves to replace control when that is not attainable or is repugnant to the nature of the investigator? (Wilson, 1926, p. 52).

(b)   *Statistical stability.* As we have seen under (a), this can be interpreted as an abstract, formalized form of replication stability, in which the model takes over the burden of replication from the investigator. A great deal of classical statistical theory can be translated as a study of the stability of data analysis techniques the concept of standard error is a good example. If a statistician uses 'optimal' it can often be interpreted as 'optimally stable' (over independent replications). Bayesian statistics, of course, is different because it does not want to use the framework of repetitions. It is probably possible to develop and study Bayesian analogues of stability, because it is also possible to talk about Bayesian versions of the related concept of robustness, but we are quite happy to leave this job to somebody else.

(c)   *Stability under data selection.* This covers a multitude of procedures, all of them very objectionable from a Bayesian point of view. The first one is postexperimental randomization, for example to derive permutation distributions of statistics. Another form of data-selection stability is the jackknife (a good review is Miller, 1974), the bootstrap (Efron, 1979), or subsampling (Hartigan, 1969). All these approaches construct random mechanisms to perturb the given data in a probabilistic sense (the classical jackknife uses deterministic perturbations, but it is easy to construct probabilistic versions). Because the random mechanisms are all introduced conditional on the data, it is not necessary, at least for some questions, to assume any probabilistic model for the data themselves. This is an extremely useful feature in many situations. The concept of the influence curve (Hampel, 1974) in robust statistics is also directly related to this form of stability. Rejection of outliers is also an interesting form of data-selection which occurs very frequently. It may be true that these data-selection techniques cannot be fitted into any one of the formal approaches to statistical inference. From our point of view this does not cause any inconvenience:

> A sort of question that is inevitable is: 'Someone taught my students exploratory, and now (boo, hoo!) they want me to tell them how to assess significance or confidence for all these unusual functions of the data. (Oh, what can we do?)' To this there is an easy answer: TEACH them the JACKKNIFE." (Tukey, 1980, p. 25).

In fact we think that the answer is a bit too easy, because there are more forms of stability than data-selection stability and there are more techniques similar to the jackknife, but we agree with the spirit of the answer.

(d) *Stability under model selection.* A small change in the model that we fit must result in a small change in the estimates of the free parameters, and consequently also in a small change in the interpretation of the results. The study of predictor-selection techniques, of multicollinearity, and of specification errors comes under this heading. Much of the work in sociology on fitting 'causal models', in psychometric genetics on fitting 'genotype–environment models', and in criminology on fitting 'bio-social models' should pay more attention to this form of stability than they usually do.

(e) *Numerical stability.* This is an underrated but very important form of stability in data analysis. It studies the influence of rounding errors and of computation with limited precision on the results given by the techniques. Study of numerical stability has profoundly influenced the field of linear least squares regression, but more complicated techniques in MVA and scaling are much more careless in this respect.

(f) *Analytical stability.* If the possible data structures and the possible representations have enough mathematical structure, then the idea that 'a small change in the input should lead to a small change in the output' can be made precise in terms of continuity or differentiability. Statistical large-sample theory, for instance, concentrates on consistency (a continuity condition) and asymptotic normality (a differentiability condition). The main difference between analysis and statistics is that in analysis we often derive results in the form of inequalities and bounds, in statistics we derive similar results in terms of expected values. Thus analysis is often unduly pessimistic according to statistical criteria.

(g) *Algebraic stability.* In techniques based on the procedures of linear algebra it is often feasible to derive perturbation results by algebraic means. The effect of deleting a predictor or a variable in component analysis, for example, can often be bounded by an inequality. This is closely related to (f), but often the algebra of the problem gives us simpler, more general, and more precise results.

(h) *Stability under selection of technique.* If we apply a number of techniques that roughly tries to answer the same question to the same data, then the result should give us roughly the same information. As the use of 'roughly' indicates, this form of stability is somewhat complicated to study. However, if nine out of ten techniques point to the same important characteristic of a data set, then the tenth technique is disqualified if it does not show this characteristic.

It is clear that the study of stability can be made to include large parts of statistics by only changing the emphasis somewhat. An implication of our approach to data analysis is that no data analytic technique is complete without some gauging and without an investigation of its stabilities. Consequently, some of these results are discussed in this book too. In the past, data analytic or psychometric techniques were often considered to be justified because they 'performed well in practice', because they provided 'insight', or because the users were 'satisfied'. It is, of course, always a good thing if the customers do not complain, but from our point of view it is certainly not sufficient. Good advertising can sell bad products. We agree with Harris (Section 1.1.9) that we need quality control.

1.6 SPECIFIC PROBLEMS OF MVA

1.6.1 The multinormal distribution

We have seen in the discussion of the MVA books that mathematical statisticians often assume from the start that multivariate data are multinormally distributed. This seems to result in a considerable loss of generality and applicability of the techniques. Why then is this assumption so common? In Section 1.1.3 we briefly mentioned the reasons given by Anderson for using this assumption. We now discuss them in more detail.

(a) *Usually a good description in practice.* Is it really? Anderson uses the classical Galton data on the distribution of length of fathers and sons, and in fact more examples of this sort can be found in anthropometry. However, we must not forget that Pearson discovered in studying the equally classical shrimp data of Weldon that the normal distribution is certainly not universally valid in biometry and that as a consequence of this discovery he constructed his famous system of skew frequency curves. Pearson is more careful than Anderson in this respect:

> On the basis of a very large experience of frequency curves and surfaces we have no hesitation in saying that up to the present time no distribution has been proposed which roundly represents experience so effectively as the Gaussian frequency. One of the present writers has indicated over and over again how it fails, and he has measured the significance of its failure, but has always recognized that he must put against this the large percentage of cases in which it gives reasonable results, close enough for all practical purposes (Pearson and Heron, 1913, p. 162).

Of course we must remember that Pearson's practical purposes were descriptive and not inferential and that subsequent research on robustness has been more pessimistic. Moreover, considerable efforts of Pearson's school to construct a system of bivariate or multivariate frequency surfaces comparable to the univariate system have not been successful, which means that there are no systematic alternatives to multivariate normality. A second argument against Anderson's multinormal optimism is that 'goodness-of-fit' tests for the multinormal distribution are still fairly primitive, although it is true that they get a lot of attention in recent times. In any case normality of the marginals is not sufficient to conclude that the data are multivariate normal, and the assumption does not make sense if the variables are categorical or ordinal.

(b) *The central limit theorem.* The assumption that the data can be interpreted as resulting from summation of a large number of independent effects is not very natural in many situations. It may be sensible to assume something like this in biometrical genetics or in astronomical error theory, but it seems very far-fetched in sociology or economics. We also know that convergence to the normal distribution can be very slow if the components in the summation are skew. We know that different normalizations can lead to different limiting distributions and we know that in some cases it is more natural to think in terms of the product or the maximum of a large number of independent effects.

(c) *Simple formulas and many theoretical results.* This seems to be the most important argument. The fact that there are many theoretical results is not only the consequence, however, of the simplicity of the formulas but also of the large amount of interest and energy that has been invested in the multinormal because of reasons (a) and (b). Simplicity is only a relative matter; computing tetrachoric correlations, for example, is not simple; computing the exact distribution of the eigenvalues of a Wishart matrix is not simple either. However, because the multinormal distribution is so popular, much of the complicated computing has already been done. This includes Monte Carlo work, which could in principle of course have been applied equally well in nonnormal situations.

It is undoubtedly true, however, that the multinormal distribution has a number of very attractive theoretical properties, which simplify the job of the statistician considerably. We mention the ones that are most important for our purposes.

(a) If a vector of random variables is multinormal, then any linear transformation of this vector is also multinormal. Because MVA often uses linear transformations of random vectors this property is extremely important.

(b)   If the joint distribution of two vectors of random variables is multinormal, then the conditional distribution of the first vector given the values of the second vector is again multinormal, with a dispersion matrix that is independent of the value of the second vector and with a mean vector that is a linear function of the value of the second vector. The linearity of the mean is called linear regression, the constancy of the dispersions is called homoscedasticity.

(c)   For samples from a multivariate normal distribution the sample mean and the sample dispersion matrix are independent. The sample mean and sample dispersion are also complete sufficient statistics for the multinormal parameters, and they are their maximum likelihood estimates. To put it differently: in most other distributions, moments and product moments are complicated functions of the parameters of the distribution, which implies that it is difficult to compute 'optimal' estimators and that it is also difficult to interpret the parameters. The multinormal distribution is exceedingly simple in this respect, and the first-order moments and second-order product moments contain all information in the sample.

(d)   Multinormally distributed variates are independent if and only if they are uncorrelated, which is true if and only if the covariance matrix is diagonal. In general zero correlation is only a necessary condition for independence; for the multinormal distribution we find again that interdependence of the variates can be described completely in terms of the second-order product moments.

(e)   Properties of the multinormal distribution are closely connected with properties of Euclidean geometry. Points with equal probability density are located on ellipsoids with the vector of mean values as the centre, which implies that probability density and weighted Euclidean distance can easily be translated into each other. If, for example, two multinormal distributions have equal dispersions, then the points where the first density is larger than the second one are separated from the other points by a hyperplane.

These properties are not only statistically interesting, but it is also clear that simple properties like these are important from a data analytical point of view. This makes the multinormal distribution both very important and very interesting as a gauge. However, in many situations the assumption of multivariate normality is not very natural and is difficult to test rigorously. This depends to a large extent on the properties of some commonly used statistical procedures. In MVA we generally test a parametric hypothesis *within* a more general hypothesis, and the more general hypothesis almost always contains the assumption of multinormality. The usual procedures do not test the more general hypothesis or model; they assume this to be true, and test the additional specification on the parameters. This can be quite dangerous, of course. In summary we must agree with the verdict: 'Theorists of multivariate analysis clearly need to venture away from multivariate normal models' (Dempster, 1971, p. 317).

In univariate statistics the normal distribution, which once was all-powerful, has already been abandoned to a much larger extent. It has been shown for several procedures (such as the t-test) that they are moderately robust, in the sense that their properties remain approximately valid under moderate deviations from normality. In addition, a large number of non-parametric statistical procedures have been developed that do not assume normality, or indeed any parametric model. The two have also been combined recently. In MVA both approaches have not been used systematically. Little is known about robustness of MVA procedures, and the things that are known are not very encouraging. Sample covariances, for example, are very sensitive to departures from normality in the heavy tails direction. Also, as we have seen, for example in our discussion of Kendall's book, nonparametric MVA has not been developed sufficiently. There are a number of versions of some of the more simple multivariate tests, relying quite heavily on univariateness of dependent variables (compare, for example, Puri and Sen, 1971). The properties of these procedures are less satisfactory than those of univariate nonparametric statistics and (above all) their data analytical value is limited, because the quantities that are computed do not have straightforward geometrical interpretations.

1.6.2 Tabular analysis

In sociology, political science, and related sciences, surveys (also called observational studies) are very important. A large number of people have to respond to a large number of questions; the data are the answers to the questions, sometimes combined with background information about the respondents. Variations on this theme are questionnaires for clinical diagnostic purposes, attitude studies, panel studies, multiple choice tests, and so on. Continuously varying numerical variables are rare in investigations like these. If the variables are numerical they are usually categorized in fairly broad categories; other variables are ordinal, and background information such as religion or party affiliation can easily be nominal. Good surveys of the problems in the analysis and interpretation of observational studies are Hirshi and Seivin (1973), Cochran (1972), and McKinley (1975). We concentrate on one particular aspect.

It was conceded very soon that assuming multinormality for complex sample surveys made no sense; instead of using Pearson's measures of association people preferred those of Yule (compare MacKenzie, 1978). According to Pearson everything in this world varied continuously on a scale; discrete variables are always discreticized continuous variables and measures of association have the purpose to estimate the underlying correlation between the continuous variables. Pearson made these assumptions because he was convinced that a unified conception of science was possible starting from the concept of correlation, instead of causality. The unified conception implied the idea that biology and anthropology had the same kind of lawlike relationships as physics, i.e. functional relationships between measurable variables. The theory is outlined in the various editions of Pearson's classical *The Grammar of Science* (Pearson, 1892, 1900, 1910). In the last analysis this is, of course, a metaphysical point of view, which also explains why it was very difficult for Pearson to accept the essentially 'discrete' doctrine of inheritance of Mendel (Norton, 1976, 1978). Pearson's metaphysics has had a considerable influence in psychometrics, because many people thought that only continuous variation was 'measurable' and that only the assumption of underlying continuous variation could affect the ascent of psychology to the level of the real sciences such as physics. Tetrachorical correlation, for example, is used almost exclusively in psychometrics, and there are still articles published that tell us how to compute this coefficient faster or better, without paying any attention to the fact that the assumption of underlying bivariate normal variation is extremely contrived for most binary data.

Yule did not subscribe to Pearson's metaphysical ideas. For him something was a measure of association if it was +1 with perfect positive relationship, −1 with perfect negative relationship, and zero in the case of independence. On the basis of this axiomatic point of view he proposed a number of measures that satisfied these assumptions. They are discussed, with many modifications and an avalanche of interesting details, in the famous papers of Goodman and Kruskal (1954, 1959, 1963, 1972), now reprinted in the book by Goodman and Kruskal (1979). Sociologists have never been bothered to the same extent as psychologists by the idea that their discipline should be constructed after the model and with the methods of the exact sciences. They also never had the uncompromising empiricism and the corresponding correlation mania of the Pearson–Spearman school. The technique that became popular in sociological data analysis was making contingency tables with corresponding measures of association. This worked satisfactory at first, for obvious reasons: 'Many tried and tested techniques of multivariate data analysis were invented at a time when ten was a typical number of variables in an ambitious data collection program' (Dempster, 1971, p. 336). With ten variables we have 45 contingency tables, which is still a manageable amount. However, in contemporary surveys 100 variables are quite common, the MMPI has approximately 700 questions, and in longitudinal studies even more variables can occur. The computer came to the aid of the sociologists; packages such as CROSSTABS were constructed and all cross tables were printed, each table with a long list of association measures. This evidently produces enormous amounts of output, and it actually still happens that you find these mountains of paper on people's desks, gathering dust and looking desperate:

> That one still sees these long lists of little tables coming out of the printers, is because too many scientists, especially in the social sciences, have not adapted their methods to the power of these new computing tools. They are like an engineer who builds a bridge by designing blocks of concrete in the form of bricks (Benzécri, 1973, p. 11).

A number of problems can be mentioned. In the first place 5000 cross tables cannot be presented in a research report or paper, and consequently one must select. Usually one selects what seems interesting, which is a rather subjective criterion. It is perfectly possible that others, with different (and perhaps opposite) interests, will find other relationships in the material. In the second place a very long list of cross tabulations at the very least suggests the question of how these tables are related to each other. They are usually presented as independent findings, but it is clear that they certainly are not independent. It seems that reporting selected cross tables gives the impression that the relationships are stronger than they actually are, and gives moreover a very unsystematic presentation of these relationships (compare Hirshi and Selvin, 1973). We can compare it with the following procedure. Suppose we have a large number of numerical variables and compute their large matrix of intercorrelations. The methods of tabular analysis now suggest that we discuss each of the individual correlations in the table that is interesting from our point of view. It seems more natural to us to look for techniques that give a compact description of the correlation matrix, which can consequently be described and discussed independently. Another habit is to discuss only significant relationships. Of course we should take into account here that with 5000 tables we expect 250 tables with 5 per cent significant relationships on the basis of chance alone.

The sociologists have found two different ways out of the ruins of tabular analysis. The first one is the analysis of multidimensional contingency tables, usually by using the so-called loglinear models. Books describing these techniques are Haberman (1974), Bishop, Fienberg and Holland (1975), and

Gokhale and Kullback (1978). This is the discrete MVA we have encountered earlier in our discussion of the books of Roy and Kendall. A second way out is the so-called 'causal analysis', extensively discussed in Blalock (1964) and Boudon (1967). We discuss these two recent developments in separate sections.

1.6.3 Discrete MVA

One of the main disadvantages of tabular analysis is that it is not at all clear how the various tables are related to each other. This makes it possible to find various relationships of the well-known 'spurious' sort, such as the relation between the number of imported bananas and the number of illegal births. The solution in discrete MVA is the analysis of multidimensional contingency tables, which means that we consider three or more variables at the same time and analyse the corresponding multidimensional array of frequencies. We first mention some of the more important disadvantages of this approach.

If we have a large number of variables, then we can make an extremely large number of multidimensional tables. The problem of selection of tables thus becomes more serious. With ten variables there are 45 two-dimensional tables, 120 three-dimensional ones, 210 four-dimensional ones, and so on. There is of course only one ten-dimensional table, and it is possible in principle to analyse only this table, which contains all information in the data. But then, unfortunately, we encounter the second disadvantage of discrete MVA. If every variable has four categories, then the ten-dimensional table has $4^{10} = 1048576$ cells. The number of observations will in general be much smaller than that, and consequently most of the cells will be empty. Because the inferential aspects of discrete MVA are based on the asymptotic normality of the frequencies it is necessary that the table is reasonably well filled. According to the classical, although somewhat arbitrary, prescription of Cochran we want on the average about five observations in each cell, which means that even in this small example we need more than five million observations. If there are more variables then analysis of the complete table is not possible at all, and selection of subtables can be done in an enormous number of different ways. In this sense discrete MVA is useful if we want a fairly exhaustive analysis of the relationships between three or four variables, either because there are only three or four variables or because there are reasons to find three or four variables extremely important. Discrete MVA can not be used for the simultaneous analysis of a large number of categorical variables.

We have seen that Roy (1957) was the first author to discuss discrete MVA in a handbook, and that he thought that this approach was more realistic than the usual multinormal one. What are the most important differences? In MVA we are generally interested in dependence and interdependence of variables. Dependence and interdependence are properties of the probability distribution of the variables, which can be defined in various ways. For the multinormal distribution there is not much choice; all relationships can be defined in terms of the covariance matrices and derived marginal and conditional covariance matrices. For more general distributions more general definitions are needed. Roy uses, following Fisher and Bartlett, definitions in terms of conditional probabilities and in terms of the product rule for combining independent events. This is the so-called *multiplicative* approach, also used in the more recent work on loglinear analysis. There is also an *additive* approach, used mainly by Lancaster and his school. The relationship between the two different systems is easy to describe: multiplicative analysis is the same thing as additive analysis on the logarithms of the probability measures. In general the usual probability-based definitions of independence, interdependence, and interaction can more easily be investigated in the multiplicative approach. Additive analysis has other advantages. The two techniques are compared in Darroch (1974) and Lancaster (1971, 1975a).

In general the discreteness of the variables in discrete MVA is not essential for definitions of the interactions; it is only essential in the subsequent statistical analysis. It is possible to define a completely general system of nonlinear multivariate analysis, valid for both discrete and continuous variables, of which discrete MVA and multinormal MVA are just special cases. In this general system we start by defining complete sets of orthogonal functions on the marginals, and we decompose the probability distribution (or its logarithm) in terms of the tensor product of the functions from the various sets. In discrete MVA the complete sets on the marginals are finite, which makes it possible to handle the analysis in practice; in multinormal MVA only linear functions contribute to the interdependence and the complete sets consist of a single function for each variable. In the general continuous case we need bases that consist of an infinite number of functions on each variable, and consequently this case is only interesting theoretically, not in practice. In several other places in this book we shall discuss the relationships of general nonlinear MVA with the techniques we prefer. ➻

1.6.4 Causal analysis

Causal analysis has a different historical origin than tabular analysis. In biometrical genetics (which used to be almost identical with what we now call statistics) the dominant philosophy of science was the descriptive and empiristic system of Pearson's *Grammar*. One of the fundaments was Pearson's doctrine that correlation is more fundamental than causation, because causality is merely the (theoretical) limit of perfect correlation. It is not necessary to look for causal relationships, you only have to compute correlation coefficients. The theory then comes automatically, because a scientific theory is merely a short and simple summary of a large number of empirical observations (for example correlations). This interpretation of science is not very popular these days, except perhaps in some isolated psychometric and biometric centres. There are at least three reasons for this. In the first place it does not work if you actually try it, as psychometric theories about intelligence or heredity or criminality have clearly shown. The number of correlations that has been computed since Pearson must run in the zillions, but no theory has as yet come out of this mountain of numbers. In the second place Yule clearly showed that correlations have their limitations. If we correlate time series, for example, we often find nonsense correlations. And there are many, many examples that show that correlation does not imply causality (such as the income of Presbyterian ministers and the import of rum from Jamaica). The third reason is more philosophical. Causality is asymmetric, implies a direction and a temporal order. Correlation is symmetric. Nonsense correlations made some people believe that correlations can only be interpreted within an assumed causal model. Sociologists in particular, who never cared in the first place for Pearson's empiricism, find this an attractive point of view. It is, of course, only logical that a field in which there is ten times as much theory as empirical data reacts differently to correlation coefficients than a field in which there is ten times as much data as theory.

Causal analysis was defined originally (by the geneticist Sewall Wright) for continuous multivariate data. The postulated interrelations between the variables were pictured in an arrow diagram, the arrows are interpreted as linear relations between the variables and clearly the arrows indicated a direction in which causality operated. It is obvious that even for a small number of variables it is already possible to draw a very large number of different arrow diagrams. This is the major problem of causal analysis. Instead of the problem of table selection we now have the problem of model selection. Once again this problem is more serious when the number of variables is large. We also must not forget that the model more or less automatically implies its causal interpretation (all arrows have a direction, some possible arrows are not there) and that the estimated correlation and regression coefficients are always interpreted within the postulated model. If we had chosen another model, then the same statistics would have been interpreted differently. The problem of interpretation, which corresponds to relating the different tables in tabular analysis, has been shifted to the *a priori* level, but it has not been solved by this clever move. It is true that, with some additional assumptions, we can also test the goodness-of-fit of the model, but these additional assumptions are often not very realistic and the power of these tests for complicated models with many variables is usually very low. In biometrical genetics the theory of Mendel imposes many restrictions on the choice of model, at least in relatively simple situations under direct experimental control. In studying the inheritance of intelligence, for example, genetical theory does not tell us anything useful, and consequently the choice of model is quite arbitrary, with the unpleasant effect that the same data can lead to very different interpretations.

In modern versions of causal models, inspired by biometrical genetics, by psychometrics, and by econometrics, we even use 'latent' or 'unmeasurable' variables to extend the model. Genotype, for example, is almost always unmeasured, as is general intelligence. The latent variables are only defined in terms of the relationships they have with each other, and with the measured variables; postulating latent variables only has consequences through the structure they impose on the covariances of the observed variables. This clearly makes the problem of model choice even more complicated than it already is, and consequently makes interpretation even trickier. It is now even true that if we choose another *name* for the unobserved variables then the same statistics are interpreted differently. The goodness-of-fit test now becomes even more overburdened and is not of much help in the selection of an appropriate model. It is always possible that 'better' models exist, more so because most investigators only look at relatively minor variations within a model with fixed global structure. Interpretability of the results in this context (as in others) is a poor criterion of success, because the interpretation has already been largely determined at the moment of model choice.

Nevertheless, we think that causal analysis is a useful attempt to incorporate prior information about the variables (for example their natural order in time) in MVA. As such it is an interesting generalization of the simple distinction between analysis of dependence and interdependence. We think that it is appropriate to incorporate some rationalistic conjectures into the untenable empiricistic optimism of the psychometricians. However, there are many situations in the social sciences in which the choice of a causal model is very arbitrary because the necessary *a priori* knowledge is either absent or extremely fragmentary. This is especially the case if there is a large number of variables. In situations like these it can be misleading to interpret results in terms of the parameters of the causal model. Because the choice of the model was arbitrary the same thing must be true for the interpretation. It does not make sense to use highly structured models in highly unstructured situations. The subjective choices of tabular analysis are assumed away, and are not questioned any more. A program package such as LISREL is in many respects much more satisfactory than CROSSTABS: there is some indication of statistical stability of the results and there is considerably less output and thus more data reduction. However, the alleged statistical respectability of the approach (compared with some of the earlier alternatives) invites uncritical acceptance of the results by the uninitiated.

1.7 DEFINITION OF MVA

1.7.1 Asymmetric role of rows and columns

On the basis of the discussion in the preceding sections we can now try to give a definition of MVA. The oldest, and the most restrictive, definition was that MVA is the analysis of random samples from a multinormal distribution. The data are collected in an $n \times m$ matrix, the rows of the matrix are the n independent replications of the same multinormal m-vector. We have seen that both the assumption of independence between rows and the assumption of multinormality are too restrictive to construct a general theory of MVA.

The first, and most far-reaching, generalization (suggested by the work of Van de Geer, Cailliez and Pagès, and Green and Carroll) is that MVA is the analysis of an arbitrary rectangular matrix, with the explicit purpose of describing the matrix in terms of a smaller number of parameters and of making pictures of this representation. We do not make any assumptions on the origin of the matrix. If we define MVA like this, then multidimensional scaling and various forms of cluster analysis are also included in the definition. Indeed, the French followers of Benzécri include these techniques in their definition of data analysis, and the group around Krishnaiah and the *Journal of Multivariate Analysis* also pays attention to these 'nonstatistical' forms of MVA. Nevertheless, we think that this definition is somewhat too general. The name multi*variate* analysis implies that a number of entities are involved that we call variables or variates. In the matrix definition of MVA the rows and columns of the matrix are treated symmetrically, variables are not mentioned. It may be better to use the term *multidimensional analysis* for this field.

We consequently need to preserve more elements from the classical definition, notably the asymmetric treatment of rows and columns. We also want to drop the restriction that the data are real numbers, because this is not necessarily true in discrete MVA. As the starting point we do not use an arbitrary $n \times m$ matrix, but m random variables defined on a common probability space, not necessarily real-valued. In the simplest case the space on which the variables are defined has n elements, and the probability is defined by counting the number of elements in the subset and dividing by n. The m variables can be defined by making a list of all n elements, with the corresponding m values of the functions for each element. This list can be organized in an $n \times m$ matrix. The probabilistic component in this case is essentially irrelevant, and it is consequently still possible to analyse $n \times m$ matrices, but the approach in terms of m functions on the same space has introduced asymmetry and has made it possible to think of generalizations. In statistical terminology we study the population in this case, and the population is finite and completely observed. We can also study a finite population with a more general discrete probability distribution over the n elements, but now the probability content is no longer trivial and we have information that is not represented in the $n \times m$ matrix.

It is also possible that the probability space on which the variables are defined has an infinite number of elements and that we consequently can not define our functions by giving an explicit list of values. We are still studying the population, but the population is now infinite and can not be observed. Instead of listing the functions explicitly we now state their properties mathematically, for example by assuming that the joint distribution of the variables is multivariate normal. It is clear that this situation is of theoretical interest only; nothing is observed, there is no data matrix. We are not doing data analysis, we are not doing statistics. The situation is of interest in the process of gauging our techniques or in the study of various forms of stability.

In the third situation we have a data matrix again, but we now assume that the rows of the matrix are a random sample of size n or the rows are independent *realizations* of the same population model. The situation is completely different from the previous two, in which the stochastic variables were defined completely, but not necessarily observed. We can also state this by saying that we now have observations on the probability measure defined on the probability space; the probability measure itself now defines the basic random variable. We have to work with the empirical probability measure, on which we have observations and which estimates the theoretical one.

Assumptions about the theoretical measure (i.e. about the population) have consequences for the possible empirical measures we can observe, in the same way as the nature of the sample has consequences for what we observe.

On the basis of this analysis we now give the following definition: *MVA studies systems of correlated random variables or random samples from such systems.* We do not specify in this definition that the number of random variables is finite. This will always be the case in this book, but we do not want to exclude the analysis of continuous time stochastic processes from the definition, although an infinite number of random variables will also only occur in theory. We have also incorporated the stochastic element explicitly in our definition, but we have seen that it can be made trivial in the case of a finite population with counting measure. Thus it causes no real loss of generality. Statistics only becomes important, of course, in the special case that we actually have random samples.

1.7.2 Linear, monotone, and nonlinear MVA

We now define some specific forms of MVA which will be important in this book. MVA is *linear* if the results are invariant under one-to-one linear transformations of the random variables, it is *monotone* if the results are invariant under one-to-one monotone transformations, and it is *nonlinear* if they are invariant under all one-to-one nonlinear transformations of the random variables. These definitions are somewhat vague, because we do not specify what we mean by 'the results'. The results can be formulas, which are the result of derivations; they can be idealized numbers, which result from substituting values for the variables in the formulas; and they can be actual computer output, influenced by rounding errors, choice of initial configuration, or stopping criteria. It is possible that a part of the output of the actual computer programs changes and another part does not change, or even that all the results change, but in a simple way. It will become clear in the rest of the book, for all the techniques we discuss, where this distinction is important and what remains invariant. Our definitions are also idealized, because they do not take inevitable shortcomings of computer programs into account. It is possible that transformations of the variables result in slower convergence to the desired solution, or even to convergence to an undesirable local minimum. Nevertheless the distinction is a very useful one, and using it we can now state that one of the main purposes of this book is to discuss monotone and nonlinear versions of some of the more familiar linear multivariate techniques.

Up to now we have discussed the classical $n \times m$ matrix of MVA, but we have also seen that tabular analysis and discrete MVA use cross tabulation or contingency tables. The relationship between the two representations will be explained more formally in Chapter 2; in this introduction it is merely observed that the relationship is one between the values of a random variable and its distribution. In the discrete case, in which the m random variables map the space into m finite sets, every possible combination of values corresponds with a *cell* of the m-dimensional cross tabulation. The data can be represented by indicating how many times each of these possible *profiles* occurs in the data matrix. This is true in the case of a population (the cell values then are probabilities), but it is also true in the case of a sample, in which the cell values are observed frequencies. If the variables have values in a *range* with an infinite number of elements, then we replace the cross tabulation with the product of these m ranges and we replace the probabilities in the cells by a multivariate probability distribution or density. Again we see that the three special cases to which our definition of MVA applies are all covered.

It is, of course, true that a sample is finite. Continuous variables, if they exist at all, are always measured with finite precision, which leads to a representation with a finite number of decimals. Thus infinite ranges only occur in a theoretical analysis of continuous population models. One of the basic ideas in this book is that all *data* are discrete (or categorical) and that continuous models are used for gauging and to simplify calculations and approximations in some cases. This is, in fact, the way in which the normal distribution was introduced into the history of probability theory. Only much later, after Galton and especially Pearson have made continuous variation the norm, do we find the point of view that discrete variables are in some sense degenerate or rounded continuous variables. This last point of view is still very important in most of the classical books on MVA, although usually implicit.

1.7.3 Bivariate and multivariate MVA

The m stochastic variables define an m-dimensional probability distribution. This distribution has univariate marginals, bivariate marginals, and so on. There are still people whose approach to MVA is essentially univariate, by which we mean that they apply techniques that give the same univariate marginals. Most people agree that such an approach can be extremely misleading (compare Rao, 1960). We have seen that multinormal MVA is typically bivariate; the techniques give the same results if we apply them to another multivariate distribution with the same bivariate marginals. In the multinormal distribution this does not lead to loss of information, because multinormal distributions are completely determined by their bivariate marginals, but in other distributions we do ignore information. Tabular analysis is also bivariate, but tabular analysis is nonlinear, while multinormal analysis is linear. Consequently multinormal analysis gives the same results on different multivariate distributions that merely have the same variances and covariances. Moreover tabular analysis looks at all bivariate distributions separately while multinormal analysis looks at them jointly. Thus if two multivariate distributions have the same bivariate marginals except for one, then tabular analysis; will give the same results for all other bivariate marginals while multinormal analysis will give different results for the complete analysis.

The techniques in this book are largely *joint bivariate,* although we discuss some extensions in the multivariate direction. Thus they can be considered to be somewhere in between multinormal and general multivariate analysis; we combine nonlinearity and bivariateness, hoping that the bivariate marginals give sufficient information on the interdependencies in the multivariate distribution. By concentrating on bivariate marginals only we circumvent the problem of the many empty cells and the problem of difficult interpretation of higher order interactions (also familiar from the analysis of variance). By allowing for nonlinear transformations we drop many of the restrictions of multinormal analysis. By analysing all bivariate distributions jointly and by concentrating on low-structure models we circumvent the model-selection problem. In addition, by applying as much data reduction as possible we avoid the problem of having to cope with ten pounds of output from each analysis. The emphasis in this book is consequently on large data sets with many variables, on efficient computation, and on nonlinear transformations. The idea of a 'random sample' does not play a prominent role, and the multinormal distribution is nowhere used as a starting point. Nevertheless, for gauging purposes, we are very interested in the performance of our techniques if we apply them to multinormal populations and samples. Wherever possible we use statistical stability techniques, mainly asymptotic perturbation methods and versions of the jackknife. It is difficult to indicate the position of the book in Figure 1.1, because in a sense we are close to tabular analysis and discrete MVA, which are not represented in the figure.

1.8 SOME IMPORTANT INGREDIENTS

1.8.1 Join and meet problems

The techniques discussed in this book can be classified into two different groups. In the first place there are various generalizations of principal components analysis and in the second place similar generalizations of canonical analysis. This distinction corresponds with the already familiar distinction between internal and external MVA, or between the analysis of interdependence and the analysis of dependence. In this section we discuss a more general distinction on a verbal level. The distinction will be explained more formally in Chapter 11.

Techniques such as principal components analysis try to find a subspace of the space spanned by the variables, which has minimum dimensionality and yet contains all the variables. Canonical analysis tries to find a subspace of maximum dimensionality which is contained in all groups of variables. Principal components analysis approximates from the outside and tries to find the 'least common multiple' of all the variables; canonical analysis approximates from the inside and tries to find the 'greatest common divisor' of all groups of variables. Although our techniques are nonlinear, in the sense that the results are (often) invariant under nonlinear transformations of the variables, we use computational tools from linear analysis and algebra. This is because nonlinear transformations of a variable themselves define a linear space, of which the linear transformations form a subspace. We consequently work in a larger space, but the space is still a linear space in the technical sense of the word. General nonbivariate MVA works in even larger linear spaces.

Let us introduce some terminology to generalize the distinction between components analysis and canonical analysis. Assume for the moment that we are dealing with an ordinary $n \times m$ data matrix. The m variables are partitioned into K sets of variables. Each set of variables defines a subset of n-dimensional space. In the linear case this is the set of all linear combinations of the variables in the set; the dimensionality is not larger than the number of variables in the set. In the nonlinear case it can be the set of all nonlinear transformations of the variables in the set, whose dimensionality is not larger than the total number of different values assumed by the variables in the set (which is equal to the number of nonempty cells in the corresponding multidimensional cross tabulation). It can also be the space of all linear combinations of separate nonlinear transformations on each of the variables, in this case the dimensionality does not exceed the sum of the numbers of values assumed by each of the variables separately. These K subsets, once they are defined, can be combined in various ways. In the first place they have an *intersection* which is the largest subspace contained in all K subspaces, and they have a linear *sum*, which is the space of all *linear* combinations of K vectors, one from each of the subspaces. The linear sum is the smallest subspace that contains each of the K subspaces we started with. We borrow some terminology from lattice theory and call the intersection the *meet* of the K subspaces and the linear sum their *join*. It is important to observe that the join of a number of subspaces generated by linear combinations of nonlinear transformations of separate variables is the same as the join of the m subspaces defined by the nonlinear transformations of each variable. Thus in this case the partitioning of variables into subsets is irrelevant in computing the join. The same thing is true in the linear case.

We translate some familiar MVA problems into this terminology. In principal components analysis we try to find p orthogonal vectors in n-space, in such a way that each variable is a linear combination of these p components. This is possible if and only if the join of the variables has a dimension not exceeding p. Thus principal components analysis is a *join problem*, and we can say that it tries to compute the smallest subspace containing all variables (which is equivalent to computing the dimensionality of the join, or the *join rank*). In canonical analysis there are usually only two sets of variables, and consequently only two subspaces. We try to find p orthogonal vectors in n-space that belong to both subspaces. This is possible if and only if the meet of the two subspaces has a dimension of at least p. Thus canonical analysis is a *meet problem*, in which we compute the largest subspace contained in both sets of variables (which is equivalent to computing the dimensionality of the meet, or the *meet rank*). The restriction of this purely algebraic formulation of two subspaces is in no way essential, and we can extend it directly to K subspaces. The restriction of meet and join problems to finite dimensional space is also not essential, and neither is it necessary on this abstract level to consider only a finite number of variables or subspaces.

Our formulation does not use any specific coordinate system in the space. For numerical purposes, however, it is necessary that coordinates are used and that the problem is translated into matrix algebra. Moreover, we cannot expect in general that perfect solutions exist: in general the join rank of m variables will be m and the meet rank of K sets will be zero. We consequently must introduce *loss functions* which measure the departure from perfect fit: *join loss* measures the departure from 'join rank $= p$' and *meet loss* measures the departure from 'meet rank $= p$'. The dimensionality p is chosen by the user; the theory in this section tells us that we want to choose p as small as possible in a join problem and as large as possible in a meet problem. The extensions 'as large as possible' and 'as small as possible' will not be defined here, because their interpretation depends not only on the value of a meet loss or join loss but also on various other properties of the data and on other data analytic considerations. The definitions and properties of the loss function will be discussed in more mathematical detail in Chapter 11.

The definitions of meet and join problems above assume that all variables are linear, or at least that the possible transformations of a variable define linear subspaces. This formulation is not quite general enough for some purposes, because ordinal variables, for example, cannot be fitted into this framework. We now give a more general discussion, starting with a more precise analysis of the concept of a join rank. The partitioning into subsets is irrelevant here, so we let each variable define a subspace. In the linear case this is the subspace of linear transformations, but we now extend the analysis and merely assume that for each variable we can choose from a set of possible transformations (for nominal variables transformations are also called, perhaps more appropriately, *quantifications*) that is not necessarily a subspace. Each choice of transformations makes it possible to compute the correlation matrix of the transformed variables; different transformations lead to different correlation matrices. If the join rank is equal to p, then a correlation matrix will have a rank less than or equal to p, *no matter how we choose the transformations*. Because the correlation matrix is invariant under linear transformation of each of the variables the join rank of linear variables is p if and only if their ordinary correlation matrix has rank p.

It is now possible to use two different practical approaches to join problems, and both approaches have been used by previous authors. If we have a class of nonlinear transformations at our disposal we can suppose that we actually are in the linear situation and that for some unfortunate reason the precise values of the numerical variables are unknown. In this case we can still speak of *the* correlation matrix of the variables, which is also unfortunately unknown, except for the fact that it has rank p. We choose our transformation in such a way that the correlation matrix is as close as possible to a rank p matrix. This is called the *single* approach to the join problem, because we only find a single transformation for each variable and only a single correlation matrix. In the *multiple* approach we take the nonlinear situation as our starting point, and we look for a number of different transformations, all of which give correlation matrices of rank p. The definition of the join rank tells us that the number of linear independent solutions for the transformations must be equal to the dimensionality of the space of possible quantifications. Thus the single approach computes a single solution (because it believes in a true hidden transformation), while the multiple approach computes more than one solution.

The two approaches to the join problem are implemented in two different computer programs, which are both discussed in this book. We mention them here because discussing their properties shows clearly that nice theoretical distinctions do not necessarily lead to equally nice distinctions in the implementation. HOMALS and PRINCALS are discussed in Chapters 3 and 4 of this book. We must emphasize in the first place that interpreting these programs in terms of solving join problems is just one possible interpretation, and not necessarily the most illuminating one. Other geometrical and algebraic interpretations of especially HOMALS are possible, and will be discussed in detail in Chapters 3 and 8. HOMALS implements the approach to nonlinear component analysis, also known as Guttman's principal components of scale analysis, Hayashi's third method of quantification, or Benzécri's correspondence analysis of multiple disjoint tables. If interpreted as a program for solving a join problem it takes the multiple approach, and aims at solutions with a join rank of one. HOMALS accepts only nominal variables, not numerical or ordinal ones. PRINCALS generalizes the approach used in Kruskal and Shepard's nonlinear factor analysis, in PRINCIPALS of Takane, Young, and De Leeuw, and in other similar programs by Roskam or Guttman and Lingoes. PRINCALS in its simplest versions uses the single approach, and aims at solutions with a join rank equal to any given p. The variables can be either nominal, ordinal, or numerical. Thus, briefly, HOMALS computes p solutions of join rank one, PRINCALS computes one solution of join rank p. If a perfect fit is not possible we merely have to replace 'computes' by 'approximates' in this last sentence.

Things become more complicated because PRINCALS also has the possibility of mixing the multiple and single approaches. We interpret this as HOMALS with restrictions: we use the multiple approach aiming at join rank one, but for single variables we use the restriction that the transformations in the different solutions must be proportional. It is also possible to extend PRINCALS for the case with a single treatment of variables in order to arrive at a multiple solution for general p. This would amount to computing an ordinary single PRINCALS solution and then to compute another one, taking care somehow that the second solution is actually different from the first one. All this will be explained in more detail in Chapters 8 and 10.

The situation is more or less the same for the meet problem. For each transformation of the variables and for each linear combination of the variables within the K sets we can compute a correlation matrix of order K, i.e. between sets. The meet rank of K sets is equal to p if there are p different transformations and combinations with a correlation matrix of rank one or, to put it differently, if there are p different transformations and combinations with a generalized canonical correlation equal to one. Again we can choose for each variable, if we desire, a multiple or a single treatment. Now suppose that $K = m$, i.e. all sets of variables consist of exactly one variable. Then the meet rank is equal to p if there are p transformations that give a correlation matrix of rank one. Remember that the join rank is equal to one if *all* transformations give a correlation matrix of rank one. The multiple approach to the join rank one problem is to compute p transformations with a rank one correlation matrix, and consequently the solutions can also be used as a solution for the meet problem with rank p. Thus join problems can be identified in practice with meet problems in which $K = m$. Again this will be explained in more detail in the later chapters.

Meet problems with $K < m$ occur in many disguises in linear MVA. Our programs implement various generalizations of these linear problems. Thus CANALS has $K = 2$ and generalizes canonical correlation analysis, CRIMINALS has $K = 2$ and generalizes multiple group discriminant analysis, PATHALS has $K = 2$ and generalizes path analysis, and MORALS has $K = 2$ and generalizes multiple regression. OVERALS is a program for general K, which generalizes K-set canonical analysis. If we choose $K = m$ and all variables are both nominal and multiple then we are back to HOMALS. By choosing $K = 2$ in OVERALS we recover CANALS and the less general programs MORALS, PATHALS, CRIMINALS, which have only been written because the special structure of the problem leads to more efficient algorithms or to more specialized output.

1.8.2 Optimal scaling and alternating least squares

The basic computational ingredients in our computer programs are the *alternating least squares* method and the concept of *optimal scaling*. The loss functions we are minimizing have two different sets of parameters: in the first place a basis for the meet or the join (corresponding with scores for the individuals or objects) and in the second place parameters for the transformations of the variables (or for the quantifications of the categories). The loss functions are all of the least squares type, and they have the obvious property that they are zero for a particular choice of the parameters if and only if that choice defines a perfect solution for the corresponding meet or join problem. The algorithms we use are usually (but not always) of the alternating least squares type, by which we mean that in each iteration two substeps are alternated. In the first substep we compute the optimum basis for given values of the transformations, in the second substep we compute new values for the optimum transformations for the given basis computed in the first substep. Alternating these substeps obviously produces a decreasing sequence of loss function values, which always converges because loss is bounded below by zero. Under some mild regularity conditions we can also prove that the basis and transformations converge to values corresponding with a stationary value of the loss function.

This particular way of computing the transformation is called optimal scaling, because the transformations are chosen in such a way that they minimize the loss function. In other forms of MVA the transformations are chosen on *a priori* grounds, after which ordinary linear MVA is applied.

Of course, optimality must not be interpreted in any wider sense. We do not pretend that our procedures always give better transformations than other procedures; they are only better in terms of the particular loss function we choose. Whether they are better in any wider sense must in principle be decided by the gauging process.

Our choice of least squares loss function could be considered as old-fashioned. The results of robust statistics can be interpreted as showing that least squares loss functions often fail, and can almost always be replaced by more appropriate ones. Our basic join and meet philosophy is formulated in purely algebraic terms, and least squares only enters the picture if we start making use of the fact that most of the linear spaces we study in practice can be made into inner product spaces, with a corresponding (weighted) least squares distance function. However, other norms for the linear spaces could in principle be used. They lead to unpleasant complications in the computational process and, even more importantly, they destroy most of the geometrical interpretations of our procedures which play an important role in the making and interpreting of pictures. It is quite clear, however, that in some cases the least squares loss function is a rather poor choice, for example if outliers are quite common. We explicitly study the consequences of our choice of loss function in our analysis of stability and our gauging of the techniques on various theoretical and practical examples. Thus it is conceivable that in later versions of this book other norms will be studied and used, but for the moment we have to be satisfied with least squares.

We also do not wish to maintain that alternating least squares always gives the best algorithm. In fact it is clear that in some situations much more efficient computation is possible. Solving a join or meet problem is in the simplest cases (discussed mainly in Chapters 3 and 8) equivalent to finding the partial or truncated singular value decomposition of a given matrix, and sometimes alternating least squares (which is equivalent to the familiar power method in these cases) is not a good algorithm for computing the singular value decomposition. Other methods are available which are faster, use less storage, and provide us with all singular values and/or singular vectors. In our ANACOR and ANAPROF programs, discussed in Chapter 8, for example, we use singular value algorithms not based on alternating least squares. The major advantage of alternating least squares is its generality; it can also be applied if the problem is not of the singular value type. The method is useful because the transformations of the variables are often restricted in various ways. We have already seen that single variables in the multiple approach are restricted by proportionality constraints, another very common restriction is imposing monotonicity when analyzing ordinal variables. A consequence of these restrictions is that the resulting problems are no longer equivalent to singular value decompositions. Moreover, alternating least squares can be applied to extremely large examples, in which singular value decomposition using other methods is not practical. The various restrictions that are possible and useful in our programs will be discussed in detail in Chapters 4 and 12.

1.8.3 Dimensionality and data massage

We have been very casual so far about the choice of the dimensionality. Observe in the first place that there are two different dimensionalities involved. We have to choose the join rank or meet rank and also the number of different solutions we want to compute. These choices are usually dependent on the type of program we want to use. If we use HOMALS in a join problem, then this implies that we choose a join rank equal to one, and we only have to decide how many solutions we want to compute. If we use PRINCALS then we decide to compute only one solution, but we must decide what join rank we want to use. We have merely said so far that the dimensionality must be chosen by the user, but this is somewhat unfair because the user obviously needs some guidelines.

In almost all of the examples in this book the rank and/or dimensionality is either equal to one or equal to two. Computing just one HOMALS solution (if all variables are nominal this is the same thing as computing a PRINCALS solution with a join rank of one) is used quite often to compute 'optimal' transformations or quantifications of variables. These transformed or quantified variables are then used in a subsequent linear MVA or in any other data analytic technique that requires numerical variables. We call this using HOMALS *as a first step*. It is also possible to use PRINCALS or CANALS as a first step, and, although considerably less common, this is just as useful.

It is also very common to compute two HOMALS solutions, or a PRINCALS/CANALS/OVERALS with rank two. The basic emphasis in this case is on making low-dimensional representations. It is clear from our experience with nonlinear MVA that studying pictures in more than two dimensions is not very rewarding. The 'rotation problem' of linear MVA becomes considerably more complicated in nonlinear MVA, because we in fact often have to transform nonlinear manifolds of points to a 'simple structure'. The first two dimensions often give us a fairly clear idea of the most important effects in the data. If we interpret our programs as fitting models (for example the model that the join rank is equal to two) then the routine choice of $p = 2$ is not defensible; there is no reason whatsoever why $p = 2$ should occur more 'in the real world' than any other value of p. However, if we interpret our techniques as making transformations and preparing for a linear analysis, then

$p = 1$ is the natural choice. If we see them as techniques for making two-dimensional pictures of data, then $p = 2$ is the only reasonable choice.

There are some cases in which two-dimensional solutions indicate that the results are dominated by either a single deviating object or by the properties of a single variable. These solutions are similar to 'degenerate' solutions in multidimensional scaling, although they are not quite so degenerate, because multidimensional scaling actually shifts points to infinity and collapses clusters into single points. In such degenerate cases it often happens that the interesting structure is hidden somewhere in the higher dimensions or in the remaining solutions. We could consequently study higher dimensional solutions to find the interesting structure, but we have seen that this is very problematic in practice. Our solution is consequently a different one: we delete offending objects and/or variables and compute a new two dimensional solution. In our analysis of stability in Chapter 12 we will show that this often has the same effect as eliminating the offending dimension.

Some people call this *massaging* the data, with the implication that it is subjective, not respectable, and possibly somewhat indecent. It is true that it is difficult to formalize this massage process, because it involves what Tukey calls 'judgment' in his 1962 paper. We do not agree, however, with the point of view that the process can be used to find any conclusion or interpretation that you want to find. The main safeguard is that in reporting the analysis we should also report and motivate our deletions and other manipulations. It is also extremely naive to suppose that users of classical statistical techniques never apply massage; the major difference seems to be that exploratory data analysis encourages people to report this explicitly, while classical statistics seems to encourage the attitude that massaging should only be practised behind closed doors (compare our discussion in Section 1.4).

1.9 EPILOGUE

A number of new books on various aspects of MVA have appeared since the previous material of this chapter was written. Perhaps the one most in the spirit as the present one is the book by Everitt and Dunn (1983). It emphasizes exploratory aspects of methods that are usually described as confirmatory, and a large part of its contents is devoted to principal components analysis, multidimensional scaling, and cluster analysis (approximately 40 per cent). Another 30 per cent deals with (generalized) linear models, 10 per cent with structural covariance modelling, and a final 10 per cent with mathematical preliminaries. If we locate this book in the triangle CORR – STAT – MATH (Figure 1.1), a strategy called fitting in *supplementary points*, then we find that the text is close to Gnanadesikan (1977) and Kendall (1975), and not too far from the centroid of the triangle. This location also suggests that it might not fit very well into the plane of the other books, due to the fact that the attention for comparatively modern methods constitutes a new dimension perpendicular to the old-fashioned triangle. Similar remarks can be made about the books by Dillon and Goldstein (1984), Mardia, Kent, and Bibby (1979), and Seber (1984), where the last two are probably most interesting for the technically advanced reader.

Of course there also still appear books embedded in the classical tradition. A prime example is Muirhead (1982). As well as major new developments since Anderson (1958), it includes material on zonal polynomials and hypergeometric functions of matrix argument, through which a unified study of the noncentral distributions that arise in the classical case has become possible. The book by Press (1982), a revision of an earlier text and still quite in the classical tradition, is brought up to date in slightly different directions as well. It discusses Bayesian viewpoints, includes a small part on multidimensional scaling and clustering, and reflects a background in business applications by a very interesting chapter on portfolio analysis. Most importantly, however, a revised edition of Anderson (1958) appeared in 1984. Although the general outline of topics has been retained, there are substantial differences in treatment. We quote from the Preface to the Second Edition:

> The method of maximum likelihood has been augmented by other considerations. In point estimation of the mean vector and covariance matrix alternatives to the maximum likelihood estimators that are better with respect to certain loss functions, such as Stein and Bayes estimators, have been introduced. In testing hypotheses likelihood ratio tests have been supplemented by other invariant procedures. (...) Simultaneous confidence intervals for means and covariances are developed. A chapter on factor analysis replaces the chapter sketching miscellaneous results in the first edition. Some new topics, including simultaneous equation models and linear functional relationships, are introduced (Anderson, 1984, p. vii).

There can be no doubt that this second edition is a new landmark in the field.

Some comments are also in order with respect to the MATH corner of Figure 1.1. It turns out that, indeed, Green and Carroll (1976) do not view their *Mathematical Tools* text as a book about MVA, primarily because they published another book, called *Analyzing Multivariate Data* (1978), which makes their position less extreme. The new book by Van de Geer (1986) is different from the previous one, although perhaps only slightly so in terms of topics covered. However, it is entirely based on the analysis of operators and projectors, so it now propagates even more radically MVA as a form of applied coordinate-free linear algebra. There are now also a number of more specialized books, covering in particular principal components analysis and correspondence analysis (Lebart, Morineau, and Warwick 1984; Greenacre, 1984; Joliffe, 1986; and Van Rijckevorsel and De Leeuw, 1988).

Although the robust approach has had a healthy development within the linear model (Huber, 1981; Hampel *et al.*, 1986; Rousseeuw and Leroy, 1987), for truly multivariate problems progress has been much less. Moreover, from scattered remarks in Barnett and Lewis (1984) we get the impression that the issue of starting from the data versus starting from the model is still alive in this area as well. For a discussion of these matters in the context of multidimensional scaling, see Heiser and Meulman (1983). There is a rapidly growing literature that relates (model-based) loglinear and (data-based) correspondence analysis methods (cf. Van der Heijden, De Falguerolles, and De Leeuw, 1989).

CHAPTER 2
CODING OF CATEGORICAL DATA

Basic in MVA is a finite set of n *objects* (or *individuals*). A *variable* h_j maps the set of objects into a finite set of k_j categories; this set of *categories* is called the *range* of h_j. We shall assume that there is a finite number of m variables h_j ($j = 1, ..., m$). The Cartesian product of all categories is called the *multivariate range*. Its elements are all possible combinations of m categories; they are called *profiles*. Since variables are ordered (from 1 to m), each is an ordered m-fold. The *data matrix* $\mathbf{H}$ is an $n \times m$ matrix with elements h_{ij} giving the category of variable h_j for object i. These elements are not necessarily numbers.

An example of a data matrix $\mathbf{H}$ *is given in Table 2.1, with $n = 10$, $m = 3$, $k_j = 3$ ($j = 1, 2, 3$). Elements of $\mathbf{H}$ are 'category labels': the first variable has categories 'a', 'b', 'c'; the second 'p', 'q', 'r'; the third 'u', 'v', 'w' (with zero frequency for 'w').*

Table 2.1 Example of data matrix $\mathbf{H}$

a	p	u
b	q	v
a	r	v
a	p	u
b	p	v
c	p	v
a	p	u
a	p	v
c	p	v
a	p	v

The number of possible profiles equals Πk_j, the product of all k_j. It may happen that this number is much smaller than n. In this case the data matrix is not the most efficient way of coding. Instead one might prefer a *profile frequency matrix*. A complete profile frequency matrix would list all possible profiles and indicate for each one how often it occurs. Such a matrix has Πk_j rows and $(m+1)$ columns: the first m elements of a row give the categories of the profile and the last element shows its frequency.

> *Table 2.2 shows the complete profile frequency matrix derived from the data matrix of Table 2.1. Obviously, if many profiles have zero frequency, it becomes more economical to drop the corresponding rows from it; we then obtain a reduced profile frequency matrix, as illustrated in Table 2.3. Another possible coding is as follows. Profiles correspond with the cells of an m-dimensional $k_1 \times k_2 \times \ldots \times k_m$ array. Inserting profile frequencies in the appropriate cells, a higher-dimensional cross tabulation is obtained. For the example this is illustrated in Table 2.4.*

Table 2.2 Profile frequency matrix

a	p	u	3
a	p	v	2
a	p	w	0
a	q	u	0
a	q	v	0
a	q	w	0
a	r	u	0
a	r	v	1
a	r	w	0
b	p	u	0
b	p	v	1
b	p	w	0
b	q	u	0
b	q	v	1
b	q	w	0
b	r	u	0
b	r	v	0
b	r	w	0
c	p	u	0
c	p	v	2
c	p	w	0
c	q	u	0
c	q	v	0
c	q	w	0
c	r	u	0
c	r	v	0
c	r	w	0

Table 2.3 Reduced profile frequency matrix

a	p	u	3
a	p	v	2
a	r	v	1
b	p	v	1
b	q	v	1
c	p	v	2

Table 2.4 Higher-dimensional cross tabulation

	p	q	r		p	q	r		p	q	r
a	3	0	0	a	2	0	1	a	0	0	0
b	0	0	0	b	1	1	0	b	0	0	0
c	0	0	0	c	2	0	0	c	0	0	0
		u				v				w	

2.1 THE COMPLETE INDICATOR MATRIX AND ITS PROPERTIES

A third way of coding data will be of crucial interest for the type of analysis described in this text. For each variable h_j an $n \times k_j$ binary matrix G_j is defined, by taking

$$g_{(j)ir} = \begin{cases} 1 & \text{if the } i\text{th object is mapped in the } r\text{th category of } h_j \\ 0 & \text{if the } i\text{th object is not mapped in the } r\text{th category of } h_j. \end{cases}$$

G_j is called the *indicator matrix* of h_j. Such matrices can be collected in a partitioned matrix $G = (G_1, ..., G_j, ..., G_m)$ of dimension $n \times \Sigma k_j$, also called the 'indicator matrix'. For the example of Table 2.1 the indicator matrix G is shown in Table 2.5.

Table 2.5 Indicator matrix G for data matrix H of Table 2.1

a	b	c	p	q	r	u	v	w
1	0	0	1	0	0	1	0	0
0	1	0	0	1	0	0	1	0
1	0	0	0	0	1	0	1	0
1	0	0	1	0	0	1	0	0
0	1	0	1	0	0	0	1	0
0	0	1	1	0	0	0	1	0
1	0	0	1	0	0	1	0	0
1	0	0	1	0	0	0	1	0
0	0	1	1	0	0	0	1	0
1	0	0	1	0	0	0	1	0

The indicator matrix G_j is said to be *complete* if each row of G_j has only one element equal to unity and zeros elsewhere, so that row sums of G_j are equal to unity. The latter can be written as $G_j u = u$, where u is a vector of unit elements. If all G_j are complete, their combined matrix G is also said to be complete, and it then follows that $Gu = mu$: rows of G add up to m.

Let d_j be the vector of the column totals of G_j. Its rth element $d_{(j)r}$ corresponds to the marginal frequency of the rth category of h_j. Also, the sum of the elements of d_j must be equal to $u'd_j = n$. Write $D_j = G_j'G_j$. This matrix D_j is diagonal (columns of G_j are orthogonal), and its diagonal elements are the same as the marginal frequencies given in d_j.

Define $C_{jl} = G_j'G_l$; it is a two-dimensional cross tabulation of variables h_j and h_l. Its elements correspond to the frequency of objects characterized by a particular combination of one category in h_j and one in h_l. Define $C = G'G$. This matrix C combines all C_{jl} along its diagonal with the matrices $D_j = C_{jj}$. It is a matrix of *bivariate marginals*. In the French data analysis literature C is called the 'tableau de Burt'.

Define D as the partitioned diagonal matrix of C, in the sense that elements of D and C are identical in the diagonal submatrices $C_{jj} = D_j$, whereas D has zero elements in its off-diagonal submatrices. D is a matrix of *univariate marginals*. Although D is strictly a diagonal matrix for a complete indicator matrix, we prefer to think of it as a partitioned diagonal matrix, because of later, somewhat different, applications. For the numerical example, C and D are given in Tables 2.6 and 2.7. The 3×3 submatrices along the diagonal of D are the same as those along the diagonal of C.

Table 2.6 Matrix C of bivariate marginals

	a	b	c	p	q	r	u	v	w
a	6	0	0	5	0	1	3	3	0
b	0	2	0	1	1	0	0	2	0
c	0	0	2	2	0	0	0	2	0
p	5	1	2	8	0	0	3	5	0
q	0	1	0	0	1	0	0	1	0
r	1	0	0	0	0	1	0	1	0
u	3	0	0	3	0	0	3	0	0
v	3	2	2	5	1	1	0	7	0
w	0	0	0	0	0	0	0	0	0

Table 2.7 Matrix D of univariate marginals

	a	b	c	p	q	r	u	v	w
a	6	0	0	0	0	0	0	0	0
b	0	2	0	0	0	0	0	0	0
c	0	0	2	0	0	0	0	0	0
p	0	0	0	8	0	0	0	0	0
q	0	0	0	0	1	0	0	0	0
r	0	0	0	0	0	1	0	0	0
u	0	0	0	0	0	0	3	0	0
v	0	0	0	0	0	0	0	7	0
w	0	0	0	0	0	0	0	0	0

2.2 QUANTIFICATION

Categories of variables may be numerical values, like midpoints of intervals on some continuous variable. In that case the $n \times m$ data matrix $\mathbf{H}$ is a 'classical' multivariate data matrix and can be handled with classical techniques of linear MVA. Most of the techniques to be discussed in this text, however, do not assume such an *a priori* quantification. Even in the case where an *a priori* quantification is available, such quantification could be ignored and replaced by a 'nominal' categorization.

> *Suppose we have a variable 'age' that maps individuals into 15 age groups, each group being characterized by an interval midpoint on the age scale. The data matrix would give for age a single column with 15 possible numerical values. The corresponding indicator matrix would have 15 columns, one for each age group. We might, from then on, as it were, 'forget' that these 15 categories were a priori ordered on an 'interval scale', and interpret them as 15 nominal categories.*

Quantification of categories should, of course, follow rules, with the intention of optimizing some criterion or, in other words, with the intention of minimizing some loss function. For the moment we shall not discuss loss functions, however, but indicate in a global way how quantification of an indicator matrix is feasible. Quantification of categories of variable h_j implies that these k_j categories are mapped as the k_j numerical values of a vector $\mathbf{y}_j$. Then the quantified variable $\mathbf{q}_j = \mathbf{G}_j \mathbf{y}_j$ becomes a single vector which gives a numerical result for each object with respect to h_j. Define $\mathbf{x}$ as the mean vector of all $\mathbf{q}_j$, i.e.

$$\mathbf{x} = m^{-1} \Sigma \mathbf{q}_j. \tag{2.1}$$

This vector $\mathbf{x}$ now will contain the quantification of the objects, and we will say that for some *direct* quantification $\mathbf{y}_j$ of categories, $\mathbf{x}$ is the *induced* score of objects. On the other hand, let $\mathbf{x}$ be some direct quantification of the objects. We then define the induced category quantification of a category as the average of the scores of those objects that are mapped into that category. In formula:

$$\mathbf{y}_j = \mathbf{D}_j^{-1} \mathbf{G}_j' \mathbf{x}. \tag{2.2}$$

(The latter assumes that D_j has an inverse, which implies that there are no categories with zero frequency. If some category has zero frequency, we may as well skip its column from the indicator matrix.)

The two procedures can be connected as follows. Let y_j be a direct quantification of the categories of the jth variable. Let y be a vector that combines all y_j in a single vector with $\sum k_j$ elements. Induced object scores then are Gy/m. We now require that a solution for the direct quantification of objects, x, must be proportional to the induced object scores, and vice versa, that the direct category quantification y_j must be proportional to the induced category quantification $D_j^{-1}G_j'x$. In Chapter 3 it will be shown that this requirement not only makes solutions for x and y feasible but it also results in minimization of attractive loss functions.

The discussion above should not suggest that there is only one solution for x and y. In general, we might be interested in p different solutions. This implies that the category quantification corresponds to a matrix Y_j of dimension $k_j \times p$. Induced object scores then are given in the $n \times p$ matrix GY/m. Similarly, given an $n \times p$ quantifications of the objects, induced category quantifications appear in the $k_j \times p$ matrices $D_j^{-1}G_j'X$.

2.3 THE INCOMPLETE INDICATOR MATRIX

Thus far we have described a complete indicator matrix. Its typical feature is that each row of G_j adds up to unity. This could be stated more formally by defining M_j as the diagonal matrix of row totals of G_j. For a complete indicator matrix G_j it must then be true that $M_j = I$. Also, define $M_* = \sum M_j$. For a complete indicator matrix G this implies $M_* = mI$. An indicator matrix G_j is *incomplete* if it has rows with only zero elements.

Let G have n rows, one for each of n individual parliamentarians. Columns of G correspond to m different proposals. G has entry '1' if the corresponding parliamentarian voted 'in favour' of the proposal, and '0' otherwise. The matrix could be completed by adding, for each proposal, a second column in which '1' is registered if the individual voted 'not in favour', and '0' otherwise. This creates for each proposal a complete indicator matrix G_j with two columns, and $M_j = I$.

Another example is that of missing data. It will be discussed more fully in Section 2.4. For the moment we remark that, if an individual has missing data on the jth variable, this could be coded by registering only zeros in the corresponding row of $\mathbf{G}_j$. Then $\mathbf{G}_j$ is incomplete, but could be completed by adding a column with entries '1' for individuals with missing data.

An incomplete indicator matrix can be quantified according to the same principles as outlined in Section 2.2 for the complete case. Again we want object scores to be proportional to the vector of average category quantifications for categories that apply to the object and vice versa category quantifications proportional to the average score of objects within the category. In formulas, we require

$$\mathbf{x} \;\propto\; \mathbf{M}_*^{-1}\mathbf{G}\mathbf{y} \tag{2.3}$$

$$\mathbf{y}_j \;\propto\; \mathbf{D}_j^{-1}\mathbf{G}_j'\,\mathbf{x}. \tag{2.4}$$

A solution based on these requirements will be different from a solution based on the completed indicator matrix. The reason is that, in general, object scores will become more similar to the extent that two objects have more categories in common. In the example with missing data, for instance, if we add to $\mathbf{G}_j$ a column with '1' for each object with missing data on that variable, the effect will be that such objects will be quantified closer together, as if 'having missing data' can be interpreted as a 'positive' characteristic shared by those objects.

(a) In the example of the parliamentarians, completion of the matrix implies that parliamentarians who vote 'not in favour' will be considered as being in the 'same category' in this respect. However, parliamentarians can vote against some proposal for opposite reasons, and it follows that analysis of the incomplete indicator matrix could be more realistic than analysis of the completed matrix.

(b) A well-known method in archaeology is 'seriation'. In the graves of an ancient graveyard, fragments of pottery are found, with different decorative motives on them. If motives belonging to two different graves are quite similar, one might assume that the graves are about equally old. The seriation problem is solved if it turns out to be possible to find an order for the graves, and simultaneously an order for the motives, in such a way that the

'parallelogram' structure is found which is illustrated in Table 2.8a. For the incomplete matrix the oldest and the newest grave have nothing in common and it can be expected that they will be quantified at opposite ends of a scale. In the completed matrix (Table 2.8b), however, these two graves share the characteristics 'b–' and 'c–'. This must lead to a solution where these two graves come closer together (the numerical solution for this example is given in Section 3.12). In the context of multidimensional scaling theory the present example illustrates a Coombs scale. Typical of the representation of a Coombs scale is that it maps objects as segments of the continuum in such a way that each segment includes the categories that apply to the object. It then follows that objects which do not share some category are not necessarily close together. (We come back to this subject in Section 8.2, on unfolding theory.)

Table 2.8a Incomplete indicator matrix

	a	b	c	d	e	f
I old	1	0	0	0	0	0
II	1	1	0	0	0	0
III	0	1	1	1	0	0
IV	0	0	1	1	1	0
V new	0	0	0	1	1	1

Table 2.8b Completed indicator matrix

	a		b		c		d		e		f	
	+	–	+	–	+	–	+	–	+	–	+	–
	1	0	0	1	0	1	0	1	0	1	0	1
	1	0	1	0	0	1	0	1	0	1	0	1
	0	1	1	0	1	0	1	0	0	1	0	1
	0	1	0	1	1	0	1	0	1	0	0	1
	0	1	0	1	0	1	1	0	1	0	1	0

(c) Table 2.9a gives a typical example of a Guttman scale. Objects and categories can be ordered in such a way that zeros appear only in the upper right corner. In other words, an object that has some category, must also have all categories to the left of that category. Completing the matrix results in Table 2.9b (two adjacent columns, one registering 'presence' and the other 'absence' of the category). This completed matrix can be reordered as in Table 2.9c. This last table shows the typical Guttman parallelogram structure, and the scaling solution of this completed matrix remains consistent with the ordering implied in the incomplete matrix.

Table 2.9a Guttman scale

	a	b	c	d
1	0	0	0	0
2	1	0	0	0
3	1	1	0	0
4	1	1	1	0
5	1	1	1	1

Table 2.9b Completed indicator matrix for Guttman scale

	a +	a −	b +	b −	c +	c −	d +	d −
1	0	1	0	1	0	1	0	1
2	1	0	0	1	0	1	0	1
3	1	0	1	0	0	1	0	1
4	1	0	1	0	1	0	0	1
5	1	0	1	0	1	0	1	0

Table 2.9c Ordered completed indicator matrix for Guttman scale

	a−	b−	c−	d−	a+	b+	c+	d+
1	1	1	1	1	0	0	0	0
2	0	1	1	1	1	0	0	0
3	0	0	1	1	1	1	0	0
4	0	0	0	1	1	1	1	0
5	0	0	0	0	1	1	1	1

2.4 MISSING DATA

A special and ever recurring problem in MVA is the presence of missing data. They can occur for a variety of reasons: a subject left a blank on his response sheet, an experimental animal died, an 'impossible' code has been entered in the code book, an archaeological fragment is so damaged that one cannot decide whether a certain motive was ever present or not. In classical MVA many ways of handling missing data have been proposed, such as (a) to insert a random value selected from the range of possible values, (b) to insert the mean of the variable, (c) to insert the best prediction from other variables. Alternatively, for derived data such as a correlation matrix, one might insert values that minimize the rank of this matrix, or that maximize its largest eigenvalue, etc. Sometimes inserting values for missing data has no other purpose than making it possible to perform standard calculations. The inserted values are 'stand-ins', so to speak, and should leave the scene before results are presented. However, there are also situations where it is of special interest to make a sophisticated estimate about the missing value (as in the example of a damaged fragment, where the archaeologist will want to make the best guess as to whether or not a motive was present). Still another way of handling missing data is to throw away each object with missing data. This option has little to do with data analysis as such, and we shall not discuss it.

From the point of view of nonlinear data analysis we shall distinguish between the following three options:

(i)   the indicator matrix is left incomplete;
(ii) the indicator matrix is completed with a single additional column for each variable with missing data;
(iii) the indicator matrix is completed by adding to G_j as many additional columns as there are missing data for the *j*th variable.

Option *(i)* is called *missing data passive* or *missing data deleted*. It implies that when an object has missing data for the *j*th variable, the corresponding row of G_j is a zero row. Option *(ii)* is called *missing data single category*. It implies that one extra column is added to G_j, with entry '1' for each object with missing data on the *j*th variable. Missing data are thus treated as if they are a category by themselves. Objects with missing data are handled as if they are in the same category in this respect. Option *(iii)* is called *missing data multiple categories*. It adds to G_j as many extra columns as there are objects with missing data on the *j*th variable, and each such column has only one entry '1'. The option handles missing data as if for each individual each single missing datum forms a category of its own. With not too many missing data, distributed randomly over objects and variables, the three options will have roughly the same results. However, when missing data cluster at some individuals (or variables), results can be rather different.

Table 2.10a Incomplete indicator matrix with missing data corresponding to option *(i)*

a	b	c		p	q	r		u	v
0	0	0		1	0	0		1	0
0	1	0		0	1	0		0	1
0	0	0		0	0	1		0	1
1	0	0		1	0	0		1	0
0	1	0		1	0	0		0	1
0	0	1		1	0	0		0	1
1	0	0		1	0	0		1	0
1	0	0		1	0	0		0	1
0	0	1		0	0	0		0	1
1	0	0		1	0	0		0	1

Table 2.10b Completed indicator matrix, missing data single category (option *(ii)*)

a	b	c	?	p	q	r	?	u	v
0	0	0	1	1	0	0	0	1	0
0	1	0	0	0	1	0	0	0	1
0	0	0	1	0	0	1	0	0	1
1	0	0	0	1	0	0	0	1	0
0	1	0	0	1	0	0	0	0	1
0	0	1	0	1	0	0	0	0	1
1	0	0	0	1	0	0	0	1	0
1	0	0	0	1	0	0	0	0	1
0	0	1	0	0	0	0	1	0	1
1	0	0	0	1	0	0	0	0	1

Table 2.10c Completed indicator matrix, missing data multiple categories (option *(iii)*)

a	b	c	?	?	p	q	r	?	u	v
0	0	0	1	0	1	0	0	0	1	0
0	1	0	0	0	0	1	0	0	0	1
0	0	0	0	1	0	0	1	0	0	1
1	0	0	0	0	1	0	0	0	1	0
0	1	0	0	0	1	0	0	0	0	1
0	0	1	0	0	1	0	0	0	0	1
1	0	0	0	0	1	0	0	0	1	0
1	0	0	0	0	1	0	0	0	0	1
0	0	1	0	0	0	0	0	1	0	1
1	0	0	0	0	1	0	0	0	0	1

Suppose that in the data matrix **H** *of Table 2.1 data were missing for individuals 1 and 3 on variable 1 and for individual 9 on variable 2. The three resulting indicator matrices are shown in Table 2.10.*

Note that the options for missing data can also be chosen when an indicator matrix is incomplete for different reasons, such as in the examples of Section 2.3. For example, if an indicator matrix is incomplete because it registers only 'presence' of categories but not 'absence', so that each G_j has only a single column, we might complete G_j with a second column registering 'absence' as '1' and 'presence' as '0', or we might complete G_j by adding as many columns as there are objects with missing data.

2.5 THE REVERSED INDICATOR MATRIX

In some cases it can be useful to 'reverse' the indicator matrix, i.e. to derive the indicator matrix from the transposed data matrix. An illustration is given in Table 2.11. Table 2.11a gives the data matrix H for five objects and three variables; the classical indicator matrix is shown in Table 2.11b. Table 2.11c gives the transposed data matrix H'. If we now treat objects as variables (and variables I, II, II as objects), we obtain the reversed indicator matrix of Table 2.11d. Analysis of this reversed indicator matrix implies that we must be prepared to accept II and III as similar on the basis of the columns 2b and 4a, showing that some individuals apply some category to II and III, but not to I.

Table 2.11a Data matrix H: 5 individuals, 3 variables

	I	II	III
1	a	b	c
2	a	b	b
3	b	b	c
4	c	a	a
5	b	c	b

Table 2.11b Classical indicator matrix

	I			II			III		
	a	b	c	a	b	c	a	b	c
1	1	0	0	0	1	0	0	0	1
2	1	0	0	0	1	0	0	1	0
3	0	1	0	0	1	0	0	0	1
4	0	0	1	1	0	0	1	0	0
5	0	1	0	0	0	1	0	1	0

Table 2.11c Transposed matrix H'

	1	2	3	4	5
I	a	a	b	c	b
II	b	b	b	a	c
III	c	b	c	a	b

Table 2.11d Reversed indicator matrix data

	1a	1b	1c	2a	2b	3b	3c	4a	4c	5b	5c
I	1	0	0	1	0	1	0	0	1	1	0
II	0	1	0	0	1	1	0	1	0	0	1
III	0	0	1	0	1	0	1	1	0	1	0

The most logical application of a reversed indicator matrix is that of a sorting task: individuals are presented with a number of objects and are asked to sort the objects into as many categories as they like, where it is left to the individual's own fancy how he or she wants to define the categories. Data then consist of different groupings for each individual, on the basis of categories that are not comparable across individuals. Such an example will be given in Section 3.13.

Whether the reversed indicator matrix is also of interest in other cases depends on the question as to what extent data can be interpreted as if derived from a sorting task. Analysis of a classical indicator matrix quantifies objects and categories, but does not quantify 'variables'. Analysis of the reversed indicator matrix quantifies variables and categories per individual, but does not quantify individuals.

2.6 THE INDICATOR MATRIX FOR A CONTINGENCY TABLE

An indicator matrix for a two-dimensional $n_r \times n_c$ contingency table is obtained as follows. The matrix will have as many rows as indicated by the total frequency (the sum of all cells of the contingency table). The matrix will have

Table 2.12a Contingency table for 15 objects

	a	b	c
p	4	2	0
q	1	3	2
r	0	1	2

Table 2.12b Indicator matrix

	p	q	r	a	b	c
1	1	0	0	1	0	0
2	1	0	0	1	0	0
3	1	0	0	1	0	0
4	1	0	0	1	0	0
5	1	0	0	0	1	0
6	1	0	0	0	1	0
7	0	1	0	1	0	0
8	0	1	0	0	1	0
9	0	1	0	0	1	0
10	0	1	0	0	1	0
11	0	1	0	0	0	1
12	0	1	0	0	0	1
13	0	0	1	0	1	0
14	0	0	1	0	0	1
15	0	0	1	0	0	1

$(n_r + n_c)$ columns: the first n_r columns for row categories and the last n_c columns for column categories. Each row will have two entries '1': one for the row category that applies to the individual and one for the column category. Table 2.12 gives an example. Obviously such an indicator matrix is not an efficient way of coding data, but sometimes it will be theoretically convenient to imagine data coded in this way.

2.7 GROUPING OF CATEGORIES

Sometimes it makes sense to group categories into a single category. An example is an indicator matrix G_j for one item of a multiple choice examination, with four response categories, one of which is correct. Instead of setting up G_j with four columns (one for each response category), one might take only two columns (one for the correct answer, and one for the other answers).

Results of the two types of indicator matrices will not necessarily be similar. For example, suppose that individuals differ in response bias (some individuals systematically avoid response category 'a', others favour it). Such a response bias would not be revealed by the analysis based on two categories per item. Also, the analysis based on the four response categories will produce a scaling of the wrong answers as to their 'degree of wrongness'. This might be useful in particular when it turns out that some wrong response is chosen by the majority of individuals with many correct responses on the other items; such a result gives reason for a close inspection of the content of the deviant item. An illustration is given in Section 13.1. When categories are grouped, their quantification will not necessarily be something like the average of the quantifications they would have obtained before grouping. When one groups categories, one makes the *a priori* decision that they must have equal weights, and that their contribution to the object score must be the same. Ungrouped categories, on the other hand, obtain differential weights.

2.8 GROUPING OF VARIABLES

Sometimes the number of categories is expanded by creating a new variabl with as many categories as the number of possible combinations of categorie of the initial variables. For example, let one variable have categories 'a' and 'b' and another one categories 'p', 'q', 'r'. The combined variable will have categories 'ap', 'aq', 'ar', 'bp', 'bq', 'br'.

Grouping of variables can be useful if one suspects that the variables have *higher-order interaction* (in the sense of analysis of variance). Without grouping, the contribution of a combination of categories to the object score $\mathbf{G}\mathbf{y}/m$ is *additive*. When variables are grouped, their contribution no longer needs to be additive (which is precisely the 'interaction' effect). For more details, compare Section 5.3. Obviously, more than two variables can be grouped. The most extreme case is that all variables are grouped. The number of combined categories then becomes equal to the number of possible response patterns or profiles. Columns of the indicator matrix will add up to marginal profile frequencies. Analysis of profile data will be discussed in Section 8.5.3 on ANAPROF. The following two examples illustrate special cases of grouping.

(a) A common procedure for collecting preference data for m stimuli S_j $(j = 1,...,m)$ is the method of paired comparisons. It presents all possible pairs of stimuli (S_j, S_l) $(j \neq l)$, and for each pair the individual is asked to say which of the two is preferred. The (incomplete) indicator matrix for n individuals would have $1/2m(m-1)$ columns, one for each pair, with entry '1' if the first stimulus of the pair is preferred and '0' otherwise. The indicator matrix can be completed by interpreting each pair as a 'variable', and defining indicator matrices $\mathbf{G}_{(j,l)}$ with two columns, the first one with entry '1' if S_j is preferred, the other one with entry '1' if S_l is preferred.

(b) The method of triads is a well known method for collecting similarity data (Coombs, 1964). Given stimuli S_p $(p = 1,...,m)$ all possible $m(m-1)(m-2)/6$ triplets of stimuli are formed, and the individual is asked to tell, for each triplet, which two stimuli are most alike and which two are least alike. Obviously, there are six possible responses to each triad (S_p, S_q, S_r):

| | | Most similar | | |
		pq	pr	qr
Least	pq	–	prq	qrp
similar	pr	pqr	–	rqp
	qr	qpr	rpq	–

The entries in this table give the six possible similarity orderings (assuming that individuals are consistent and their choices transitive). A possible way of coding would be to create for each triad an indicator matrix $G_{(p,q,r)}$ with six columns, one for each response pattern.

2.9 EPILOGUE

New work on coding, developed mainly in France, has been reviewed in Van Rijckevorsel (1987). Special ways of coding are frequently motivated by smoothness considerations, but Heiser (1985) shows that choice of basis may be guided by ordination models as well. The effect of choosing various missing data options has been studied in Meulman (1982).

CHAPTER 3
HOMOGENEITY ANALYSIS

There is a *strict* sense and a *broad* sense in which one can use the term homogeneity analysis. In the strict sense homogeneity analysis is a technique for the analysis of purely categorical data, with a particular loss function that defines it and with a particular method for finding an optimal solution. This technique and its properties will be our concern in the second part of this chapter, from Section 3.8 onwards. Homogeneity analysis in the broad sense refers to a class of criteria for analysing multivariate data in general, sharing the characteristic aim of *optimizing the homogeneity of variables* under various forms of manipulation and simplification. This class of criteria will be introduced in the first part of this chapter, in a sequence of increasing complexity and with some discussion of historical antecedents and ramifications. ➤

At a number of points in the present and the next chapters, knowledge of matrix algebra is necessary for a proper understanding of the results. The interested reader who is unfamiliar with this topic is advised to study Appendix B as a first introduction, or to use the books by Green and Carroll (1976) or Van de Geer (1971) as accompanying texts.

3.1 HOMOGENEITY OF VARIABLES

Historically, the idea of homogeneity is closely related to the idea that different variables may measure 'the same thing'. If the latter were perfectly valid, the data matrix (assuming that the variables are in deviations from their mean, i.e. they are centered, and that they are identically normalized) would turn out with identical values in each row or, if we plot observations as profiles, each profile would be a straight horizontal line. If the idea of 'measuring the same thing' were imperfectly true (variables measure the same thing, but with random error), rows of the data matrix may have elements that vary somewhat (more to the extent that measurement error increases). A graph of profiles then would then show zigzag curves at different levels. Replacing such profiles by a straight line then implies some *loss of information*. Variables are homogeneous if the loss is relatively small. This is illustrated in Figure 3.1.

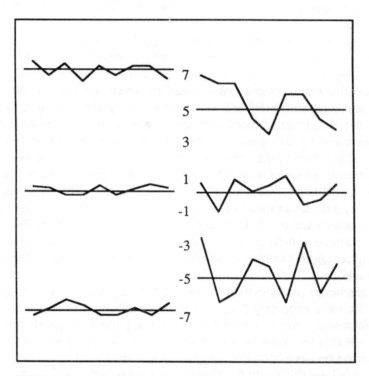

Figure 3.1   The three sets of curves on the left have very different levels, but vary little within levels. They illustrate homogeneity of variables. The three curves on the right have less variation between levels and more variation within levels. The set of curves on the right is therefore less homogeneous than the set on the left.

3.2 HISTORICAL PRELIMINARIES OF DIFFERENTIAL WEIGHTING

Since the early beginnings of quantitative social science there has been a lively interest in the problem of reduction of multivariate data to univariate scales by means of 'weighted averaging'. In 1888 Edgeworth wrote a paper on 'The statistics of examinations', in which he discusses how various parts of an examination might be weighted relative to each other. In 1913 Spearman published a paper 'Correlations of sums and differences' in which he investigates the effects of differential weighting of variables and in which he gives a basis for multiple regression and canonical analysis. In the thirties the problem obtained a new impetus from work on attitude scales by Thurstone and Likert.

In all these studies the basic problem was how to define the univariate scale: by simply adding scores on different variables? Or by some sophisticated differential weighting method? Empirical studies, especially those from the field of mental test theory (where the problem is whether we can simply add scores on separate items of a test or should give differential weights to the items), showed that differential weighting had little effect. This literature has been reviewed by Gulliksen (1950, ch. 20) and by Burt (1950a, and also 1948a, 1951). Guilford's general conclusion (in the 1936 edition of his *Psychometric Methods*) was 'weighting is not worth the trouble'. One should qualify this conclusion against the context of mental test theory, where items are usually selected in such a way that they correlate highly with each other. To substantiate Guilford's conclusion we give some results from classical psychometrics – without proof.

Let $h_1, ..., h_m$ be stochastic variates, with expected value $E(h_j) = 0$ and variance $V(h_j)=1$. Define $r_{jk} = E(h_j h_k)$. Let $v = \Sigma a_j h_j$ and $w = \Sigma b_j h_j$, so that v and w are weighted sum variates with different weights a_j and b_j. Obviously $r_{(v,w)}$, the correlation between v and w, can take any value between +1 and −1. However, if we require that all weights a_j and b_j are nonnegative, a lower bound for $r_{(v,w)}$ is the smallest correlation in the correlation matrix **R** of correlations between h_j and h_k. Suppose now that we simply add variables to obtain w (all weights b_j are equal to unity). In this case a better lower bound for $r_{(v,w)}$ is available: if all elements a_j are nonnegative and all elements b_j are equal to unity, then $r_{(v,w)}$ cannot be smaller than the smallest average correlation for the rows of **R**. This shows that if all correlations in **R** are large, the correlation between any linear compound with nonnegative weights and the simple sum variate necessarily must be large, too. This then is an argument to advocate the simple sum as a reasonable choice for a univariate compression of multivariate data. A recent example of such a plea can be found in Wainer (1976), from whose paper we quote: 'Estimating coefficients in linear models: it don't make no never mind.' (For a critical review, cf. Rozeboom, 1979.) Similar arguments as mentioned above can be derived from selection of random weights. This idea has been explored by Wilks (1938), Burt (1950a), and Gulliksen (1950). The general conclusion is that $r_{(v,w)}$ approaches unity to the extent that

(a)   variation in weights is small,
(b)   the average correlation is large,
(c)   the number of variables is large,

a very clean and acceptable result. Its practical significance, however, is limited. As with most probabilistic arguments, it is based on an 'ideal' situation. Also, in practice one never selects random weights, but one will take weights that are in some sense 'optimal'. For binary variables there are theoretical models, like those of Birnbaum or Rasch, from which optimal weights can be derived. The same is true for Spearman's 'one factor model' and for Guttman scales. We discuss such models in more detail in Chapter 9.

At any rate, in the context of mental test theory (how to define the overall test score as a differentially weighted sum of item scores) the conclusion is that the weighting problem seems trivial. However, in a different context the problem remains: to what extent can we replace a number of stochastic variates by one single variate? The problem is implied in a classical paper by Galton (1888), in which he introduced the correlation coefficient. At the end of this paper one reads:

> Neither is it necessary to give examples of a method by which the degree may be measured, in which the variables in a series, each member of which is the summed effect of n variables, may be modified by their partial correlation. After transmuting the separate variables as above, and summing them, we should find the probable error of any of them to be $\sqrt{n}$ if the variables were perfectly independent, and n if they were rigidly and perfectly co-related (pp. 144–145).

Permitting ourselves, following in Burt's steps, a liberal interpretation of this somewhat obscure quotation, it says that the average correlation is a measure for homogeneity among a number of variables. A formal proof is as follows. Assume that all $\mathbf{h}_j$ are standardized. Let $\mathbf{x}$ be the candidate for replacing all $\mathbf{h}_j$. Such a replacement implies loss of information, to be evaluated by the loss function

$$\sigma(\mathbf{x}) \equiv m^{-1} \sum_j \mathrm{SSQ}(\mathbf{x} - \mathbf{h}_j) . \tag{3.1}$$

The notation $\mathrm{SSQ}(\mathbf{v})$ is used throughout this book to denote the sum of squares of the elements of the vector $\mathbf{v}$. Clearly, $\sigma(\mathbf{x}) = 0$ only if $\mathbf{x} = \mathbf{h}_j$ for all j, which implies that all $\mathbf{h}_j$ are identical. Let

$$\sigma(*) \equiv \min \{ \sigma(\mathbf{x}) \mid \mathbf{x} \} \tag{3.2}$$

be the minimum of $\sigma(\mathbf{x})$. It is obtained by taking $\mathbf{x} = \mathbf{h}_.$ (the mean of all $\mathbf{h}_j$). The minimum value of the loss function then becomes

$$\sigma(*) = 1 - \text{SSQ}(\mathbf{h}_.) = 1 - r_{..} , \qquad\qquad (3.3)$$

where $r_{..}$ is the average correlation between all $\mathbf{h}_j$ (including $r_{jj} = 1$). This result corresponds with Galton's observation.

These concepts, as well as some additional ones, are illustrated by the following small numerical example.

$$Let\ \mathbf{H} = \begin{pmatrix} 1 & 3 & 2 \\ 3 & -1 & 1 \\ 2 & 1 & -1 \\ -6 & -3 & -2 \end{pmatrix}.$$

Can we replace the columns of $\mathbf{H}$ *by a single vector* $\mathbf{x}$ *without re-scaling the original columns? The best solution for* $\mathbf{x}$ *is the vector of row means*

$$\mathbf{x} = 1/3 \begin{pmatrix} 6 \\ 3 \\ 2 \\ -11 \end{pmatrix}.$$

The total sum of squares T for $\mathbf{H}$ *is the trace of*

$$\mathbf{H'H} = \begin{pmatrix} 50 & 20 & 15 \\ 20 & 20 & 20 \\ 15 & 10 & 10 \end{pmatrix}, \quad with\ T = (50+20+10) = 80.$$

If we replace each column of $\mathbf{H}$ *by* $\mathbf{x}$*, the resulting sum of squares becomes* $B = 3\mathbf{x'x} = 56.67$*. It is called B from Between, because it depends only on differences between rows (elements within a row are identical). A direct expression is* $B = \mathbf{u'H'Hu}/m$*; a direct expression for the total sum of squares T is* $T = \mathbf{u'Du}$*, where* $\mathbf{D}$ *is the diagonal matrix of* $\mathbf{H'H}$*:*

$$\mathbf{D} = \begin{pmatrix} 50 & & \\ & 20 & \\ & & 10 \end{pmatrix}.$$

Define $W = T - B$. The symbol W comes from Within, since W gives the sum of squares of deviations from row means within rows. W is a measure of absolute loss; in the example $W = -56.67 = 23.33$. We could also define $W/T = 1 - B/T$ as a measure of relative loss; in the example $W/T = 0.29$. Suppose now, as a further step, that columns of H are equally normalized (to unity). This can be expressed as $HD^{-1/2}$ (divide h_1 by $\sqrt{50}$, h_2 by $\sqrt{20}$, and h_3 by $\sqrt{10}$). The result is

$$HD^{-1/2} = \begin{pmatrix} 0.14 & 0.67 & 0.63 \\ 0.42 & -0.22 & 0.32 \\ 0.28 & 0.22 & -0.32 \\ -0.85 & -0.67 & -0.63 \end{pmatrix}, \ with$$

$$D^{-1/2}H'HD^{-1/2} = \begin{pmatrix} 1.00 & 0.63 & 0.67 \\ 0.63 & 1.00 & 0.71 \\ 0.67 & 0.71 & 1.00 \end{pmatrix}$$

(the correlation matrix R). Row means of $HD^{-1/2}$ now become

$$x = \begin{pmatrix} 0.48 \\ 0.17 \\ 0.06 \\ -0.72 \end{pmatrix},$$

and we find $B = u'D^{-1/2}H'HD^{-1/2}u/m = u'Ru/m = 2.34$, whereas $T = m = 3$ and $W = T - B = 0.66$. Relative loss becomes $W/T = 0.22$. The average correlation (averaged over all nine elements of R) becomes $r_{..} = 7.02/9 = 0.78$, which illustrates $W/T = 1 - r_{..}$. Correlations between x and the columns of H are

$$r_{(x,h)} = \begin{pmatrix} 0.87 \\ 0.88 \\ 0.90 \end{pmatrix},$$

with average $\bar{r}_{(x,h)} = 0.88 = (0.78)^{1/2}$, which illustrates the equality

$$W/T = 1 - (\bar{r}_{(x,h)})^2.$$

3.3 MAXIMIZING HOMOGENEITY BY LINEAR WEIGHTING: PRINCIPAL COMPONENTS ANALYSIS

The average correlation of the variables provides an estimate of how well they can be reduced to one score vector if we keep them exactly in their original form; for obtaining the new scores we have to do nothing more than averaging the variables. Suppose now that it is allowed to *rescale* the variables before averaging, i.e. to allocate *weights* to $\mathbf{h}_j$, in an attempt to further increase homogeneity. Let $\mathbf{x}$ again be an arbitrary score vector of dimension n and with zero mean. Let $\mathbf{a}$ be a vector of m weights. Rescaling columns of $\mathbf{H}$ comes to the same as replacing $\mathbf{h}_j$ by $a_j\mathbf{h}_j$. The problem thus becomes that of maximizing homogeneity by a suitable choice of $\mathbf{x}$ *and* $\mathbf{a}$. More explicitly, we could minimize the *departure from homogeneity*, as measured by the loss function

$$\sigma(\mathbf{x},\mathbf{a}) \equiv m^{-1} \sum_j \mathrm{SSQ}(\mathbf{x} - a_j\mathbf{h}_j). \tag{3.4}$$

Obviously, this loss function attains a trivial absolute minimum of zero if we take $\mathbf{x} = \mathbf{0}$ and $\mathbf{a} = \mathbf{0}$. To exclude this trivial solution it is necessary to normalize $\mathbf{x}$ so that $\mathbf{x'x} = c$, where c is any prechosen constant different from zero, or to normalize $\mathbf{a}$ in a similar fashion.

The objective of choosing scores and weights so as to maximize homogeneity or to minimize departure from homogeneity (3.4) is one of the possible definitions of finding the (first) *principal component* of $\mathbf{H}$. It involves *linear* weighting, since $a_j\mathbf{h}_j$ can be viewed as a linear transformation of $\mathbf{h}_j$. Actually calculating $\mathbf{x}$ and $\mathbf{a}$ is more involved than simple averaging. Yet we shall see in the next section that a *sequence of weighted averages* is sufficient to approximate the solution as closely as desired.

3.4 ALTERNATING LEAST SQUARES ALGORITHMS FOR LINEAR WEIGHTING

In the following we shall give algorithms for finding optimal scores $\mathbf{x}^*$ and weights $\mathbf{a}^*$ for the linear weighting problem. These algorithms are based on the principle of *alternating least squares*. This means that the algorithms proceed in alternating steps, where in one step the loss function is minimized with respect to $\mathbf{x}$ for fixed $\mathbf{a}$ and in the other step the loss function is minimized with respect to **a** for fixed **x**. We shall first describe two varieties of these algorithms, corresponding to the two ways of normalization mentioned in connection with the trivial solutions for the loss function (3.4). In one **x** is normalized whereas the scale of **a** is left free and in the other **a** is normalized whereas the scale of **x** is left free. In order to keep the notation simple, we shall first assume that the columns of the data matrix **H**, in addition to being centered, are normalized to unity. This normalization implies that $\mathbf{H'H} = \mathbf{R}$ (the correlation matrix).

3.4.1 Normalized scores algorithm

In the normalized scores algorithm the object scores are constrained to satisfy $\mathbf{x'x} = 1$. The algorithm requires an initial arbitrary choice of $\tilde{\mathbf{a}}$ ($\tilde{\mathbf{a}} \neq \mathbf{0}$) and then proceeds with the following steps:

(1) *Update scores:* $\tilde{\mathbf{x}} \leftarrow \mathbf{H}\tilde{\mathbf{a}}/m$
(2) *Normalization:* $\mathbf{x}^+ \leftarrow \tilde{\mathbf{x}}(\tilde{\mathbf{x}}'\tilde{\mathbf{x}})^{-1/2}$
(3) *Update weights:* $\mathbf{a}^+ \leftarrow \mathbf{H'x}^+$
(4) *Convergence test:* Go back to step 1, setting $\tilde{\mathbf{a}} \leftarrow \mathbf{a}^+$, as long as the values of $\mathbf{x}^+$ and $\mathbf{a}^+$ are not sufficiently stabilized (according to some preselected criterion of accuracy).

Step 1 corresponds to the unconstrained, conditional minimum of loss function (3.4) for fixed $\tilde{\mathbf{a}}$. Note that $\mathbf{H}\tilde{\mathbf{a}}/m$ is a vector of row means of the rescaled matrix with columns $\tilde{a}_j\mathbf{h}_j$. The updated scores $\tilde{\mathbf{x}} = \mathbf{H}\tilde{\mathbf{a}}/m$ therefore also minimize the relative loss W/T for a rescaled **H** with fixed weights $\tilde{\mathbf{a}}$. Step 2 is the projection of $\tilde{\mathbf{x}}$ onto the hypersphere of all unit normalized **x**; it transfers the unconstrained minimizer to the *feasible region* (the region containing all solutions that satisfy the constraints). Step 3 corresponds to the unconstrained, conditional minimum of loss function (3.4) for fixed $\mathbf{x}^+$. Because $\mathbf{x}^+$ and the columns of **H** are centered and unit normalized, $\mathbf{a}^+$ is a vector of correlations. The algorithm converges monotonically, since steps 1 and 2 together, and step 3, always yield a smaller value of the loss function, which is bounded below by zero.

Appendix B discusses extensively how $\tilde{\mathbf{x}} = \mathbf{H}\tilde{\mathbf{a}}/m$ can be interpreted as an *image* of $\tilde{\mathbf{a}}$ (and $\mathbf{a}^+ = \mathbf{H'x}^+$ is an image of $\mathbf{x}^+$). The appendix also demonstrates that a stationary point is reached when the image $\mathbf{Ha}^+ = \mathbf{HH'x}^+$ is proportional to $\mathbf{x}^+$. The normalized scores algorithm requires that **x** is a radius of a hypersphere, so that $\mathbf{H'x}$ becomes a pseudo-radius of a hyper-ellipsoid. The algorithm converges to a so-called *invariant direction*, or *principal axis* of the latter hyperellipsoid. In the appendix the algorithm is related to the singular value decomposition of $\mathbf{H}$. Let $\mathbf{H} = \mathbf{K\Lambda L'}$ be this singular value decomposition (SVD); thus $\mathbf{K}$ is the matrix of left singular vectors (satisfying $\mathbf{K'K} = \mathbf{I}$), $\mathbf{L}$ is the matrix of right singular vectors (satisfying $\mathbf{L'L} = \mathbf{I}$), and $\mathbf{\Lambda}$ is the diagonal matrix of singular values. The algorithm converges to $\mathbf{x}^*$ as the first left vector $\mathbf{k}_1$, with $\mathbf{a}^* = \lambda_1 \mathbf{l}_1$, where λ_1 is the first diagonal element of $\mathbf{\Lambda}$ and $\mathbf{l}_1$ is the first column of $\mathbf{L}$.

This implies the equalities (or *stationary equations*):

$$\mathbf{H'x}^* = \mathbf{L\Lambda K'x}^* = \mathbf{L\Lambda K'k}_1 = \lambda_1 \mathbf{l}_1 = \mathbf{a}^*, \tag{3.5}$$

$$\mathbf{Ha}^* = \mathbf{K\Lambda L'a}^* = \mathbf{K\Lambda L'}\lambda_1 \mathbf{l}_1 = \lambda_1^2 \mathbf{k}_1 = \lambda_1^2 \mathbf{x}^*. \tag{3.6}$$

In addition, at the stationary point the between and total sums of squares are

$$B = \mathbf{a}^{*'}\mathbf{H'Ha}^* / m = \lambda_1^2 \mathbf{l}_1'\mathbf{L\Lambda}^2\mathbf{L'l}_1 / m = \lambda_1^4 / m, \tag{3.7}$$

$$T = \mathbf{a}^{*'}\mathbf{Da}^* = \mathbf{a}^{*'}\mathbf{a}^* = \lambda_1^2 \mathbf{l}_1'\mathbf{l}_1 = \lambda_1^2, \tag{3.8}$$

with $\mathbf{D} = \text{diag}(\mathbf{H'H}) = \text{diag}(\mathbf{R}) = \mathbf{I}$. The sum of squares of the optimal weights is equal to the dominant eigenvalue of $\mathbf{H'H} = \mathbf{R}$. The relative loss is

$$W / T = 1 - B/T = 1 - \lambda_1^2 / m. \tag{3.9}$$

The matrix $\mathbf{L\Lambda}$ is called the *matrix of loadings* of the principal components analysis, so that $\mathbf{a}^* = \lambda_1 \mathbf{l}_1$ is the first vector of loadings, corresponding to $\mathbf{x}^*$ as the vector of component scores on the first principal component. It has already been remarked that $\mathbf{a}^+$ is a vector of correlations between $\mathbf{x}^+$ and the vectors $\mathbf{h}_j$. The solution $\mathbf{a}^*$ maximizes the sum of the squared correlations $\mathbf{a}^{*'}\mathbf{a}^* = \lambda_1^2$.

For a miniature example,

$$let\ \mathbf{H} = \begin{pmatrix} 0.707 & 0.000 \\ -0.707 & -0.707 \\ 0.000 & 0.707 \end{pmatrix},\ with\ \mathbf{R} = \mathbf{H'H} = \begin{pmatrix} 1.0 & 0.5 \\ 0.5 & 1.0 \end{pmatrix},$$

where $\mathbf{R}$ is the correlation matrix. Table 3.1 gives the results for the first two iterations and the final solution. Figures 3.2 and 3.3

*give the geometry of the solution (also compare Appendix B).
Figure 3.2 shows the plane of the two column vectors $\mathbf{h}_j$; they are
radii of the unit circle. Figure 3.3 gives the image of Figure 3.2*

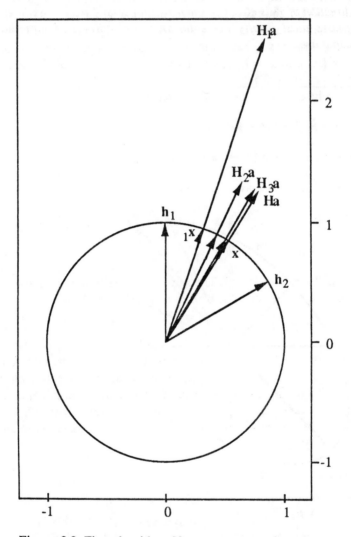

Figure 3.2 First algorithm. Vectors $_t\mathbf{x}$ are on the unit
circle. Images $\mathbf{H}_t\mathbf{a}$ of $_t\mathbf{a}$ with respect to H are shown; they
converge to $\mathbf{H}\mathbf{a}$ (t is the iteration index). $_t\mathbf{x}$ is the unit
length version of $\mathbf{H}_t\mathbf{a}$ and converges to $\mathbf{x}$.

with respect to **H'**. *The image of the unit circle now becomes an ellipse. Figure 3.3 also shows the arbitrary* vector $_1\tilde{a}$ *(the initial choice for* **a***); its image* $H_1\tilde{a}$ *is in Figure 3.2, where* $_1x^+$ *is obtained by giving* $H_1\tilde{a}$ *unit length. The image of* $_1x^+$ *becomes* $_2\tilde{a}$*, where the direction of* $_2\tilde{a}$ *is closer to the principal axis of the ellipse. This continues until finally the solution for* a^* *coincides with the principal axis.*

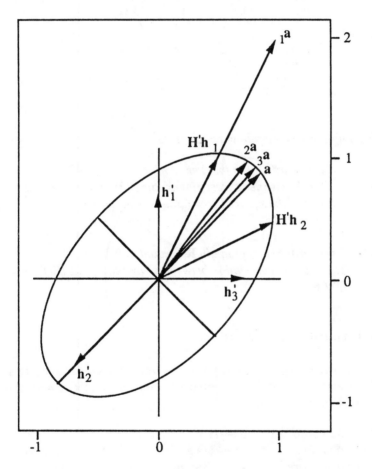

Figure 3.3 Companion to Figure 3.2. $_1a$ is an arbitrary initial vector. The dotted ellipse is the image of the unit circle in Figure 3.1. $_ta$ $(t = 2,...)$ is the image of $_tx$. $_ta$ converges to **a**, the long principal axis of the ellipse.

Table 3.1 Abridged history of iterations of the
normalized scores algorithm

Iteration	$\tilde{\mathbf{a}}$	$\mathbf{H}\tilde{\mathbf{a}}$	$\mathbf{x}^+$	$\mathbf{H'x}^+$
1	2	10.414	0.535	0.945
	1	–20.121	–0.802	0.756
		0.707	0.267	
2	0.945	0.668	0.453	0.896
	0.756	–10.203	–0.815	0.832
		0.535	0.362	
Final	0.866	0.612	0.408	0.866
	0.866	–10.225	–0.816	0.866
		0.612	0.408	

One might verify the following properties:

(a) $\mathbf{a}^*$ *is proportional to the dominant eigenvector of* **R**.
(b) $\mathbf{a}^{*'}\mathbf{a}^* = 1.5$ *is the largest eigenvalue of* **R**, *corresponding to
         the sum of the squared correlations between* $\mathbf{x}^*$ *and* $\mathbf{h}_j$.
(c)   *The ratio* $B/T = 0.75$; *the relative loss is* $1 - B/T =$
         $1 - \lambda_1^2 / m = 0.25$.
(d)   *The average correlation in* **R** *is equal to* $B/T = 0.75$.
(e)   *The average correlation between* $\mathbf{x}^*$ *and* $\mathbf{h}_j$ *equals* $(B/T)^{1/2} =$
         0.866.

3.4.2 Normalized weights algorithm

In the normalized weights algorithm the weights are constrained to satisfy
$\mathbf{a'a} = 1$. The algorithm requires an initial arbitrary choice of $\tilde{\mathbf{x}}$ ($\tilde{\mathbf{x}} \neq \mathbf{0}$), and then
proceeds with the following steps:

(1) *Update weights:* $\tilde{\mathbf{a}} \leftarrow \mathbf{H'\tilde{x}}$
(2) *Normalization:* $\mathbf{a}^+ \leftarrow \tilde{\mathbf{a}}(\tilde{\mathbf{a}}'\tilde{\mathbf{a}})^{-1/2}$
(3) *Update scores:* $\mathbf{x}^+ \leftarrow \mathbf{H}\mathbf{a}^+/m$
(4) *Convergence test:* Go back to step 1, setting $\tilde{\mathbf{x}} \leftarrow \mathbf{x}^+$, as long
                          as the values of $\mathbf{x}^+$ and $\mathbf{a}^+$ are not
                          sufficiently stabilized (according to some
                          preselected criterion of accuracy).

Again the update steps correspond to the unconstrained, conditional minimum of loss function $\sigma(\mathbf{x},\mathbf{a})$. Since $\mathbf{x}^+$ is not unit normalized, $\mathbf{a}^+$ now is a vector of covariances rather than correlations. In terms of the SVD of $\mathbf{H}$ the normalized weights algorithm converges to $\mathbf{a}^* = \mathbf{l}_1$ and $\mathbf{x}^* = \lambda_1 \mathbf{k}_1/m$, which shows that the two algorithms are basically producing the same results apart from normalization. The images of the optimal score and weight vectors are

$$\mathbf{H}'\mathbf{x}^* = \lambda_1 \mathbf{L}\Lambda\mathbf{K}'\mathbf{k}_1/m = \lambda_1^2 \mathbf{l}_1/m = \lambda_1^2 \mathbf{a}^*/m , \tag{3.10}$$

$$\mathbf{H}\mathbf{a}^* = \mathbf{K}\Lambda\mathbf{L}'\mathbf{l}_1 = \lambda_1 \mathbf{k}_1 = m\mathbf{x}^*, \tag{3.11}$$

and for the between and total sums of squares we obtain

$$B = \mathbf{a}^{*\prime}\mathbf{H}'\mathbf{H}\mathbf{a}^* / m = \mathbf{l}_1\mathbf{L}\Lambda^2\mathbf{L}'\mathbf{l}_1 / m = \lambda_1^2/m \tag{3.12}$$

$$T = \mathbf{a}^{*\prime}\mathbf{D}\mathbf{a}^* = \mathbf{a}^{*\prime}\mathbf{a}^* = 1 \tag{3.13}$$

so that again $W/T = 1 - B/T = 1 - \lambda_1^2/m$. The scores and weights obtained from the normalized weights algorithm can be rescaled to obtain the corresponding quantities in the normalized scores algorithm, and vice versa. Therefore, in practice we may execute either one of them (whatever the final normalization should be), depending on additional criteria such as accuracy and efficiency.

Table 3.2 Abridged history of iterations of the normalized weights algorithm

Iteration	$\tilde{\mathbf{a}}$	$\mathbf{H}\tilde{\mathbf{a}}/2$	$\mathbf{H}'\mathbf{x}^+$	$\mathbf{a}^+$
1	0.894	0.316	0.559	0.781
	0.447	−0.474	0.447	0.625
		0.158		
2	0.781	0.276	0.547	0.733
	0.625	−0.497	0.508	0.680
		0.221		
3	0.733	0.259	0.537	0.716
	0.681	−0.500	0.523	0.698
		0.241		
Final	0.707	0.250	0.530	0.707
	0.707	−0.500	0.530	0.707
		0.250		

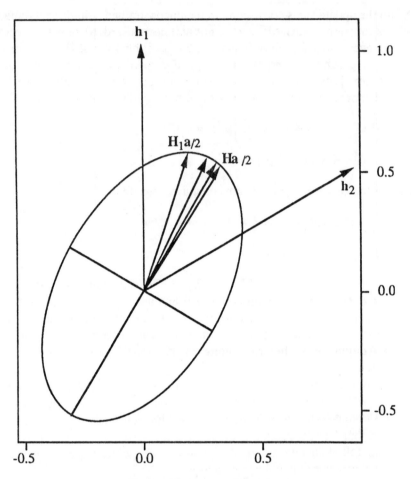

Figure 3.4 Second algorithm. $H_t a/2$ is the image of $_t a$ with respect to $H/2$. It converges to $Ha/2$, the long axis of the ellipse. This ellipse is the image of the unit circle in Figure 3.5.

For the miniature example used above, Table 3.2 gives the results for the normalized weights algorithm. Figures 3.4 and 3.5 show the geometry of the algorithm for this example. Figure 3.5 shows the unit circle of which $_1\tilde{a}$ is a radius. Its image is $_1x^+ = H_1\tilde{a}$ in Figure 3.4 and appears as the pseudo-radius of an ellipse. Successive iterations move $_tx^+$ towards the principal axis.

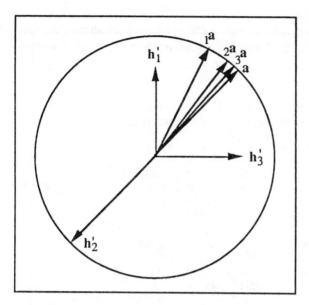

Figure 3.5 $_i$a is on the unit circle and converges to a.

3.4.3 Adaptation for heterogeneous variances

We now drop the assumption that the columns of the data matrix **H** are normalized to unity (but retain the assumption that they have zero means). Let **D** be the diagonal matrix containing the diagonal elements of **H'H**. It then follows that $HD^{-1/2}$ again has unit normalized columns. The two algorithms need the following adaptations. In the normalized scores algorithm the weights are updated according to the new step:

(3#) *Update weights:* $a^+ \leftarrow D^{-1}H'x^+$.

In terms of the SVD $HD^{-1/2} = K\Lambda L'$, the algorithm converges to a stationary point with $x^* = k_1$ and $a^* = \lambda_1 D^{-1/2}l_1$. The W/T ratio remains the same. For the normalized weights algorithm we also need to adjust the normalization step:

(2#) *Update weights:* $\tilde{a} \leftarrow D^{-1}H'x^+$
(3#) *Normalization:* $a^+ \leftarrow \tilde{a}(\tilde{a}'D\tilde{a})^{-1/2}$

Note that **a** is now scaled so that $\mathbf{a'Da} = 1$ (the total sum of squares equal to unity). The normalized weights algorithm converges to $\mathbf{a}^* = \mathbf{D}^{-1/2}\mathbf{l}_1$ and $\mathbf{x}^* = \lambda_1\mathbf{k}_1 / m$.

Table 3.3 History of iterations of the normalized scores algorithm, adapted for unequal variances of **H**

Iteration	$\tilde{\mathbf{a}}$	$\mathbf{H}\hat{\mathbf{a}}$	$\mathbf{x}^+$	$\mathbf{D}^{-1}\mathbf{H'x}^+$
1	1	6	0.460	0.130
	1	3	0.230	0.192
	1	2	0.153	0.286
		−11	−0.843	
2	0.130	1.242	0.535	0.124
	0.192	0.467	0.202	0.197
	0.268	0.184	0.079	0.283
		−1.894	−0.816	
3	0.124	1.279	0.547	0.123
	0.197	0.458	0.196	0.197
	0.283	0.162	0.069	0.284
		−1.899	−0.811	
4	0.123	1.283	0.548	0.123
	0.197	0.455	0.194	0.197
	0.284	0.159	0.068	0.284
		−1.897	−0.811	

We take the same example as in Section 3.2, with

$$\mathbf{H} = \begin{pmatrix} 1 & 3 & 2 \\ 3 & -1 & 1 \\ 2 & 1 & -1 \\ -6 & -3 & -2 \end{pmatrix}, \mathbf{H'H} = \begin{pmatrix} 50 & 20 & 15 \\ 20 & 20 & 10 \\ 15 & 10 & 10 \end{pmatrix}, \mathbf{D} = \begin{pmatrix} 50 & & \\ & 20 & \\ & & 10 \end{pmatrix}.$$

The history of the two adjusted algorithms is shown in Tables 3.3 and 3.4. The optimally rescaled data matrix becomes **Q**, *with columns* $\mathbf{q}_j = a_j\mathbf{h}_j$:

$$Q = \begin{pmatrix} 0.080 & 0.387 & 0.372 \\ 0.240 & -0.129 & 0.186 \\ 0.160 & 0.129 & -0.186 \\ -0.480 & -0.387 & -0.372 \end{pmatrix}$$

the row means of which are equal to $\mathbf{x}^*$ *(normalized weights algorithm). Also,* $\mathbf{D}^{1/2}\mathbf{a}^*$ *(normalized weights algorithm) is the first eigenvector of the correlation matrix of* $\mathbf{H}$ *(or* $\mathbf{Q}$*), which has the largest eigenvalue* $\lambda_1^2 = 2.341$*. This illustrates the close relationship between differential weighting and the singular value decomposition. For this numerical example the relative loss equals* $W/T = 0.220$*, which is not appreciably better than the result for the unweighted solution given in Section 3.2. In fact, the elements of* $(\mathbf{D}^{1/2}\mathbf{a}^*)' = (0.567 \quad 0.577 \quad 0.588)$ *are almost identical, showing that* $\mathbf{H}\mathbf{a}^*$ *will be almost proportional to* $\mathbf{H}\mathbf{D}^{-1/2}\mathbf{u}$*.*

Table 3.4 History of iterations of the normalized weights algorithm, adapted for unequal variances of $\mathbf{H}$ and started with arbitrary object scores

Iteration	$\mathbf{x}^+$	$\mathbf{D}^{-1}\mathbf{H}'\mathbf{x}^+$	$\mathbf{a}^+$	$\mathbf{x}^+$
1	1	0.48	0.096	0.258
	1	0.60	0.120	0.109
	1	0.80	0.160	0.051
	−3			−0.418
2	0.258	0.604	0.083	0.277
	0.109	0.098	0.127	0.101
	0.051	0.141	0.183	0.037
	−0.418			−0.415
3	0.258	0.063	0.081	0.279
	0.101	0.100	0.129	0.099
	0.037	0.145	0.185	0.035
	−0.415			−0.413
4	0.279	0.063	0.080	0.280
	0.099	0.101	0.129	0.099
	0.035	0.145	0.186	0.035
	−0.413			−0.413
Final	0.280	0.063	0.080	0.280
	0.099	0.101	0.129	0.099
	0.035	0.145	0.186	0.034
	−0.413			−0.413

3.4.4 Multiple solutions

The two algorithms as described find one set of weights and scores. In fact, the introductory discussion in terms of homogeneity emphasized the idea of a one-dimensional definition. The variables are weighted in such a way that their rescaled versions are as homogeneous as possible, so that they can be replaced by one new variable: the object scores. In some cases, however, it can be useful to find several of such weighting schemes at the same time. For instance, some of the variables may get small – close to zero – weights in the best one-dimensional solution, so that they do not contribute very much to the analysis; then we may be interested in finding out whether a second best solution would yield larger weights for these variables, which would indicate the existence of multiple forms of homogeneity.

We have to make sure that characteristics of the first weighting scheme are not repeated in the second weighting scheme (and perhaps in further ones), or at least that such reoccurrences are avoided as much as possible. One way to translate this objective into the optimization problem is by requiring that the object scores and the weights form an orthogonal system. Thus the multidimensional version of loss function (3.4) becomes

$$\sigma(\mathbf{X},\mathbf{A}) \equiv m^{-1} \sum_j \mathrm{SSQ}(\mathbf{X} - \mathbf{h}_j \mathbf{a}_j'), \qquad\qquad (3.14)$$

where $\mathbf{X}$ is now an $n \times p$ matrix and $\mathbf{A}$ is an $m \times p$ matrix. The vectors $\mathbf{a}_j'$ are the rows of $\mathbf{A}$. The number p is the number of weighting schemes we are interested in (the *dimensionality*).

Minimizing the departure-from-homogeneity function (3.14) can be done by at least two different algorithms again, precisely as in the one-dimensional case. The algorithms differ because we either normalize $\mathbf{X}$ or we normalize $\mathbf{A}$. We shall discuss the normalized scores algorithm in which $\mathbf{X}$ is normalized. This means here that we require $\mathbf{X}'\mathbf{X} = \mathbf{I}$. In the case of multiple dimensions the normalized scores algorithm starts with an arbitrary full rank $\tilde{\mathbf{A}}$, and consists of the following steps:

(1) *Update scores:* $\tilde{\mathbf{X}} \leftarrow \mathbf{H}\tilde{\mathbf{A}}/m$

(2) *Normalization:* $\mathbf{X}^+ \leftarrow \tilde{\mathbf{X}}(\tilde{\mathbf{X}}'\tilde{\mathbf{X}})^{-1/2}$

(3) *Update weights:* $\mathbf{A}^+ \leftarrow \mathbf{H}'\mathbf{X}^+$

(4) *Convergence test:* Go back to step 1, setting $\tilde{\mathbf{A}} \leftarrow \mathbf{A}^+$, as long as the values of $\mathbf{X}^+$ and $\mathbf{A}^+$ are not sufficiently stabilized (according to some preselected criterion of accuracy).

Step 1 corresponds again to the unconstrained, conditional minimum of loss function (3.14) for fixed $\tilde{A}$. Step 3 corresponds to the unconstrained, conditional minimum of loss function (3.14) for fixed X^+. Because X^+ and the columns of H are centered and unit normalized, A^+ is a matrix of correlations.

The normalization step 2 involves the matrix $(\tilde{X}'\tilde{X})^{-1/2}$. This matrix, which *orthonormalizes* $\tilde{X}$, can be defined in a number of ways, since any resulting orthonormal matrix will remain orthonormal under rotations. In terms of the singular value decomposition $\tilde{X} = M \Xi N'$ we may write $\tilde{X}'\tilde{X} = N\Xi^2N'$ and define $(\tilde{X}'\tilde{X})^{-1/2} = N\Xi^{-1}N'$. We obtain $\tilde{X}(\tilde{X}'\tilde{X})^{-1/2} = M\Xi N'N\Xi^{-1}N' = MN'$. When p is large this method could become quite expensive from a computational point of view. It can be replaced in the master algorithm by the cheaper *Gram–Schmidt method*. The Gram–Schmidt method starts with unit normalizing the first column of $\tilde{X}$, then projects the second column of $\tilde{X}$ onto the space orthogonal to the first column, replaces the second column by the unit normalized antiprojection, next projects the third column of $\tilde{X}$ onto the space orthogonal to the first and second column, and so on. This process can be summarized by stating that $\tilde{X}$ is decomposed as $\tilde{X} = UT$, with $U'U = I$ and T an upper triangular matrix. Thus we can replace step 2 by two substeps:

(2a) *Gram–Schmidt decomposition*: $(U,T) \leftarrow \tilde{X}$,

(2b) *Basis selection*: $X^\# \leftarrow U$.

The adjusted algorithm is not an alternating least squares method any more, strictly speaking, because steps 1, 2a, and 2b do *not* minimize $\sigma(X,A)$, for fixed A, over X that satisfies $X'X = I$. In fact, while X^+ is the conditionally optimal X when computed through the singular value decomposition as $X^+ = MN'$, steps 2a and 2b give a suboptimal solution $X^\#$. Both X^+ and $X^\#$ are bases for the column space of X, and thus $X^\# = X^+P$ for some rotation matrix P satisfying $P'P = PP' = I$. It follows that step 3 of the algorithm computes either $A^+ = H'X^+$ or $A^\# = H'X^\# = H'X^+P$, but now $\sigma(X^\#,A^\#) = m^{-1} \sum_j$ $SSQ[(X^+ - h_j a_j^{+\prime})P] = \sigma(X^+,A^+)$, because the sum of squares operator is invariant under rotation. Thus after performing step 3 the suboptimality of steps 2a and 2b is corrected again. In terms of the value of the loss function, the net result of steps 2 and 3 is the same as that of steps 2a, 2b, and 3. However, when using the Gram–Schmidt method we have to take care of finding the exact principal axes orientation of X^* and A^* after convergence, since inaccuracies may accumulate in this process.

3.5 LINEAR WEIGHTING FOR K SETS OF VARIABLES

Let the data matrix $\mathbf{H}$ be partitioned into K sets $\mathbf{H} = (\mathbf{H}_1,...,\mathbf{H}_k,...,\mathbf{H}_K)$ where $\mathbf{H}_k$ has m_k columns, with $\sum m_k = m$. The partitioning in sets usually has a specific data analytical purpose beyond mere convenient grouping of variables. One example of such a purpose is to keep the results of the analysis unchanged when adding, in any one set $\mathbf{H}_k$, a new variable that is linearly predictable from the existing variables in $\mathbf{H}_k$. Other possible objectives in K sets analysis will be discussed more extensively in Chapter 5.

In this section we merely want to introduce a version of the K sets problem that naturally follows from the concept of homogeneity of variables as developed so far. Let $\mathbf{a}_k$ be a vector of weights for $\mathbf{H}_k$ and define the weighted combination $\mathbf{z}_k \equiv \mathbf{H}_k\mathbf{a}_k$. Collect the K new variables $\mathbf{z}_k$, called *canonical variables* in this context, in the $n \times K$ matrix $\mathbf{Z}$. Define the diagonal matrix $\mathbf{D} \equiv \text{diag}(\mathbf{Z}'\mathbf{Z})$ so that the diagonal elements of $\mathbf{D}$ are $\mathbf{a}_k'\mathbf{H}_k'\mathbf{H}_k\mathbf{a}_k$. Our objective is to find those sets of weights that make the canonical variables as homogeneous as possible. In order to prevent all $\mathbf{z}_k$ from vanishing simultaneously, we normalize on their total sum of squares. In terms of relative loss, maximum homogeneity will be obtained by solving for the vectors $\mathbf{a}_k$ in such a way that $1 - B/T$ attains its minimum, with

$$B = \mathbf{u}'\mathbf{Z}'\mathbf{Z}\mathbf{u} \,/\, K = (\textstyle\sum_k \mathbf{H}_k\mathbf{a}_k)'(\textstyle\sum_k \mathbf{H}_k\mathbf{a}_k), \qquad (3.15)$$

$$T = \mathbf{u}'\mathbf{D}\mathbf{u} = \textstyle\sum_k \mathbf{a}_k'\mathbf{H}_k'\mathbf{H}_k\mathbf{a}_k. \qquad (3.16)$$

A slight reformulation directly shows the relationship with a generalized eigenvalue problem. Collect all $\mathbf{a}_k$ in a single vector $\mathbf{a}$ (of dimension m). Define $\mathbf{E}$ as the partitioned diagonal matrix of $\mathbf{H}'\mathbf{H}$ (i.e. $\mathbf{E}$ is identical to $\mathbf{H}'\mathbf{H}$ in its diagonal submatrices $\mathbf{H}_k'\mathbf{H}_k$, but the off-diagonal submatrices $\mathbf{H}_k'\mathbf{H}_l$ with $k \neq l$ are replaced by zero submatrices). Then

$$B = \mathbf{u}'\mathbf{Z}'\mathbf{Z}\mathbf{u}/K = \mathbf{a}'\mathbf{H}'\mathbf{H}\mathbf{a}/K , \qquad (3.17)$$

$$T = \mathbf{u}'\mathbf{D}\mathbf{u} = \mathbf{a}'\mathbf{E}\mathbf{a}, \qquad (3.18)$$

so that $\mathbf{a}$ must be found in such a way that $\mathbf{a}'\mathbf{H}'\mathbf{H}\mathbf{a}/K\mathbf{a}'\mathbf{E}\mathbf{a}$ is maximized.

We give a numerical example with $K = 2$, $m_1 = m_2$, $m = 4$.

$$H = \begin{pmatrix} 1 & -1 & 3 & 0 \\ 3 & -3 & 1 & 1 \\ 2 & 1 & 2 & -1 \\ -5 & 0 & -1 & 2 \\ -1 & 3 & -5 & -2 \end{pmatrix},$$

$$H'H = \begin{pmatrix} 40 & -11 & 20 & -7 \\ -11 & 20 & -19 & -10 \\ 20 & -19 & 40 & 7 \\ -7 & -10 & 7 & 10 \end{pmatrix},$$

$$E = \begin{pmatrix} 40 & -11 & 0 & 0 \\ -11 & 20 & 0 & 0 \\ 0 & 0 & 40 & 7 \\ 0 & 0 & 7 & 10 \end{pmatrix}.$$

A solution can be obtained with the same algorithms as described in Section 3.4.3 for nonnormalized H*; the only difference is the definition of* E *(a partitioned diagonal block matrix instead of a diagonal matrix). Using the normalized weights algorithm with normalization* $a'Ea = 1$*, the result becomes*

$$a = \begin{pmatrix} 0.089 \\ 0.157 \\ 0.010 \\ 0.230 \end{pmatrix},$$

$$Ha = \begin{pmatrix} -0.038 \\ -0.422 \\ 0.584 \\ -0.914 \\ 0.791 \end{pmatrix},$$

$$Z = \begin{pmatrix} -0.068 & 0.029 \\ -0.203 & -0.220 \\ 0.334 & 0.249 \\ -0.445 & -0.469 \\ 0.381 & 0.410 \end{pmatrix}$$

with $Ha = Zu$. *Since* $a'Ea = 1$, *we have* $B/T = 1.982/2 = 0.991$, *so that relative loss is equal to* $1 - B/T = 0.009$. *Appendix B* *shows that this solution, for $K = 2$, is the solution of canonical correlation analysis, dependent on the generalized eigenvector equation $\mathbf{H'Ha} = \lambda^2 \mathbf{Ea}$, with $\lambda^2 = 1 + \rho$, where ρ is the canonical correlation coefficient (the correlation between z_1 and z_2). For the example we find $\rho = \lambda^2 - 1 = 1.982 - 1 = 0.982$. The example also shows that $z_1'z_1 = z_2'z_2 = 0.50$. This result is typical for the special case $K = 2$, and has no simple equivalent when $K \geq 3$.*

3.6 MORE HISTORICAL COMMENTS ON PCA

The argument that weighting is unnecessary implies the supposition that the largest eigenvalue λ_1^2 of the correlation matrix $\mathbf{R}$ is close to the value $mr_{..}$ (where $r_{..}$ is the average correlation). In the example in Section 3.4.3 we found this eigenvalue to be equal to 2.341. For the same example $r_{..} = 0.780$ with $mr_{..} = 2.340$, which is very close to the eigenvalue. In fact, the example did show that differential weighting gives no appreciable improvement with respect to relative loss. Somewhat more generally, it can be shown that, if $\mathbf{R}$ is a matrix of *positive* correlations, and if $r_{j.}$ denotes the average of the jth row, then

$$\min r_{j.} \leq \lambda_1^2 / m \leq \max r_{j.}. \tag{3.19}$$

In words, λ_1^2 / m will be in the range between the smallest and the largest row average of $\mathbf{R}$. It follows that for variables with large and not very much distinct correlations (such as items of the same achievement test) it will be true that 'weighting is not worth the trouble' (cf. Section 3.2), or, as Rozeboom phrased it: 'To put it bluntly, second digit precision in item weighting is generally a waste of effort' (Rozeboom, 1979, p. 296). On the other hand, in cases where row means of a correlation matrix are different, weighting can be rewarding enough. The solution sketched in Section 3.4.3, which related differential weighting to singular value decomposition, was, according to Burt, first mentioned by MacDonell (1901, p. 209). The relevant quotation is 'Professor Pearson has pointed out to me that the ideal index characters would be given if we calculated the seven directions of uncorrelated variables, that is the principal axes of the correlation "ellipsoid".' And, further on the same page: 'I propose to return in a later paper to this calculation.' As far as we know the latter promise never has been honoured. The quotation shows that Pearson was the actual inventor of the technique. His name is often mentioned in this connection, usually with reference to Pearson (1901). However, in this paper Pearson indicated an approach in which SSQ(**Ha**) is made as *small* as possible, which implies a maximum relative loss rather than a minimum.

In somewhat modernized terminology Pearson (1901) describes a model that assumes **Ha** = **0**, but where the actual observations only satisfy **Ha** = **e**, with **e** a vector of (supposedly small) 'model-disturbances'. It follows that an estimate of **a** should minimize **e'e**. The solution for **a** then becomes the eigenvector with the *smallest* associated eigenvalue. This result is not so surprising as it may look at first sight.

If there is linear dependence among the columns of **H**, there will be a solution for **a** with **Ha** = **0**. Such a solution demonstrates that the observations in **H** agree with some linear 'functional relation' between the variables. In this case the smallest eigenvalue of **H'H** is zero. Furthermore, recall that in multiple regression theory one fits a hyperplane for given predictor variables h_j and a single criterion variable **z** in such a way that the sum of the squared differences between **z** and some linear compound **Ha** is minimized. This corresponds to a minimum sum of squared distances between the observation points and the fitted hyperplane, where these distances are measured in the direction of **z**. Pearson's model, on the other hand, for a given data matrix **H** fits a hyperplane, again in such a way that the sum of the squared distances is minimized, but now the distances are measured in the direction *orthogonal to the hyperplane*.

The invention of linear weighting is often attributed to Hotelling (1933). This is not correct, as shown in the quotation above: Pearson was earlier. In addition, in Hotelling's paper PCA is introduced as a form of factor analysis, with the aim to minimize a loss function $\sum_j \text{SSQ}(h_j - a_j x)$. In terms of the concepts introduced in Section 1.8.1 this definition of loss is the *join* rather than the *meet* version of the problem (as we have discussed it so far). Hotelling shows that PCA solves the join problem. He adds: 'An easily verified property of the method is that the first of our principal components has a greater mean square correlation with the tests than does any other variable' (Hotelling, 1933, p. 422). In other words, the first principal component has a maximum for its squared factor loadings. This implies that the first principal component also gives the best weights (for a matrix with equally normalized columns), but Hotelling did not make this explicit. Later, the idea was explicitly mentioned by Horst (1936), Edgerton and Kolbe (1936), and Wilks (1938).

3.7 MAXIMIZING HOMOGENEITY BY NONLINEAR
TRANSFORMATION: NONLINEAR PCA

Up to now we have studied the homogeneity of a set of variables in two different forms. In the first one, the variables were simply replaced by their average. In the second one, differential weighting of the variables was applied, which constituted linear principal components analysis. Differential weighting is equivalent to linear transformation of the variables. In this section nonlinear transformations are introduced, and it is shown how HOMALS, the technique that will be discussed in the remainder of this chapter, incorporates nonlinear transformations of a specific type.

Nonlinear transformations enter in the PCA framework described in the previous sections by allowing for transformations in the form $\phi_j(\mathbf{h}_j)$, where ϕ_j may be any nonlinear function of the variable $\mathbf{h}_j$. In order to attain an optimal solution we might minimize the loss function

$$\sigma(\mathbf{x};\mathbf{a};\phi) \equiv m^{-1} \sum_j SSQ(\mathbf{x} - a_j\,\phi_j(\mathbf{h}_j)) \tag{3.20}$$

over the object scores $\mathbf{x}$, the weights a_j, and the nonlinear transformations ϕ_j. Note that the weights a_j can only be identified when $\phi_j(\mathbf{h}_j)$ is normalized. Because these two quantities are confounded in (3.20), we might as well subsume a_j in $\phi_j(\mathbf{h}_j)$ and write (3.20) simply as

$$\sigma(\mathbf{x};\phi) \equiv m^{-1} \sum_j SSQ(\mathbf{x} - \phi_j(\mathbf{h}_j)), \tag{3.21}$$

to be minimized under the condition $\mathbf{x}'\mathbf{x} = 1$ or, alternatively, under the condition $SSQ(\phi_j(\mathbf{h}_j)) = 1$. In this form, however, the minimization of $\sigma(\mathbf{x};\phi)$ is *trivial*. By admitting all nonlinear transformations, the space of admissible transformations is enlarged to an n-dimensional space, and thus *any* quantification of the variables will suffice in order to obtain a perfect fit, though a trivial solution.

This situation indicates that we have to impose *restrictions* on the quantifications. In Chapter 4 restrictions will be discussed that limit the admissible transformations to monotonic transformations. However, as we shall see, sometimes these restrictions are not sufficient in order to prevent trivial solutions. In these cases, as well as when no *a priori* order of the categories is available, we have to diminish the dimensionality of the transformation space; in other words, we have to choose a *basis* whose rank is much smaller than n. One possible way to do this is to restrict the transformations to be polynomials of low order; another one is to use splines (these two approaches are discussed in Section 11.3). In the remainder of this chapter we will limit ourselves to those cases in which it is reasonable to construct an indicator matrix G_j for each variable h_j (cf. Chapter 2), where the number of columns of G_j, indicated by k_j, must be *much smaller* than n. This means that the variables are either categorical from the start, containing a lot of ties, or that they are discretized into a limited number of categories. In this way the k_j vectors in G_j span a k_j-dimensional subspace of the n-dimensional space of all possible nonlinear transformations. After optimal quantification the ties in the data will still be tied. The trivial solution to (3.21) mentioned in the previous paragraph is prevented by this treatment of ties; the corresponding quantification of the categories is called nominal. ↠

Incorporating our choice of basis G_j implies that we may write $\phi_j(h_j) = G_j y_j$, for some vector of coefficients y_j, so that the loss function (3.21) for nonlinear PCA becomes

$$\sigma(x;y) = m^{-1} \sum_j SSQ(x - G_j y_j), \qquad (3.22)$$

where the role of the *quantifications* y_j is comparable to the role of the weights a_j in the K sets problem of Section 3.5. This notation does not mean that the y_j cannot be interpreted as weights; rather, it is chosen to emphasize that, in contrast to the a_j the y_j are weights in a space of larger dimensionality, spanning all feasible transformations, i.e. G_j.

We immediately generalize (3.22) to its multidimensional form, in the same way as (3.4) in Section 3.3 for linear PCA was generalized to (3.14) in Section 3.4.4. The loss function for nonlinear PCA with multiple solutions is written as

$$\sigma(X;Y) = m^{-1} \sum_j SSQ(X - G_j Y_j). \qquad (3.23)$$

Table 3.5 Overview of the loss functions in the search for homogeneity

Loss function	Description
$\sigma(x) = m^{-1} \sum_j SSQ(x - h_j)$	Just averaging
$\sigma(x;a) = m^{-1} \sum_j SSQ(x - a_j h_j)$	Linear PCA
$\sigma(X;A) = m^{-1} \sum_j SSQ(X - h_j a_j')$	Linear PCA: multiple solutions
$\sigma(x;a;\phi) = m^{-1} \sum_j SSQ(x - a_j \phi_j(h_j))$	Nonlinear PCA
$\sigma(x;\phi) = m^{-1} \sum_j SSQ(x - \phi_j(h_j))$	Nonlinear PCA: weights incorporated
$\sigma(x;y) = m^{-1} \sum_j SSQ(x - G_j y_j)$	HOMALS: single solution
$\sigma(X;Y) = m^{-1} \sum_j SSQ(X - G_j Y_j)$	HOMALS: multiple solutions

This loss function will in fact turn out to be the basic loss function in the rest of this book. For the time being, however, we call it the HOMALS loss function because, with $\mathbf{Y}_j$ unrestricted and $\mathbf{X}$ restricted to satisfy $\mathbf{X'X} = \mathbf{I}$, it is minimized by the HOMALS program. HOMALS is the acronym for HOMogeneity analysis by Alternating Least Squares. Before proceeding to the next sections, which will elaborate on this particular form of nonlinear principal components analysis, an overview is given of the history of loss functions discussed so far (Table 3.5).

3.8 CATEGORICAL PCA: HOMALS

Categorical PCA is that particular form of nonlinear PCA that is based on a categorical coding of the variables in indicator matrices. In Section 2.2 it was suggested that quantifications of objects and of categories for a set of complete indicator matrices $\{\mathbf{G}_1,...,\mathbf{G}_j,...,\mathbf{G}_m\}$ should satisfy the proportionalities

$$\mathbf{x} \ \propto m^{-1} \textstyle\sum_j \mathbf{G}_j \mathbf{y}_j, \qquad\qquad (3.24)$$

$$\mathbf{y}_j \ \propto \mathbf{D}_j^{-1} \mathbf{G}_j' \mathbf{x}, \qquad\qquad (3.25)$$

where $\mathbf{x}$ is the vector of object scores and $\mathbf{y}_j$ the vector of the quantifications of the categories of variable j. After some thought it will be evident that the stationary equations of the loss function in (3.22), obtained by setting the partial derivatives with respect to $\mathbf{x}$ equal to zero and by setting the partial derivatives with respect to $\mathbf{y}_j$ equal to zero as well, satisfy these proportionalities in fact as equalities. Thus (3.24) and (3.25) are the two basic quantification mechanisms for categorical PCA. The object scores $\mathbf{x}$ are proportional to the mean category quantification of all categories to which each object is associated (3.24), and the category quantifications $\mathbf{y}_j$ are proportional to the mean object score of objects falling into the same category of variable j (3.25). As before, it depends on the normalization chosen as to which one of the two proportionalities can be replaced by a straight equality and which one can be expressed as an equality with an explicit rescaling factor.

The computer program HOMALS performs categorical PCA by using the normalized scores algorithm adapted for heterogeneous variances, as discussed in Section 3.4.3. Using the concatenations $\mathbf{G} = \{\mathbf{G}_1,...,\mathbf{G}_j,...,\mathbf{G}_m\}$, and $\mathbf{y'} = \{\mathbf{y}_1',...,\mathbf{y}_j',...,\mathbf{y}_m'\}$, the proportionalities (3.24) and (3.25) can be written in a condensed way as

$$\mathbf{x} \propto \mathbf{G}\mathbf{y}/m, \tag{3.26}$$

$$\mathbf{y} \propto \mathbf{D}^{-1}\mathbf{G}'\mathbf{x}, \tag{3.27}$$

which is of the same form as the update steps 1 and 3# of the ALS schemes from Section 3.4. In the normalized scores algorithm the object scores are constrained to satisfy $\mathbf{x}'\mathbf{x} = 1$. The HOMALS implementation of the algorithm uses $\mathbf{G}$ in the role of $\mathbf{H}$ and $\mathbf{y}$ in the role of $\mathbf{a}$. It starts with a uniformly random choice of $\mathbf{x}$ ($\mathbf{x} \neq \mathbf{0}$), with a mean of zero, normalizes it to the sum of squares n (rather than 1, so that the scores have *variance* 1), and computes a first set of category quantifications $\tilde{\mathbf{y}}$ by applying (3.27) as an equality. Then HOMALS proceeds with the following steps:

(1) *Update object scores:* $\tilde{\mathbf{x}} \leftarrow \mathbf{G}\tilde{\mathbf{y}}/m$
(2) *Normalization:* $\mathbf{x}^+ \leftarrow \sqrt{n}\ \tilde{\mathbf{x}}(\tilde{\mathbf{x}}'\tilde{\mathbf{x}})^{-1/2}$
(3#) *Update category quantifications:* $\mathbf{y}^+ \leftarrow \mathbf{D}^{-1}\mathbf{G}'\mathbf{x}^+$
(4) *Convergence test:*   as usual.

Note that $\mathbf{G}\tilde{\mathbf{y}}/m$ is a vector of means of selected category quantifications; thus explicit matrix multiplication involving $\mathbf{G}$ can be avoided and is done implicitly in HOMALS through averaging. Reciprocally, $\mathbf{D}^{-1}\mathbf{G}'\mathbf{x}^+$ is a vector of means of selected object scores; thus finding the conditionally optimal category quantifications also merely needs to involve averaging – hence the alternative name *reciprocal averaging*. By capitalizing on the sparseness of $\mathbf{G}$, HOMALS is capable of handling very large data sets with reasonable accuracy.

3.8.1 Properties in terms of the SVD

In order to distinguish the SVD of $\mathbf{H}$ from the SVD related to $\mathbf{G}$, let $\mathbf{G}\mathbf{D}^{-1/2} = \mathbf{V}\boldsymbol{\Psi}\mathbf{W}'$. The ALS algorithm will reach the stationary pair of vectors $\mathbf{x}^* = \mathbf{v}_1$ and $\mathbf{y}^* = \psi_1\mathbf{D}^{-1/2}\mathbf{w}_1$, where ψ_1 is the first singular value of $\mathbf{G}\mathbf{D}^{-1/2}$ and $\mathbf{v}_1$ and $\mathbf{w}_1$ are the corresponding singular vectors. As an aside, note that although in practice (i.e. in HOMALS) the normalization of $\mathbf{x}$ is on n, it is more convenient to use $\mathbf{x}'\mathbf{x} = 1$ in theory; of course, when $\mathbf{x}$ is rescaled with a factor $\sqrt{n}$, then the $\mathbf{y}_j$ are automatically scaled with that same factor in step 3#. We now first write the departure from homogeneity for categorical PCA (3.22) in terms of the category quantifications only, using the stationary equations $\mathbf{D}_j\mathbf{y}_j = \mathbf{G}_j'\mathbf{x}$ and the normalization condition on $\mathbf{x}$:

$$\sigma(\mathbf{x},\mathbf{y}) = \mathbf{x'x} + m^{-1} \sum_j \mathbf{y}_j'\mathbf{D}_j\mathbf{y}_j - 2\,m^{-1} \sum_j \mathbf{y}_j'\mathbf{G}_j'\mathbf{x}$$

$$= 1 - m^{-1} \sum_j \mathbf{y}_j'\mathbf{D}_j\mathbf{y}_j = 1 - m^{-1}\,\mathbf{y'Dy}. \tag{3.28}$$

Upon substitution of $\mathbf{y}^* = \psi_1 \mathbf{D}^{-1/2}\mathbf{w}_1$ it follows from (3.28) that at the minimum the loss is

$$\sigma(*,*) = 1 - \psi_1^2/m. \tag{3.29}$$

This equals the relative loss W/T defined for numerical data in (3.9). In the HOMALS program the quantity ψ_1^2/m is called 'the eigenvalue'. Of what matrix is it the (largest) eigenvalue?

If $\mathbf{GD}^{-1/2} = \mathbf{V\Psi W'}$, then we correspondingly have the eigenvalue decomposition

$$\mathbf{D}^{-1/2}\mathbf{G'GD}^{-1/2} = \mathbf{W\Psi}^2\mathbf{W'}. \tag{3.30}$$

So ψ_1^2 (m times the HOMALS eigenvalue) is the largest eigenvalue of the matrix formed by rescaling $\mathbf{C} = \mathbf{G'G}$ by $\mathbf{D}^{-1/2}$. In Section 2.1 $\mathbf{C}$ was introduced as the matrix of bivariate marginals, while $\mathbf{D}$ contains the univariate marginals. Formulated in this way the categorical PCA technique is sometimes called 'factorial analysis of the Burt table', crediting Burt (1950), even though this formulation was already fully discussed in Guttman (1941). When we rework (3.30) a little, we get

$$\mathbf{CD}^{-1/2}\mathbf{W} = \mathbf{D}^{1/2}\mathbf{W\Psi}^2. \tag{3.31}$$

Thus the identification $\mathbf{y} = \psi\mathbf{D}^{-1/2}\mathbf{w}$, with $\mathbf{w}$ *any* column of $\mathbf{W}$ and ψ the corresponding singular value, yields the combined proportionalities (3.26) and (3.27) in the form of a generalized eigenvalue problem:

$$\mathbf{Cy} = \psi^2\mathbf{Dy}. \tag{3.32}$$

By itself the generalized eigenvalue problem does not tell us which eigenvector to take, and how to scale it. Answers to these questions follow from the loss function and the normalization conditions chosen.

The proportionalities (3.26) and (3.27) also imply

$$\mathbf{x} \propto (\mathbf{GD}^{-1}\mathbf{G'}/m)\,\mathbf{x}. \tag{3.33}$$

So the object scores are eigenvectors of the matrix $\mathbf{GD}^{-1}\mathbf{G'}/m$, which has the eigenvalue decomposition $\mathbf{GD}^{-1}\mathbf{G'}/m = \mathbf{V}(\mathbf{\Psi}^2/m)\mathbf{V'}$, and ψ_1^2/m is the largest eigenvalue of this matrix, which can alternatively be expressed as

$m^{-1} \sum_j G_j D_j^{-1} G_j'$ (for more discussion of this *average similarity table*, see Chapter 8). ↠

When we consider multiple solutions, based on (3.23), result (3.28) generalizes to

$$\sigma(X,Y) = \text{tr } X'X - m^{-1} \sum_j \text{tr } Y_j' D_j Y_j$$

$$= p - m^{-1} \sum_s y_s' D y_s. \qquad (3.34)$$

The stationary pair of multiple quantifications now is $X^* = V_p$ for the object scores and $Y^* = D^{-1/2} W_p \Psi_p$ for the category quantifications, where the subscript p denotes the selection of the first p singular values and vectors; the minimum loss becomes

$$\sigma(*,*) = p - \sum_s \psi_s^2/m. \qquad (3.35)$$

The derivations in this section were based on the full SVD of $GD^{-1/2}$. In the next section we will see that one of the singular vectors is *extraneous*, i.e. it does not correspond to a sensible categorical PCA solution.

3.8.2 Centering

In Section 3.4 the ALS algorithms were applied to a data matrix H for which it was assumed that its columns have zero means. This is not true for G.

We will now show that this does not really matter when appropriate precautions are taken. For the present section only, let S be the matrix of deviations from column means of G.

The argument can perhaps be more easily followed with the help of a numerical illustration.
To this end we take G of Table 2.5 (from which we drop the last column since category 'w' has zero frequency). The result for S is given in Table 3.6a, whereas Table 3.6b shows the result for $S'S$, which is the doubly centered Burt table.

Equation (3.32) has a trivial solution $y = u$ with eigenvalue m, since $Cu = m \, Du$. This solution is trivial, or extraneous to the categorical PCA problem, because it follows from (3.26) and from the constant row sum property of a complete partitioned indicator matrix that x has constant elements as well. This fact in turn implies complete lack of discrimination, independently of the data. Thus we have to disregard the solution $y = u$ from consideration. It now follows from the orthogonality properties of generalized eigenvectors that any other solution y_s must satisfy $u'Dy_s = 0$, i.e. y_s has zero weighted mean. However, equation (3.32) also implies $G_v'Cy_s = \psi_s^2 G_v'Dy_s$. Here G_v is the indicator matrix that links categories to variables as illustrated in Table 3.7; it has m columns and Σk_j rows, and in each column it contains unit elements corresponding to the partitioning of G. For a complete indicator matrix G we find $G_v'Cy_s = 0$, and therefore we may also conclude, for any $0 < \psi_s^2 < m$, that $G_v'Dy_s = 0$.

The algebraic expression for S is $S = G - uu'G/n$. Since $GG_v = uu'$, we also have $SG_v = GG_v - uu'GG_v/n = uu' - uu' = 0$, which shows that not only the column totals of S are zero, but also its row totals, and even the row totals of each submatrix, S_j, as one may verify in Table 3.6a. The latter result also shows that S has rank $\Sigma k_j - m$, so that for y_s there generally will be at most $p_{max} = \Sigma k_j - m$ nontrivial solutions.

Table 3.6a Indicator matrix of Table 2.5 in deviations from means

a	b	c	p	q	r	u	v
0.4	−0.2	−0.2	0.2	−0.1	−0.1	0.7	−0.7
−0.6	0.8	−0.2	−0.8	0.9	−0.1	−0.3	0.3
0.4	−0.2	−0.2	−0.8	−0.1	0.9	−0.3	0.3
0.4	−0.2	−0.2	0.2	−0.1	−0.1	0.7	−0.7
−0.6	0.8	−0.2	0.2	−0.1	−0.1	−0.3	0.3
−0.6	−0.2	0.8	0.2	−0.1	−0.1	−0.3	0.3
0.4	−0.2	−0.2	0.2	−0.1	−0.1	0.7	−0.7
0.4	−0.2	−0.2	0.2	−0.1	−0.1	−0.3	0.3
−0.6	−0.2	0.8	0.2	−0.1	−0.1	−0.3	0.3
0.4	−0.2	−0.2	0.2	−0.1	−0.1	−0.3	0.3

Table 3.6b Sums of squares and cross products for Table 3.6a (doubly centered Burt table)

a	b	c	p	q	r	u	v
2.4	−1.2	−1.2	0.2	−0.6	0.4	1.2	−1.2
−1.2	1.6	−0.4	−0.6	0.8	−0.2	−0.6	0.6
−1.2	−0.4	1.6	0.4	−0.2	−0.2	−0.6	0.6
0.2	−0.6	0.4	1.6	−0.8	−0.8	0.6	−0.6
−0.6	0.8	−0.2	−0.8	0.9	−0.1	−0.3	0.3
0.4	−0.2	−0.2	−0.8	−0.1	0.9	−0.3	0.3
1.2	−0.6	−0.6	0.6	−0.3	−0.3	2.1	−2.1
−1.2	0.6	0.6	−0.6	0.3	0.3	−2.1	2.1

Table 3.7 Matrix G_v for numerical example

1	0	0
1	0	0
1	0	0
0	1	0
0	1	0
0	1	0
0	0	1
0	0	1

In order to relate the analysis of S to the analysis of G, we proceed as follows. The algebraic expression for $S'S$ is $S'S = G'G - Duu'D/n$; next define the partitioned block diagonal matrix $D_{S'S} \equiv D - DG_vG_v'D/n$. Then equation (3.32), reformulated for $S'S$ and $D_{S'S}$, becomes

$$S'Sy = \psi^2 D_{S'S}y \qquad\qquad (3.36)$$

or

$$(G'G - Duu'D/n)y = \psi^2(D - DG_vG_v'D/n)y . \qquad\qquad (3.37)$$

Since $u'Dy = 0$ and $G_v'Dy = 0$ for all nontrivial eigenvectors, equation (3.36) must have the same solutions for y and ψ^2 as equation (3.32). So G will also have at most $p_{max} = \Sigma k_j - m$ nontrivial solutions. The way in which the undesired vector is removed from the generalized eigenvector problem in (3.37) is called *Hotelling deflation*. In the ALS algorithms no deflation is used (in order to fully exploit the sparseness of G), but the extraneous solution is avoided by keeping the *quantifications* centered, rather than G itself. For the HOMALS algorithm we see that:

$$y^+ \leftarrow D^{-1}G'x^+ \quad \text{yields} \quad u'Dy^+ = u'G'x^+ = u'x^+, \qquad (3.38)$$

$$\tilde{x} \leftarrow G\tilde{y}/m \quad \text{yields} \quad u'\tilde{x} = u'G\tilde{y}/m = u'D\tilde{y}/m. \qquad (3.39)$$

So the process does not alter the mean quantifications, and it is sufficient to set the (weighted) means equal to zero at the beginning in order to stay away from the extraneous solution. Yet centering of quantifications may have to be repeated once in a while, due to the imperfect precision of machine calculations.

3.8.3 Normalization

There are two basic normalization options in categorical PCA (cf. the algorithms of Section 3.4):

(a) $\mathbf{y}$ is normalized so that $\mathbf{y'Dy}$ is some constant. The induced object scores then are obtained by $\mathbf{x} = \mathbf{Gy}/m$, which makes the object score equal to the average of its category quantifications. In a plot of multiple dimensions, an object point will be the centre of gravity of the points for categories that apply to the object.

(b) $\mathbf{x}$ is normalized to some constant. The induced category quantifications are then obtained by $\mathbf{y}_j = \mathbf{D}_j^{-1}\mathbf{G}_j'\mathbf{x}$, where each category is quantified as the average of the objects within the category. In a plot of multiple dimensions, a category point will be the centre of gravity of the points for objects within the category.

The standard HOMALS program takes normalization (b), with $\mathbf{x'x} = n$, so that $\mathbf{x}$ becomes a 'standard score'. There are two practical reasons for this choice. The first one is that elements of $\mathbf{x}$ now can be interpreted with the help of all the familiar properties of standard scores. The second is that in HOMALS applications it happens very often that n is much larger than $\sum k_j$. For plots, normalization (b) then gives the nicest arrangement of the picture, with object points equally spread in all directions and category points indicating the means of subgroups of objects (objects sorted into the respective categories of a variable). The normalization above implies, in terms of the SVD solution $\mathbf{GD}^{-1/2} = \mathbf{V\Psi W'}$, that for each of p solutions $\mathbf{x}_s$ and $\mathbf{y}_s$ ($s = 1,...,p$) the following relations will be valid:

$$\mathbf{x}_s = \sqrt{n}\,\mathbf{v}_s \tag{3.40}$$

$$\mathbf{y}_s = \sqrt{n}\,\psi_s\,\mathbf{D}^{-1/2}\mathbf{w}_s \tag{3.41}$$

$$\mathbf{y}_s'\mathbf{D}\mathbf{y}_s = n\psi_s^2 \tag{3.42}$$

In the output of the HOMALS program the eigenvalues are reported in the form ψ_s^2/m. The scale of the category quantifications is *not* one of standard scores, but we may derive upper and lower bounds for getting an impression of its range. The category quantifications always satisfy

$$- [(n - d_{(j)r}) / d_{(j)r}]^{1/2} \le y_{(j)rs} \le [(n - d_{(j)r}) / d_{(j)r}]^{1/2}, \tag{3.43}$$

where $d_{(j)r}$ denotes the marginal frequency of category r of variable j.

Proof:
We first write the category quantifications in the form

$$y_{(j)rs} = 1/d_{(j)r} \sum_i g_{(j)ir} x_{is} = 1/d_{(j)r} \sum_i (g_{(j)ir} - d_{(j)r}/n) x_{is}. \qquad (3.44)$$

For any two vectors **a** and **b**, the Cauchy-Schwartz inequality tells us that

$$[\sum_i a_i b_i]^2 \le \sum_i a_i^2 \sum_i b_i^2. \qquad (3.45)$$

Applying this inequality to the inner product on the right-hand side of (3.44) we find

$$-c_{(j)r} \le 1/d_{(j)r} \sum_i (g_{(j)ir} - d_{(j)r}/n) x_{is} \le c_{(j)r}, \qquad (3.46)$$

with

$$c_{(j)r} = [\sum_i (g_{(j)ir} / d_{(j)r} - 1/n)^2 \sum_i x_{is}^2]^{1/2}$$

$$= [(n - d_{(j)r}) / d_{(j)r}]^{1/2}. \qquad (3.47)$$

Substituting (3.44) and (3.47) into (3.46) now gives the desired result.

Q.E.D.

The bounds on the category quantifications depend on the univariate marginals $d_{(j)r}$. Thus the maximal range of very frequent categories is much smaller than the maximal range of very infrequent categories. For instance, if category r of variable j contains 10 per cent of the objects, $c_{(j)r} = 3$; if it contains half of the objects, $c_{(j)r} = 1$; and if it contains 90 per cent of the objects, $c_{(j)r} = 1/3$. This differential restriction of range is one of the reasons to be careful when the variables have widely distinct numbers of categories, or univariate marginals with very thin tails.

3.8.4 Contribution of variables: discrimination measures

The HOMALS program also calculates *discrimination measures*, one for each variable and each dimension, defined as $\eta_{js}^2 \equiv y'_{(j)s} D_j y_{(j)s}/n$ (where $y_{(j)s}$ is the quantification for h_j in the sth dimension of the solution). The discrimination measures add up across variables to $y'_s D y_s/n = \psi_s^2$, so that the reported eigenvalue ψ_s^2/m is the average of the discrimination measures in the sth dimension. When a variable does not contribute to the sth dimension of the solution, the discrimination measure is zero (its category quantifications coincide with the origin). It can be shown that the discrimination measures are equal to the squared correlations between x_s and the optimally scaled variables $q_{(j)s} = G_j y_{(j)s}$.

Proof:
In this proof we omit the index s. The squared correlation between x and q_j equals

$$r^2_{(x,q_j)} = (x'G_j y_j)^2 (y_j'G_j'G_jy_j)^{-1} (x'x)^{-1}. \tag{3.48}$$

Using the equality $D_jy_j = G_j'x$, (3.48) can be rewritten as

$$r^2_{(x,q_j)} = (y_j'D_jy_j)^2 (y_j'D_jy_j)^{-1} (n)^{-1}$$

$$= y_j'D_jy_j/n, \tag{3.49}$$

which is the discrimination measure η_j^2.   Q.E.D.

In Section 3.9 it will be shown that the discrimination measure also has an interpretation as a *squared component loading*. Before proceeding with a discussion of the geometrical properties of the HOMALS quantifications, a small numerical example is given.

For the data matrix of Table 2.1, with the corresponding indicator matrix in Table 2.5 (the example was also used in Section 3.8.2), we give the HOMALS results, in two versions.

(a) *With normalization* $y'Dy = 1$, $x = Gy/m$.
(b) *With the standard HOMALS normalization* $x'x = n$ *and* $y_j = D_j^{-1}G_j'x$.

(a) *Normalization* $y'Dy = 1$.
The generalized eigenvector equation $Cy = \psi^2 Dy$ *(C and D are given in Tables 2.6 and 2.7) has the three largest eigenvalues:*

$$\psi_1^2 = 1.886, \quad \psi_1^2/m = 0.629, \quad \text{relative loss } 1 - \psi_1^2/m = 0.371,$$

$$\psi_2^2 = 1.277, \quad \psi_2^2/m = 0.426, \quad \text{relative loss } 1 - \psi_2^2/m = 0.574,$$

$$\psi_3^2 = 1.167, \quad \psi_3^2/m = 0.389, \quad \text{relative loss } 1 - \psi_3^2/m = 0.611.$$

The category quantifications y *are given in Table 3.8. Using the first quantification, the optimally scaled data matrix is given by* Q_1 *in Table 3.9, with columns* $q_{(j)1} = G_j y_{(j)1}$. *The object scores based on* $x = Gy/m$ *are given, for the three solutions, in Table 3.10. The equality* $y'Dy = 1$ *implies that the sum of squares of all elements in* Q_1 *is unity. The sums of squares for the separate columns of* Q_1 *are* 0.429, 0.338, *and* 0.233, *respectively. The larger this value, the better the categories of the variables discriminate between objects.*

Figure 3.6 gives a plot for the first two dimensions of the solution. The first two columns of Table 3.8 give coordinates of

Table 3.8 Solutions for y, with $y'Dy = 1$

	y_1	y_2	y_3
a	−0.14	−0.11	0.06
a	0.39	−0.19	−0.09
c	0.04	0.51	−0.10
p	−0.06	0.05	−0.09
q	0.55	−0.34	−0.09
r	−0.05	−0.06	0.85
u	−0.23	−0.21	−0.18
v	0.10	0.09	0.08

Table 3.9 Optimally scaled data matrix Q_1 with $y'Dy = 1$

−0.14	−0.06	−0.23
0.39	0.55	0.10
−0.14	−0.05	0.10
−0.14	−0.06	−0.23
0.39	−0.06	0.10
0.04	−0.06	0.10
−0.14	−0.06	−0.23
−0.14	−0.06	0.10
0.04	−0.06	0.10
−0.14	−0.06	0.10

Table 3.10 Solution for X, first three dimensions, normalization $y'Dy = 1$

	x_1	x_2	x_3
1	−0.15	−0.09	−0.07
2	0.35	−0.14	−0.04
3	−0.03	−0.03	0.33
4	−0.15	−0.09	−0.07
5	0.14	−0.01	−0.04
6	0.03	0.22	−0.04
7	−0.15	−0.09	−0.07
8	−0.04	0.01	0.02
9	0.03	0.22	−0.04
10	−0.04	0.01	0.02

category points. The first two columns of Table 3.10 give those of object points. One should verify that each object point is the centre of gravity of its categories. In the figure this is illustrated for objects 6 and 9 with categories 'c', 'p', and 'v'. Relative loss for the first two dimensions is visible in the figure as the sum of

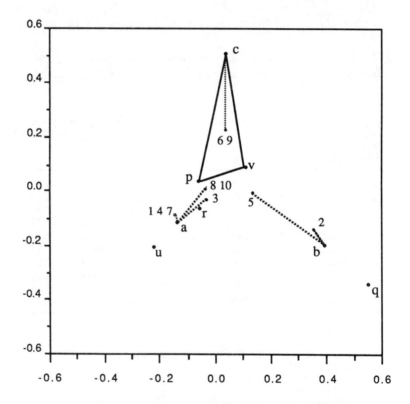

Figure 3.6 Nonstandard HOMALS solution. The figure illustrates for objects 6 and 9 that object points are the centre of gravity of category points (triangle). The figure also illustrates what variable 1 contributes to the loss function (sum of squared lengths of dotted lines).

the squared distances between a category point and the object points for objects belonging to the category. In the figure, dotted lines are drawn between object points and their categories for variable 1. The sum of the squared lengths of these dotted lines is what variable 1 contributes to relative loss. Since a variable discriminates better to the extent that its category points have larger spread, the sum of the squared distances between category points for some variable and the origin of the plot visualizes how well the variable discriminates.

(b) Normalization $x'x = n$; standard HOMALS solution.
The solution for X (first three dimensions) is given in Table 3.11 and the corresponding solution for $Y = D^{-1}G'X$ in Table 3.12. The optimally scaled data matrix, using y_1 for the category quantification, becomes Q_1 in Table 3.13. The difference with the previous solution is only in normalization. The correlation matrix for Q_1 of Table 3.13 is the same as that for Q_1 in Table 3.9.

Table 3.11 Object scores X, normalization $x'x = n$

	x_1	x_2	x_3
1	-1.01	-0.77	-0.62
2	2.40	-1.20	-0.32
3	-0.21	-0.22	2.89
4	-1.01	-0.77	-0.62
5	0.98	-0.12	-0.32
6	0.18	1.83	-0.34
7	-1.01	-0.77	-0.62
8	-0.25	0.09	0.14
9	0.18	1.83	-0.34
10	-0.25	0.09	0.14

Table 3.12 HOMALS category quantifications

	y_1	y_2	y_3
a	-0.62	-0.39	0.22
b	1.69	-0.66	-0.32
c	0.18	1.83	-0.34
p	-0.27	0.18	-0.32
q	2.40	-1.20	-0.32
r	-0.21	-0.22	2.89
u	-1.01	-0.77	-0.62
v	0.43	0.33	0.27

Table 3.13 Optimally scaled data matrix, based on the first HOMALS dimension

1	-0.62	-0.27	-1.01
2	1.69	2.40	0.43
3	-0.62	-0.21	0.43
4	-0.62	-0.27	-1.01
5	1.69	-0.27	0.43
6	0.18	-0.27	0.43
7	-0.62	-0.27	-1.01
8	-0.62	-0.27	0.43
9	0.18	-0.27	0.43
10	-0.62	-0.27	0.43

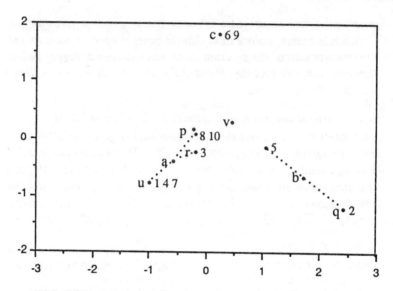

Figure 3.7 Standard HOMALS plot, with category points as the centre of gravity of their objects.

Figure 3.7 gives a plot for the first two dimensions. Category points now are the centre of gravity of object points, as illustrated for the first variable, where category 'c' coincides with objects 6 and 9, category 'b' is midway between 2 and 5, and category 'a' is at the centre of gravity of the cloud of object points 1, 3, 4, 7, 8, 10. The plot also illustrates that category 'u' coincides with objects 1, 4, 7, category 'q' with object 2, and category 'r' with object 3.

For the first dimension of the solution, discrimination measures are 0.809, 0.638, and 0.439, respectively. They add up to $\psi_1^2 = 1.886$ and their mean is 0.629. Under this normalization, the discrimination measures are squared correlations between x_1 and the columns of Q_1.

3.8.5 Geometrical properties

One should note that HOMALS solutions are *nested*, which means that the first dimension of the solution of a higher dimensional solution is identical with the one-dimensional solution; the second dimension of a two-dimensional solution is also the second dimension of a higher-dimensional solution, and so on. Some of the techniques of nonlinear MVA to be discussed later (e.g. PRINCALS, Chapter 4) do not have this property.

We now summarize some geometrical aspects of the plot in Figure 3.7, which are typical for all HOMALS plots.

(a) Categories and objects are represented as points in a joint space.
(b) Category points are the centre of gravity of the object points that share the same category. For each variable, categories of that variable partition the object points into 'subclouds', and the category points are the means of those subclouds.
(c) A variable discriminates better to the extent that the category points are farther apart.

   In the figure, this is shown (with dotted lines) for the first variable, where category 'a' brings objects 1, 3, 4, 7, 8, 10 together, category 'b' objects 2 and 5, and category 'c' objects 6 and 9.

(d) The sum of the squared distances between object points in a subcloud and their mean point (the category point) is related to relative loss, in the sense that, if we project the displacements on a dimension and take the sum of squares over all variables, this sum equals $1 - \psi_s^2/m$. This implies that for a 'perfect' solution, without any loss at all, all object points must coincide with their category points.
(e) The spread of the category points for each variable is an indication of how much the variable contributes to relative loss, for the dimensions displayed in the plot.
(f) The distance between two object points is related to the 'similarity' between their response patterns or profiles. In particular, objects with the same response pattern will be plotted as identical points. However, the reverse is not necessarily true. If object points are close together in a plot of the first two HOMALS dimensions, they might be far apart in the third, fourth, etc., dimension.

   For the example, aspect (f) is illustrated by objects 8 and 10 or 1, 4, 7. Object 3 is plotted roughly midway between these two groups. However, on the third dimension (Table 3.11), we can see that object 3 is distant from all other objects.

(g) If a category applies uniquely to only one object, the object point and this category point will coincide.

   Example in the illustration: object 2 and category 'q'. The same is true when a category applies uniquely to a group of objects with an identical response pattern (category 'u' for objects 1, 4, 7).

(h) A category point with low marginal frequency will be plotted farther towards the periphery, whereas a category with high marginal frequency will be plotted nearer to the centre of the plot.

For the example this is illustrated by category 'q' with marginal frequency of 1 versus category 'b' with marginal frequency of 8. However, category 'r', also with marginal frequency of 1, is not very far from the centre. This category point becomes peripheral in the third dimension.

(i) Objects with a response pattern similar to the 'average' response pattern will be plotted more towards the centre, whereas objects with a 'unique' response pattern appear in the periphery. This is mainly a corollary of aspect (h). Again, this statement will be true for a plot in *all* relevant dimensions, and not necessarily for a plot of the first two dimensions only.

For the example, objects 8 and 10, which for each variable have the most frequently selected category, are near the centre of the plot, whereas objects 2, or 6, or 9, with unique response patterns, are at the periphery. However, object 3, also with a unique response pattern, is not far from the centre of the plot; its uniqueness appears in the third dimension.

3.9 RELATIONS BETWEEN HOMALS AND LINEAR PCA

In this section we start with having a look at the first HOMALS dimension. It is related to linear PCA in the following way. Let Q_1 be the optimally scaled data matrix and let the correlation matrix between the transformed variables in Q_1 be denoted by R_1. It is assumed that the columns of Q_1 are unit normalized, so that $R_1 = Q_1'Q_1$. We write the singular value decomposition of Q_1 as $Q_1 = K_1 \Lambda_1 L_1'$ and the eigenvalue decomposition of R_1 as $R_1 = L_1 \Lambda_1^2 L_1'$. Then it can be shown that the normalized object scores in the first HOMALS dimension x_1 are proportional (with respect to a factor $n^{1/2}$) to the normalized component scores on the basis of Q_1, which are obtained by taking k_1. The discrimination measures are equal to the squares of the component loadings in the first PCA dimension, which are obtained by $a_1 = \lambda_1 l_1$.

To illustrate, we take the unit-normalized version of the optimally quantified data matrix Q_1 from Table 3.13. The corresponding correlation matrix R_1 becomes

$$R_1 = \begin{pmatrix} 1.000 & 0.622 & 0.453 \\ 0.622 & 1.000 & 0.223 \\ 0.453 & 0.223 & 1.000 \end{pmatrix}.$$

R_1 has eigenvalues $\lambda_{11}^2 = 1.886$, $\lambda_{12}^2 = 0.789$, and $\lambda_{13}^2 = 0.325$. Here double subscripts are used to remind us that these are three ordered eigenvalues of R_1. The eigenvectors are

$$L_1 = \begin{pmatrix} 0.655 & -0.104 & 0.748 \\ 0.582 & -0.563 & -0.587 \\ 0.482 & 0.820 & -0.308 \end{pmatrix},$$

the matrix of component loadings is given by

$$A_1 = L_1 \Lambda_1 = \begin{pmatrix} 0.900 & -0.092 & 0.427 \\ 0.799 & -0.500 & -0.335 \\ 0.662 & 0.728 & -0.176 \end{pmatrix},$$

and the matrix of component scores is obtained as

$$K_1 = Q_1 L_1 \Lambda_1^{-1} = \begin{pmatrix} -.0320 & -.0352 & .0085 \\ .0757 & -.0479 & -.0310 \\ -.0068 & .0270 & -.0312 \\ -.0320 & -.0352 & .0085 \\ .0310 & .0190 & .0778 \\ .0057 & .0252 & .0083 \\ -.0320 & -.0352 & .0085 \\ -.0078 & .0285 & -.0288 \\ .0057 & .0252 & .0083 \\ -.0078 & .0285 & -.0288 \end{pmatrix}.$$

The first column of the matrix of component loadings contains correlations between the columns of Q_1 and the first column of the component scores. The HOMALS discrimination measures in the first dimension are the squares of these component loadings. Figure 3.8 depicts the component loadings in two-dimensional

*space. The plot is based on A_1: the horizontal axis gives the first
dimension and the vertical axis the second dimension.*

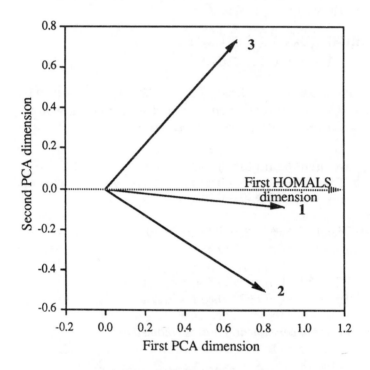

Figure 3.8 PCA solution in two dimensions on the
basis of the correlation matrix R_1 between the
variables in the optimally quantified data matrix Q_1.
The projections on the horizontal axis (the first principal component) are equal to the square root of the
HOMALS discrimination measures.

One of the ways to prove that the first HOMALS dimension is the first
principal component of the optimally scaled data matrix Q_1 is the following.

Proof:
In this proof we omit the index for the dimension. Let Q_y be a $\Sigma k_j \times m$ matrix
with elements y_j in its jth column, where each set of k_j rows corresponds to
the position for the jth variable, so that $Q_y u = y$. The optimally scaled data
matrix then can be written as $Q = GQ_y$. The diagonal elements of $Q_y'G'GQ_y/n$
are the discrimination measures $\eta_j^2 = y_j' D_j y_j/n$. Now let b be the $m \times 1$ vector

with elements η_j. Let D_b be the $m \times m$ diagonal matrix with diagonal elements η_j. Then $R = D_b^{-1}Q_y'G'GQ_yD_b^{-1}$ is the matrix of correlations between the transformed variables in Q. Now b is an eigenvector of R associated with (m times) the HOMALS eigenvalue ψ^2 because

$$D_b^{-1}Q_y'G'GQ_yD_b^{-1}b = D_b^{-1}Q_y'G'GQ_yu$$

$$= D_b^{-1}Q_y'G'Gy$$

$$= \psi^2 D_b^{-1}Q_y'Dy$$

$$= \psi^2 b, \qquad (3.50)$$

where we have used the Burt table formulation of HOMALS (3.32) in the middle step. Thus $\psi^2 = \lambda^2$, and since $b'b = \sum_j \eta_j^2 = y'Dy/n = \psi^2$, it follows that b is the eigenvector of R normalized to have the sum of squares equal to the associated eigenvalue, so that $b = \lambda l = a$: the vector of component loadings on the first principal component. The discrimination measures η_j^2 are the squared elements of b, and therefore they are equal to the squared component loadings.   Q.E.D.

We summarize some results. When Q_1 denotes the optimally scaled data matrix on the basis of the category quantifications in the first HOMALS dimension y_1 and when R_1 denotes the correlation matrix between the transformed variables in Q_1, then the largest HOMALS eigenvalue will always be proportional to the largest eigenvalue of R_1, so that the first HOMALS dimension corresponds to the first principal component of Q_1. In addition, the first HOMALS dimension guarantees that Q_1 is optimally scaled in such a way that the first eigenvalue of R_1 is maximized.

The procedure sketched above justifies applying HOMALS in one dimension in order to find an optimal quantification of the categories, whereafter the analysis of the data is continued on the basis of the optimally scaled data matrix. Such an approach can be useful not only when variables have no prior quantification (the categories are purely nominal), but sometimes also when there is prior quantification but one suspects that the relations between the variables cannot be optimally described in terms of linear relations. For example, if one variable is 'age' of individuals and one suspects that the other variables have a curvilinear relation with age, then the HOMALS quantification might confirm that suspicion. Also, the HOMALS quantification might in some cases suggest a well-defined transformation of the prior quantification of a variable, e.g. the HOMALS quantification corresponds to a logarithmic transformation.

The first HOMALS solution can also be described in the following way. The indicator matrix as it were *expands*, or *blows up*, the data matrix in the sense that a column h_j of the data matrix becomes a *set* of k_j binary variables in G_j, one for each category. The HOMALS object scores provide a *meet solution* for these m sets of binary variables, which means that HOMALS solves for m linear compounds $q_j = G_j y_j$ in such a way that these m vectors q_j are as close as possible to their sum vector Gy, where Gy is proportional to x. Geometrically, this means that the vectors q_j form a bundle in m-dimensional space around their sum vector. The HOMALS solution minimizes the sum of the squared cosines of the angles between the q_j and Gy (where Gy and x coincide as to their direction). A different way of formulating the HOMALS criterion geometrically is: if we give all q_j unit length, then HOMALS maximizes the sum of the squared projections of such unit-length vectors on Gy (or on x).

PCA on the optimally scaled data matrix Q_1 solves a join problem. It means that in Q_1 the indicator matrix is again *compressed* into a set of m variables q_j. PCA investigates how these m vectors are located in their m-dimensional space, and the PCA solution describes the bundle of vectors q_j as well as possible with as few dimensions as possible. HOMALS gives a one-dimensional approximation of the bundle while PCA gives additional dimensions.

The second HOMALS dimension compresses the indicator matrix in a different way, so that we obtain a different set of vectors $q_{(j)2}$, a second bundle of vectors around their sum vector Gy_2. It is important to note that all subsequent solutions Gy_s may no longer correspond to the *first* principal component of Q_s, but to another eigenvector of Q_s instead. Some of the implications of PCA on the optimally scaled data matrix Q_2 are shown in the following example, based on the standard HOMALS solution from Section 3.8.

Results are given below: the correlation matrix R_2, its eigenvector matrix L_2, and its principal components $A_2 = L_2 \Lambda_2$, where Λ_2^2 is the matrix of eigenvalues of R_2. For the example, these eigenvalues are $\lambda_{21}^2 = 1.279$, $\lambda_{22}^2 = 1.277$, and $\lambda_{23}^2 = 0.445$. Note that it is not the first but the second eigenvalue that is equal to the second HOMALS eigenvalue $\psi_2^2 = 1.277$: the second HOMALS dimension must now be identified with the second principal component of Q_2. The HOMALS discrimination measures (they are 0.850, 0.175, and 0.252, respectively) are the squares of the second column of A_2.

$$R_2 = \begin{pmatrix} 1.000 & 0.277 & 0.277 \\ 0.277 & 1.000 & 0.279 \\ 0.277 & 0.279 & 1.000 \end{pmatrix}.$$

The eigenvectors are obtained as

$$L_2 = \begin{pmatrix} -0.043 & 0.816 & -0.577 \\ -0.728 & 0.370 & 0.578 \\ 0.685 & 0.445 & 0.578 \end{pmatrix},$$

the matrix of component loadings is given by

$$A_2 = L_2\Lambda_2 = \begin{pmatrix} -0.049 & 0.992 & -0.385 \\ -0.823 & 0.418 & 0.385 \\ 0.774 & 0.502 & 0.385 \end{pmatrix},$$

and the matrix of component scores is obtained as

$$K_2 = Q_2 L_2 \Lambda_2^{-1} = \begin{pmatrix} -0.374 & 0.243 & -0.186 \\ 0.719 & -0.380 & -0.412 \\ 0.238 & -0.070 & 0.150 \\ -0.374 & -0.243 & -0.186 \\ 0.047 & -0.038 & 0.492 \\ 0.015 & 0.579 & -0.247 \\ -0.374 & -0.243 & -0.186 \\ 0.044 & 0.029 & 0.412 \\ 0.015 & 0.579 & -0.247 \\ 0.044 & 0.029 & 0.412 \end{pmatrix}.$$

Figure 3.9 gives a graph for the first two dimensions of the PCA of R_2. One should realize, however, that if we continue in this way for successive HOMALS dimensions, things become redundant. In general, there are $\sum k_j - m$ (cf. Section 3.8) possible HOMALS dimensions (we denote this number by p_{max}). Each of them creates m vectors $q_{(j)s}$ ($s=1,...,p_{max}$), so that a *complete* HOMALS solution, with a PCA for each dimension, creates a situation with $m p_{max}$ dimensions (a clear example of 'data production' instead of 'data reduction'). The p_{max}th solution, however, is special. It will correspond to that principal component of the p_{max}th correlation matrix that is associated with the *smallest* eigenvalue. In addition, the data matrix is quantified in such a way that this smallest eigenvalue is minimized (and therefore the sum of the other eigenvalues is maximized). This is further explained in Section 11.3. ↠

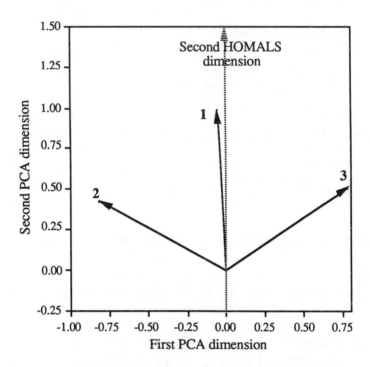

Figure 3.9 PCA solution in two dimensions on the basis of the correlation matrix R_2. Compared to Figure 3.8 this second solution is not as good as the first because the vectors are not so close to their sum vector (the second dimension). The projections on this dimension (the second principal component) are equal to the square root of the HOMALS discrimination measures in the second dimension.

HOMALS *as a first step* need not necessarily be followed by a principal components analysis. For example, suppose that the data matrix contains two different sets of variables, e.g. a set of subjects scoring on test items and a set of characteristics of the subjects. In linear multivariate analysis one would probably resort to a form of canonical analysis. Applying HOMALS as a first step would imply that this canonical analysis is performed on optimally scaled data matrices. There are two different strategies: HOMALS as a first step could be applied to the data matrix as a whole or to the two data matrices for the two sets separately.

3.10   THE RELATIONSHIP BETWEEN HOMALS AND TOTAL CHI-SQUARE

The association between any pair of categorical variables can be expressed in terms of the classical chi-square statistic. For a set of m variables we may study the relationship between the sum of such chi-square statistics and HOMALS most easily through analysis of the Burt table. Recall that the Burt table C contains all pairwise contingency tables and the univariate marginals. In Section 3.8.1 we have seen that the category quantifications can be obtained from the eigenvalue decomposition

$$D^{-1/2}(C - Duu'D/n)D^{-1/2} = W\Psi^2W',   \tag{3.51}$$

where the trivial solution $w = u / \sqrt{n}$ cannot occur due to the centering of C. In Section 3.8.2 it was shown that the matrix in (3.51) has at most a rank of $p_{max} = \Sigma k_j - m$. The matrix on the left in equation (3.51) has trace p_{max}, and it is immediately seen that the matrix on the right has trace $\Sigma \psi_s^2$. It follows that $\Sigma \psi_s^2 = p_{max}$. When $k_j = 2$ for all j, then $p_{max} = m$, and the sum of the eigenvalues also equals m, as is the case with an eigenanalysis of a correlation matrix R. An off-diagonal submatrix of (3.51) has the form

$$D_j^{-1/2}(C_{jl} - D_juu'D_l/n)D_l^{-1/2},   \tag{3.52}$$

for variables j and l. This submatrix has the following interpretation. $D_juu'D_l/n$ contains elements equal to the *expected values under independence*, based on the univariate marginals for variables j and l. Since C_{jl} is the matrix of bivariate marginals (Section 2.1) it follows that $C_{jl} - D_juu'D_l/n$ is a matrix of differences between observed and expected values. The matrix of (3.52), after multiplication with $\sqrt{n}$, then has elements equal to the discrepancy between the observed and expected values, divided by the square root of the expected value. The sum of the squares of those elements equals the chi-square statistic X_{jl}^2.

For diagonal submatrices the expression in (3.52), with $j = l$, becomes

$$D_j^{-1/2}(D_j - D_juu'D_j/n)D_j^{-1/2} = I - D_j^{1/2}uu'D_j^{1/2}/n.   \tag{3.53}$$

This is an idempotent matrix, so that its trace is equal to the sum of its squared elements. Its trace equals $k_j - 1$. Combining the results for diagonal and off-diagonal submatrices we find that the sum of the squared elements of the matrix on the left side of equation (3.51) must be equal to $p_{max} + \sum_j \sum_l X_{jl}^2 / n$. This sum of squares must be equal to the sum of squares of the elements of the matrix on the right in equation (3.51). The latter sum of squares is the trace of $W \Psi^2 W' W \Psi^2 W' = W \Psi^4 W'$, with trace $\sum \psi_s^4$. We already know that $p_{max} = \sum \psi_s^2$. It then follows that

$$1/2 \, n \sum_s (\psi_s^2 - 1)^2$$

$$= 1/2 \, n \sum_s \psi_s^4 + 1/2 \, n \, p_{max} - n \sum_s \psi_s^2 =$$

$$= 1/2 \, n \, (p_{max} + \sum_j \sum_l X_{jl}^2/n) + 1/2 \, n \, p_{max} - n \, p_{max}$$

$$= \sum_{j<l} X_{jl}^2. \tag{3.54}$$

Under the assumption that all variables are independently distributed the latter quantity converges to χ^2 with $1/2[(p_{max})^2 - \sum (k_j - 1)^2]$ degrees of freedom. Note that (3.54) also shows that the total chi-square can be split up dimension-wise; we could compute a 'proportion-of-chi-square-accounted-for' for each HOMALS dimension (also note that, with respect to the present notation, the computer program HOMALS reports ψ_s^2/m as the eigenvalue).

3.11 AN ILLUSTRATION: HARTIGAN'S HARDWARE

The following illustration is obtained by performing homogeneity analysis on data taken from Hartigan (1975, p. 228). A number of bolts, nails, screws, and tacks are classified according to a number of criteria. Table 3.14 explains the symbols used and gives the basic data matrix. HOMALS was applied in two dimensions. The results are given in Table 3.15a (object scores X) and in Table 3.15b (category quantification Y), while Table 3.15c gives the discrimination measures for the various criteria. Figure 3.10 depicts the object scores. We see in the figure that the first dimension discriminates screws and bolts (which have a thread) from nails and tacks (which do not have a thread). It also separates bolts (with flat bottom) from screws, tacks, and nails (with sharp bottom). It does not separate nails from tacks. The second dimension separates SCREW1 and NAIL6, both being very long, from the rest.

Table 3.14a Hartigan's hardware: variables and categories

Variables	Categories and codes		
1 Thread	Yes = Y	No = N	
2 Head	Flat = F	Cup = C	Cone = O
	Round = R	Cylinder = Y	
3 Head indentation	None = N	Star = T	Slit = L
4 Bottom	Sharp = S	Flat = F	
5 Length	(in half inches)		
6 Brass	Yes = Y	No = N	

Table 3.14b Hartigan's hardware

Object	1	2	3	4	5	6
TACK	N	F	N	S	1	N
NAIL1	N	F	N	S	4	N
NAIL2	N	F	N	S	2	N
NAIL3	N	F	N	S	2	N
NAIL4	N	F	N	S	2	N
NAIL5	N	F	N	S	2	N
NAIL6	N	U	N	S	5	N
NAIL7	N	U	N	S	3	N
NAIL8	N	U	N	S	3	N
SCREW1	Y	O	T	S	5	N
SCREW2	Y	R	L	S	4	N
SCREW3	Y	Y	L	S	4	N
SCREW4	Y	R	L	S	2	N
SCREW5	Y	Y	L	S	2	N
BOLT1	Y	R	L	F	4	N
BOLT2	Y	O	L	F	1	N
BOLT3	Y	Y	L	F	1	N
BOLT4	Y	Y	L	F	1	N
BOLT5	Y	Y	L	F	1	N
BOLT6	Y	Y	L	F	1	N
TACK1	N	F	N	S	1	Y
TACK2	N	F	N	S	1	Y
NAILB	N	F	N	S	1	Y
SCREWB	Y	O	L	S	1	Y

These results are confirmed in Figure 3.11 which depicts the discrimination measures. This plot, too, shows that the first dimension is related to variables 1 (THREAD) and 4 (BOTTOM). Closest to the second dimension is variable 5 (LENGTH). Variables 2 and 3 are in between, whereas variable 6 discriminates very poorly in the first two dimensions. In fact, we would expect that BRASS and LENGTH do not discriminate very well, because in general there are screws, nails, bolts and tacks of any length, and either in brass or not. If such variables still discriminate, it must be because of peculiarities in the sample. Figure 3.12 shows the category quantifications; the points in Figure 3.12 are the centres of gravity of the object points associated with each category.

Table 3.15a Object scores

Object	Dimension 1	Dimension2
TACK	0.75	0.46
NAIL1	0.68	0.47
NAIL2	0.96	0.52
NAIL3	0.96	0.52
NAIL4	0.96	0.52
NAIL5	0.96	0.52
NAIL6	1.00	-1.69
NAIL7	1.25	-0.74
NAIL8	1.25	-0.74
SCREW1	-0.38	-3.96
SCREW2	-0.85	0.23
SCREW3	-0.91	0.26
SCREW4	-0.57	0.28
SCREW5	-0.63	0.31
BOLT1	-1.31	0.38
BOLT2	-1.18	-0.51
BOLT3	-1.30	0.40
BOLT4	-1.30	0.40
BOLT5	-1.30	0.40
BOLT6	-1.30	0.40
TACK1	0.93	0.67
TACK2	0.93	0.67
	0.93	0.67
SCREWB	-0.54	-0.44

Table 3.15b Category quantifications

Category		Dimension 1	Dimension2
1	yes	-0.96	-0.15
1	no	0.96	0.15
2	flat	0.90	0.56
2	cup	1.16	-1.05
2	cone	-0.70	-1.64
2	round	-0.91	0.30
2	cylinder	-1.12	0.36
3	no	0.96	0.15
3	star	-0.38	-3.96
3	slit	-1.02	0.19
4	sharp	0.43	-0.08
4	flat	-1.28	0.25
5	0.5 inch	-0.34	0.31
5	1 inch	0.44	0.44
5	1.5 inch	1.25	-0.74
5	2 inch	-0.60	0.34
5	2.5 inch	0.31	-2.82
6	yes	0.56	0.39
6	no	-0.11	-0.08

Table 3.15c Hartigan's hardware: discrimination measures

Variables	Dimension 1	Dimension 2
Thread	0.93	0.02
Head	0.95	0.64
Head indentation	0.95	0.68
Bottom	0.55	0.02
Length	0.29	0.82
Brass	0.06	0.03
Eigenvalues	0.62	0.37

A more precise and detailed analysis is possible by studying the six plots in Figure 3.13. Here the object scores are plotted again, but now labelled for each variable separately, using the labels of Table 3.14a. From these plots we see that variables 5 and 6 have categories that cannot be separated very well (at least in the first two dimensions). For the other variables the objects with the

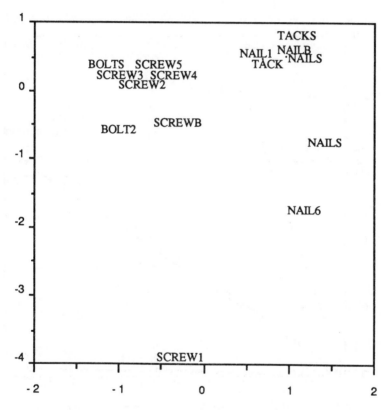

Figure 3.10 Hartigan's hardware: object scores labelled by type.

same label form fairly homogeneous clumps, with some exceptions (such as that for variable 2 the cylinders cannot be separated from the rounds). There is another important point. For variable 1 the categories 'Y' and 'N' are not very homogeneous, in the sense that objects in the same category are not necessarily very close together. On the other hand, it is clear that the first dimension separates the categories 'Y' and 'N' perfectly. We consider this result satisfactory, although the HOMALS loss might not be small in situations like this. To put it differently, in situations like this we essentially want categories to be well separated, but we do not necessarily want that objects in a category to form compact clumps. The HOMALS definition of what is the 'best' or 'optimal' solution is frequently stronger than the definition we really want to use. Another point illustrated by this example is that the interesting separation is almost completely along the first dimension.

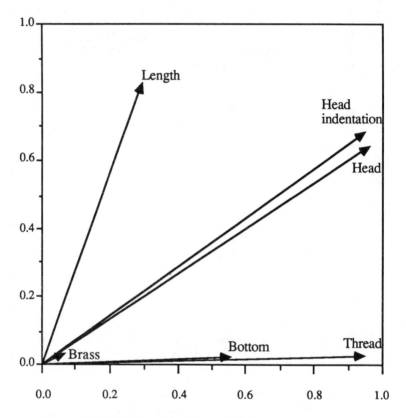

Figure 3.11 Hartigan's hardware: discrimination measures.

The second dimension primarily capitalizes on the fact that there is one special screw in the sample: SCREW1 (very long, cone head, star indentation). If we delete SCREW1 from the data, the HOMALS solution changes considerably with respect to the second dimension. The second dimension then contrasts the U heads (variable 2) with the four brasses (variable 6). The link between these two variables is LENGTH (variable 5), because the brasses all have length 1 and the U heads are the only objects with length 3 or 5. It thus turns out that the second dimension depends strongly on the particular choice of objects in the sample. (The results of this second analysis are not presented.)

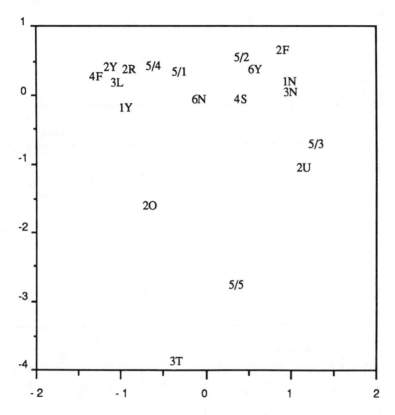

Figure 3.12 Hartigan's hardware: category quantifications.

All tables and plots shown so far can be produced by the standard HOMALS program. This is not true for the plot in Figure 3.14. It illustrates the history of the HOMALS algorithm. The plot maps the 24 objects in two dimensions on the basis of results for **X** in successive iteration steps. Lines have been drawn that connect successive positions of the same object, showing that some objects gradually move towards their eccentric position whereas others move away from their eccentric position towards the centre. Obviously,

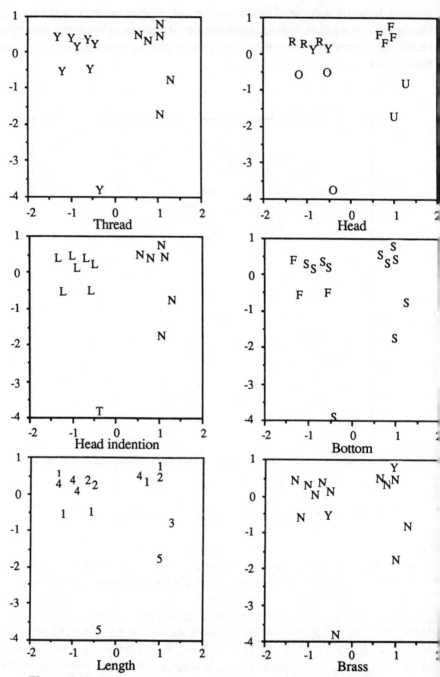

Figure 3.13 Hartigan's hardware: object scores labelled by variables.

the initial positions (first iteration) very much depend on the arbitrary initial identifications of **X** or **Y**. Nevertheless, the figure shows rather nicely how HOMALS starts with big moves, which gradually become smaller, until finally they become very small indeed.

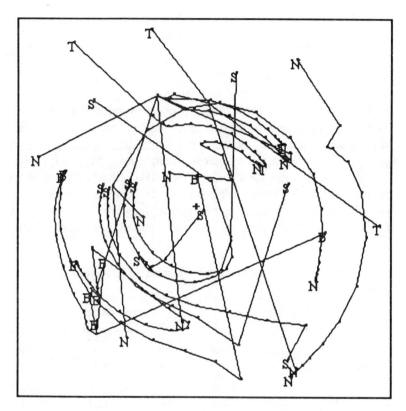

Figure 3.14 Hartigan's hardware: iteration history (point 10 omitted).

3.12 HOMALS WITH INCOMPLETE INDICATOR MATRIX

The discussion in Section 2.3 suggests for the incomplete indicator matrix a quantification that obeys the proportionalities

$$y \propto D^{-1}G'x, \qquad\qquad (3.55)$$

$$\mathbf{x} \propto \mathbf{M}_*^{-1}\mathbf{Gy}. \tag{3.56}$$

Such a quantification is consistent with a solution based on the SVD

$$\mathbf{M}_*^{-1/2}\mathbf{GD}^{-1/2} = \mathbf{V\Psi W'}, \tag{3.57}$$

from which we may form

$$\mathbf{x} = \mathbf{M}_*^{-1/2}\mathbf{v} \ \sqrt{(mn)}, \tag{3.58}$$

$$\mathbf{y} = \mathbf{D}^{-1/2}\mathbf{w}\psi \ \sqrt{(mn)} \ , \tag{3.59}$$

so that $\mathbf{x'M}_*\mathbf{x} = mn$ and $\mathbf{y'Dy} = mn\psi^2$. It remains true that $\mathbf{u'Dy} = 0$, but we no longer have $\mathbf{u'D}_j\mathbf{y}_j = 0$, because in this weighted summation of means with respect to the row scores $\mathbf{x}$ some of these scores are skipped. It remains true that $\mathbf{u'M}_*\mathbf{x} = 0$. In the complete indicator matrix case, $\mathbf{M}_* = m\mathbf{I}$, so that there we have $\mathbf{u'x} = 0$. A consequence is that the HOMALS solution no longer can be interpreted in such a simple way as a principal components analysis on the optimally scaled data matrix. Since an incomplete indicator matrix has some rows that are completely zero, the optimally scaled data matrix obtains zero

Table 3.16 Results for numerical example with option *(i)* 'missing values passive'

Object scores X			Category quantifications Y		
1	−1.33	−0.10	a	−0.82	0.03
2	1.59	−2.01	b	1.07	−1.39
3	1.63	2.67	c	0.62	0.59
4	−1.29	−0.04			
5	0.56	−0.78	p	−0.53	−0.04
6	0.34	0.46	q	1.59	−2.01
7	−1.29	−0.04	r	1.63	2.67
8	−0.36	0.11			
9	0.89	0.72	u	−1.30	−0.06
10	−0.36	0.11	v	0.61	0.18
Eigenvalues Ψ^2	0.688	0.534			

entries on its corresponding places. Such zeros are not the quantification of a category that was missing. Apart from that, columns of the optimally scaled data matrix will not add up to zero. This has, among other things, the consequence that PCA on the optimally scaled matrix produces a different result.

Another consequence is that the discrimination measure $y_j'D_jy_j/n$ can no longer be interpreted as the variance of the elements in a column of the optimally scaled data matrix, since such a column G_jy_j does not have zero mean. Also, the discrimination measure no longer equals the squared correlation between x and G_jy_j, for the same reason. Finally, it now may happen that a discrimination measure becomes larger than unity.

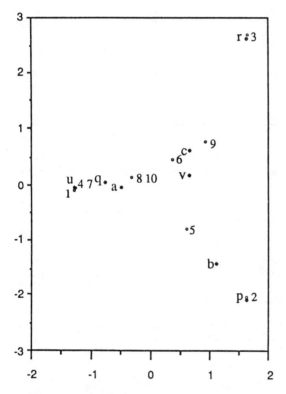

Figure 3.15 HOMALS solution with option 'missing values passive'.

One special case of incomplete indicator matrix is that of *missing data*, as discussed in Section 2.4, for the option 'missing data passive'. In the present section we will compare results for the numerical example of Section 2.4, Table 2.10, for the three different options mentioned there.

Table 3.17 Results for numerical example with option *(ii)* 'missing values single category'

Object scores X			Category quantifications Y		
1	−1.11	−0.00	a	−0.77	0.00
2	1.59	−2.07	b	1.08	−1.46
3	−0.11	−0.00	c	1.08	1.46
4	−1.19	0.00	?	−0.61	0.00
5	0.57	−0.86			
6	0.57	−0.86	p	−0.44	−0.00
7	−1.19	0.00	q	1.59	−2.07
8	−0.36	−0.00	r	−0.11	−0.00
9	1.58	2.07	?	1.59	2.07
10	−0.36	−0.00			
			u	−1.16	0.00
			v	0.50	−0.00
Eigenvalues Ψ^2	0.665	0.569			

Table 3.18 Results for numerical example with option *(iii)* 'missing values multiple categories'

Object scores X			Category quantifications Y		
1	−1.36	−0.70	a	−0.69	−0.26
2	0.82	1.36	b	0.52	0.98
3	2.06	−2.16	c	0.52	0.98
4	−1.06	−0.47	?1	−1.36	−0.70
5	0.22	0.60	?3	2.06	−2.16
6	0.22	0.60			
7	−1.06	−0.47	p	−0.53	−0.08
8	−0.32	−0.06	q	0.82	1.36
9	0.82	1.36	r	2.06	−2.16
10	−0.32	−0.06	?9	0.82	1.36
			u	−1.16	−0.54
			v	0.50	0.23
Eigenvalues Ψ^2	0.747	0.631			

(i)   *Missing values passive.* This is the example with an incomplete indicator matrix. Results are given in Table 3.16, for $p = 2$ dimensions, graphically displayed in Figure 3.15.

(ii) *Missing values single category.* Here the indicator matrix is completed. Results are given in Table 3.17 and Figure 3.16.

(iii) Missing values multiple category. Here, too, the indicator matrix is completed. Results are given in Table 3.18 and Figure 3.17.

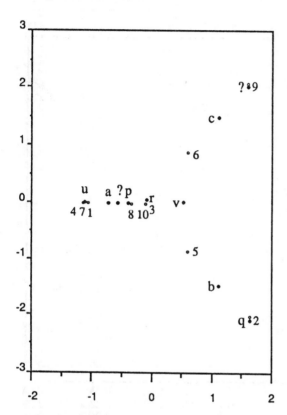

Figure 3.16 HOMALS solution with option 'missing values single category'.

Comparing these results, one may note that in solution *(ii)* objects 1 and 3 are brought closer together (because they share the single category of missing data on variable 1). In solution *(iii)* objects 1 and 3 are again farther apart. On the whole, solution *(iii)* seems to be largely affected by this particular treatment of missing data, which, of course, in this miniature example with 3 out of 30 data missing, is not too surprising. For actual data matrices, if the number of missing data is relatively small and if missing data are randomly divided over objects and categories, differences between the three options will be small (and the interpretation of discrimination measures will be almost the same as for the complete case).

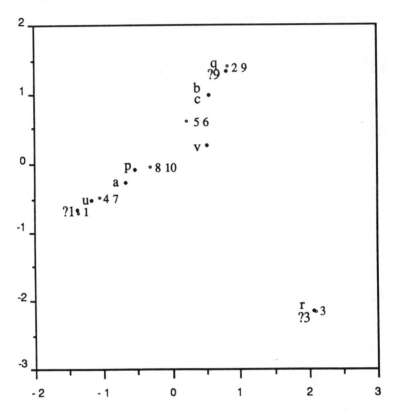

Figure 3.17 HOMALS solution with option 'missing values multiple categories'.

Options *(ii)* and *(iii)* do produce a quantification of the missing data. In option *(ii)* it is the average object score for objects with missing data on a variable. In option *(iii)* each missing value obtains the quantification of the associated object. In option *(i)* missing values are not quantified. We cannot think of them as having quantification equal to 0, since this would be inconsistent with the idea that category quantification is the average of the objects within that category.

Suppose an object has missing values on all variables. Option *(i)* gives scores equal to 0 to this object on all dimensions. Option *(ii)* will quantify this object as the average of all other objects with missing data. Option *(iii)* will quantify this object on a separate dimension of its own.

As an additional example of comparing the incomplete versus the completed indicator matrix, Figures 3.18 and 3.19 give results for the example used in Section 2.3 (the seriation problem), both for the first two dimensions.

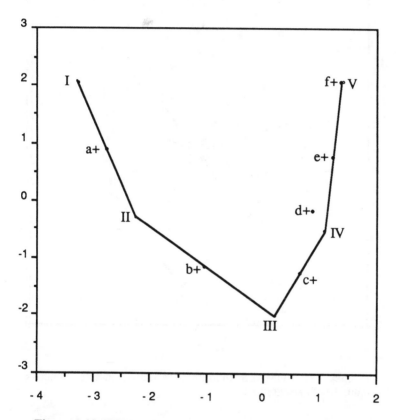

Figure 3.18 HOMALS solution for seriation example, based on incomplete indicator matrix.

Although for both solutions the order for the objects (or for the categories) as given in Table 2.8a 'curls up', the first dimension of Figure 3.18 still shows objects (and categories) ordered in the same way as in Table 2.8a, whereas in Figure 3.19 there is no longer any direction in the plot on which objects can be projected in their order of the table. Figure 3.19 therefore can be said to make objects I and V (the extremes of the scale) more similar. Note also that in Figure 3.19 category points for + and – of each category are the opposites of each other with respect to the origin of the plot.

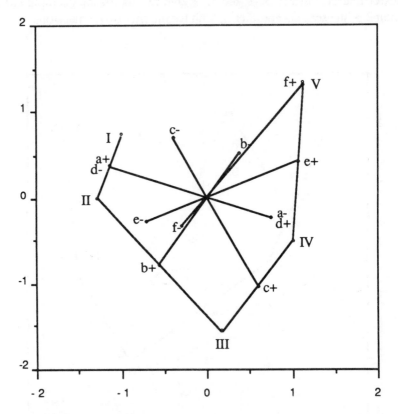

Figure 3.19 HOMALS solution for seriation example, based on completed indicator matrix.

3.13 REVERSED INDICATOR MATRIX

In Section 2.5 we observed that the most typical application of an analysis of a reversed indicator matrix would be in a sorting task. As a miniature example, suppose we have four items. They are the names 'Washington', 'Lincoln', 'Churchill', and 'de Gaulle' (W, L, C, G). Three judges are asked to sort these names into categories. Suppose judge 1 makes the groups (W L C) and

(G), perhaps related to whether the items were English speaking or not. Judge 2 sorts (W L) (C G), perhaps implicating that he sorts 'before 1900' versus 'after 1900'. Judge 3 sorts (W L) (C) (G), perhaps by country of birth. In the reversed indicator matrix approach the objects of analysis are the items and the variables of analysis are the judges. The data matrix and the reversed indicator

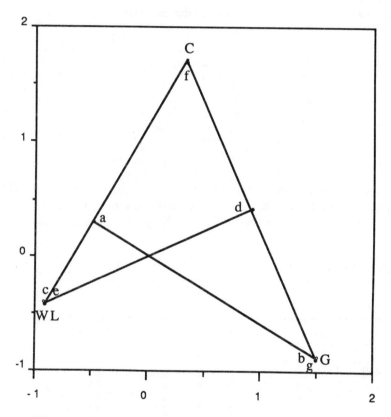

Figure 3.20 HOMALS solution for mini sorting task, illustrating solution for reversed indicator matrix.

matrix are given in Table 3.19. A HOMALS solution on this reversed indicator matrix is given in Figure 3.20. Objects W and L coincide (they are always sorted together), and they also coincide with categories 'c' and 'e' (unique for W and L). Object C coincides with category 'f', object G with categories 'b' and 'g'. Category 'a' is in the centre of gravity of W, L and C, category 'd' in the centre of gravity of C and G. Assuming that our interpretation of categories was correct, the picture shows that W and L are both born in the USA and lived before 1900, whereas G is French and lived after 1900, and C, W, and L are English speaking.

Table 3.19 Example of sorting task

Data matrix				Reversed indicator matrix							
					1		2			3	
	W	L	C	G		a	b	c	d	e	f	g
1	a	a	a	b	W	1	0	1	0	1	0	0
2	c	c	d	d	L	1	0	1	0	1	0	0
3	e	e	f	g	C	1	0	0	1	0	1	0
					G	0	1	0	1	0	0	1

A second miniature example is given to show that the results for the analysis of the classical and the reversed indicator matrix can be quite different. Consider the data matrix in Table 3.20. Suppose the eight rows are

Table 3.20 Data matrix

					Items				
		1	2	3	4	5	6	7	8
	1	1	1	1	1	1	2	2	2
	2	1	1	2	2	2	2	3	3
	3	1	2	2	2	3	3	3	4
Subjects	4	2	3	3	3	3	4	4	4
	5	3	3	3	4	4	4	4	5
	6	4	4	4	4	4	5	5	5
	7	1	2	2	2	5	5	5	5
	8	1	1	1	2	2	2	5	5

subjects that gave ratings for eight items on a five point scale. There is a lot of variability among the subjects in the way they use the scale. Consider, for instance, subjects 1 and 6. They give exactly the same pattern, except for the fact that subject 1 uses the left side of the scale, whereas subject 6 uses the right side. The pattern for subject 4 resembles the pattern of both subjects 1 and 6, only this subject uses the middle of the scale.

Table 3.21 Transposed data matrix

		Subjects							
		1	2	3	4	5	6	7	8
	1	1	1	1	2	3	4	1	1
	2	1	1	2	3	3	4	2	1
	3	1	2	2	3	3	4	2	1
Items	4	1	2	2	3	4	4	2	2
	5	1	2	3	3	4	4	5	2
	6	2	2	3	4	4	5	5	2
	7	2	3	3	4	4	5	5	5
	8	2	3	4	4	5	5	5	5

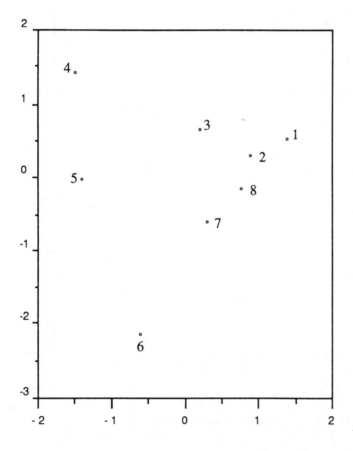

Figure 3.21 HOMALS solution for miniature example: subjects.

When this data matrix is analysed with HOMALS, we obtain the plots for the subjects and the categories as given in Figures 3.21 and 3.22. We notice that the subjects are located on a triangle, with the subjects 1, 4, and 6 on the corners. The same applies to the category points. How do we interpret the fact that the categories of the rating scales have obtained quantifications that are in both dimensions, far from monotone with the original scale values? It might be clear that this analysis tells us little about the original items. On the other hand, it tells us quite a lot about the response behaviour of the subjects.

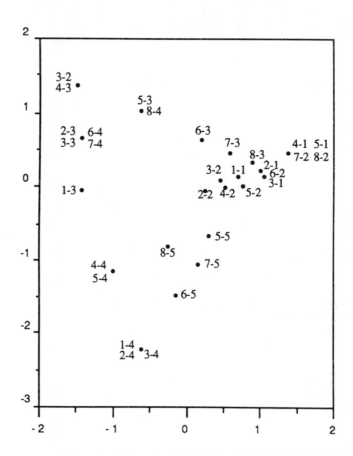

Figure 3.22 HOMALS solution for miniature example: category points. Each first number refers to an item, each second number to a category.

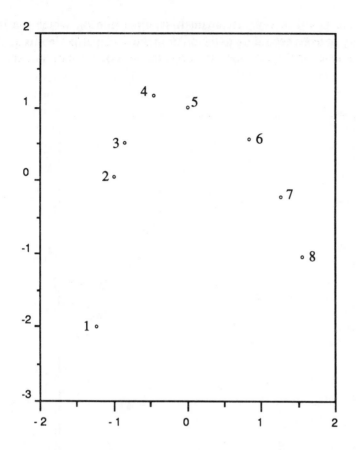

Figure 3.23 HOMALS solution for transposed data matrix of miniature example: items.

Now consider the situation for the reversed indicator matrix, which comes to the analysis of the transposed data matrix given in Table 3.21. Here the subjects act as the variables and the items as the objects. It is immediately clear that the fact that subjects 1 and 4 use different sides of the scale means that they cannot play a role in the analysis any longer, because the HOMALS results are invariant under one-to-one nonlinear transformations of the categories per variable. This implies, for instance, that column 6 can be replaced by column 1 without changing the solution.

The results of the analysis of Table 3.21 are depicted in Figures 3.23 and 3.24. Now we obtain object points for the items: the horseshoe indicates that we are dealing with a very dominant first dimension on which the items are perfectly ordered according to the order in the data matrix. In this analysis we acquire category points for the subjects: we notice that, as expected,

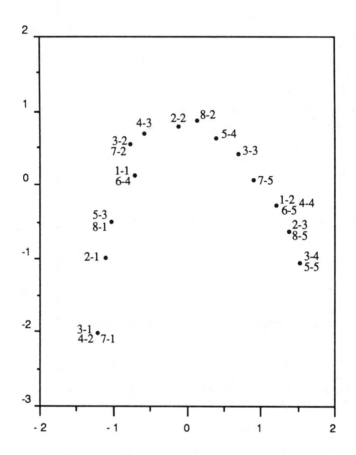

Figure 3.24 HOMALS solution for transposed data matrix of miniature example: category points. Each first number refers to a subject, each second number to a category.

categories 1 and 2 for subject 1 have obtained exactly the same quantifications as categories 4 and 5 for subject 6. Moreover, all categories are monotonically transformed in the first dimension.

Whether or not we have to analyse a reversed indicator matrix is in general difficult to decide. It depends on considerations that go beyond the formal appearance of the data. Yet in any particular situation the choice can be crucial for the results (see also the example in Section 13.1).

3.14 EPILOGUE

Unfortunately, the term homogeneity analysis, perhaps especially in the strict sense, did not really gain acceptance. Much better known are 'multiple correspondence analysis' and 'analysis of the Burt table'. Yet it is our conviction that homogeneity and her twin brother heterogeneity are the basic concepts that appear in different guises (cf. Tenenhaus and Young, 1985).

For an extended discussion of choosing a suitable space of transformations in canonical analysis, see Mallet (1982). It should be remarked that one special feature of the basis of indicator functions is the property of invariance of analysis results under one-to-one nonlinear transformations of the original data. This property was used as the defining characteristic of nonlinear MVA in Chapter 1. Homogeneity analysis viewed as eigenanalysis of the average similarity table is treated extensively in Meulman (1982). A new idea in this area is to drop the aim of complete homogeneity and to switch to three-way analyses of the m separate similarity matrices (Marchetti, 1988; Kiers, 1989).

The reader is referred to Bekker and De Leeuw (1988) for a thorough discussion of the interplay between the two types of dimensionality mentioned in Section 3.9. In Van de Geer (1985) more practical comments and examples of the use of HOMALS can be found.

CHAPTER 4
NONLINEAR PRINCIPAL
COMPONENTS ANALYSIS

4.1 METRIC PRINCIPAL COMPONENTS ANALYSIS

In Sections 3.2 and 3.6 we reviewed some of the history of *homogeneity analysis*, which constitutes one particular way of introducing principal components analysis. We have seen that Pearson introduced a 'dual' form of homogeneity analysis in 1901 and that ordinary PCA, in terms of using the principal axes of the correlation ellipsoid to compute optimal index characters, was also invented by Pearson around the same time (cf. MacDonell, 1901). The first systematic account of principal components analysis as a data analytic technique was by Hotelling (1933), and his particular form of 'factor analysis' was introduced to statisticians by Girshick (1936). Both Hotelling and Girshick used homogeneity ideas as their starting point. They looked for the linear composite with the largest variance, which is the same as the linear composite with a maximum sum of squared correlations with the variables. Additional components were introduced by looking for the second best solution under orthogonality constraints. This line of reasoning was taken up by Horst (1936), Edgerton and Kolbe (1936), Wilks (1938), and Guttman (1941). It leads directly to our form of homogeneity analysis and to the computer program HOMALS.

At about the same time Eckart and Young (1936) introduced a natural way to define principal components analysis for a general p, i.e. to formulate optimality properties that hold for the first p components simultaneously. This is a more natural way to introduce multidimensional solutions than the earlier successive procedures, although it turns out that in the simple situations investigated by these early authors simultaneous and successive multidimensionality lead to the same solution (or, as we sometimes say, the solutions for different dimensionalities are *nested*). Eckart and Young used the least squares properties of the singular value decomposition (also discussed in Appendix B), which were already described in a more general context by

Schmidt (1906). The singular value decomposition itself was discovered, independently, by Beltrami (1873), Jordan (1874), and Sylvester (1889). Simultaneous and successive optimality conditions of principal components are reviewed by Rao (1964), Okamoto (1968), Roux and Rouanet (1979), and Rao (1980). The Roux and Rouanet paper is a beautiful 'French' introduction to principal components analysis. Rao (1980) extends the approximation properties of the singular value decomposition to a more general class of matrix norms.

The Eckart–Young theorem tells us that principal components analysis of an $n \times m$ data matrix $\mathbf{H}$ can be formulated in terms of the loss function

$$\sigma_J(\mathbf{X},\mathbf{A}) \equiv SSQ(\mathbf{H} - \mathbf{X}\mathbf{A}'), \tag{4.1}$$

where $\mathbf{X}$ is $n \times p$ of rank p, and where $\mathbf{A}$ is $m \times p$. We use the symbol σ_J (join loss) because $\sigma_J(\mathbf{X},\mathbf{A}) = 0$ for some $\mathbf{X}$ and $\mathbf{A}$ means, in the terminology of Section 1.8.1, that the join rank of $\mathbf{H}$ is less than or equal to p. Appendix B explains that the optimal $\mathbf{X}$ and $\mathbf{A}$ that minimize $\sigma_J(\mathbf{X},\mathbf{A})$ can be found by computing the singular value decomposition of $\mathbf{H}$ and by retaining the largest p singular values with corresponding singular vectors. Another possible computational procedure is alternating least squares, in which we alternate the steps

$$\mathbf{X} \leftarrow \mathbf{HA}(\mathbf{A'A})^{+}, \tag{4.2}$$

$$\mathbf{A} \leftarrow \mathbf{H'X}(\mathbf{X'X})^{+}, \tag{4.3}$$

where the superscript + indicates the Moore–Penrose inverse of a matrix. This computational procedure is described by Daugavet (1968).

We emphasize that principal components analysis is usually not interpreted in terms of fitting a statistical model. The model $\mathbf{h}_j = \sum \mathbf{x}_s a_{js}$, implicit in the Eckart–Young formulation, is equivalent with the model that the rank of the covariance matrix of the $\mathbf{h}_j$ is less than or equal to p. If we assume a multinormal sample, then statistical theory tells us to reject this model whenever the observed sample dispersion matrix has rank larger than p. The performance characteristics of this test are very satisfactory. It is impossible to reject the model if it is true, and the probability of accepting it when it is false tends to zero exponentially fast. In practical situations, however, the test means that we shall always reject the model, which is not a very satisfactory situation. A more realistic model is possible in which we suppose that the smallest $m - p$ eigenvalues of the covariance matrix are equal (but not necessarily equal to zero). This corresponds with $\mathbf{h}_j = \sum \mathbf{x}_s a_{js} + \mathbf{e}_j$, where the errors $\mathbf{e}_j$ are independent and have equal variances. This model, however, i better interpreted as a special case of the factor analysis model, and not as principal components analysis.

4.2 NONMETRIC PRINCIPAL COMPONENTS ANALYSIS

The word 'nonmetric' in nonmetric principal components analysis could be somewhat confusing. In the history of psychological scaling we can distinguish three different uses of nonmetric. In the first place there are nonmetric *data*. Preference rank orders are, for example, usually thought of as nonmetric data, although it is of course possible to *code* them metrically, for example by using rank numbers or order statistics. The idea is, however, that we are supposed to use only the ordinal properties of the data in arriving at our representations. This does not mean that it is forbidden, by some mysterious central directorate for the administration of scale levels, to correlate rank numbers or paired comparisons. Of course such a prohibition would not make sense, correlating rank numbers gives Spearman's ρ and correlating paired comparisons gives Kendall's τ, both of which are perfectly respectable. The idea is that if the data 'are' nonmetric, then we should first transform them to rank numbers or paired comparisons or some other set of invariant conventional numbers before we correlate them, and thus make the correlation coefficient reflect the ordinal properties of the data only. We have used nonmetric here in almost the same sense as ordinal. This is somewhat restrictive, although it is quite common. Clearly nominal data are nonmetric too. We are not sure whether missing data are nonmetric. Observe that this use of the word is consistent with our definition of nonlinear and ordinal MVA in Chapter 1.

The second use of the word nonmetric is somewhat more specialized. The older scaling programs for nonmetric data produced nonmetric representations. Guttman scaling, for example, provided a rank order of the items and individuals; the unfolding theory of Coombs gave a rank order of objects only (the common J scale). These older algorithms are very ably reviewed in the book of Coombs (1964), who also presents similar algorithms for multidimensional unfolding and nonmetric multidimensional scaling, which rank the objects on a number of dimensions. In what is usually called the 'nonmetric breakthrough' or the 'computational breakthrough' in psychological scaling, Shepard, Guttman, and Kruskal showed that it was actually possible in some circumstances to derive *metric* representations from *nonmetric* data, or, more precisely, stable metric representations. This breakthrough produced a lot of enthusiasm at the time, but after twenty years we start to find out that it does not always work very well (compare Section 4.3 in this chapter). This makes even older approaches due to Thurstone and Guttman, which already derived metric representations from nonmetric data by introducing stronger requirements, more interesting. As we shall see in Section 8.2 this is basically how HOMALS fits into the scaling tradition.

Finally, nonmetric is sometimes used for error theories and loss functions. If we have a model that is formulated in terms of a finite number of inequalities that must be true (or even a finite number of equations that must be true), then we can measure loss by counting the number of inequalities and/or equations that are not true for a particular representation, and we can find our representation in such a way that the number of violations is minimized. Such nonmetric error theories are particularly attractive in combination with non-metric data, but they present us with considerable computational problems. ➤ The algorithms outlined in Coombs (1964) were nonmetric in all three senses of the word, and could consequently be called purely nonmetric. Since the computational breakthrough, the word is used almost exclusively in the first sense we discussed.

We start our brief account of the history of *nonmetric* principal components analysis with a quotation from Thurstone:

> One of the principal assumptions underlying factorial theory is that the scores are monotonic increasing or decreasing functions of the scores on the primary factors or parameters. The fundamental observation equation of factor analysis makes the further assumption that these functions can be expressed in linear form as a first approximation. It would be possible to start with a second-degree observation equation and to develop factorial methods on that basis. Instead of developing factorial theory more completely with observation equations of higher degree, it would probably be more profitable to develop nonmetric methods of factor analysis. An idea for such a development would be to determine the number of independent parameters of a score matrix by analyzing successive differences in rank order on the assumption that they are monotonic functions of a limited number of independent parameters. A score would then be regarded as merely an index of rank order, and that is essentially what we are now doing. The raw scores are transmuted into a normal distribution of unit standard, and these transmuted scores are used for the correlations. Instead of dealing with the transmuted scores in this manner, one might deal with the rank orders directly or with some equivalent indices of rank order (Thurstone, 1947, p. xiii–xiv).

There are no further contributions to this problem in Thurstone's book. The next contribution was Bennett (1956). This is in the purely nonmetric tradition of Coombs and his school. He determines the number of distinct ways in which n subjects can be ranked by linear functions of p factors, and uses these combinatorial results to bound the number of factors in the complete case, when all possible linear functions (tests) have been used. This is clearly not very relevant for practical data analysis, although it does tell us some interesting theoretical properties of nonmetric components analysis.

Guttman (1959) makes another important contribution. He adds the additional restriction that the regressions between transformed variables must be linear, and shows that there is at most one possible transformation of the variables which linearizes regressions *and* is monotonic. This is easy to see if there are just two variables. Linearity of regression defines the stationary equations of correspondence analysis (Hirschfeld, 1935). We shall see in Section 8.3 that the number of solutions to these equations is equal to the number of categories of the variable with the smallest number of categories and that the solutions are orthogonal. Orthogonality implies that if one of them is monotone, then the others certainly are not monotone. Guttman discusses one of the possible extensions of this result to more than two variables. His conclusion is interesting:

> Since the metrics of observed test scores are usually arbitrary, Thurstone posed the problem of how to 'factor' them by using only rank-order considerations. One form of solution is to seek transformations that will yield new scores with a correlation matrix that is best from some point of view of factor analysis. But analyzing data via their correlation matrix is justified stochastically only if the regressions are linear. Assuming only the linearity restriction on regression, it is shown that – in general – at most one set of new scores can be found to maintain the observed rank-orders. The factor analyst has no freedom to mould the new correlation matrix by further considerations (Guttman, 1959, p. 257).

In the early sixties the computational breakthrough in nonmetric scaling was accomplished by Shepard and Kruskal. Shepard showed that ordinal restrictions, at least in some situations, constrain the solution almost as tightly as linear restrictions, and he showed that metric solutions can be obtained from ordinal data by an iterative algorithm. Both ideas were revolutionary at the time; the easiest way to appreciate this is to compare them with the techniques discussed in the book by Coombs (1964). Kruskal emphasized explicit loss functions, showed that any differentiable loss function could be minimized by gradient methods, and introduced monotone regression. The introduction of monotone regression made it possible to construct nonmetric versions of all metric scaling methods, and the stability and gauging results of Shepard seemed to indicate that these new programs actually amounted to getting something (stable metric representations) from merely nothing (ordinal data).

A working program for nonmetric principal components analysis existed in 1962; in Shepard (1966) we find the first published stability results, but the basic paper was published only in 1974 by Kruskal and Shepard. The earliest published account is Roskam (1968, ch. 5). Both Kruskal and Shepard and Roskam use basically the same approach. They fit what they call the 'linear model' (actually 'bilinear model' is a better name) to a rectangular data matrix **H**. The model is defined by

$$q_{ij} = \sum_s x_{is} a_{js}, \tag{4.4}$$

$$h_{ij} > h_{kj} \rightarrow q_{ij} \geq q_{kj}, \tag{4.5}$$

where $i, k = 1,...,n$ and $j = 1,...,m$ and $s = 1,...,p$. Thus each column of the data matrix (each variable) must be transformed monotonically in such a way that the model with p components optimally fits the transformed data. The dimensionality p is important here, because choosing a different dimensionality will lead to different transformations; thus we can no longer expect the solutions to be nested. Another point that we must remember is that the data are interpreted as *column conditional*, by which we mean that only order relations within columns are imposed. In general the elements in a column may be only partially ordered; thus there may be missing data that do not impose constraints. The model does not specify what should be done if $h_{ij} = h_{kj}$; thus equalities in the data do not impose restrictions. This is called the primary approach to ties in Kruskal (1964a, 1964b) or De Leeuw (1977); it is called continuous ordinal data in De Leeuw, Young, and Takane (1976). If we add the restrictions

$$h_{ij} = h_{kj} \rightarrow q_{ij} = q_{kj}, \tag{4.6}$$

then this defines the *secondary approach to ties*, or *discrete ordinal data*. We shall come back to this distinction later; for the moment we write the restrictions imposed by the order relations in the data in the simple form $q_j \in C_j$, with C_j a convex cone in n-space (see Appendix C).

The loss function used by both Kruskal and Shepard and by Roskam is, except for some irrelevant details,

$$\sigma_J(\mathbf{Q},\mathbf{X},\mathbf{A}) \equiv m^{-1} \sum_j SSQ(\mathbf{q}_j - \mathbf{X}\mathbf{a}_j) / SSQ(\mathbf{q}_j - AVE(\mathbf{q}_j)), \tag{4.7}$$

which is minimized over all **X**, **A** and over all $\mathbf{q}_j \in C_j$ ($j = 1,...,m$).

The notation in definition (4.7) deserves some attention here. The matrix **A** is $m \times p$, the vector $\mathbf{a}_j$ has p elements, and there are m such vectors. Thus we mean that $\mathbf{a}_j$ is *row j* of **A**, *written as a column*. This sounds complicated, but no confusion is possible, and the notation is considerably simpler than something less ambiguous such as $\mathbf{a}'_{j \rightarrow}$. AVE (.) is the mean of a vector or the expected value of a random variable (cf. Appendix A). Because of the form of σ_J we call it a *normalized loss function*; the denominator is the *normalization factor*. For each j the corresponding component of the loss is the variance of the residuals divided by the variance of the transformed data. The loss function is minimized by gradient methods, combined with monotone regression. A very detailed account of the algorithms can also be found in Hartmann (1979, ch. 4).

Subsequent contributions to nonmetric/nonlinear principal components analysis are Roskam (1977), Tenenhaus (1977), and Young, Takane, and De Leeuw (1978). They all use the same loss function, but they differ in the types of data they can handle or in the algorithm. Tenenhaus assumes that variables are either numerical or nominal; Young, Takane, and De Leeuw can handle any mixture of nominal, ordinal, and numerical variables. All three programs (called MNNFAEX, PRINQUAL, and PRINCIPALS, respectively) use alternating least squares methods to minimize the loss function. They alternate transformations of the data with fitting of the PCA model to these transformed data; the latter is done by computing a partial or complete singular value decomposition. Another difference with the earlier programs is that PRINQUAL and PRINCIPALS use *explicit normalizations*. They minimize

$$\sigma_J(Q,X,A) \equiv m^{-1} \sum_j SSQ(q_j - Xa_j), \tag{4.8}$$

over X and A and over q_j satisfying $q_j \in C_j$, $AVE(q_j) = 0$, $SSQ(q_j) = 1$. The homogeneity of the bilinear model proves that the problem with explicit normalizations and the problem with normalized loss functions (or *implicit normalizations*) are equivalent. Using explicit normalizations leads to more compact formulas, and generally seems slightly more elegant.

It must be emphasized that all programs can also be applied to row-conditional data by simply transposing the data matrix. A very important application is to *preference rank orders*, in which a number of persons rank a number of objects with respect to preference (or utility, or beauty, or whatever). Tucker (1960) proposed a classical model which assumes that each person defines a direction in p-dimensional space, each object defines a point in the same space, and preference strength for an object–person combination is the length of the projection of the object point on the person direction. In Coombs (1964) the obvious nonmetric extension of this model is discussed. If the person direction is a_j, with $SSQ(a_j) = 1$, and the object point is x_i, then the model says that

$$q_{ij} = \sum_s x_{is}a_{js} \tag{4.9}$$

is the preference strength of person j for object i. If h_{ij} is the observed preference strength, then the model requires in addition

$$h_{ij} > h_{kj} \rightarrow q_{ij} \geq q_{kj}. \tag{4.10}$$

This is clearly exactly identical to the nonmetric principal components model, but the comparisons are now within individuals. If the data are collected in an individual × objects matrix, then they are row-conditional, and we have to transpose the matrix first before we can apply the programs discussed above. We shall return to this important application in our examples (Section 4.8).

4.3 THEORY OF JOIN LOSS

In this section we study some properties of $\sigma_J(Q,X,A)$, using explicit normalization. The easiest way to do this is to define $\sigma_J(Q,*,*)$, which is the minimum of $\sigma_J(Q,X,A)$ over X and A for fixed Q. By the Eckart–Young theorem (cf. Appendix B)

$$\sigma_J(Q,*,*) = m^{-1} \Sigma_s \lambda_s^2(R(Q)), \quad s = p+1,...,m \tag{4.11}$$

where $R(Q)$ stands for the correlation matrix of the m transformed variables q_j and λ_s^2 for its eigenvalues, in decreasing order. So the smallest $m - p$ eigenvalues are summed in (4.11). Clearly

$$0 \leq \sigma_J(Q,*,*) \leq 1 - p/m, \tag{4.12}$$

with $\sigma_J(Q,*,*) = 0$ if and only if $\text{rank}(R(Q)) \leq p$, and $\sigma_J(Q,*,*) = 1 - p/m$ if and only if $R(Q)$ is the identity matrix.

Minimizing $\sigma_J(Q,*,*)$ over $q_j \in C_j$ with $\text{AVE}(q_j) = 0$ and $\text{SSQ}(q_j) = 1$ means that we transform or quantify our variables in such a way that the sum of the m–p smallest eigenvalues is minimized or, equivalently, that the sum of the p largest eigenvalues of $R(Q)$ is maximized.

If we want to compare this form of components analysis with homogeneity analysis discussed in Chapter 3, we must suppose in the first place that all variables are discrete and nominal. Thus the cones C_j are subspaces defined by

$$C_j = \{q_j \mid q_j = G_j y_j\}, \tag{4.13}$$

with the G_j complete indicator matrices. If $p = 1$ then the theory of the previous paragraph tells us that component analysis amounts to finding the y_j in such a way that the largest eigenvalue of the correlation matrix of the $q_j = G_j y_j$ is maximized, and we have already seen in Chapter 3 that this is one of the ways in which (one-dimensional) homogeneity analysis can be defined. Thus for $p = 1$ and all variables discrete and nominal the two techniques are equivalent. If $p = m - 1$, then we want to minimize the smallest eigenvalue of the correlation matrix. This is related to the 'dual' form of homogeneity analysis introduced by Pearson in 1901, and discussed briefly in Section 3.6. We know from Chapter 3 that the generalized eigenvalue problem in homogeneity analysis has one trivial eigenvalue equal to one, $m - 1$ trivial eigenvalues equal to zero, and $p_{max} = \Sigma k_j - m$ nontrivial solutions. In the same way as was shown that the largest nontrivial eigenvalue in homogeneity analysis corresponds with the components solution for $p = 1$, we can also show that the smallest nontrivial eigenvalue in homogeneity analysis corresponds with the principal components solution for $p = m - 1$.

For intermediate values of p the situation is more complicated. One useful way of looking at the problem is as follows. We distinguish between the *dimensionality p*, which we have defined already in terms of the number of columns of X and A in the loss function, and the *number of successive solutions r*. In homogeneity analysis we have $p = 1$. We compute an optimal solution q^*; the next step is the compute a second solution, still minimizing σ_J with $p = 1$, but imposing the additional *orthogonality* requirement that

$$\Sigma_j q_j' q_j^* = 0. \tag{4.14}$$

In general we can compute additional solutions by requiring that they must be orthogonal with all previous solutions. Thus homogeneity analysis has $p = 1$ and $r \geq 1$, and it yields *multiple quantifications* of the variables. Components analysis, on the other hand, has $p \geq 1$ and $r = 1$; we minimize σ_J for a given value of p, and then we are done. Thus components analysis gives a *single quantification*.

We can illustrate some of these points with the example of Section 3.9. – not all of them, however, because in this example m = 3. Consequently we can only choose p = 1 or p = m–1 = 2 here. There are five nontrivial solutions to the generalized eigenvalue problem of homogeneity analysis. The r successive solutions to the components analysis problem with p = 1 correspond with the r largest nontrivial eigenvalues of the homogeneity problem; the r successive solutions to the components analysis problem with

$p = m-1$ correspond with the r smallest nontrivial eigenvalues. In Table 4.1 we give the solution for $\mathbf{Q}$ with $p = 1$ and $r = 1$, with the corresponding correlation matrix, and the eigenvalues of $m^{-1}\mathbf{R}(\mathbf{Q})$. In Table 4.2 similar results for $p = 2$, $r = 1$ are given.

Table 4.1a Matrix $\mathbf{Q}_1$ for $p = 1$ normalized by diag($\mathbf{Q}'_1\mathbf{Q}_1$) = $n\mathbf{I}$

1	0.69	0.34	1.53
2	-1.88	-3.00	-0.65
3	0.69	0.26	-0.65
4	0.69	0.34	1.53
5	-1.88	0.34	-0.65
6	-0.20	0.34	-0.65
7	0.69	0.34	1.53
8	0.69	0.34	-0.65
9	-0.20	0.34	-0.65
10	0.69	0.34	-0.65

Table 4.2a Matrix $\mathbf{Q}_1$ for $p = 2$, normalized by diag($\mathbf{Q}'_1\mathbf{Q}_1$) = $n\mathbf{I}$

1	0.75	-0.04	-1.51
2	-1.73	2.35	0.65
3	0.75	-2.12	0.65
4	0.75	-0.04	-1.51
5	-1.73	-0.04	0.65
6	-0.53	-0.04	0.65
7	0.75	-0.04	-1.51
8	0.75	-0.04	0.65
9	-0.53	-0.04	0.65
10	0.75	-0.04	0.65

Table 4.1b $\mathbf{R}(\mathbf{Q}_1)$ for $p = 1$

1.00		
0.62	1.00	
0.45	0.22	1.00

Table 4.2b $\mathbf{R}(\mathbf{Q}_1)$ for $p = 2$

1.00		
-0.57	1.00	
-0.49	0.02	1.00

Table 4.1c Eigenvalues of $m^{-1}\mathbf{R}(\mathbf{Q}_1)$

0.63
0.26
0.11

Table 4.2c Eigenvalues of $m^{-1}\mathbf{R}(\mathbf{Q}_1)$

0.59
0.33
0.09

Note on Tables 4.1 and 4.2.
The eigenvalues of the corresponding homogeneity analysis are:

0.63
0.43
0.39
0.14
0.09

The largest one is the largest one in Table 4.1c, the smallest one is the smallest one in Table 4.2c.

It is, of course, possible to combine the two approaches and construct a technique with both $p \geq 1$ and $r \geq 1$. This amounts to applying components analysis r times, each time imposing an additional orthogonality constraint. We do not have any experience with this combined technique, however. It is of some interest that the definition of successive solutions implies automatically that they are *nested* with respect to different values of r, although components analysis is usually *not nested* with respect to different values of p. ➤

In De Leeuw, Young and Takane (1976) and De Leeuw and Van Rijckevorsel (1980) a system of measurement and process levels is discussed, which can be used to define many different types of cones C_j. If the data are numerical and we require that the transformations are all linear, then

$$C_j = \{ \mathbf{q}_j \mid \mathbf{q}_j = \alpha_j \mathbf{h}_j + \beta_j \}. \tag{4.15}$$

Without loss of generality we suppose that $SSQ(\mathbf{h}_j) = 1$ and $AVE(\mathbf{h}_j) = 0$ for all j. Because we require that $SSQ(\mathbf{q}_j) = 1$ and $AVE(\mathbf{q}_j) = 0$, it follows that $\sigma_J(\mathbf{Q},\mathbf{X},\mathbf{A}) = m^{-1} \sum_j SSQ(\mathbf{h}_j - \mathbf{X}\mathbf{a}_j) = m^{-1} SSQ(\mathbf{H} - \mathbf{X}\mathbf{A}')$. There is no freedom for choosing $\mathbf{Q}$ different from $\mathbf{H}$, and consequently the analysis amounts to computing eigenvalues and eigenvectors of $\mathbf{R}(\mathbf{H})$.

In this book we assume generally that data are categorical and that the number of categories of a variable is usually much less than the number of observations. This has some consequences. Suppose, for example, that we define all C_j by the continuous ordinal option (formula 4.5), which poses no constraints within blocks of tied data values. Let us investigate under what condition we can choose the $\mathbf{q}_j \in C_j$ in such a way that $rank(\mathbf{R}(\mathbf{Q})) = 1$, which implies that a perfect solution exists. Assume for convenience that the h_{ij} are integers, satisfying $1 \leq h_{ij} \leq k_j$. Suppose the n individuals are a random sample from a multivariate population, in which

$$\pi_0 \equiv prob(\underline{h}_1 = 1 \ \& \ ... \ \& \ \underline{h}_m = 1), \tag{4.16}$$

$$\pi_1 \equiv prob(\underline{h}_1 = k_1 \ \& \ ... \ \& \ \underline{h}_m = k_m), \tag{4.17}$$

with $\pi_0 + \pi_1 > 0$. If the variables are binary items, with wrong–correct interpretation, for example, then this supposes that there are individuals *in the population* that will answer all items correctly or all items incorrectly. Suppose $\underline{n}_0$ is the number of individuals in the sample with $h_{ij} = 1$ for all j and $\underline{n}_1$ is the number of individuals in the sample with $h_{ij} = k_j$ for all j. If $\underline{n}_0 + \underline{n}_1 > 0$, then a perfect solution can easily be constructed. (Choose three points on a straight line; allocate all $\underline{n}_0$ individuals who answer all items incorrectly to the left-most point, all $\underline{n}_1$ individuals who answer all items correctly to the right-most point, and all other individuals to the point in the middle.) Thus p_n, the probability that a perfect solution exists in a random sample of size n, satisfies

$$p_n \geq \text{prob}(n_0 + n_1 > 0) = 1 - (1 - \pi_0 - \pi_1)^n. \tag{4.18}$$

Thus $p_n \to 1$ if $n \to \infty$, which implies that the minimum of σ_J converges almost surely to zero if $n \to \infty$. Because all measurements are discrete in the last analysis, this result applies quite generally. It shows that the continuous ordinal option only works if the sample is small; in large samples it will almost surely find a 'trivial' perfect solution. This is not satisfactory at all, especially because p_n is generally much larger than the lower bound we have derived. Similar objections can be raised against the other continuous options mentioned in De Leeuw and Van Rijckevorsel (1980). The discrete options, which require in addition that tied values remained tied (see formula 4.6), are much more restrictive, and have the additional advantage that the transformation $h \to q$ is really a function in the mathematical sense.

The option of discrete ordinal data makes it possible to use indicator matrices in the obvious way. Then it is more convenient to write σ_J as a function of the category quantifications y_j. Thus formula (4.8) becomes

$$\sigma_J(Y,X,A) = m^{-1} \sum_j \text{SSQ}(G_j y_j - X a_j), \tag{4.19}$$

which must be minimized under the conditions

$$u'G_j y_j = 0, \tag{4.20}$$

$$y_j' D_j y_j = 1, \tag{4.21}$$

$$y_j \in C_j, \tag{4.22}$$

where C_j is now a cone in k_j–dimensional space, usually $k_j \ll n$.

We can now describe the algorithms of MNNFAEX, PRINQUAL, and PRINCIPALS mentioned in Section 4.2 in more detail. They are of the alternating least squares type, with two different steps. Suppose we start an iteration with an estimate of the y_j satisfying the constraints. We then compute $q_j = G_j y_j$ and minimize $\text{SSQ}(Q - XA')$ over X and A by using the Eckart–Young theorem. This is the first step. In the second step we compute new y_j for given X and A. This can be done for each j separately. Define

$$\tilde{y}_j \equiv D_j^{-1} G_j' X a_j. \tag{4.23}$$

This particular choice of the auxiliary category quantifications $\tilde{\mathbf{y}}_j$ is obtained by minimizing (4.19) over $\mathbf{y}_j$ *unrestricted*, for given $\mathbf{X}$ and $\mathbf{A}$. Then the loss for variable j can be partitioned as

$$\mathrm{SSQ}(\mathbf{G}_j\mathbf{y}_j - \mathbf{X}\mathbf{a}_j) = \mathrm{SSQ}(\mathbf{G}_j\tilde{\mathbf{y}}_j - \mathbf{X}\mathbf{a}_j) + (\mathbf{y}_j - \tilde{\mathbf{y}}_j)'\mathbf{D}_j(\mathbf{y}_j - \tilde{\mathbf{y}}_j), \qquad (4.24)$$

and consequently we have to minimize the second term on the right over all $\mathbf{y}_j$ satisfying the constraints. This is a normalized cone projection problem, for which the relevant theory is treated in Appendix C. We then go back to the first step, and so on. In Table 4.3 we illustrate three iterations of the algorithm applied to the example in Section 3.9.

Alternatively, we can also use the Daugavet algorithm mentioned in Section 4.1 to construct a three-step alternating least squares method. The algorithm starts each iteration with an estimate of the $\mathbf{y}_j$ and of $\mathbf{X}$. In the first step we compute $\mathbf{A} = \mathbf{Q}'\mathbf{X}(\mathbf{X}'\mathbf{X})^+$. In the second step we compute a new $\mathbf{X}$, by $\mathbf{X} = \mathbf{Q}\mathbf{A}(\mathbf{A}'\mathbf{A})^+$. The third step is the same as the second step of the previous algorithm, it computes new $\mathbf{y}_j$ by normalized cone regression. We also illustrate three iterations of this algorithm in Table 4.4. In general, of course, this alternative algorithm has less work in each iteration, and will as a consequence often need more iterations. Many other variations are possible, because we can change the order of the three steps, and we can perform a number of iterations of the two Daugavet steps before computing new $\mathbf{y}_j$. We have not experimented seriously with these different versions to find out which is more efficient.

We have already discussed two of the more important choices of C_j earlier in this section. A variable is treated as *single nominal* if $\mathbf{y}_j$ can be anywhere in k_j-space (only the normalization requirements are used). Adjustment for $\mathbf{y}_j$ in the alternating least squares algorithm consists of computing $\tilde{\mathbf{y}}_j$ and normalizing it. A variable is *single numerical* if $\mathbf{y}_j$ is restricted to be a linear function of a given k_j-vector. Thus $\mathbf{y}_j = \alpha_j \mathbf{t}_j + \beta_j$, where $\mathbf{t}_j$ is such that $\mathbf{h}_j = \mathbf{G}_j \mathbf{t}_j$. The normalization requirements then imply that $\mathbf{y}_j$ does not change at all during the iterations; $\mathbf{y}_j$ remains equal to the normalized $\mathbf{t}_j$. We can consequently skip the $\mathbf{y}_j$ adjustment step. The third type of cone is defined by monotonicity requirements (for *single ordinal* variables)

$$y_1 \leq \dots \leq y_{k_j}. \qquad (4.25)$$

The $\mathbf{y}_j$ adjustment steps apply weighted monotone regression to the vector $\tilde{\mathbf{y}}_j$ and normalize the result (see Appendix C).

Table 4.3 Three iterations of the two-step algorithm to minimize σ_J

	Iteration 1			Iteration 2			Iteration 3		
Y	−0.237	0.158	0.558	−0.240	0.178	0.543	−0.243	0.197	0.532
	−0.148	0.346	0.839	−0.146	0.316	0.854	−0.144	0.285	0.868
	−0.483	0.207		−0.483	0.207		−0.483	0.207	
R	1.000			1.000			1.000		
	−0.156	1.000		−0.158	1.000		−0.161	1.000	
	0.491	0.307	1.000	0.498	0.303	1.000	0.504	0.298	1.000

Eigenvalues

	Iteration 1			Iteration 2			Iteration 3		
	0.506	0.379	0.114	0.508	0.379	0.113	0.509	0.380	0.112
A	0.625	−0.498		0.631	−0.486				
	0.249	0.856		0.237	0.863				
	0.739	0.132		0.738	0.139				
X	−0.542	−0.073		−0.543	−0.076				
	0.338	0.245		0.340	0.215				
	0.214	0.865		0.203	0.882				
	−0.542	−0.073		−0.543	−0.076				
	0.215	−0.178		0.231	−0.184				
	0.462	−0.375		0.461	−0.361				
	−0.542	−0.073		−0.543	−0.076				
	−0.032	0.019		−0.034	0.020				
	0.462	−0.375		0.461	−0.361				
	−0.032	0.019		−0.034	0.020				

The situation becomes more complicated if there are missing data. We have already seen in Section 2.4 that there are at least three ways in which we can proceed (compare also Section 3.12). Options *ii* and *iii* ('missing values single category' and 'missing values multiple categories') continue to work with complete indicator matrices G_j, but the cones C_j change, because we do not require monotonicity or linearity for the missing values. For single nominal variables nothing changes; for single ordinal variables monotone regression is simply performed over the nonmissing categories; for the missing categories we copy the corresponding elements of $\tilde{y}_j$; and afterwards we normalize. This is also no real complication. For single numerical variables the situation is slightly more complicated. For the nonmissing part of y_j, denoted by y_j^o where ^{o} stands for observed, we require

$$y_j^o = \alpha_j t_j + \beta_j. \tag{4.26}$$

Table 4.4 Three iterations of the three-step algorithm to minimize σ_J

	Iteration 1			Iteration 2			Iteration 3		
Y	−0.237	0.158	0.558	−0.241	0.179	0.543	−0.243	0.197	0.532
	−0.148	0.346	0.839	−0.146	0.316	0.854	−0.144	0.285	0.868
	−0.483	0.207		−0.483	0.207		−0.483	0.207	
X	−0.540	−0.088		−0.542	−0.085		−0.543	−0.082	
	−0.330	0.256		0.337	0.220		0.342	0.184	
	0.188	0.872		0.189	0.885		0.191	0.899	
	−0.540	−0.088		−0.542	−0.085		−0.543	−0.082	
	0.221	−0.173		0.234	−0.180		0.246	−0.188	
	0.473	−0.360		0.467	−0.354		0.460	−0.347	
	−0.540	−0.088		−0.542	−0.085		−0.543	−0.082	
	−0.032	0.014		−0.034	0.020		−0.035	0.022	
	0.473	−0.360		0.467	−0.354		0.460	−0.347	
	−0.032	0.014		−0.034	0.020		−0.035	0.022	
A	0.639	−0.478		0.639	−0.477		0.638	−0.474	
	0.223	0.865		0.223	0.866		0.223	0.868	
	0.736	0.152		0.736	0.151		0.737	0.149	
σ_J	0.114			0.113			0.112		

We do not restrict the missing part y_j^m, where ^{m} denotes missing. Suppose that t_j (which contains the *a priori* numerical category values and has the number of elements equal to the number of nonmissing categories) is normalized in such a way that

$$\mathbf{u}'\mathbf{D}_j^0 t_j = 0, \tag{4.27}$$

$$t_j'\mathbf{D}_j^0 t_j = 1, \tag{4.28}$$

with, of course, $\mathbf{D}_j^0$ the part of $\mathbf{D}_j$ corresponding with the nonmissing categories. We compute

$$\beta_j = \mathbf{u}'\mathbf{D}_j^0 \tilde{\mathbf{y}}_j / \mathbf{u}'\mathbf{D}_j^0 \mathbf{u}, \tag{4.29}$$

$$\alpha_j = t_j'\mathbf{D}_j^0 \tilde{\mathbf{y}}_j^0, \tag{4.30}$$

$$\bar{\mathbf{y}}_j^0 = \alpha_j t_j + \beta_j, \tag{4.31}$$

$$\bar{\mathbf{y}}_j^m = \tilde{\mathbf{y}}_j^m, \tag{4.32}$$

and we then compute y_j^+ by normalizing $\bar{y}_j$. This adjustment process has the obvious consequence that y_j changes during the iterations and that the principal component problem with missing data is no longer equivalent to a simple singular value decomposition problem.

Option i ('missing values deleted') has even more far-reaching consequences (as already shown in the HOMALS context in Section 3.12). Missing values are not quantified and we cannot interpret our results any more in terms of $R(Q)$ and its eigenvalues. This is a major disadvantage, but as we have already seen option ii often does not make sense from the interpretational point of view, while option iii may lead to the situation that individuals or variables with missing data completely dominate the solution. Thus option i may be the only viable alternative. However, it also leads to very unpleasant computational complications. The loss function changes to

$$\sigma_J(Y,X,A) = m^{-1} \sum_j (G_j y_j - X a_j)' M_j (G_j y_j - X a_j), \tag{4.33}$$

where M_j is a binary diagonal matrix, indicating which observations are missing (if i is missing for j, the diagonal element i of M_j is zero otherwise it is one). Because G_j is now an incomplete indicator matrix we have the convenient relationship $M_j G_j = G_j$ we also have $G_j u = M_j u$. The y_j adjustment step in the alternating least squares algorithm is now simpler than in options ii and iii, because the vectors y_j are shorter. It is also true in the single numerical case that y_j does not change during the iterations. However, unfortunately, adjusting for X and A for fixed y_j is not a singular value problem any more. We can generalize the Daugavet three-step algorithm, but it becomes more complicated and computationally much more expensive. Using $q_j = G_j y_j$ we find that the optimal A for fixed X and Y must be computed row-wise. Row j, written as a column a_j, is

$$a_j = (X'M_j X)^+ X' q_j. \tag{4.34}$$

Thus it will take approximately m times as long to update A, compared to $A = QX(X'X)^+$. Updating X must also be done row-wise, and takes approximately n times as long. Define N_i as a binary diagonal matrix of order m (if i is missing for j, then diagonal element j of N_i is zero; otherwise it is one). Then row i of X, written as a column x_i, becomes

$$x_i = (A'N_i A)^+ A' q_i. \tag{4.35}$$

The programs MNNFAEX, PRINQUAL, and PRINCIPALS handle missing data by using option iii. Since they rely on explicit computation of the singular value decomposition, they cannot incorporate option i.

4.4 THEORY OF MEET LOSS

Principal components analysis based on σ_J has some inconvenient features. In the first place multiple quantifications must be computed *successively*, while in homogeneity analysis we can compute them *simultaneously*. Secondly, missing data option i gives us computational troubles. We now present an alternative approach to components analysis, which does not have these disadvantages. Recall the homogeneity loss function

$$\sigma_M(X,Y) \equiv m^{-1} \sum_j SSQ(X-G_jY_j), \qquad (4.36)$$

with normalization $AVE(x_s) = 0$ for $s = 1, ..., p$ and $X'X = I$. We use the notation σ_M (meet loss) here because $\sigma_M(X,Y) = 0$ implies that the meet rank of the G_j is at least p. As explained in Chapter 3 the alternating least squares algorithm for minimizing σ_M is

$$Y_j \leftarrow D_j^{-1}G_j'X, \qquad (4.37)$$

$$Z \leftarrow \sum_j G_jY_j, \qquad (4.38)$$

$$X \leftarrow GRAM(Z), \qquad (4.39)$$

and it computes the first p dimensions of a homogeneity analysis simultaneously. We use the abbreviation GRAM for a Gram–Schmidt orthogonalization (cf. Section 3.4.4 and Appendix A). These three steps are identical with the HOMALS algorithm (without missing data).

As we have seen in the previous section it is characteristic for ordinary components analysis to have single rather than multiple quantifications. The step from multiple to single is achieved by imposing *rank one restrictions* These are written as

$$Y_j = y_ja_j' \qquad (4.40)$$

with the additional requirements

$$u'D_jy_j = 0, \qquad (4.41)$$

$$y_j'D_jy_j = 1, \qquad (4.42)$$

$$y_j \in C_j. \qquad (4.43)$$

Some simple computation, using the normalization conditions, gives

$$\sigma_M(X,Y,A) = \sigma_J(Y,X,A) + (p - 1),\tag{4.44}$$

where for all variables single meet loss is written as

$$\sigma_M(X,Y,A) = m^{-1} \sum_j \text{SSQ}(X - G_j y_j a_j').\tag{4.45}$$

This relationship between meet loss for single variables and join loss holds because we can write

$$\sigma_M(X,Y,A) = p + m^{-1} \sum_j a_j' a_j - 2m^{-1} \sum_j a_j' X' G_j y_j,\tag{4.46}$$

where we have used the conditions $X'X = I$, from which it follows that $\text{tr}(X'X) = p$ and $y_j' D_j y_j = 1$. Similarly we obtain for join loss

$$\sigma_J(Y,X,A) = 1 + m^{-1} \sum_j a_j' a_j - 2m^{-1} \sum_j a_j' X' G_j y_j,\tag{4.47}$$

from which (4.44) follows easily. (The fact that we use Y_j for the multiple category quantifications and y_j for the single quantifications may be confusing. On the other hand, a variable is either single or multiple, not both, and has either a Y_j or a y_j.) It follows from equation (4.44) that minimizing $\sigma_M(X,Y,A)$ and $\sigma_J(Y,X,A)$ are equivalent problems, with the same solutions.

The first advantage of using σ_M is that we can impose the conditions $Y_j = y_j a_j'$ for some variables and not for others. If we impose them for *all* variables, then we minimize σ_J, and we are doing components analysis, as before. If we do not impose them at all, then we are doing homogeneity analysis as in Chapter 3. It may be interesting to mix the two options and to give some variables single quantifications and others multiple quantifications. Especially for nominal variables single quantification often is not very natural, because it seems to suggest that the categories can be ordered on a single scale. Consequently we might compute single quantifications for ordinal and numerical variables, and multiple quantifications for nominal variables.

Another advantage becomes clear if we analyse the implementation of option *i* for missing data. Now

$$\sigma_M(X,Y) = m^{-1} \sum_j \text{tr}(X - G_j Y_j)' M_j (X - G_j Y_j),\tag{4.48}$$

with normalization

$$u' M_* X = 0,\tag{4.49}$$

$$X' M_* X = mI,\tag{4.50}$$

where

$$\mathbf{M}_* = \sum_j \mathbf{M}_j. \tag{4.51}$$

The simple relation between σ_M and σ_J (equation 4.44) is no longer true with this treatment of missing data. The alternating least squares algorithm for minimizing the generalized σ_M (equation 4.48) is quite simple, however. We start with an $\mathbf{X}$ satisfying the constraints. Define

$$\tilde{\mathbf{Y}}_j \equiv \mathbf{D}_j^{-1}\mathbf{G}_j'\mathbf{X}. \tag{4.52}$$

As we know these are the unrestricted conditional minimizers of σ_M of the category quantifications of variable j. The relevant part of σ_M can now be partitioned as

$$\mathrm{tr}(\mathbf{X} - \mathbf{G}_j\mathbf{Y}_j)'\mathbf{M}_j(\mathbf{X} - \mathbf{G}_j\mathbf{Y}_j)$$

$$= \mathrm{tr}(\mathbf{X} - \mathbf{G}_j\tilde{\mathbf{Y}}_j)'\mathbf{M}_j(\mathbf{X} - \mathbf{G}_j\tilde{\mathbf{Y}}_j) + \mathrm{tr}(\mathbf{Y}_j - \tilde{\mathbf{Y}}_j)'\mathbf{D}_j(\mathbf{Y}_j - \tilde{\mathbf{Y}}_j). \tag{4.53}$$

Thus minimizing over $\mathbf{Y}_j$ can be done by minimizing the second term on the right. If the variable is *multiple nominal* we simply set $\mathbf{Y}_j = \tilde{\mathbf{Y}}_j$, and we are home.

We can also introduce at this point a new type of variable, which fits naturally into the general framework. For a multiple ordinal variable we require that all columns of $\mathbf{Y}_j$ are either increasing or decreasing. As explained by Guttman (1959), we cannot require them all to be increasing. Computing $\mathbf{Y}_j$ amounts to solving two monotone regression problems for each column, the increasing and the decreasing one, and to keep the best one. *Multiple numerical* variables require that all columns are linear functions of a given vector. It is easy to see, however, that this amounts to requiring that $\mathbf{Y}_j = \mathbf{y}_j\mathbf{a}_j'$, with $\mathbf{y}_j$ known, and that consequently multiple numerical is identical with single numerical. When a variable is single, the second part of equation (4.53) becomes

$$\mathrm{tr}(\mathbf{y}_j\mathbf{a}_j' - \tilde{\mathbf{Y}}_j)'\mathbf{D}_j(\mathbf{y}_j\mathbf{a}_j' - \tilde{\mathbf{Y}}_j), \tag{4.54}$$

which must be minimized over $\mathbf{y}_j \in C_j$ and over $\mathbf{a}_j$. For *single nominal* variables $\mathbf{y}_j$ is unrestricted, and we can minimize over $\mathbf{y}_j$ and $\mathbf{a}_j$ simultaneously by computing the dominant singular vectors of $\mathbf{D}_j^{1/2}\tilde{\mathbf{Y}}_j$. In general, however, we prefer alternating least squares *inner iterations* to solve for $\mathbf{y}_j \in C_j$, given $\mathbf{a}_j$, and for $\mathbf{a}_j$, given $\mathbf{y}_j$. Solving for $\mathbf{a}_j$, given $\mathbf{y}_j$, can be done by defining:

$$\tilde{\mathbf{a}}_j = \tilde{\mathbf{Y}}_j'\mathbf{D}_j\mathbf{y}_j / \mathbf{y}_j'\mathbf{D}_j\mathbf{y}_j, \tag{4.55}$$

which gives the decomposition of formula (4.54) into

$$\text{tr}(y_j\tilde{\mathbf{a}}_j' - \tilde{\mathbf{Y}}_j)'\mathbf{D}_j(y_j\tilde{\mathbf{a}}_j' - \tilde{\mathbf{Y}}_j) + y_j'\mathbf{D}_j y_j \cdot \text{SSQ}(\mathbf{a}_j - \tilde{\mathbf{a}}_j) \tag{4.56}$$

Because $\mathbf{a}_j$ is unrestricted in general, this means that we can simply set $\mathbf{a}_j = \tilde{\mathbf{a}}_j$. It is not difficult to generalize this to restricted $\mathbf{a}_j$ (which occurs in various forms of factor analysis). If we require $\mathbf{a}_j \in \Gamma_j$, for some convex set Γ_j, then this substep of the inner iterations means that we must project $\tilde{\mathbf{a}}_j$ on Γ_j.

Solving for $\mathbf{y}_j \in C_j$ for fixed $\mathbf{a}_j$ (the other half of one inner iteration) is done through using the auxiliary vector

$$\tilde{\mathbf{y}}_j = \tilde{\mathbf{Y}}_j \mathbf{a}_j / \mathbf{a}_j'\mathbf{a}_j, \tag{4.57}$$

i.e. the conditionally optimal, single category quantifications without further restrictions, and uses the alternative partitioning of the sum of squares (formula 4.54) given by

$$\text{tr}(\tilde{\mathbf{y}}_j\mathbf{a}_j' - \tilde{\mathbf{Y}}_j)'\mathbf{D}_j(\tilde{\mathbf{y}}_j\mathbf{a}_j' - \tilde{\mathbf{Y}}_j) + \mathbf{a}_j'\mathbf{a}_j \cdot (\mathbf{y}_j - \tilde{\mathbf{y}}_j)'\mathbf{D}_j\,(\mathbf{y}_j - \tilde{\mathbf{y}}_j). \tag{4.58}$$

Thus finding $\mathbf{y}_j$ means projecting the vector $\tilde{\mathbf{y}}_j$ on C_j, which is a monotone regression problem in the ordinal case, a linear regression problem in the numerical case, and a question of simply setting $\mathbf{y}_j = \tilde{\mathbf{y}}_j$ in the nominal case.

Several comments are in order here. In the first place we can choose how many inner two-step alternating least squares iterations we want to make, before we go on to compute a new $\mathbf{X}$. In the second place we do *not* use or assume here that $\mathbf{y}_j'\mathbf{D}_j\mathbf{y}_j = 1$ or that $\mathbf{u}'\mathbf{D}_j\mathbf{y}_j = 0$. In minimizing σ_J it was not necessary to normalize $\mathbf{X}$ and $\mathbf{A}$, because $\mathbf{Y}$ was normalized. In minimizing σ_M it is not necessary to normalize $\mathbf{Y}$ and $\mathbf{A}$, because $\mathbf{X}$ is normalized. In the third place the situation without missing data is a special case in which $\mathbf{M}_j = \mathbf{I}$ for all j. From the mean preserving property of the regression algorithms we find that after regression

$$\mathbf{u}'\mathbf{D}_j\mathbf{y}_j = \mathbf{u}'\mathbf{D}_j\tilde{\mathbf{y}}_j = \mathbf{u}'\mathbf{D}_j\tilde{\mathbf{Y}}_j\mathbf{a}_j / \mathbf{a}_j'\mathbf{a}_j$$

$$= \mathbf{u}'\mathbf{D}_j\mathbf{D}_j^{-1}\mathbf{G}_j'\mathbf{X}\mathbf{a}_j / \mathbf{a}_j'\mathbf{a}_j$$

$$= \mathbf{u}'\mathbf{G}_j'\mathbf{X}\mathbf{a}_j / \mathbf{a}_j'\mathbf{a}_j = \mathbf{u}'\mathbf{M}_j\mathbf{X}\mathbf{a}_j / \mathbf{a}_j'\mathbf{a}_j. \tag{4.59}$$

Thus if $\mathbf{M}_j = \mathbf{I}$ for all j (no missing data, or option *ii* or *iii*), then $\mathbf{u}'\mathbf{D}_j\mathbf{y}_j = 0$ follows from $\mathbf{u}'\mathbf{X} = 0$, and our method can be interpreted in terms of $R(Q)$.

The second step in the outer iteration is to compute a new $\mathbf{X}$ for given $\mathbf{Y}_j$, on the conditions that $\mathbf{u}'\mathbf{M}_*\mathbf{X} = 0$ and $\mathbf{X}'\mathbf{M}_*\mathbf{X} = m\mathbf{I}$. This amounts to the following three steps. We first compute

$$\mathbf{Z} = \Sigma_j \, \mathbf{M}_j\mathbf{G}_j\mathbf{Y}_j, \tag{4.60}$$

then

$$\tilde{\mathbf{Z}} = \{\mathbf{M}_* - (\mathbf{M}_*\mathbf{u}\mathbf{u}'\mathbf{M}_* / \mathbf{u}'\mathbf{M}_*\mathbf{u})\}\mathbf{Z}, \tag{4.61}$$

and

$$\mathbf{X} = m^{1/2}\mathbf{M}_*^{-1/2} \, \text{GRAM}(\mathbf{M}_*^{-1/2}\tilde{\mathbf{Z}}). \tag{4.62}$$

Upon checking these computations, it will be found out that this is *not* the least squares solution for $\mathbf{X}$ given the $\mathbf{Y}_j$. We find the least squares solution by solving a Procrustes problem (Cliff, 1966); our Gram–Schmidt solution is a *rotation* of the optimal solution. It is easy to see, however, that this implies that the $\mathbf{Y}_j$ in the next iteration will be rotated in the same way and that consequently σ_M will decrease as much in a major iteration if we use Gram–Schmidt. Using Gram–Schmidt gives a smaller decrease in the step that updates $\mathbf{X}$, but a larger decrease in the next step that updates the $\mathbf{Y}_j$. The total decrease is the same, and because Gram–Schmidt is even cheaper than Procrustes we prefer it (this is the same reasoning as the one used in Section 3.4.4).

We remark finally that there is another way in which homogeneity analysis can be fitted into the framework of component analysis. Suppose we consider each of the Σk_j categories in a complete homogeneity analysis as a new variable, indexed $l = 1,...,L$. For each l we define $\mathbf{M}_l$ as the diagonal matrix with the corresponding column $\mathbf{g}_l$ of $\mathbf{G}$ on the diagonal. Clearly $\mathbf{M}_* = m\mathbf{I}$. We also define the cone C_l in n-space as the set of those vectors $\mathbf{q}$ with the property that if objects i and k are in the category corresponding with l, then $q_i = q_k$. Alternatively, if $\mathbf{q} \in C_l$ then $\mathbf{M}_l\mathbf{q}$ is proportional to $\mathbf{g}_l$; for normalization purposes we can simply require $\mathbf{M}_l\mathbf{q} = \mathbf{g}_l$. However, under these conditions some computation gives

$$\Sigma_l \, \text{tr}(\mathbf{X} - \mathbf{q}_l\mathbf{a}_l')'\mathbf{M}_l(\mathbf{X} - \mathbf{q}_l\mathbf{a}_l') = \Sigma_j \, \text{SSQ}(\mathbf{X} - \mathbf{G}_j\mathbf{A}_j), \tag{4.63}$$

which transfers the meet loss function with rank one restrictions in L variables to the usual loss function for homogeneity analysis.

4.5 GEOMETRY OF MEET LOSS

The conditions for minimum meet loss for each variable have interesting geometrical interpretations. As in homogeneity analysis we represent object scores as points in p-dimensional space. If a variable is multiple nominal, its loss contribution is

$$\mathrm{tr}(\mathbf{X} - \mathbf{G}_j\mathbf{Y}_j)'\mathbf{M}_j(\mathbf{X} - \mathbf{G}_j\mathbf{Y}_j), \tag{4.64}$$

which vanishes if and only if $\mathbf{M}_j\mathbf{X}_j = \mathbf{G}_j\mathbf{Y}_j$, i.e. if and only if all objects in the same nonmissing category have the same object score, which is then of

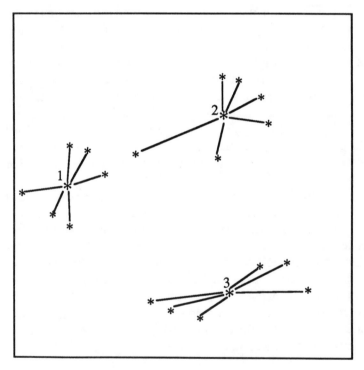

Figure 4.1 Meet loss for a multiple nominal variable.

course equal to the corresponding category quantification. As in homogeneity analysis we expect objects in a category to be close together. This is shown in

Figure 4.1. If a variable is multiple ordinal we partition the loss contribution as (cf. equation 4.53)

$$\text{tr}(X - G_j\tilde{Y}_j)'M_j(X - G_j\tilde{Y}_j) + \text{tr}(Y_j - \tilde{Y}_j)'D_j(Y_j - \tilde{Y}_j), \tag{4.65}$$

where (cf. equation 4.52)

$$\tilde{Y}_j = D_j^{-1} G_j'X. \tag{4.66}$$

Here the loss component vanishes if and only if all object scores corresponding with a nonmissing category are the same *and* all category quantifications project on the dimensions in an increasing or a decreasing order.

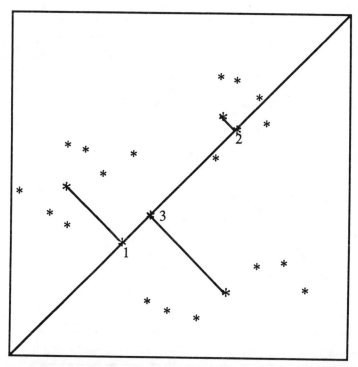

Figure 4.2 Meet loss for a single nominal variable: additional loss because the quantifications have to be on a straight line through the origin.

Observe that for multiple ordinal data there generally is no complete freedom of rotation of the axes. For single variables there are at least two loss components for each variable. A single additive partitioning of the loss is not possible (for fixed $\mathbf{a}_j$ we can partition with respect to j (cf. equation 4.58), and vice versa (cf. equation 4.56), but not both at the same time). Geometrically, however, we want all object scores corresponding with a nonmissing category to be the same and we want category quantifications (centroids of object scores) to be on a line through the origin. For single nominal variables these requirements give two loss components; this situation is pictured in Figure 4.2. For single ordinal and numerical variables we want the category quantifications on a line through the origin *and* we want them to be on this line in the correct way (defined by some cone C_j). Meet loss for a single ordinal

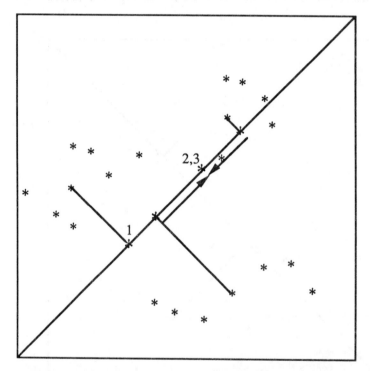

Figure 4.3a Join loss for a single ordinal variable: additional loss because the quantifications have to be in the right order on the line.

variable is depicted in Figure 4.3a. Here the third loss component arises by moving categories 2 and 3 so that the order of the quantifications is monotonically increasing with the category number. Finally, in Figure 4.3b we see meet loss for a single numerical variable: the quantifications should be in the right order *and* equally spaced.

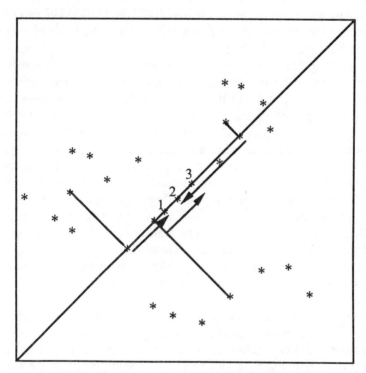

Figure 4.3b Meet loss for a single numerical variable: additional loss because the quantifications have to be equally spaced on the line.

Observe that for single variables the sum of all three loss components vanishes only if all objects are on the line through the origin. However, this can never occur because of the normalization of $\mathbf{X}$, which forces it to be of full rank. This explains the relation between meet loss and join loss, which is $\sigma_M = \sigma_J + (p - 1)$; consequently $p > 1$ implies $\sigma_M > 0$. Then the contribution to meet loss of variable j is minimized (but generally not zero) if the corresponding component of join loss is zero, which means that $\mathbf{G}_j \mathbf{y}_j = \mathbf{X}\mathbf{a}_j$. Geometrically this condition means that the object points corresponding with a category must be *on* parallel hyperplanes perpendicular to the direction defined by $\mathbf{a}_j$.

It is not very natural, in many situations, that the objects should be *on* parallel hyperplanes. It seems more satisfactory, from a geometrical point of view, to require that there exist $k_j - 1$ of these hyperplanes which *separate* the categories of a variable, in the sense that all objects in a category should be between two of these parallel hyperplanes. This particular way of representing categorical data was investigated by Lingoes (1968) and by De Leeuw (1969). It corresponds to minimizing the join loss with all variables being continuous ordinal, and we have seen in Section 4.3 that such a technique will not work in many interesting situations. The other techniques discussed by Lingoes and De Leeuw, all based on the idea of separation, basically have the same problems. Guttman's MSA-I technique, for example, can be interpreted as an attempt to find representations similar to HOMALS, but with a far more realistic way of measuring loss. Unfortunately loss is defined in such a way that it is too easy to find a solution with perfect fit. MSA-I, in fact, uses a technique like HOMALS as an initial configuration, and experience seems to indicate that the sophisticated iterative procedure either can not or does not need to get away from its starting point. The argument is the same as in Section 8.2, in which we compare HOMALS with nonmetric unfolding. By making the requirements on the representation unrealistically strong we prevent degeneracy and instability. This is also the reason, presumably, why sometimes nonmetric methods do not work and metric methods do work, even when the data are clearly ordinal (Woodward and Overall, 1976; Bentler and Weeks, 1978).

There is one special situation in which the distinction between discrete and continuous becomes irrelevant which is when $k_j = n$ for all j. This situation arises, for instance, in the analysis of preference rank orders (mentioned at the end of Section 4.2; an example will be given in Section 4.8). If we analyse preference rank orders, then each individual defines a variable, and the objects that are compared with respect to preference are also the objects in the components analysis. If the rank orders are complete, then each variable has n categories when n objects are ranked, and thus each G_j is a permutation matrix, $D_j = I$, and $G_j \tilde{Y}_j = G_j D_j^{-1} G_j' X = X$ for all $j = 1,...,m$. Thus multiple loss is always equal to zero, and homogeneity analysis would give a completely arbitrary result. In fact HOMALS stops after one iteration, with X^* equal to the random initial configuration, and with $Y_j^* = G_j' X^*$. If the variable is single nominal, then we can minimize $SSQ(X - G_j y_j a_j')$ by setting $y_j^+ = G_j' X b_j$ and $a_j^+ = b_j / b_j' b_j$, with b_j completely arbitrary, and X arbitrary except for the usual normalization condition $X'X = nI$. The minimum of σ_M is $(p - 1)$ if all variables are single nominal, which is another way of saying that σ_J is trivially equal to zero.

It follows that we can exclude all variables that are either multiple or single nominal in this case, and analyse only those variables that are single numerical or ordinal. In the numerical case the optimal X^* and A^* can be computed, a usual, from the matrix with columns $G_j y_j$. ➤ If the y_j are all equal to the centered rank numbers, for example, then the technique becomes identical to the one proposed by Guttman (1946), Slater (1960), Carroll and Chang (1964), and Benzécri (1965). Compare also Bechtel, Tucker, and Chang (1971), De Leeuw (1968, 1973, ch. 4), and Nishisato (1978). If all variables are ordinal, then we require that the projections of the objects on the direction corresponding with individual j (remember that the usual variables are here individuals) must be in the appropriate order. There is no multiple 'within-category' loss; all the loss is single. In fact even the rank one restrictions do not lead to nontrivial loss; only the ordinal or numerical restrictions give a contribution.

4.6 THE PRINCALS PROGRAM

The PRINCALS program minimizes σ_M and incorporates option i ('missing data passive') for missing data. This means that the loss function contains the matrices M_j and that the interpretation in terms of $R(Q)$ is not possible if there are missing data. Of course it is always possible to use options ii ('missing data single category') and iii ('missing data multiple categories') in PRINCALS as well, but in that case the user must recode the data in the appropriate way. Because PRINCALS without missing data (or with missing data under options ii and iii) can only perform linear regression or monotone regression over *all* categories, this recoding will only work properly if the recoded data are treated as single nominal or multiple nominal. Otherwise we have to impose order restrictions or linearity on missing data, which is clearly undesirable.

PRINCALS accepts multiple nominal, single nominal, ordinal, and numerical variables. Multiple ordinal is not (yet) implemented, there are no continuous options. We can compare these possibilities with some of the other programs we mentioned. MNNFAEX of Roskam and Lingoes and NMFA of Kruskal and Shepard accept single discrete ordinal and single continuous ordinal. PRINQUAL of Tenenhaus accepts single discrete ordinal and single discrete numerical. ➤ PRINCIPALS of Young, Takane, and De Leeuw accepts single discrete and continuous, nominal, ordinal, and numerical, in any combination, but not multiple nominal. The reason that none of the other programs accepts multiple nominal is that they are based on join loss whereas PRINCALS is based on meet loss.

The normalization of X is the same as in the HOMALS program. Thus we normalize X by $X'M_*X = mnI$ and $u'M_*X = 0$, which implies that X is in standard scores if $M_j = I$ for all j. If the variables are single, we require for purposes of identification $y_j'D_jy_j = n$. If there are no missing data (or with options ii and iii) this implies that the elements of a_j are correlations and can be interpreted as ordinary component loadings. For multiple nominal variable the category quantifications and the discrimination measures are comparable to those in HOMALS. PRINCALS also prints multiple quantification $\tilde{Y}_j = D_j^{-1}G_j'X$ for single variables, and for comparison purposes in addition y_ja_j'. In some of the programs we mentioned above other normalizations are in use. Sometimes the a_j are interpreted as direction cosines, with $a_j'a_j = 1$ which means that if we represent them as points in the joint plot they are all on a circle. This can be useful in the context of preference rank orders, but we prefer not to use this normalization in general.

PRINCALS uses the partitioning of the loss for variable j given by

$$\text{tr}(X'M_jX - \tilde{Y}_j'D_j\tilde{Y}_j) + \text{tr}(\tilde{Y}_j'D_j\tilde{Y}_j - a_ja_j') . \tag{4.67}$$

The first component is called the *multiple loss*, the second component the *single loss*. For multiple variables there is no single loss. The quantities called *multiple fit* are the diagonal elements of $\tilde{Y}_j'D_j\tilde{Y}_j/n$, which are the discrimination measures of HOMALS; the quantities called *single fit* are the a_{js}^2, the square component loadings. PRINCALS prints two $m \times p$ tables, one for multiple and one for single fit.

If there are no missing data (or with option ii or iii) the eigenvalues printed by PRINCALS are those of $m^{-1}R(Q)$. When it is also true that all variables are single, or that $p = 1$, in which case single is the same as multiple, $R(Q)$ itself is also printed. If there are missing data, and we use option i, then the eigenvalues are those of the matrix with elements $y_j'G_j'M_*^{-1}G_ly_l$, which is not necessarily a correlation matrix, although it is of course always positive semidefinite.

PRINCALS computes its solution in two phases. In the first phase all multiple (nominal) variables remain as they are and single variables are treated as numerical. After the iterations in this phase have converged, we use the result as the starting configuration for the second phase, where the appropriate measurement levels are taken into account. The program can print both solutions. The idea behind the two phases is that the first phase will give us a good start, because PRINCALS with only multiple nominal and single numerical variables amounts to a singular value problem (when there are no missing data), and consequently in the first phase the algorithm always converges to a global minimum. On the one hand, this *may* avoid local minima in the second phase (at least some local minima). It is easy to assess from the two-phase results how far the nonlinear solution deviates from and improves upon the linear one. However, we have not investigated systematically how frequently local minima occur in PRINCALS with single ordinal and single nominal variables. ➳ On the other hand, if the original linear coding of the categories used in the first phase is quite different from the optimal coding found in the second phase, then the two-phase procedure could very well tend to *introduce* local minima. Because the number of iterations and the precision in the two phases can be chosen independently, this can be avoided to some extent. If we only want local improvements on the linear solution we find precise solutions in both phases; if we expect dramatic changes, or if we do not have the faintest idea about an optimal ordering, then we only perform a single iteration in the first phase. Recent numerical experience suggests that in cases with a poor fit single ordinal is more robust with respect to local minima than single nominal.

4.7   AN EXAMPLE COMPARING MULTIPLE AND SINGLE NOMINAL TREATMENT

Table 4.5 Guttman–Bell data

Description of the variables		
Variable 1 = Intensity of interaction:	1 = slight, 2 = low, 3 = moderate, 4 = high
Variable 2 = Frequency of interaction:	1 = slight, 2 = nonrecurring, 3 = infrequent, 4 = frequent
Variable 3 = Feeling of belonging:	1 = none, 2 = slight, 3 = variable, 4 = high
Variable 4 = Physical proximity:	1 = distant, 2 = close
Variable 5 = Formality of relationship:	1 = no relationship, 2 = formal, 3 = informal

Classification of the objects					
Crowd	1	1	1	2	2
Audience	2	2	2	2	2
Public	1	1	2	1	1
Mob	4	2	4	2	3
Primary group	4	4	4	2	3
Secondary group	3	3	3	1	2
Modern community	2	3	3	2	2

The present example was discussed earlier in Guttman (1968) and Lingoes (1968), mainly to illustrate the Guttman–Lingoes MSA programs. The data were adapted by Guttman from a sociology text (Bell and Sirjamaki, 1967); they characterize seven different types of social groups in terms of five variables. These classifications can be found in Table 4.5.

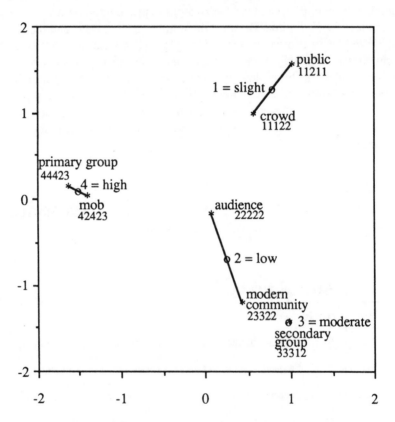

Figure 4.4 HOMALS solution of the Guttman–Bell data. Length of the lines indicate loss for variable 1; * = object, o = category.

Guttman and Lingoes apply their four MSA programs to this small matrix, which is basically not such a good idea because all MSA programs impose only a small number of constraints, and can fit small examples quite perfectly. In particular the MSA-I solution is of interest to us, because it uses Lingoes' MAC as an initial estimate, and Lingoes' MAC implements Guttman (1941) and consequently yields the same solution as HOMALS. The HOMALS solution for the object scores is virtually the same as the MSA-I solution given by Lingoes, which is not surprising because it satisfies the contiguity requirements of MSA-I perfectly. As a matter of fact, it also satisfies the requirements of MSA-III perfectly, as well as those of MSA-IV. These last two methods are the same as nonmetric principal components analysis with continuous single nominal and continuous single ordinal variables. We can compare the HOMALS solution plotted in Figure 4.4 with the PRINCALS solution, plotted in Figure 4.5, with all variables single nominal. In both figures the loss for variable 1 is indicated. The numerical results for the HOMALS and the PRINCALS analysis for this example are given in Table 4.6. We do not think this example shows us anything deep or unexpected about the structure of social groups, the data matrix is so small that

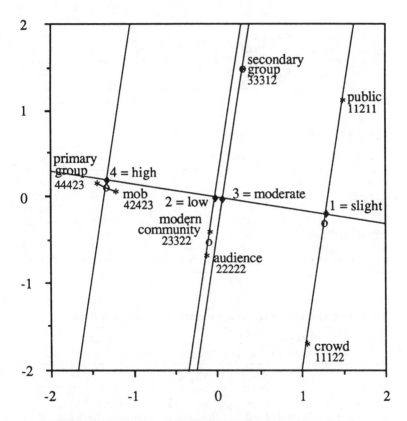

Figure 4.5 Single nominal PRINCALS solution of the Guttman–Bell data with the loss for variable 1; * = object, o = multiple category point, ♦ = single category point.

data reduction is quite unnecessary. We merely use the example to show some of the plots that can be made and some of the statistics that can be computed.

Table 4.6a Guttman–Bell data: category quantifications/coordinates

HOMALS, $p = 2$, multiple nominal

1		2		3		4		5	
0.80	1.28	0.80	1.28	0.56	0.98	0.99	0.09	1.03	1.59
0.24	−0.69	−0.67	−0.04	0.54	0.71	−0.39	−0.04	0.49	−0.45
0.94	−1.41	0.68	−1.31	0.68	−1.31			−1.50	0.11
−1.50	0.11	−1.62	0.14	−1.50	0.11				

PRINCALS, $p = 2$, single nominal

1		2		3		4		5	
1.29	−0.19	1.29	−0.12	1.54	−0.87	0.88	1.29	1.65	0.34
0.01	−0.00	−0.62	0.06	0.44	−0.25	−0.35	−0.52	0.22	0.05
0.05	−0.01	0.06	−0.01	−0.15	0.08			−1.26	−0.26
−1.32	0.20	−1.45	0.13	−1.06	0.60				

Table 4.6b Guttman–Bell data: object scores

	HOMALS, $p = 2$		PRINCALS, $p = 2$	
Crowd	0.56	0.98	1.06	−1.72
Audience	0.05	−0.16	0.08	−0.69
Public	1.03	1.59	1.48	1.11
Mob	−1.39	0.08	−1.22	0.08
Primary group	−1.62	0.14	−1.45	0.15
Secondary group	0.94	−1.41	0.27	1.47
Modern community	0.42	−1.22	−0.07	−0.40

Table 4.6c Guttman–Bell data: discrimination measures and fit

	HOMALS, $p = 2$ Discrimination measures		Multiple fit		PRINCALS, $p = 2$ Component loadings		Single fit	
	0.97	0.89	0.98	0.42	−0.99	0.15	0.98	0.02
	0.82	0.97	0.89	0.14	−0.94	−0.09	0.89	0.01
	0.91	0.78	0.81	0.52	−0.85	0.48	0.72	0.23
	0.39	0.00	0.31	0.67	−0.56	−0.82	0.31	0.67
	0.94	0.48	0.87	0.25	−0.93	−0.19	0.87	0.04
Eigenvalues:	0.81	0.63			0.75	0.19		

4.8 AN EXAMPLE WITH PREFERENCE RANK ORDERS

Table 4.7 gives preference rank orders of 39 psychologists for ten psychological journals (source: Roskam, 1968, p. 152). By convention a low element in the table indicates a high preference for the journal. The ten journals are:

Table 4.7 Roskam's journal preference rank orders

	JEXP	JAPP	JPSP	MVBR	JCLP	JEDP	PMET	HURE	BULL	HUDE	
1.	7	4	1	8	10	9	5	2	3	6	(S)
2.	7	6	2	9	3	8	10	1	4	5	(S)
3.	10	5	1	7	4	6	8	2	3	9	(S)
4.	6	5	3	7	4	8	9	2	1	10	(S)
5.	6	3	5	10	4	2	9	7	8	1	(D)
6.	8	7	4	9	2	5	10	6	3	1	(D)
7.	5	9	4	8	6	2	10	7	3	1	(D)
8.	6	7	4	9	5	3	10	8	2	1	(D)
9.	2	3	6	4	5	8	9	7	10	1	(D)
10.	5	8	2	9	1	7	10	6	4	3	(D)
11.	7	2	6	10	5	1	9	8	4	3	(D)
12.	8	7	2	9	1	6	10	5	3	4	(C)
13.	10	7	1	9	4	6	8	2	3	5	(C)
14.	5	2	3	4	1	8	7	9	6	10	(C)
15.	6	5	2	7	1	10	9	8	4	3	(C)
16.	4	7	5	2	8	9	1	6	3	10	(M)
17.	4	7	5	3	9	8	1	6	2	10	(M)
18.	5	4	7	3	9	8	1	10	2	6	(M)
19.	1	5	6	7	10	9	3	8	2	4	(E)
20.	1	5	8	7	9	3	6	10	2	4	(E)
21.	3	7	6	2	8	4	5	9	1	10	(E)
22.	1	3	8	6	6	7	4	10	2	5	(E)
23.	1	4	6	5	9	10	2	8	3	7	(E)
24.	1	7	5	4	10	9	3	8	2	6	(E)
25.	1	8	6	5	9	4	3	10	2	7	(E)
26.	1	2	5	6	10	4	7	9	3	8	(E)
27.	1	5	6	4	8	7	2	9	3	10	(E)
28.	4	6	5	1	7	10	3	8	2	9	(S)
29.	8	7	1	2	9	10	6	3	4	5	(R)
30.	7	4	1	2	9	10	8	6	3	5	(R)
31.	9	8	2	7	1	4	10	5	6	3	(R)
32.	7	1	5	8	2	6	3	9	4	10	(I)
33.	2	3	7	8	10	9	1	6	4	5	(I)
34.	10	4	2	9	3	5	6	8	1	7	(I)
35.	3	2	10	6	8	4	7	9	1	5	(I)
36.	6	1	3	9	4	7	10	2	5	8	(I)
37.	2	1	6	4	10	9	5	7	3	8	(I)
38.	2	3	6	5	7	8	4	9	1	10	(P)
39.	2	6	7	3	10	8	4	9	1	5	(P)

1. *JEXP:*   *Journal of Experimental Psychology*
2. *JAPP:*   *Journal of Applied Psychology*
3. *JPSP:*   *Journal of Personality and Social Psychology*
4. *MVBR:*   *Multivariate Behavioral Research*
5. *JCLP:*   *Journal of Consulting Psychology*
6. *JEDP:*   *Journal of Educational Psychology*
7. *PMET:*   *Psychometrika*
8. *HURE:*   *Human Relations*
9. *BULL:*   *Psychological Bulletin*
10. *HUDE:*   *Human Development*

Two PRINCALS analyses were performed, both in two dimensions: the first with all variables single numerical and the second with all variables single

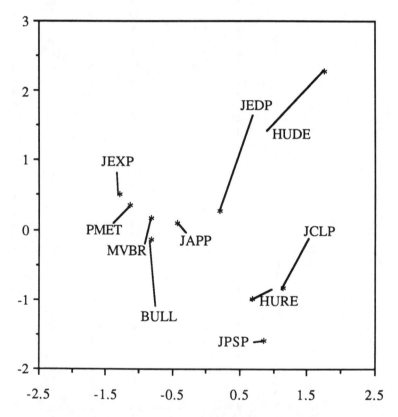

Figure 4.6 Object scores for the Roskam preference data in the single numerical solution (labelled) connected with the corresponding points in the single ordinal solution.

ordinal. The solution with all variables numerical has loss 1.43, which implies that the fit is 0.57 and the two eigenvalues are 0.41 and 0.16, respectively. The solution with all variables ordinal has loss 1.18, the fit is 0.82, and the eigenvalues are 0.48 and 0.34. Figure 4.6 gives the object points for both solutions and the labelled points indicate the linear solution, the points at the

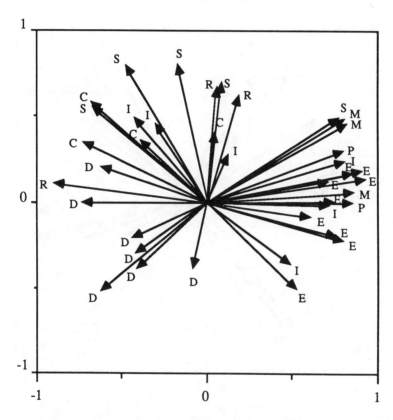

Figure 4.7 Loadings for the 39 psychologists in the single numerical solution, labelled according to their specialty.

other end of the line and the ordinal solution. Observe that in this example the psychologists are the variables and the journals are the objects. Thus we have a reversed indicator matrix situation (cf. Section 3.13). The solutions are quite different from the one given by Roskam (1968, p. 69), who used the non-metric unfolding model. In the linear solution the journals seem to cluster into a 'hard' group *(JEXP, PMET, MVBR, JAPP,* and perhaps *BULL),* a 'developmental' group *(JEDP, HUDE)*, and a 'soft' group *(JPSP, JCLP, HURE)*. It is also possible to see journals arranged on a circular structure (with *JAPP* in the middle). In the ordinal solution JAPP has moved closer to the 'hard' group and *JEDP* has moved away from *HUDE* to the middle. However, over all, the difference between the two solutions is not too big.

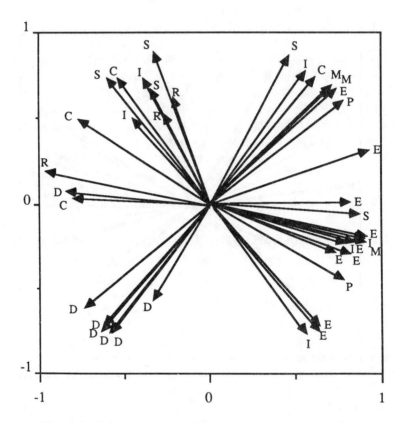

Figure 4.8 Loadings for the 39 psychologists in the single ordinal solution, labelled according to their specialty.

The loadings for the 39 psychologists are given in Figures 4.7 (numerical solution) and 4.8 (ordinal solution). Observe that the arrows point in the *least* preferred direction. For the identification of the psychologists we have used Table 8.3 from Roskam (1968, p. 152), which makes it possible to find out in which department each psychologist works. The codes are

S = Social psychology
D = Developmental and educational psychology
C = Clinical psychology
M = Mathematical psychology and psychological statistics
E = Experimental psychology
R = Cultural psychology and psychology of religion
I = Industrial psychology
P = Physiological and animal psychology

The main difference between Figures 4.7 and 4.8 is the fact that the nonlinear vectors are longer and predominantly have the same length. The phenomenon is indicative for the vastly improved fit. We also see people from the same department generally close together, with some exceptions (industrial is not very homogeneous). Experimental, mathematical, and physiological overlap completely. Social and clinical also have considerable overlap, developmental is fairly homogeneous. The nonlinear solution appears to be more tightly clustered.

4.9 AN EXAMPLE WITH DISCRETIZATION OF CONTINUOUS VARIABLES

We now discuss an example in which known nonlinear functions are simulated. A version of this problem has been used by Coombs and Kao (1960), by Kruskal and Shepard (1974), and by Young, Takane, and De Leeuw (1978) to test their approaches to principal components analysis. We made our own version with 20 objects and 10 variables. The objects varied on two dimensions, given in the first two columns of Table 4.8. These are uniform random variables on the unit interval. Each of the other eight variables was chosen as a monotonic function of these two dimensions. If the values in the first two columns of Table 4.8 are denoted by a_i and b_i, and h_{ij} is an element of the data matrix then

$$h_{i1} = a_i, \qquad\qquad h_{i6} = (2\pi)^{-1}a_i b_i^2,$$
$$h_{i2} = b_i, \qquad\qquad h_{i7} = (2\pi)a_i b_i^{-1/2},$$
$$h_{i3} = 2(\pi b_i)^{1/2}, \qquad h_{i8} = a_i b_i^{-1},$$
$$h_{i4} = 2a_i(\pi b_i)^{1/2}, \qquad h_{i9} = a_i^{-1}b_i,$$
$$h_{i5} = a_i b_i, \qquad\qquad h_{i10} = 2\pi a_i b_i^{-2}.$$

Table 4.8 Cylinder problem: data

0.63	0.76	3.08	1.95	0.48	0.06	0.29	0.84	1.20	6.95
0.99	0.37	2.14	2.12	0.36	0.02	0.65	2.71	0.37	46.64
0.25	0.98	3.51	0.87	0.24	0.04	0.10	0.25	3.98	1.61
0.72	0.75	3.08	2.22	0.54	0.07	0.33	0.96	1.04	8.00
0.65	0.07	0.96	0.62	0.05	0.00	0.06	8.96	0.11	774.63
0.63	0.88	3.33	2.11	0.56	0.08	0.27	0.71	1.40	5.07
0.27	0.44	2.34	0.64	0.12	0.01	0.16	0.62	1.60	9.00
0.77	0.48	2.45	1.88	0.37	0.03	0.44	1.60	0.62	21.10
0.24	0.27	1.86	0.44	0.07	0.00	0.18	.86	1.16	19.76
0.36	0.17	1.45	0.52	0.06	0.00	0.35	2.16	0.46	81.34
0.49	0.90	3.36	1.63	0.44	0.06	0.20	0.54	1.84	3.79
0.91	0.06	0.87	0.79	0.06	0.00	1.47	15.00	0.07	1555.76
0.90	0.50	2.52	2.28	0.46	0.04	0.51	1.79	0.56	22.32
0.52	0.32	2.00	1.03	0.16	0.01	0.36	1.62	0.62	31.86
0.99	0.49	2.49	2.46	0.49	0.04	0.56	2.00	0.50	25.40
0.27	0.09	1.07	0.28	0.02	0.00	0.35	2.93	0.34	202.87
0.95	0.07	0.96	0.91	0.07	0.00	1.39	12.84	0.08	1093.41
0.50	0.38	2.20	1.10	0.19	0.01	0.32	1.30	0.77	21.32
0.28	0.91	3.39	0.94	0.25	0.04	0.12	0.30	3.30	2.08
0.53	0.46	2.42	1.28	0.25	0.02	0.31	1.14	0.88	15.43

Table 4.9 Cylinder problem: data discreticized

3	4	4	3	4	4	2	2	3	1
4	2	2	4	3	3	4	4	1	3
1	4	4	2	2	3	1	1	4	1
3	3	3	4	4	4	2	2	3	2
3	1	1	1	1	1	4	4	1	4
3	4	4	4	4	4	2	1	4	1
1	2	2	1	2	2	1	1	4	2
3	3	3	3	3	3	3	3	2	2
1	2	2	1	1	2	1	2	3	2
2	1	1	1	1	1	3	3	2	4
2	4	4	3	3	4	1	1	4	1
4	1	1	2	1	1	4	4	1	4
4	3	3	4	4	3	3	3	2	3
2	2	2	2	2	2	3	3	2	3
4	3	3	4	4	4	4	3	2	3
1	1	1	1	1	1	3	4	1	4
4	1	1	2	2	1	4	4	1	4
2	2	2	3	2	2	2	2	3	3
1	4	4	2	3	3	1	1	4	1
2	3	3	3	3	2	2	2	3	2

Table 4.10 Correlation matrix and eigenvalues of PRINCALS ordinal

1.00									
-0.15	1.00								
-0.15	1.00	1.00							
0.60	0.60	0.60	1.00						
0.36	0.79	0.79	0.87	1.00					
0.03	0.95	0.95	0.72	0.86	1.00				
0.78	-0.61	-0.61	0.15	-0.13	-0.45	1.00			
0.56	-0.78	-0.78	-0.14	-0.40	-0.65	0.91	1.00		
-0.56	0.78	0.78	0.14	0.40	0.65	-0.91	-1.00	1.00	
0.30	-0.97	-0.97	-0.46	-0.68	-0.91	0.73	0.85	-0.85	1.00
0.669	0.280	0.023	0.010	0.009	0.005	0.004	0.002	0.000	0.000

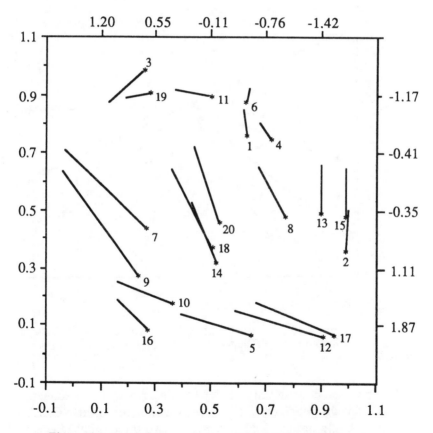

Figure 4.9 The first two columns of Table 4.7 labelled by row number, connected with the corresponding PRINCALS object scores.

This simulation example is called the *cylinder problem,* because the ten variables are used in physics and engineering to describe properties of cylinders. For our purposes it is important to note the fact that a logarithmic transformation of all variables, followed by centering of the transformed matrix, makes the resulting matrix exactly of rank two. Consequently, nonlinear components analysis should give a perfect fit in two dimensions.

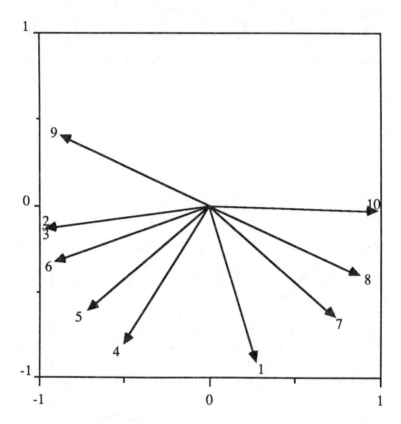

Figure 4.10 Variable loadings for the two-dimensional single ordinal solution of Thurstone's cylinder problem.

The raw data h_{ij} are given in Table 4.8; we have discreticized them in Table 4.9, and this discreticized matrix was used as input to PRINCALS with all variables single ordinal, and analysed in two dimensions. Because of the discretization perfect fit is no longer possible. The linear fit, as computed in the first phase, is 0.94 and the ordinal fit is 0.95, which is not much higher.

The correlations between the variables after the ordinal transformation are given in Table 4.10 together with the eigenvalues of $R(Q)/m$. The object scores were transformed linearly in such a way that they maximally resemble the first two columns of Table 4.8. The plot in Figure 4.9 has the numbered elements of Table 4.8 connected to the transformed object scores of ordinal PRINCALS. The loadings are given in Figure 4.10. We see that variables 1 and 2 are almost orthogonal. Variables 7, 8, and 10 make obtuse angles with variable 2 because b_i has a negative power in the formula for h_{i7}, h_{i8}, and h_{i10}. In the same way variable 9 is obtuse with variable 1, variables 2 and 3 coincide, variable 5 is between 1 and 2, 4 is closer to 1 than 2, and 6 is closer to 2 than to 1. All this is easily explained by taking logarithms of h_{ij} and looking at the coefficients of $\log a_i$ and $\log b_j$.

It is clear that PRINCALS recovers the original structure quite well, even if we have discreticized rather severely. On the other hand, the fit of the metric solution already indicates that it also performs surprisingly well.

4.10 EPILOGUE

The completely nonmetric approach mentioned in Section 4.2, which uses a nonmetric loss function for finding a nonmetric representation to nonmetric data, is becoming more tractable on present day computers, and enjoys new attention (Van Blokland-Vogelesang, 1987).

In Section 4.3 the idea of a class of quantifications generalizing the simple distinction between 'single' and 'multiple' is introduced; some elaboration can be found in De Leeuw and Van Rijckevorsel (1988). Section 4.5 mentions a number of earlier proposals of the 'vector model'. It is interesting to note that Greenacre (1984, p. 346) draws the historical line in a similar fashion: 'This particular joint display of points representing both the rows and the columns of a matrix is called a biplot (Gabriel, 1971, 1981), and is essentially the same as the vector model of preferences suggested by Tucker (1960), but applicable in a wider context of data.' Depending on the standardization, one has the covariance biplot, correlation biplot, and so on.

Segijn (1985) has studied the local minimum problem in PRINCALS; it turns out that the problem is only of some importance in the single nominal case. In Gifi (1985) more practical comments and examples of the use of PRINCALS can be found.

CHAPTER 5
NONLINEAR GENERALIZED
CANONICAL ANALYSIS

In classical canonical correlation analysis (Hotelling, 1936) one studies the relationship between two sets of variables after having removed linear dependencies within each of these sets. Any variable may contribute to the analysis in as much as it provides *independent* information with respect to the other variables within its own set *and* to the extent that it is linearly *dependent* upon other variables in the other set. The generalized canonical correlation problem involves comparing K sets of variables after having removed linear dependencies within each of the sets. Inevitably comparing K things is more intricate than comparing two things, but it turns out that the idea of optimizing homogeneity brings us a good deal further.

We have already seen in previous chapters that homogeneity analysis can be interpreted as a technique for finding p orthogonal vectors in the meet of the m indicator matrices G_j. This is not necessarily the most natural interpretation of homogeneity analysis, but it suggests an immediate extension in which we find p orthogonal vectors in the meet of K general matrices H_k, by using the loss function (cf. Section 3.5)

$$\sigma_M(X,A) = K^{-1} \sum_k SSQ(X - H_k A_k), \tag{5.1}$$

with the normalization, as in homogeneity analysis, of $u'X = 0$ and $X'X = I$. We shall suppose, throughout the following section, that the H_k are known and that their columns are in deviations from the mean. As long as the H_k are known matrices it is not necessary to refer to their individual columns (the variables). Thus in this chapter there are K sets of variables and set k consists of m_k variables. In Section 5.2 additional notation will be introduced for referring to variables within sets.

As usual we want to start with a historical overview of techniques that can be considered to be closely related to this general linear meet problem. In this case, however, we have some trouble in finding related techniques. The literature is confusing, the topic had not been systematically studied before 1960, and a great deal of the relevant papers are unpublished research reports

193

or doctoral dissertations. We shall try to bring some order into chaos; similar systematizations have been attempted by Kettenring(1971), Dauxois and Pousse (1976), Ten Berge (1977), and Van de Geer (1980). ↠

5.1 PREVIOUS WORK

Suppose the H_k are K sets of variables and $z_k = H_k a_k$ are linear composites, called *canonical variables*. The canonical variables can be collected in the $n \times K$ matrix Z, with associated correlation matrix $R(Z)$. Most optimality criteria are functions of this correlation matrix. This insight was first exploited by Kettenring (1971). In fact the most important optimality criteria are functions of the eigenvalues of $R(Z)$. Optimizing such functions means finding optimal vectors of coefficients a_k; we consequently find a *single solution* to start with. We can stop after this or proceed to solve the join problem resulting from the contraction $z_k = H_k a_k$. We can also try to find additional solutions to arrive at a *multiple solution*. There are two ways in which we can find additional solutions, which are often confused in the literature. The optimal single solution is often found by differentiating the optimality criterion and by solving the corresponding set of stationary equations (which may involve solving for undetermined multipliers). However, the stationary equations may have more than one solution and the additional solutions can be used to define a multiple solution for the canonical weights a_k. Whether this is sensible depends on the properties of the stationary equations. As we shall see in Chapter 8, the stationary equations of correspondence analysis can be interpreted both in terms of a rule called the centroid principle and in terms of scoring systems that linearize the regressions. These conceptually different interpretations make the stationary equations themselves interesting, and this interest reaches beyond the optimality criterion they were derived from.

A second way to find additional solutions is by solving the optimization problem anew, adding constraints that prevent reoccurrence of the same solution. In the context of K-set analysis these are often *orthogonality constraints*. Dauxois and Pousse (1976, pp. 197–198) have distinguished the *weak* orthogonality constraints

$$\sum_k z'_{(k)s} z_{(k)t} = \sum_k a'_{(k)s} H'_{(k)s} H_{(k)t} a_{(k)t} = 0, \tag{5.2}$$

for all previous solutions $a_{(k)t}$, with $t = 1, ..., s-1$, and the *strong* constraints

$$\mathbf{z}'_{(k)s}\mathbf{z}_{(k)t} = \mathbf{a}'_{(k)s}\mathbf{H}'_{(k)s}\mathbf{H}_{(k)t}\mathbf{a}_{(k)t} = 0, \tag{5.3}$$

which must be true for each set k and for all previous solutions $\mathbf{a}_{(k)t}$, again with $t = 1,...,s-1$. If we impose the strong constraints there can only be m_k solutions for a set with m_k variables; if we want more than m_k solutions the additional solutions will have $\mathbf{a}_{(k)s} = \mathbf{0}$ for sets with too few variables. It is a useful exercise to write down the equations of homogeneity analysis using strong orthogonality constraints and to consider what happens if in 'strong HOMALS' we have $k_j = 2$ for all j. ➤➤

The systematizations of Kettenring (1971) and Dauxois and Pousse (1976) all introduce additional solutions by using orthogonality constraints and by computing them *successively*. We have already discussed the distinction between successive and simultaneous computation in Section 4.3, and we have indicated that while in some of the simpler cases they give the same results, they do not in some of the more complicated cases. In the context of *matching configurations* with orthogonal rotations, which is closely related to multiple-set canonical analysis, Van de Geer (1968) and Ten Berge (1977) have also distinguished systematically between successive and simultaneous solutions. To introduce simultaneous computation in our context we merely have to form the matrices $\mathbf{Z}_k$ with columns $\mathbf{z}_{(k)s}$ defined as $\mathbf{z}_{(k)s} = \mathbf{H}_k\mathbf{a}_{(k)s}$, with $s = 1,...,p$ indexing the dimensionality. When we collect the weight vectors $\mathbf{a}_{(k)s}$ in the columns of the $m_k \times p$ matrix $\mathbf{A}_k$, we may also write $\mathbf{Z}_k = \mathbf{H}_k\mathbf{A}_k$. The $\mathbf{Z}_1,...,\mathbf{Z}_k,...,\mathbf{Z}_K$ are in turn collected in the $n \times (Kp)$ partitioned matrix $\mathbf{Z}$. Then the partitioned cross product matrix $\mathbf{R}(\mathbf{Z}) = \mathbf{Z}'\mathbf{Z}$ of order $(Kp) \times (Kp)$ has submatrices of the form $\mathbf{R}_{kl} = \mathbf{Z}'_k\mathbf{Z}_l = \mathbf{A}'_k\mathbf{H}'_k\mathbf{H}_l\mathbf{A}_l$ (see Figure 5.1). Usually the canonical variables are unit normalized, i.e. $\mathbf{z}'_{(k)s}\mathbf{z}_{(k)s} = \mathbf{a}'_{(k)s}\mathbf{H}'_k\mathbf{H}_k\mathbf{a}_{(k)s} = 1$. Thus the diagonal blocks of $\mathbf{R}(\mathbf{Z})$, denoted by $\mathbf{R}_{kk}$, contain the correlations among the canonical variables for set k. The off-diagonal blocks of $\mathbf{R}(\mathbf{Z})$ contain the correlations between the canonical variables of set k with those of set l.

Optimal simultaneous solutions can be computed if we now have criteria that are functions of the partitioned matrix $\mathbf{R}(\mathbf{Z})$, or of its eigenvalues. In computing these simultaneous solutions we also impose constraints, which can be strong or weak. Strong simultaneous orthogonality constraints require that $\mathbf{R}_{kk} = \mathbf{I}$ for all k, the weak version merely requires that $\sum \mathbf{R}_{kk} = K\mathbf{I}$.

Let us now consider some of the criteria proposed, first in one dimension. Steel (1951) proposes to *minimize* the *determinant* of the $K \times K$ correlation matrix $\mathbf{R}(\mathbf{Z})$, which is equal to the *product of its eigenvalues*. Clearly $0 \le \det(\mathbf{R}(\mathbf{Z})) \le 1$ with $\det(\mathbf{R}(\mathbf{Z})) = 0$ if and only if $\mathbf{R}(\mathbf{Z})$ is singular and $\det(\mathbf{R}(\mathbf{Z})) = 1$ if and only if $\mathbf{R}(\mathbf{Z})$ is the identity. Steel derives stationary equations for minimizing the determinant, but he does not propose a specific algorithm. Chang and Bargmann (1974) apply Steel's criterion to quantification of categorical data, using a quasi-Newton algorithm. Kettenring (1971) discusses Steel's work, baptizes it with the acronym GENVAR, and proposes a relaxation algorithm closely related to alternating least squares. The basic idea of this algorithm is as follows.

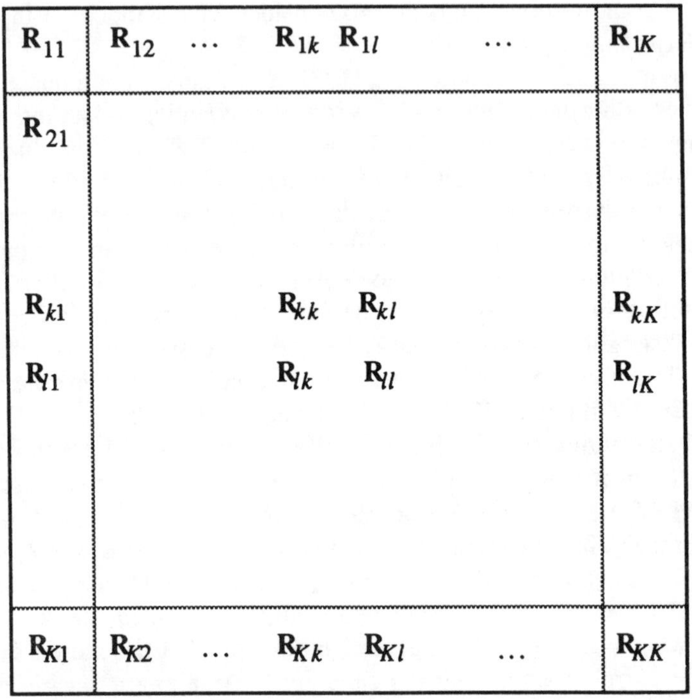

R_{11}	R_{12} ... R_{1k} R_{1l} ...	R_{1K}
R_{21}		
R_{k1} R_{l1}	R_{kk} R_{kl} R_{lk} R_{ll}	R_{kK} R_{lK}
R_{K1}	R_{K2} ... R_{Kk} R_{Kl} ...	R_{KK}

Figure 5.1 Partitioned cross product matrix $R(Z)$.

Suppose we want to find the weights a_k for forming the canonical variable of set k. For this purpose we select the kth row and column of $R(Z)$, the vector of correlations of all other canonical variables with the kth one, denoted by r_k. The remaining part of the matrix of intercorrelations is called R_{-k}. We now can apply the Schur formula for partitioned determinants in the form

$$\det(R(Z)) = (1 - r_k' R_{-k}^{-1} r_k) \cdot \det(R_{-k}). \tag{5.4}$$

It follows that minimizing the GENVAR criterion over $\mathbf{a}_k$, for fixed other $\mathbf{a}_l$, can be done by maximizing the quadratic form

$$\mathbf{r}_k' \mathbf{R}_{-k}^{-1} \mathbf{r}_k = \mathbf{a}_k' (\Sigma_{l \neq k} \Sigma_{l' \neq k} r^{ll'} \mathbf{H}_k' \mathbf{H}_l \mathbf{a}_l \mathbf{a}_l' \mathbf{H}_{l'}' \mathbf{H}_k) \mathbf{a}_k. \qquad (5.5)$$

Here l and l' are both running indices for sets, and $r^{ll'}$ denotes an element of the inverse of the correlation matrix $\mathbf{R}_{-k}$; thus

$$r^{ll'} = (\mathbf{R}_{-k}^{-1})_{ll'}. \qquad (5.6)$$

Since we want to have normalized canonical variables, $\mathbf{a}_k$ must satisfy $\mathbf{a}_k' \mathbf{H}_k' \mathbf{H}_k \mathbf{a}_k = 1$. This defines a generalized eigenvalue problem (cf. Appendix B), which is easily solved for $\mathbf{a}_k$. Observe that $\mathbf{a}_k^+$ minimizes GENVAR, for fixed other $\mathbf{a}_l$, if it maximizes the multiple correlation of z_k with the other z_l. It is also possible, with the same algorithm, to *maximize* the determinant, by finding the *smallest* generalized eigenvalue, which minimizes the multiple correlation in each step. In that case we want $\mathbf{R}(\mathbf{Z})$ to be as close as possible to the identity matrix, for example because we want to use it for prediction purposes in a *second step*. Kettenring (1971) uses GENVAR in combination with strong orthogonality constraints to compute successive solutions, but we can also use essentially the same algorithm to compute p solutions simultaneously. Maximizing or minimizing the determinant of the partitioned matrix $\mathbf{R}(\mathbf{Z})$ of order $Kp \times Kp$ can also be done by using the Schur formula. The subproblem now is maximizing (or minimizing)

$$\det\{\mathbf{A}_k' (\Sigma_{l \neq k} \Sigma_{l' \neq k} \mathbf{H}_k' \mathbf{H}_l \mathbf{A}_l \mathbf{R}^{ll'} \mathbf{A}_{l'}' \mathbf{H}_{l'}' \mathbf{H}_k) \mathbf{A}_k\} \qquad (5.7)$$

over all $\mathbf{A}_k$ satisfying the strong orthogonality constraints $\mathbf{A}_k' \mathbf{H}_k' \mathbf{H}_k \mathbf{A}_k = \mathbf{I}$, i.e. $\mathbf{R}_{kk} = \mathbf{I}$. Again this defines a generalized eigenvalue problem. Both Steel and Chang and Bargmann try to defend their choice of the GENVAR criterion by appealing to likelihood ratio tests for the multinormal distribution, but the resemblance seems merely formal to us. It is better to treat GENVAR as just a possible weighted combination of the eigenvalues, and one which tends to emphasize the smallest of the smaller ones.

The second optimality criterion was proposed by Horst (1961a, 1961b). In fact, he proposes four different methods, of which he clearly prefers the one we discuss now. The other three are treated by him as computationally convenient modifications. Kettenring uses the acronym SUMCOR, and the criterion is simply the sum of the correlations in the $K \times K$ matrix $\mathbf{R}(\mathbf{Z})$. Observe that this is *not* a function of the eigenvalues and that it depends on the sign of the elements in $\mathbf{R}(\mathbf{Z})$. The method is also discussed by Van de Geer (1968) in the matching context, by Nevels (1974), by Ten Berge (1977), also

for matching, by Dauxois and Pousse (1976, pp. 235–244), and by Van de
Geer (1980), who uses the acronym ORTHOCAN. The algorithms are
generally of the block relaxation type, in which the coefficients for each set are
the blocks. It is clear that maximizing the SUMCOR criterion over a_k, with the
other a_l fixed, amounts to maximizing the linear function

$$a_k'(\Sigma_{l \neq k} H_k'H_l a_l) \tag{5.8}$$

over $a_k'H_k'H_k a_k = 1$. This is equivalent to a multiple regression problem, with
solution proportional to

$$(H_k'H_k)^{-1}\Sigma_{l \neq k} H_k'H_l a_l. \tag{5.9}$$

Successive solutions are computed by using strong orthogonality constraints;
we can also *minimize* the sum of the correlations with the same algorithm
(simply change all signs in the previous update of a_k), and we can compute
simultaneous solutions by generalizing to the criterion

$$\Sigma_k \Sigma_l \, \text{tr}(A_k'H_k'H_l A_l) \tag{5.10}$$

which must be maximized over A_k under the restrictions $R_{kk} = I$. The
subproblems are now orthogonal Procrustes problems (Cliff, 1966).

   The third optimality criterion, also proposed by Horst, is introduced by him
as follows: 'The second model specifies that the intercorrelations of the first
transformed variables for the m sets shall give the best least square
approximation to a rank one matrix' (Horst, 1961b, p. 332). A fourth optimality criterion proposed by Carroll (1968) aims at (in our notation) an n vector
x and canonical variables $z_k = H_k a_k$ such that the sum of the squared
correlations between x and the z_k is maximized. Both formulations must
sound familiar, because they correspond to two of the ways one-dimensional
homogeneity analysis was introduced in Chapter 3. For reasons given in that
chapter both criteria yield the same solution. Carroll's contribution is of
special interest because of the introduction of the comparison scores x.

   Kettenring uses the acronym MAXVAR for both criteria, and provides the
interpretation in terms of maximizing the largest eigenvalue of $R(Z)$. He also
uses this interpretation to introduce MINVAR, which minimizes the smallest
eigenvalue of $R(Z)$. We already encountered MINVAR in Chapters 3 and 4. It
is clear from these chapters that both successive and simultaneous optimization
of this criterion is possible. It has been used there predominantly in combination with the weak orthogonality constraints. The optimization problems are
in general single eigenvalue problems, not cycles of eigenvalue problems,
which is a considerable advantage, both computationally and theoretically.

This advantage is also pointed out by Carroll (1968) and by Dauxois and Pousse, who criticize the complicatedness of the other criteria we have discussed in this section (Dauxois and Pousse, 1976, p. 194, pp. 210–212). Van de Geer (1980) discusses MAXVAR using the acronym GENCAN. Observe that multiple solutions in this case can be computed from the stationary equations without using orthogonality. In fact orthogonality follows from the fact that two solutions (with different eigenvalues) satisfy the stationary equations.

For completeness we must mention two other methods. Kettenring (1971) also introduces SSQCOR, where the criterion equals the *sum of squares* of the correlations in $R(Z)$, which is equal to the sum of squares of the eigenvalues of $R(Z)$. This leads to subproblems of the form: maximize (or minimize)

$$a_k'(\Sigma_{l \neq k}\, H_k'H_l a_l a_l' H_l' H_k)a_k \tag{5.11}$$

over $a_k' H_k' H_k a_k = 1$, which has an obvious simultaneous generalization. Computationally SSQCOR is similar to, although much simpler than, GENVAR. We expect SSQCOR to emphasize the largeness of the larger eigenvalues, while GENVAR emphasizes the smallness of the smaller ones. Clearly a similar dual relationship exists between MAXVAR and MINVAR.

The last method that we mention has not been discussed before in the literature. It is briefly mentioned in Section 4.3. If we generalize PRINCALS with single variables to situations in which the indicator matrices G_j are replaced by general, but constant, H_k, then we maximize the *sum of the p largest eigenvalues* of $R(Z)$, or, equivalently, we minimize the sum of the $K - p$ smallest ones. This generalizes MAXVAR and MINVAR at the same time and introduces some intermediate possibilities.

The different criteria have not been compared on a large scale. Horst (1961b) compares SUMCOR and MAXVAR. Kettenring uses the same example and also computes GENVAR and SSQCOR. For this example all techniques give very similar results. Haven and Ten Berge (see Ten Berge, 1977, ch. IV) made some comparisons in a matching context. It seems to us that more research is needed, either to show that canonical analysis with K matrices H_k is stable under selection of criterion or to indicate in which situation it can make a lot of difference which criterion we use. Dauxois and Pousse (1976) have already shown that if the techniques are extended to random vectors, then the solutions will all be the same if the random vectors are multinormal. As long as there are no results available which tell us how to choose a criterion or loss function, we think that mathematical and computational convenience indicate that MAXVAR and MINVAR are pretty good candidates.

Table 5.1a Correlation matrix for 11 variables, three sets

0.66									
0.46	0.49								
0.14	0.12	0.09							
0.16	0.12	0.15	0.17						
0.10	0.13	0.08	0.09	−0.05					
0.07	0.04	0.03	0.01	0.10	0.02				
0.37	0.40	0.30	0.15	0.07	0.33	0.08			
0.37	0.36	0.26	0.14	0.11	0.31	0.11	0.72		
0.46	0.47	0.35	0.19	0.09	0.34	0.09	0.80	0.71	
0.45	0.45	0.35	0.18	0.07	0.38	0.10	0.74	0.69	0.81

We illustrate this point with a small example. The correlation matrix in Table 5.1a is taken from De Leeuw and Stoop (1979, p. 142). The basic data were from the 'As years go by' study, discussed more extensively in Chapter 13. The correlation matrix is actually a correlation matrix taken from a larger matrix of order 25, in which the quantifications used to compute the correlations were derived from HOMALS. The first set consists of five variables: profession father, education father, education mother, number of children in the family, degree of urbanization. They are clearly exogenous variables describing the situation in the family. The second set, with two variables, indicates whether the child had to do one or more grades more than once and how many children there were in the child's class in the sixth grade. The third set (four variables) consists of the teacher's advice on secondary education, score on a school achievement test, choice of secondary education, and attained level of secondary education.

The resulting 11 variables, partitioned in three sets, were analysed with six canonical analysis techniques. We shall use new acronyms, which improve those of Kettenring, because they indicate that everything that can be maximized can also be minimized. First there is MAXMAX which maximizes the largest eigenvalue (previously known as MAXVAR or GENCAN or CARROLL). MINMIN (formerly MINVAR) minimizes the smallest eigenvalue of the correlation matrix. MAXSUM (Kettenring's SUMCOR) maximizes the sum of the correlations and MINSUM minimizes this sum. MINDET (used to be GENVAR) minimizes the determinant and MAXSSQ (was SSQCOR) maximizes the sum of the squared correlations (or the sum of squares of the eigenvalues, or the variance of the eigenvalues). Using our system of acronyms the reader may wonder what happened to MINMAX, MAXMIN,

MAXDET, and MINSSQ. The answer is simple. In this example, as in many examples with small m_k, we can choose the a_k in such a way that the correlation matrix becomes the identity. This identity matrix solves the four problems we have not mentioned.

Table 5.1b Canonical weights for six techniques

MAXMAX	MINMIN	MAXSUM	MINSUM	MINDET	MAXSSQ
0.41	0.48	0.40	0.52	0.45	0.43
0.48	0.43	0.49	0.35	0.44	0.46
0.21	0.24	0.21	0.25	0.23	0.22
0.27	0.18	0.29	0.09	0.21	0.24
−0.09	0.02	−0.11	0.13	−0.02	−0.06
0.78	0.57	0.96	0.99	0.97	0.96
0.21	0.10	0.25	0.10	0.23	0.25
0.03	0.01	0.04	0.03	0.01	0.02
0.09	0.11	0.09	0.11	0.08	0.08
0.48	0.58	0.39	0.39	0.45	0.43
0.63	0.67	0.56	0.55	0.53	0.54

Table 5.1c Optimality criteria (columns) for six techniques (rows)

	MAXMAX	MINMIN	MAXSUM	MINSUM	MINDET	MAXSSQ
MAXMAX	1.7555	0.3914	5.1990	1.4242	0.5862	3.9628
MINMIN	1.7482	0.3868	5.1718	1.4018	0.5849	3.9540
MAXSUM	1.7553	0.3933	5.1994	1.4294	0.5878	3.9608
MINSUM	1.7282	0.3910	5.1090	1.3934	0.5952	3.9154
MINDET	1.7530	0.3875	5.1882	1.4090	0.5838	3.9619
MAXSSQ	1.7550	0.3893	5.1958	1.4170	0.5846	3.9638

Table 5.1d As Table 5.1c, techniques ranked columnwise

	MAXMAX	MINMIN	MAXSUM	MINSUM	MINDET	MAXSSQ
MAXMAX	1	5	2	5	4	2
MINMIN	5	1	5	2	3	5
MAXSUM	2	6	1	6	5	4
MINSUM	6	4	6	1	6	6
MINDET	4	2	4	3	1	3
MAXSSQ	3	3	3	4	2	1

The solutions for the a_k are given in Table 5.1b. For MAXMAX and MINMIN we have applied the weak orthogonality constraints; for the other four techniques we have applied the strong orthogonality constraints. We do not give any substantial interpretations here; we merely remark that the solutions for the canonical weights are extremely similar. In Table 5.1c this also becomes obvious; we have computed all six criteria for all six solutions. The only problem here is that MAXSUM and MINSUM depend on the sign of the correlations between the three composites. For MAXSUM we consequently choose all correlations positive; for MINSUM we get the smallest value throughout by choosing the correlation between the third composite and the other two to be negative. In Table 5.1d the numbers in Table 5.1c are replaced by rank numbers. For this example MAXSUM and MINSUM do not perform very well, which is a nice result. MAXMAX and MINMIN are fair, MINDET and MAXSSQ are possibly better.

5.2 LOSS FUNCTION AND NORMALIZATION OF OVERALS

Our version of generalized canonical analysis, called OVERALS, is based on the MAXVAR (MAXMAX) criterion, and incorporates nonlinear transformations of the data. The notation in loss function (5.1) is appropriate enough if the matrices H_k are constant; if the variables must be quantified or transformed, however, we need more complicated notation. The notation has to take account of variables within sets and of the space of all nonlinear transformations. This is done by switching from H_k to sets of indicator matrices G_j and from weights A_k to sets of quantifications Y_j. Define

$$\sigma_M(X,Y) \equiv K^{-1} \sum_k SSQ(X - \sum_{j \in J_k} G_j Y_j), \qquad (5.12)$$

where we assume that the total number of variables is m and that the *index set* $\{1,...,m\}$ of the variables is partitioned into K sets J_k ($k = 1,...,K$). This notation has the advantage that the sets J_k do not have to consist of consecutive integers; it can also be used if the J_k do not exhaust $\{1,...,m\}$ or where the J_k are not exclusive. By constructing a partitioned indicator matrix G^k and a partitioned matrix Y^k for each of the sets, we can express the loss function as

$$\sigma_M(X,Y) = K^{-1} \sum_k SSQ(X - G^k Y^k), \qquad (5.13)$$

which makes it again very similar to the loss function (5.1). The reason why we usually prefer to use the individual G_j and not the G^k is that we prefer to use the special properties of complete indicator matrices in our algorithms and that we restrict each of the Y_j individually according to measurement level.

We have seen in Section 5.1 that other techniques would normalize the Y_j, while we have the opportunity to normalize X. The exception is Carroll (1968), who also normalizes the comparison scores, and because at least part of the French data analytic work (Masson, 1974; Saporta, 1975; Dauxois and Pousse, 1976) is inspired by this paper, it follows a similar approach. Kettenring (1971) also mentions Carroll's work as one of his inspirations. Carroll's work fits naturally into the tradition of homogeneity analysis started by Horst, Edgerton and Kolbe, Richardson, Wilks, and Guttman. It seems interesting to see how normalization on Y is related to our normalization $u'X = 0$ and $X'X = I$, even in the general case in which the Y_j can be restricted in various other ways.

The key observation here is that we shall only be interested in restrictions on Y_j that are defined in such a way that if Y_j satisfies the restrictions, then Y_jT satisfies the restrictions for all $p \times p$ matrices T. For multiple nominal variables, in which the Y_j are not restricted, this is trivially true. For single variables, in which we require $Y_j = y_ja_j'$, with $y_j \in K_j$, this condition is also true: if $Y_j = y_ja_j'$ then $Y_jT = y_ja_j'T$, which can be written as $Y_jT = y_j\tilde{a}_j'$. Observe that for multiple ordinal variables the condition is no longer true: if Y_j satisfies the constraints then Y_jT will generally only satisfy the constraints for all *diagonal* matrices. If the condition is true, then we can also minimize

$$\sigma_M(X,Y,T) = K^{-1} \sum_k SSQ(X - \sum_{j \in J_k} G_jY_jT_j) \qquad (5.14)$$

over X, Y_j restricted as before, and the new variables T_j, with the same result and solution as before. We shall use a slightly different formulation, by imposing the condition that the T_j are the same within sets and by writing them as T_k. Also let

$$Z_k = \sum_{j \in J_k} G_jY_j, \qquad (5.15)$$

and define $\sigma_M(*,Y)$ as the minimum of $\sigma_M(X,Y)$ over X satisfying $u'X = 0$ and $X'X = I$. We also write $\sigma_M(*,Y,*)$ for the minimum of $\sigma_M(X,Y,T)$ over X with $u'X = 0$ and $X'X = I$ and over T, where

$$\sigma_M(X,Y,T) = K^{-1} \sum_k SSQ(X - Z_kT_k). \qquad (5.16)$$

Our theory so far tells us that minimizing $\sigma_M(*,\mathbf{Y})$ over restricted $\mathbf{Y}$ is equivalent to minimizing $\sigma_M(*,\mathbf{Y},*)$ over restricted $\mathbf{Y}$, where the restrictions of both problems are the same. Minimizing $\sigma_M(*,\mathbf{Y})$ is our original problem, and minimizing $\sigma_M(*,\mathbf{Y},*)$ can be shown to be equivalent to maximizing the sum of the p largest eigenvalues of the matrix

$$\Sigma_k \, \mathbf{Z}_k \, (\mathbf{Z}_k'\mathbf{Z}_k)^+ \mathbf{Z}_k', \tag{5.17}$$

where $(\mathbf{Z}_k'\mathbf{Z}_k)^+$ denotes the generalized inverse. The proof follows from the Eckart–Young theorem. The eigenvalues of this matrix are also equal to the eigenvalues of the generalized eigenvalue problem with the partitioned matrices $\mathbf{R}(\mathbf{Z})$ and $\mathbf{D}(\mathbf{Z})$, where $\mathbf{R}(\mathbf{Z})$ has submatrices $\mathbf{R}_{kl} = \mathbf{Z}_k'\mathbf{Z}_l$ and where $\mathbf{D}(\mathbf{Z})$ is a block diagonal matrix with submatrices $\mathbf{R}_{kk} = \mathbf{Z}_k'\mathbf{Z}_k$. The formulation and the notation have been chosen in such a way that the problem looks as similar as possible to the linear meet problem, explored in the previous section. Of course, we must always remember that $\mathbf{Z}_k$ is not a constant matrix because it is a function of the $\mathbf{Y}_j$. It is also clear from this formulation that we now can suppose without loss of generality that $\Sigma \, \mathbf{R}_{kk} = K\mathbf{I}$, which is more explicitly

$$\Sigma_k \, (\sum_{j \in J_k} \sum_{l \in J_k} \mathbf{Y}_j'\mathbf{G}_j'\mathbf{G}_l\mathbf{Y}_l) = K\mathbf{I}, \tag{5.18}$$

or

$$\Sigma_k \, (\mathbf{G}^k\mathbf{Y}^k)'\mathbf{G}^k\mathbf{Y}^k = K\mathbf{I}, \tag{5.19}$$

which are the weak simultaneous orthogonality constraints. Some familiar special cases can be recovered easily. If J_k contains only a single variable j then $\mathbf{Z}_k = \mathbf{G}_j\mathbf{Y}_j$. If the $\mathbf{Y}_j$ are not further restricted (multiple nominal) we recover homogeneity analysis. If all variables are single then $\mathbf{Z}_k = \mathbf{q}_j\mathbf{a}_j'$ and

$$\mathbf{Z}_k \, (\mathbf{Z}_k'\mathbf{Z}_k)^+ \mathbf{Z}_k' = \mathbf{q}_j\mathbf{q}_j' \,/\, \mathbf{q}_j'\mathbf{q}_j, \tag{5.20}$$

which gives principal components analysis.

5.3 OVERALS AS A SPECIAL CASE OF HOMALS: THE USE OF INTERACTIVE VARIABLES

We have seen in Chapter 2 that it is sometimes useful to combine m variables with $k_1,...,k_m$ categories into a single variable with $k_1 \times ... \times k_m$ categories

The m indicator matrices G_j, which are $n \times k_j$, are replaced by a single indicator matrix G, which is $n \times (k_1 \times \ldots \times k_m)$. Now suppose we replace all indicator matrices in set k by the interactive indicator matrix G_k (not the same as G^k in Section 5.2, because this is the 'sum' of the G_j, which is $n \times (k_1 + \ldots + k_m)$). We rewrite the loss function as

$$\sigma_M(X,Y) = K^{-1} \sum_k SSQ(X - G_k Y_k), \tag{5.21}$$

where we must remember that the number of rows of Y_k is now equal to the number of possible profiles for all variables in set k. The matrix G_k is a proper indicator matrix, whereas G^k is not, because it consists of a number of proper indicator matrices, one for each variable in the set. It is not difficult to prove that there always exists a binary *design matrix* S_k such that $G_k S_k = G^k$. In fact it is easy to construct S_k explicitly: S_k is simply the matrix that assigns to each profile in G_k the appropriate row in G^k. Thus S_k is $(\Pi k_j) \times (\Sigma k_j)$, and is explicitly given by

$$S_k = (G_k' G_k)^{-1} G_k' G^k. \tag{5.22}$$

If there are m_k variables in the set k, then $S_k u = m_k u$. By using $u' G_k S_k = u' G^k$ we also see that S_k transforms profile frequencies in marginal frequencies. Table 5.2 gives a small example. The data matrix for a set of variables is expanded to the partitioned indicator matrix (Table 5.2b) and the interactive indicator matrix (Table 5.2c). The design matrix is given in Table 5.2d. This construction so far merely recapitulates material from previous chapters, but it is especially important at this point, because using G_k shows us that a group of variables can be reduced to a single variable and that

Table 5.2a Data matrix

a	p	u
b	q	v
a	r	v
a	p	u
b	p	v
c	p	v
a	p	u
a	p	v
c	p	v
a	p	v

Table 5.2b Partitioned indicator matrix

a	b	c	p	q	r	u	v
1	0	0	1	0	0	1	0
0	1	0	0	1	0	0	1
1	0	0	0	0	1	0	1
1	0	0	1	0	0	1	0
0	1	0	1	0	0	0	1
0	0	1	1	0	0	0	1
1	0	0	1	0	0	1	0
1	0	0	1	0	0	0	1
0	0	1	1	0	0	0	1
1	0	0	1	0	0	0	1

consequently all K-set problems can be reduced to homogeneity analysis problems (while, conversely, we have seen that all homogeneity problems are K-set problems with only one variable in each set).

Table 5.2c Interactive indicator matrix

apu	apv	aqu	aqv	aru	arv	bpu	bpv	bqu	bqv	bru	brv	cpu	cpv	cqu	cqv	cru	crv
1	0	0	0	0	0	0	0	0	0	0	0	0	0	0	0	0	0
0	0	0	0	0	0	0	0	0	1	0	0	0	0	0	0	0	0
0	0	0	0	0	1	0	0	0	0	0	0	0	0	0	0	0	0
1	0	0	0	0	0	0	0	0	0	0	0	0	0	0	0	0	0
0	0	0	0	0	0	0	1	0	0	0	0	0	0	0	0	0	0
0	0	0	0	0	0	0	0	0	0	0	0	1	0	0	0	0	0
1	0	0	0	0	0	0	0	0	0	0	0	0	0	0	0	0	0
0	1	0	0	0	0	0	0	0	0	0	0	0	0	0	0	0	0
0	0	0	0	0	0	0	0	0	0	0	0	1	0	0	0	0	0
0	1	0	0	0	0	0	0	0	0	0	0	0	0	0	0	0	0

Table 5.2d Design matrix recording additivity restrictions

	a	b	c	p	q	r	u	v
apu	1	0	0	1	0	0	1	0
apv	1	0	0	1	0	0	0	1
aqu	1	0	0	0	1	0	1	0
aqv	1	0	0	0	1	0	0	1
aru	1	0	0	0	0	1	1	0
arv	1	0	0	0	0	1	0	1
bpu	0	1	0	1	0	0	1	0
bpv	0	1	0	1	0	0	0	1
bqu	0	1	0	0	1	0	1	0
bqv	0	1	0	0	1	0	0	1
bru	0	1	0	0	0	1	1	0
brv	0	1	0	0	0	1	0	1
cpu	0	0	1	1	0	0	1	0
cpv	0	0	1	1	0	0	0	1
cqu	0	0	1	0	1	0	1	0
cqv	0	0	1	0	1	0	0	1
cru	0	0	1	0	0	1	1	0
crv	0	0	1	0	0	1	0	1

Now, in general, using interactive variables must be done with care, because it may create far too many categories (the arguments against the discrete MVA approaches connected with the empty cell problems discussed in Chapter 1 are also relevant here). Yet the use of interactive variables has proven to be useful in some examples (cf. Section 8.9). In any case, the matrix S_k is conceptually important, because $G_k S_k = G^k$ implies that $G_k S_k Y^k = G^k Y^k$, and therefore $G_k Y_k = G^k Y^k$ if $Y_k = S_k Y^k$. The restrictions $Y_k = S_k Y^k$ can be interpreted as requiring that the category quantification of a profile must be *additive*, i.e. it must be the sum of quantification of the categories in the profile. Thus S_k records the required additivity, and we can now say that K-set canonical analysis can be interpreted as homogeneity analysis with additivity restrictions. Observe that for sets of single variables the restrictions are $Y_k = S_k Y^k$ with rank one restrictions on the submatrices of Y^k. Thus for single variables it is better to write

$$Y_k = \sum_{j \in J_k} S_{(k)j} Y_j = \sum_{j \in J_k} S_{(k)j} y_j a_j', \qquad (5.23)$$

where the $S_{(k)j}$ are the submatrices of S_k, which are $(\Pi k_j) \times k_j$ (compare Table 5.2d). These are proper indicator matrices, which satisfy $G_k S_{(k)j} = G_j$. ↠

5.4 MISSING DATA

Since the options missing data single and missing data multiple both involve complete indicator matrices, their incorporation in the OVERALS framework does not create new problems compared to HOMALS and PRINCALS. In contrast, the option missing data passive, where indicator matrices are incomplete, presents an additional problem. HOMALS and PRINCALS are special cases of OVERALS when each set contains only one variable. Consequently, if an observation on a variable is missing, it is missing in the whole set. For OVERALS with m_k variables in set k, this is not necessarily true. However, the simplest generalization we can think of is

$$\sigma_M(X,Y) = K^{-1} \sum_k \text{tr}(X - \sum_{j \in J_k} G_j Y_j)' M_k (X - \sum_{j \in J_k} G_j Y_j), \qquad (5.24)$$

with M_k the elementwise minimum of the M_j in set k. This means that if an object has a missing observation on some variable, then the set that contains this variable does not contribute to the loss for this object. In HOMALS and PRINCALS $M_k = M_j$. This gives the loss function used with option i in Chapters 3 and 4.

5.5 ALGORITHM CONSTRUCTION

In this section we treat the general aspects of the algorithm construction of OVERALS, without going into too much detail. ➹ The general case is considered, i.e. indicator matrices may be incomplete and therefore we have weighting matrices M_k for missing data. If all M_j are equal to the identity matrix, which means that all G_j are complete, then we treat this as a special case of the formulas. The fitting of X for given Y_j does not present any new problems, and we refer to Section 4.4. The fitting of Y^k from the loss component

$$\text{tr}(X - G^k Y^k)' M_k (X - G^k Y^k) \tag{5.25}$$

does present new problems, because G^k is a partitioned indicator matrix and not a simple indicator matrix. This characteristic implies that in general $(G^k)'G^k$ is not diagonal, which complicates the construction of

$$\tilde{Y}^k = ((G^k)'M_k G^k)^+ (G^k)'M_k X. \tag{5.26}$$

Although it is possible, at least in principle, to base an algorithm on computing $\tilde{Y}^k$ first, and then proceeding as in Chapter 4, this will tend to become expensive. It will also only be useful for multiple variables, because for single variables we use rank one and cone restrictions defined in terms of the Y_j, which do not look very natural in terms of the Y^k. For these reasons it is preferred to fit the Y_j separately. Therefore we return to our original loss function (5.12). At this point we have to take into account that the transformation of variable j is dependent on the other variables in the set. Therefore we define the auxiliary matrix

$$X_j \equiv X - (\sum_{l \neq j} G_l Y_l - G_j Y_j), \tag{5.27}$$

where both j and l are variables in set k, in which the contributions of the other variables are removed from X.

Now the minimization of $\sigma_M(X,Y)$ with X and Y_l fixed can be done by minimizing

$$\text{tr}(X_j - G_j Y_j)' M_k (X_j - G_j Y_j) \tag{5.28}$$

over Y_j. The conditional unconstrained minimum is attained for

$$\tilde{\mathbf{Y}}_j = (\mathbf{G}_j'\mathbf{M}_k\mathbf{G}_j)^+\mathbf{G}_j'\mathbf{M}_k\mathbf{X}_j. \tag{5.29}$$

The matrix which must be inverted is now small (its order is the number of nonmissing categories) and diagonal. From this point, rank one and order restrictions can be applied completely analogously to the PRINCALS algorithm, with $\mathbf{X}_j$ in the role of $\mathbf{X}$. The algorithm does not have any more special features. Still worth mentioning is the update formula for going from variable j in set k to variable $j+1$ in set k. Then

$$\mathbf{X}_{j+1} = \mathbf{X}_j - (\mathbf{G}_j\mathbf{Y}_j - \mathbf{G}_{j+1}\mathbf{Y}_{j+1}). \tag{5.30}$$

5.6 AN EXAMPLE: EFFECTS OF RADIOACTIVITY ON FISH

The data for this example were taken from Cailliez and Pagès (1976, pp. 277–293). Twenty-four fish were placed in three aquariums, which were contaminated with radioactive strontium. The three aquariums were the same, but the fish stayed for a short period in the first aquarium (fish numbers 1–8), for a longer time in the second aquarium (fish numbers 9–17), and for the longest time in the third aquarium (fish numbers 18–24). Variables 1–9 are measures of radioactivity of various body parts of the fish after the experiment, variables 10–16 are size measurements of the fish, and variable 17 indicates the aquarium (Table 5.3). We have used a discreticized version of the data matrix, also given by Cailliez and Pagès (1976, pp. 280–284). They apply metric component analysis to the first 16 variables, and find that in the plot of the object scores it is easy to separate the fish from the different aquaria. Within the aquarium clusters we find fish of the same size close together. Fish 21 and 24, which are in aquarium 3, are close to the fish in aquarium 2. There are two clusters of variables in the plot of the loadings. The first cluster consists of variables 10–14; these all measure the size of the fish rather directly. The second cluster are the variables 1, 2, 3, 4, 8; these are measurements of radioactivity of hard tissues. The two clusters are almost at right angles in the plane of the first two principal axes, which explains about 70 per cent of the variance. The same data were also analysed by Bourouche and Saporta (1980, pp. 190–221). They used discriminant analysis with 15 predictors (variable 7 was not used) to identify aquarium membership. This could be done very well; the three centroids are in the corners of an equilateral triangle and the fish are close to their aquarium. Again the same two groups of variables can be distinguished.

Table 5.3 Fish data from an experiment by Amiard (data categorized by Cailliez and Pagès, 1976)

Fish	Set 1 1	2	3	4	5	6	7	8	9	Set 2 10	11	12	13	14	15	16	Set 3 17
1	2	2	2	2	1	1	10	1	1	10	9	10	10	10	10	10	1
2	2	1	1	1	4	1	5	1	2	9	10	10	8	8	8	7	1
3	1	1	2	2	1	2	4	1	2	10	10	10	8	9	9	10	1
4	1	2	1	1	1	3	5	1	2	10	10	10	9	10	7	10	1
5	1	1	2	2	2	3	1	3	1	2	1	1	1	4	1	4	1
6	1	1	1	2	2	1	4	1	1	2	3	3	2	3	3	4	1
7	1	1	1	1	2	1	1	2	1	2	2	2	3	3	3	4	1
8	2	2	1	2	2	1	5	1	4	1	2	4	2	1	2	1	1
9	3	3	2	4	6	2	4	2	3	3	3	4	3	5	4	7	2
10	6	6	4	3	7	2	4	2	5	4	6	7	6	5	4	4	2
11	3	3	3	3	3	6	4	2	2	4	4	4	4	6	4	10	2
12	4	5	2	3	3	3	4	8	5	3	2	3	3	5	5	7	2
13	4	6	3	3	5	5	4	2	10	4	5	5	4	5	10	7	2
14	7	3	2	3	7	5	4	2	2	1	1	1	1	3	4	4	2
15	3	3	2	3	2	3	4	2	4	5	6	6	3	5	8	7	2
16	4	4	2	2	4	1	4	3	6	5	8	5	6	6	9	7	2
17	?	?	?	?	?	?	?	?	?	?	?	?	?	?	?	?	2
18	10	10	10	9	5	1	7	6	3	3	3	4	3	3	5	4	3
19	7	7	9	9	3	9	1	7	2	2	2	3	2	3	4	4	3
20	10	9	8	10	10	1	1	10	10	1	1	2	2	2	4	1	3
21	4	5	4	5	3	1	3	1	2	7	7	8	7	9	7	10	3
22	7	7	10	8	1	1	6	9	3	5	5	6	4	8	6	4	3
23	7	7	9	9	7	10	10	7	2	5	5	5	4	5	6	7	3
24	6	2	6	6	2	1	4	4	1	6	5	5	5	7	5	10	3

The categorized data matrix is given in Table 5.3. Observe that there is no information on fish number 17, because it died during the experiment. All variables, except the last one, have ten categories, but many categories are empty. We first analyse the three sets 1–9, 10–16, and 17, with the first 16 variables single numerical and the last one multiple nominal, in two dimensions. We then change the measurement level of the first 16 variables to single ordinal and repeat the analysis. Observe that with 23 observations 10 categories for each variable is quite a lot, and the ordinal option could very well behave in the way the continuous ordinal option is expected to behave. The loss of the numerical solution is 0.401, which is the average of the loss per set, which is 0.250, 0.625, and 0.329, respectively. The loss of the ordinal solution is 0.014, which is the average of 0.000, 0.022, and 0.21.

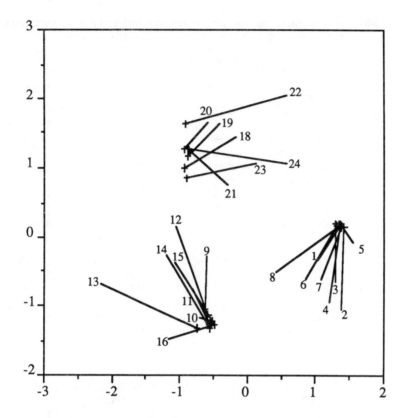

Figure 5.2 Object scores for the fish data in the numerical solution (labelled) connected with the corresponding points in the ordinal solution.

In Figure 5.2 the object scores for the 23 fish are given, the numbers correspond with the numerical solution, the crosses at the end of the line with the ordinal solution. The clustering of the aquaria is quite clear in the numerical case; in the ordinal case it is even clearer. The canonical weights (A_k) are not so informative because the within-set correlations are very high, and this often results in instability and difficulty in the interpretation. It is more informative to look at the canonical loadings (the correlations of the canonical variables with the original variables or the quantified variables). There are three sets of variables, and consequently three pairs of canonical variables, and a fourth pair, which are the comparison scores **X**. Table 5.4 gives the loadings for the numerical solution, i.e. correlations between the original variables and the canonicalvariables/object scores, and Table 5.5 gives them

Table 5.4 Canonical loadings for single numerical analysis (correlations of 16 original variables and four sets of two canonical variables/object scores)

	Canonical variables							
	Set 1		Set 2		Set 3		Object scores	
Variables								
1	−0.36	0.77	−0.33	0.36	−0.45	0.74	−0.45	0.72
2	−0.41	0.71	−0.41	0.32	−0.44	0.68	−0.51	0.66
3	0.05	0.92	−0.08	0.41	−0.16	0.90	−0.14	0.88
4	−0.06	0.95	−0.19	0.47	−0.21	0.91	−0.22	0.91
5	−0.57	0.22	−0.44	0.06	−0.49	0.22	−0.53	0.16
6	−0.27	0.22	−0.23	0.11	−0.29	0.19	−0.32	0.19
7	0.15	0.04	0.21	−0.11	0.07	0.08	0.11	0.03
8	−0.07	0.85	−0.17	0.57	−0.23	0.71	−0.21	0.81
9	−0.72	0.05	−0.59	0.00	−0.48	0.01	−0.67	0.01
10	0.34	−0.25	0.60	−0.46	0.31	−0.08	0.40	−0.24
11	0.22	−0.36	0.47	−0.67	0.21	−0.17	0.28	−0.39
12	0.25	−0.28	0.52	−0.50	0.28	−0.10	0.33	−0.27
13	0.24	−0.33	0.52	−0.57	0.27	−0.13	0.33	−0.33
14	0.35	−0.19	0.51	−0.35	0.17	−0.04	0.31	−0.15
15	−0.15	−0.20	0.03	−0.42	−0.10	−0.08	−0.15	−0.20
16	0.15	−0.23	0.24	−0.42	−0.06	−0.10	0.07	−0.23

Table 5.5 Canonical loadings for single ordinal analysis (correlations of 16 quantified variables and four sets of two canonical variables/object scores)

	Canonical variables							
	Set 1		Set 2		Set 3		Object scores	
Variables								
1	−0.98	0.02	−0.98	0.02	−0.98	0.03	−0.98	0.02
2	−0.54	0.34	−0.56	0.30	−0.52	0.37	−0.54	0.34
3	−0.62	0.64	−0.62	0.61	−0.61	0.66	−0.62	0.64
4	−0.66	0.74	−0.66	0.72	−0.66	0.75	−0.66	0.74
5	−0.34	0.09	−0.33	0.06	−0.34	0.12	−0.34	0.09
6	−0.29	0.15	−0.29	0.07	−0.30	0.22	−0.29	0.15
7	−0.06	0.26	−0.06	0.18	−0.06	0.34	−0.06	0.26
8	−0.28	0.45	−0.29	0.51	−0.28	0.38	−0.28	0.45
9	−0.35	−0.38	−0.35	−0.37	−0.34	−0.39	−0.35	−0.38
10	0.45	0.18	0.45	0.18	0.44	0.18	0.41	0.18
11	0.45	−0.01	0.44	−0.01	0.44	−0.00	0.44	−0.01
12	0.03	0.07	0.03	0.07	0.03	0.07	0.03	0.07
13	0.42	0.02	0.42	0.03	0.42	0.03	0.42	0.02
14	0.23	0.04	0.22	0.04	0.22	0.05	0.22	0.04
15	−0.58	−0.06	−0.58	−0.05	−0.59	−0.05	−0.59	−0.05
16	−0.06	−0.29	−0.06	−0.30	−0.07	−0.28	−0.06	−0.29

for the ordinal solution, i.e. correlations between transformed variables and canonical variables and object scores, respectively. The four subtables of 5.5 do not differ very much, because of the very good fit. We have plotted the component loadings of the numerical solution in Figure 5.3 and the loadings of the ordinal solution in Figure 5.4. We can see from the figures that there are four groups of variables. The first one consists of the size variables, of which

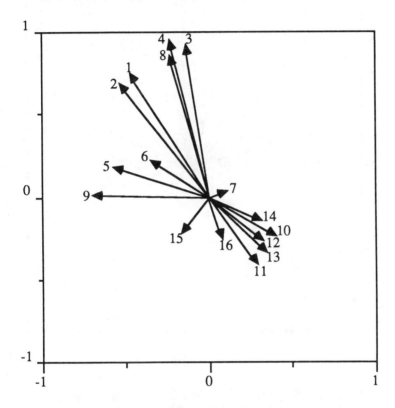

Figure 5.3 Single numerical OVERALS solution for the fish data: canonical loadings of the 16 variables.

especially 10–14 are important, the second group consists of variables 3, 4, 8 (radioactivity of gill-covers, fines, scales), the third group of variables 1 and 2 (radioactivity of eyes and gills), and the fourth group of variables 5, 6, 9 (radioactivity of liver, gullet, and muscles). Variable 7 (radioactivity of kidneys) does not belong with the other variables 1–9, which is probably why

Bourouche and Saporta eliminated it. Clearly our classification of the variables refines the one given by Cailliez and Pagès, and our technique gives results that are intermediate between discriminant analysis and principal components analysis.

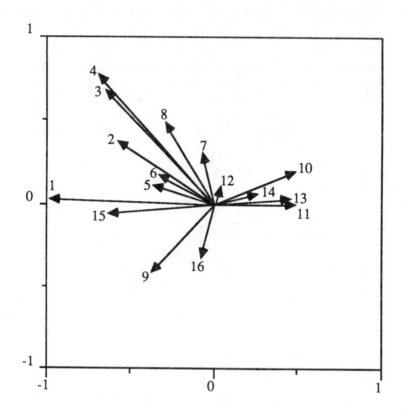

Figure 5.4 Single ordinal OVERALS solution for the fish data: canonical loadings of the 16 variables.

5.7 EPILOGUE

Further systematizations of K-set analysis can be found in Van de Geer (1984) and Gower (1984). For other recent work see Ten Berge and Knol (1984), Van de Geer (1986), Ten Berge (1988), and Peay (1988). A whole range of predominantly European contributions in this area is contained in Coppi and Bolasco (1989), which also includes recent work on three-way methods. These methods focus on the special case in which the variables of each set are the same (or at least for which we are prepared to identify K variables, one from each set, with each other, and to repeat this m times). A lot of other interesting work on multimode and multiway data analysis is contained in Kroonenberg (1983) and Law *et al.* (1984).

The useful exercise mentioned in Section 5.1 is among the things studied by Van de Geer (1987a). Viewing OVERALS as a special case of HOMALS, as is done in Section 5.3, forms the basis of Van der Burg, De Leeuw, and Verdegaal (1988). For other contributions to OVERALS, including algorithm construction, algebraic and geometric properties, and examples of applications, the reader is referred to Van der Burg (1988), Verdegaal (1985, 1986), Van de Geer (1987b, 1988), and Meulman (1988a).

CHAPTER 6
NONLINEAR CANONICAL
CORRELATION ANALYSIS

In the previous chapter canonical analysis was discussed for the general case of K sets of variables. In this chapter we study the situation where two sets of variables are involved, which have a symmetric role. The latter characteristic is associated with the technique that is known in the literature as *canonical correlation analysis*. When we have two sets of variables the meet problem simplifies considerably, at least in the linear case. In Section 5.1 various criteria were studied that were functions of the $K \times K$ correlation matrix $\mathbf{R(Z)}$ between the canonical variables. When we have only two sets, $\mathbf{R(Z)}$ is of order 2×2, and it can be shown that the various criteria are all monotonic functions of the single correlation $r_{12} = \mathbf{z}_1' \mathbf{z}_2 = \mathbf{a}_1' \mathbf{H}_1' \mathbf{H}_2 \mathbf{a}_2$.

6.1 PREVIOUS WORK

There is considerably more historical material on the problem of two sets. The problem was probably formulated for the first time by Spearman, who discussed the correlation between sums and differences in a famous paper of 1913. He observed that it was interesting to investigate how weights n_j should be computed such that the correlation between the weighted sum $n_1 a_1 + ... + n_p a_p$ and a single given variable b is as large as possible (notation adapted from Spearman). Spearman observed that Yule had already described completely how this problem should be solved, but he adds: 'By the aid of the present theorems, the still more general case of the maximum of $r_{(n_1 a_1 + ... + n_p a_p)(m_1 b_1 + ... + m_q b_q)}$ can be obtained, though sometimes through rather complicated differentiation' (Spearman, 1913, p. 420, footnote). Indeed, because Spearman does not use matrix algebra and the theory of canonical forms of matrices, his expressions already become very complicated in the simple special cases that he discusses. Hotelling (1935) was a

substantial improvement in this respect. In this paper an *asymmetric* approach
to the two sets is adopted: one set of variables consists of *predictors*, the other
set of *criteria*. The problem here is that there is more than one criterion. This
cannot be avoided in many situations because constructs such as college
success or general price level cannot be measured by a single variable.
Hotelling suggests various possible ways to combine the criteria into a composite one. As soon as this is accomplished we apply multiple regression and
we are done. One of the methods he proposes is to replace the criteria by their
principal component, which is quite similar to 'first-step' procedures we use
in some examples. Also several forms of *a priori* weighting are considered.
"In spite of this variety of grounds for choice of a variate to be predicted, the
problem of finding a linear function of the criterion variates which can *most
accurately* be predicted from given observations, in the sense of least squares,
admits a definite solution, which we shall set forth' (Hotelling, 1935, p. 139).
It seems that this particular formulation of the problem (the 'most predictable
criterion') has confused a lot of psychometricians. In their search for the
criterion, they thought (correctly) that there is no reason to prefer a particular
criterion simply because it can be predicted most accurately. They overlooked
the fact that this is only one way to formulate canonical analysis and that even
if canonical analysis does not give us *the* criterion, it does at least give us
upper and lower bounds.

In Hotelling (1936) a more symmetrical approach is adopted:

> The relations between two sets of variates with which we shall be concerned
> are those that remain invariant under internal linear transformations of each set
> separately. Such invariants are not affected by rotations of axes in the study of
> wind or of hits on a target, or by replacing mental test scores by an equal
> number of independently weighted sums of them for comparison with physical
> measurement (Hotelling, 1936, p. 332).

This symmetric formulation seems far less objectionable, but it has been
ignored by psychometricians for a fairly long time. This is despite the
excellent papers by Thomson (1947), Bartlett (1948), and Burt (1948b), who
explained Hotelling's work to psychometricians familiar with factor analysis.
Canonical correlation analysis never became as popular as multiple regression
and principal components analysis.

Thorndike and Weiss (1973) mention four possible reasons for this relative
unpopularity. The first one is *difficulty in computation*. This reason was only
valid until the early sixties, because during the sixties efficient computer
programs for performing canonical correlation analysis became widely
available. The second reason could be the availability of *other more familiar
methods* for studying the relationships between two sets of variables. Again
this does not seem to be a valid reason any more: review papers trying to

convert people to using canonical correlation analysis have been written in marketing (Green, Halbert, and Robinson, 1966), in counselling psychology (Weiss, 1972), and in educational research (Darlington, Weinberg, and Walberg, 1973). Application-oriented papers have also been published by Wood and Erskine (1976) and Thorndike (1977). We have also seen in Chapter 1 that canonical correlation analysis is covered in almost all MVA textbooks, presented as the appropriate technique to investigate relations between two sets of numerical variables. Thus there must be other reasons for its unpopularity. The third one mentioned by Thorndike and Weiss is the *difficulty in interpreting the results*. The problem here seems to be that investigators often did not know exactly what to compute, what to plot, and what to look at. This problem has been studied quite extensively recently, and the outcome indicates that the canonical weights should be supplemented by the correlations between the original variables and the canonical variables (cf. the example in Section 5.6). The final reason mentioned by Thorndike and Weiss is the *instability of results* (the 'bouncing beta' problem of multiple regression analysis becomes twice as serious). Studies by Thorndike and Weiss (1973), Barcikowski and Stevens (1975) and Thorndike (1977) indicate that indeed stability under selection of variables or objects may be low. These studies use cross validation and Monte Carlo methods, and their findings are extremely difficult to generalize, which shows that the stability of results investigating stability may also be very low. It is possible, however, to derive some analytical results which indicate that canonical correlation analysis is less stable in general than, for example, principal components analysis. Some of these stability results will be reviewed in Chapter 12.

A fifth reason, not mentioned by Thorndike and Weiss, is perhaps that the situation in which we have two sets of numerical variables with a *symmetric* role does not occur very often in practice. Asymmetric versions of canonical correlations have been introduced by Stewart and Love (1968), and asymmetric versions of canonical correlation analysis have been proposed by Rao (1964) and Van den Wollenberg (1977). ⇥ We still have the impression, however, that many interesting analysis problems with two sets of variables are even more severely asymmetric, for example because the second set is a categorical variable (as in discriminant analysis) or because the first set is a design matrix (as in analysis of variance). The fact that canonical correlation analysis is a useful computational technique that unifies a large number of seemingly very different MVA problems was already emphasized by Bartlett (1948) and Tintner (1946). These more asymmetric versions will be discussed in the next chapter, because it is often advantageous to fit them by using specialized algorithms.

6.2 THEORY

Of course it is possible to fit canonical analysis with two sets of variables with the more general algorithm for K sets outlined in Chapter 5. Strictly speaking we do not need a separate chapter for two sets. We can use the special properties of $K = 2$, however, to simplify both theory and algorithm somewhat.

We start with a normalization result similar to that in Section 5.2. Consider

$$\sigma_M(\mathbf{X},\mathbf{Y}) = K^{-1} \Sigma_k \text{SSQ}(\mathbf{X} - \mathbf{G}^k\mathbf{Y}^k), \qquad (6.1)$$

which must be minimized under the condition $\mathbf{X'X} = \mathbf{I}$ (we suppose $\mathbf{u'X} = 0$ is automatically taken care of by the definition of the restrictions on $\mathbf{Y}^k$). We assume again that the restrictions on $\mathbf{Y}^k$ are such that $\mathbf{Y}^k\mathbf{T}_k$ is feasible whenever $\mathbf{Y}^k$ is feasible, no matter how we choose $\mathbf{T}_k$. We have shown in Section 5.2 that minimizing $\sigma_M(\mathbf{X},\mathbf{Y})$ over $\mathbf{X'X} = \mathbf{I}$ means maximizing the sum of the p largest eigenvalues of

$$\Sigma_k \mathbf{Z}_k(\mathbf{Z}_k'\mathbf{Z}_k)^+\mathbf{Z}_k', \qquad (6.2)$$

with $\mathbf{Z}_k = \mathbf{G}^k\mathbf{Y}^k$. We have also shown that minimizing $\sigma_M(\mathbf{X},\mathbf{Y})$, applying the weak orthogonality constraint $\Sigma_k \mathbf{R}_{kk} = \Sigma_k \mathbf{Z}_k'\mathbf{Z}_k = K\mathbf{I}$, is equivalent to maximizing the sum of the p largest eigenvalues of the generalized eigenvalue problem with matrices $\mathbf{R}(\mathbf{Z})$ and $\mathbf{D}(\mathbf{Z})$, which have submatrices $\mathbf{R}_{kl} = \mathbf{Z}_k'\mathbf{Z}_l$ and $\mathbf{R}_{kk} = \mathbf{Z}_k'\mathbf{Z}_k$, respectively. Because the eigenvalues of the second problem and the eigenvalues of the first problem are the same, the two problems are equivalent, and we can use either one of the normalization conditions (either on $\mathbf{X}$ or on $\mathbf{Y}$). Normalizing $\mathbf{Y}$, in the case $K = 2$, shows that the problem is equivalent to minimizing

$$\sigma_M(\mathbf{Y}^1,\mathbf{Y}^2) = \text{SSQ}(\mathbf{G}^1\mathbf{Y}^1 - \mathbf{G}^2\mathbf{Y}^2), \qquad (6.3)$$

where $\mathbf{X}$ has been eliminated, and in which we require that $\mathbf{Z}_1'\mathbf{Z}_1 + \mathbf{Z}_2'\mathbf{Z}_2 = 2\mathbf{I}$. Because we can choose $\mathbf{T}_1$ and $\mathbf{T}_2$ independently to transform $\mathbf{Y}^1$ into $\mathbf{Y}^1\mathbf{T}_1$ and $\mathbf{Y}^2$ into $\mathbf{Y}^2\mathbf{T}_2$, we obtain an equivalent solution if we require $\mathbf{Z}_1'\mathbf{Z}_1 = \mathbf{Z}_2'\mathbf{Z}_2$ $= \mathbf{I}$ (strong orthogonality). Finally, we can also require that either $\mathbf{Z}_1'\mathbf{Z}_1 = \mathbf{I}$ *or* $\mathbf{Z}_2'\mathbf{Z}_2 = \mathbf{I}$, which again gives essentially the same solution (except for a possibly different scaling of the dimensions). It is clear that these normalization results generalize corresponding results from Chapter 3 to the case in which there are restrictions on the $\mathbf{Y}_j$. The important condition that makes these generalizations possible is that $\mathbf{Y}^k\mathbf{T}_k$ is feasible, whenever $\mathbf{Y}^k$ is. This condition is satisfied by sets of single and multiple nominal variables.

Incorporating optimal scaling in the simplified case for two-set canonical analysis was done for the first time in Young, De Leeuw, and Takane (1976), but unfortunately the algorithm proposed in that paper does not work. The problem is caused by the normalization. Young, De Leeuw, and Takane require that *both* $Z_1'Z_1$ and $Z_2'Z_2 = I$ (actually, their CORALS algorithm is limited to computing one-dimensional solutions, but this is not essential). These requirements have as a consequence that the constraints on Y become very complicated. If we modify one of the Y_j (by an algorithm as in Section 5.5) then we must impose both *measurement constraints* (rank one and order restrictions) on Y_j and *normalization constraints* on the whole set, and the latter involves the other Y_l in the set. Young, De Leeuw, and Takane first ignore the normalization constraints while modifying the Y_j, and impose them afterwards. This is against the general principles of alternating least squares algorithm construction and consequently their procedure does not work; CORALS diverges.

Fortunately, there is a way out of this normalization problem. As we have seen in the previous paragraph the strong orthogonality constraints are equivalent to the requirement that either $Z_1'Z_1 = I$ *or* $Z_2'Z_2 = I$. Which one of these two constraints we impose at any particular time depends on the set that we are modifying. If we modify the Y_j of the first set, we impose $Z_2'Z_2 = I$. If all Y_j in the first set are modified, we find a linear transformation T_1 and another linear transformation T_2 in such a way that $\sigma_M(Y^1, Y^2, T_1, T_2) = \sigma_M(Y^1, Y^2)$ *and* $Z_1'Z_1 = I$. This *transfer of normalization* guarantees that the loss does not change and that after rescaling the first set is normalized. We then proceed to modify the Y_j in the second set, without bothering about set normalization. When the Y_j in the second set are modified, we rescale again. This scheme is repeated until convergence. Convergence is assured because in each step the loss decreases. The present procedure exploits the fact that a complicated normalization problem can be replaced by a simpler one; this replacement is possible because we know from theory that their solutions are equivalent. ⮞

6.3 THE CANALS PROGRAM

The CANALS program does not fit naturally into the series HOMALS – PRINCALS – OVERALS, because it does not incorporate multiple variables and does not use indicator matrices. The latter is explained by the fact that CANALS dates back to a period in which we still wanted the possibility to incorporate continuous variables (a continuous variable is associated with $G_j = I$). Practical experience and some theory given in Section 4.3 has suggested, however, that incorporating continuous variables is not urgent, to put it mildly. CANALS is much more similar to the ALS programs resulting from the cooperation between Young, De Leeuw, and Takane. In fact, it is a fairly straightforward extension of CORALS to $p \geq 1$ dimensions, in which the transfer of normalization mentioned in Section 6.2 is used to correct the error made in CORALS. ➤

The new aspect in CANALS can be described as follows. We have Z_1 and Z_2 such that $Z_2'Z_2 = I$. We want to find T_1 and T_2 such that $SSQ(Z_1T_1 - Z_2T_2) = SSQ(Z_1 - Z_2)$ and $T_1'Z_1'Z_1T_1 = I$. The solution is simple: we find any T_2 such that $Z_1'Z_1 = T_2T_2'$ and we set $T_1' = T_2^{-1}$. CANALS uses the Gram–Schmidt decomposition to find T_2 from Z_1, but the singular value decomposition could also have been used (cf. Section 3.4).

The CANALS output will be illustrated in the examples, but we mention some of the more important features. In CANALS all variables are treated single, which means that we have a vector y_j of category quantifications (scaled in such a way that $u'D_jy_j = 0$ and $y_j'D_jy_j = n$) for each j. Moreover, for each j we have a vector a_j of (canonical) *weights*. Finally, we have correlations between the canonical variables $Z_1 = G^1Y^1 = Q_1A_1$ and $Z_2 = G^2Y^2 = Q_2A_2$ and the optimally transformed variables $q_j = G_jy_j$. These correlations are called *canonical loadings* (they are called *canonical components* by Thorndike and Weiss, 1973). The CANALS program also gives the following four correlation matrices:

$Q_1'Z_1$: the correlations between the transformed variables of the first set and the canonical variables of the first set,

$Q_1'Z_2$: the correlations between the transformed variables of the first set and the canonical variables of the second set,

$Q_2'Z_1$: the correlations between the transformed variables of the second set and the canonical variables of the first set,

$Q_2'Z_2$: the correlations between the transformed variables of the second set and the canonical variables of the second set.

The correlations between the canonical variables are called the *canonical correlations* and are given in the diagonal matrix $Z_1'Z_2 = \Phi$. We then have the relationships $Q_1'Z_2 = Q_1'Z_1\Phi$ and $Q_2'Z_1 = Q_2'Z_2\Phi$, while $Z_1'Z_1 = Z_2'Z_2 = I$.

We have to emphasize another difference between CANALS and the other programs: it does not have provisions to handle missing data according to option i; instead, missing data are handled according to option iii. Because CANALS does not use indicator matrices this simply means that the missing data approach must be incorporated in the definitions of the cones C_j. Option *ii* can be simulated if one codes missing data in one extra category per variable. It would be sensible to use the nominal option in this case when one does not know where to place this additional category among the nonmissing ones.

6.4 EXAMPLE 1: ECONOMIC INEQUALITY AND POLITICAL STABILITY

The data for this example are taken from a paper by Russett (1964), which has been reprinted in Rowney and Graham (1969). The basic hypothesis in Russett's paper is that economic inequality leads to political instability; the basic problem is how to measure these complicated constructs:

> We shall be concerned with information on the degree to which agricultural land is concentrated in the hands of a few large landholders. Information on land tenure is more readily available, and is of more dependable comparability, than are data on the distribution of other economic assets like current income or total wealth (Russett, 1964, p. 444).

Three variables are used to measure the inequality of land distribution. We discuss them, and mention their codes in the tables and figures. GINI is the Gini index of concentration, which measures the deviation of the Lorenz curve from the line of equality. FARM is the percentage of farmers that own half of the land, starting with the smallest ones. Thus if FARM is 90 per cent, then 10 per cent of the farmers own half of the land. The third indicator is RENT, which is the percentage of farm households that rent all their land. Russett adds two extra economic variables, the gross national product per capita, GNPR, and the percentage of the labour force employed in agriculture, LABO. There are four measures of political stability. The first is a function of the number of chief executives in office and the number of years a country has been independent in the period between 1945 and 1961. This index is called INST; it runs from 0 (very stable) to 17 (very unstable). The second index is ECKS, the total number of violent internal war incidents in 1946–1961. The third one is DEAT, the number of people killed as a result of civic group violence. The fourth one is DEMO, which is a threefold classification of countries in 'stable democracies', 'unstable democracies', and 'dictatorships'. The original data are in Table 6.1 and the discretized data are in Table 6.2.

Table 6.1 Russett's original data

	GINI	FARM	RENT	GNPR	LABO	INST	ECKS	DEAT	DEMO
Argentina	86.3	98.2	32.9	374	25	13.6	57	217	2
Australia	92.9	99.6	–	1215	14	11.3	0	0	1
Austria	74.0	97.4	10.7	532	32	12.8	4	0	2
Belgium	58.7	85.8	62.3	1015	10	15.5	8	1	1
Bolivia	93.8	97.7	20.0	66	72	15.3	53	663	3
Brazil	83.7	98.5	9.1	262	61	15.5	49	1	2
Canada	49.7	82.9	7.2	1667	12	11.3	22	0	1
Chile	93.8	99.7	13.4	180	30	14.2	21	2	2
Colombia	84.9	98.1	12.1	330	55	14.6	47	316	2
Costa Rica	88.1	99.1	5.4	307	55	14.6	19	24	2
Cuba	79.2	97.8	53.8	361	42	13.6	100	2900	3
Denmark	45.8	79.3	3.5	913	23	14.6	0	0	1
Dominican Republic	79.5	98.5	20.8	205	56	11.3	6	31	3
Ecuador	86.4	99.3	14.6	204	53	15.1	41	18	3
Egypt	74.0	98.1	11.6	133	64	15.8	45	2	3
El Salvador	82.8	98.8	15.1	244	63	15.1	9	2	3
Finland	59.9	86.3	2.4	941	46	15.6	4	0	2
France	58.3	86.1	26.0	1046	26	16.3	46	1	2
Guatemala	86.0	99.7	17.0	179	68	14.9	45	57	3
Greece	74.7	99.4	17.7	239	48	15.8	9	2	2
Honduras	75.7	97.4	16.7	137	66	13.6	45	111	3
India	52.2	86.9	53.0	72	71	3.0	83	14	1
Iraq	88.1	99.3	75.0	195	81	16.2	24	344	3
Ireland	59.8	85.9	2.5	509	40	14.2	9	0	1
Italy	80.3	98.0	23.8	442	29	15.5	51	1	2
Japan	47.0	81.5	2.9	240	40	15.7	22	1	2
Libya	70.0	93.0	8.5	90	75	14.8	8	0	3
Luxembourg	63.8	87.7	18.8	1194	23	12.8	0	0	1
Netherlands	60.5	86.2	53.3	708	11	13.6	2	0	1
New Zealand	77.3	95.5	22.3	1259	16	12.8	0	0	1
Nicaragua	75.7	96.4	–	254	68	12.8	16	16	3
Norway	66.9	87.5	7.5	969	26	12.8	1	0	1
Panama	73.7	95.0	12.3	350	54	15.6	29	25	3
Peru	87.5	96.9	–	140	60	14.6	23	26	3
Philippines	56.4	88.2	37.3	201	59	14.0	15	292	3
Poland	45.0	77.7	0.0	468	57	8.5	19	5	3
South Vietnam	67.1	94.6	20.0	133	65	10.0	50	1000	3
Spain	78.0	99.5	43.7	254	50	0.0	22	1	3
Sweden	57.7	87.2	18.9	1165	13	8.5	0	0	1
Switzerland	49.8	81.5	18.9	1229	10	8.5	0	0	1
Taiwan	65.2	94.1	40.0	132	50	0.0	3	0	3
United Kingdom	71.0	93.4	44.5	998	5	13.6	12	0	1
United States	70.5	95.4	20.4	2343	10	12.8	22	0	1
Uruguay	81.7	96.6	34.7	569	37	14.6	1	1	1
Venezuela	90.9	99.3	20.6	762	42	14.9	36	111	3
West Germany	67.4	93.0	5.7	762	14	3.0	4	0	2
Yugoslavia	43.7	79.8	0.0	297	67	0.0	9	0	3

Table 6.2 Discreticized variables

	Economic variables					Political variables				Label in Figs. 6.3 and 6.4
Argentina	5	5	4	4	2	5	3	4	2	Arg
Australia	6	5	9	7	1	4	1	1	1	Aus
Austria	4	5	2	5	2	5	2	1	2	Aut
Belgium	2	3	5	6	1	6	2	2	1	Belg
Bolivia	6	5	3	1	4	6	3	5	3	Bol
Brazil	5	5	2	4	4	6	3	2	2	Bra
Canada	1	2	2	7	1	4	2	1	1	Can
Chile	6	5	3	3	2	5	2	2	2	Chil
Colombia	5	5	3	4	3	5	3	4	2	Col
Costa Rica	5	5	2	4	3	5	2	3	2	Cos
Cuba	4	5	5	4	3	5	3	5	3	Cuba
Denmark	1	1	2	6	2	5	1	1	1	Den
Dominican Republic	4	5	3	3	3	4	2	3	3	Dom
Ecuador	5	5	3	3	3	6	3	2	3	Ecu
Egypt	4	5	3	2	4	6	3	2	3	Egy
El Salvador	5	5	3	3	4	6	2	2	3	ElS
Finland	2	3	2	6	3	6	2	1	2	Fin
France	2	3	4	6	2	6	3	2	2	Fra
Guatemala	5	5	3	3	4	5	3	3	3	Gua
Greece	4	5	3	3	3	6	2	2	2	Gre
Honduras	4	5	3	2	4	5	3	4	3	Hon
India	2	3	5	1	4	2	3	2	1	Ind
Iraq	5	5	5	3	5	6	2	4	3	Iraq
Ireland	2	3	2	5	2	5	2	1	1	Ire
Italy	5	5	4	5	2	6	3	2	2	Ita
Japan	1	2	2	3	2	6	2	2	2	Jap
Libya	3	4	2	1	4	5	2	1	3	Lib
Luxembourg	3	3	3	7	2	5	1	1	1	Lux
Netherlands	3	3	5	6	1	5	2	1	1	Neth
New Zealand	4	5	4	7	1	5	1	1	1	NwZ
Nicaragua	4	5	9	4	4	5	2	2	3	Nic
Norway	3	3	2	6	2	5	2	1	1	Nor
Panama	4	4	3	4	3	6	2	3	3	Pan
Peru	5	5	9	2	3	5	2	3	3	Peru
Philippines	2	3	4	3	5	5	2	4	1	Phil
Poland	1	1	1	5	3	3	2	2	3	Pol
South Vietnam	3	4	3	2	4	4	3	5	3	SVi
Spain	4	5	5	4	3	1	2	2	3	Spa
Sweden	2	3	3	7	1	3	1	1	1	Swe
Switzerland	1	2	3	7	1	3	1	1	1	Swi
Taiwan	3	4	4	2	3	1	2	1	3	Tai
United Kingdom	4	4	5	6	1	5	2	1	1	UnK
United States	4	5	3	8	1	5	2	1	2	USA
Uruguay	5	5	4	5	2	5	2	2	1	Uru
Venezuela	6	5	3	6	3	5	2	4	3	Ven
West Germany	3	4	2	6	1	2	2	1	2	WGe
Yugoslavia	1	1	1	4	4	1	2	1	3	Yug

Russett's analysis is somewhat primitive; he simply correlates everything with everything and reports some of the higher correlations (tabular analysis for interval scale variables, or for supposedly interval scale variables). The two economic variables GNPR and LABO were introduced to make a more refined analysis possible. Russett's discussion seems to suggest that he has a partial correlation analysis in mind, but he says that he has used multiple regression and does not give any details. We have chosen to use canonical correlation analysis for the two sets of variables GINI, FARM, RENT, GNPR, LABO versus INST, ECKS, DEAT, DEMO. This is not necessarily the most rational choice. Both Russett's discussion and the definition of the variables suggest that it may be preferable to use three-set canonical analysis (program OVERALS, explained in Chapter 5) or a partial canonical correlation analysis (program PARTALS, explained in Chapter 7). Because both programs were not yet available in exportable FORTRAN versions we prefer to use this example to demonstrate some of the possibilities of CANALS. All variables are treated as single ordinal, we use $p = 2$ dimensions. There are two

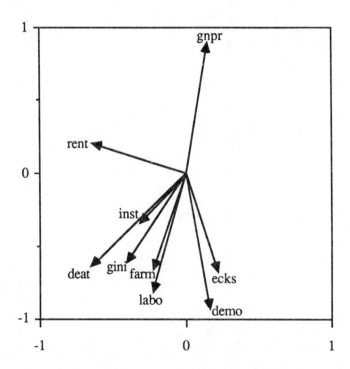

Figure 6.1 Russett's original data: correlations in the canonical space of economic variables.

different analyses; the first one uses Table 6.1, the second one uses a discreticized version of this table, given in Table 6.2; the two solutions can be compared to indicate how stable they are. Of course DEMO is already discrete, which prevents Russett from computing correlations, but which does not bother CANALS.

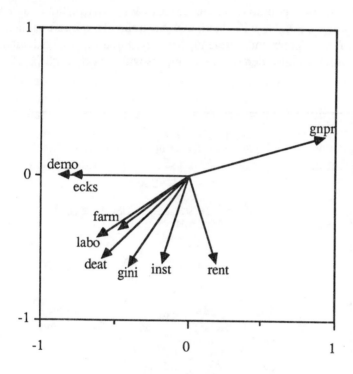

Figure 6.2 Discreticized variables: correlations in the canonical space of economic variables.

Figures 6.1 and 6.2 give the canonical loadings in the space of the economic variables, i.e. the correlations between the nine quantified variables and the optimal linear combinations of the first set of quantified variables that define the two canonical variables of the first set. The solutions for the rankings and the discreticized data are not very different, except for the fact that the two dimensions are interchanged. For the rankings the canonical correlations are 0.97 and 0.92, for the discrete data they are 0.90 and 0.81. Figures 6.1 and 6.2 show that the first dimension of the discrete data, which is the second dimension of the rankings, contrasts poor dictatorships having a large percentage of agricultural labour (variables DEMO, LABO, and GNPR) with rich industrialized democracies. The other dimension is more difficult to interpret. It helps to study Figures 6.3 and 6.4, which depict the object scores on the two canonical variables of the first set. For the discrete data matrix we see a number of clusters. On the right we see a number of countries with high GNPR, classified by Russett in his final table as stable democracies with greater than median equality. These are Canada, Switzerland, Sweden, Luxembourg, the USA, New Zealand, and Australia. To the left of them we find Belgium, the Netherlands, Ireland, Norway, Finland, United Kingdom, and France which are also democratic countries with a fairly high GNPR and

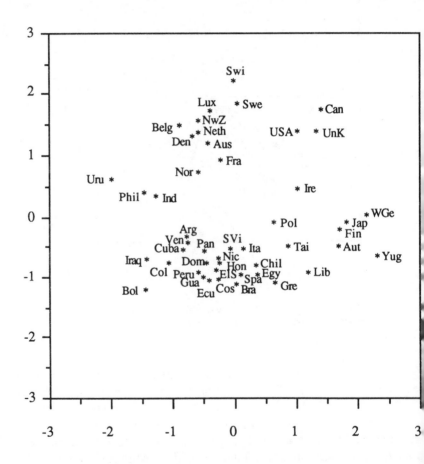

Figure 6.3 Russett's original data: canonical scores in the space of economic variables.

with greater than median equality. Only France and Finland are unstable democracies. Below to the left we find dictatorships with less than median equality, in the left upper corner we find dictatorships with greater than median equality. It is possible to find a refinement of Russett's classification of countries, although some countries such as West Germany and South Vietnam are not where they should be. The same clusters can be found, with much more trouble, in the figure for the rankings. It is more striking that the solutions are quite different. They are also different from the solution reported in Gifi (1980, p. 227–236), which seems to indicate that the stability is far from satisfactory for this example. The two new analyses are, as far as the

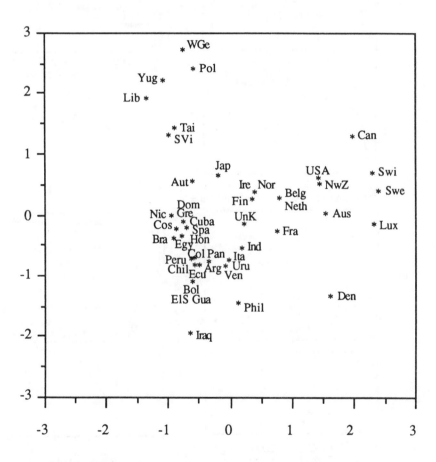

Figure 6.4 Discreticized variables: canonical scores in the space of economic variables.

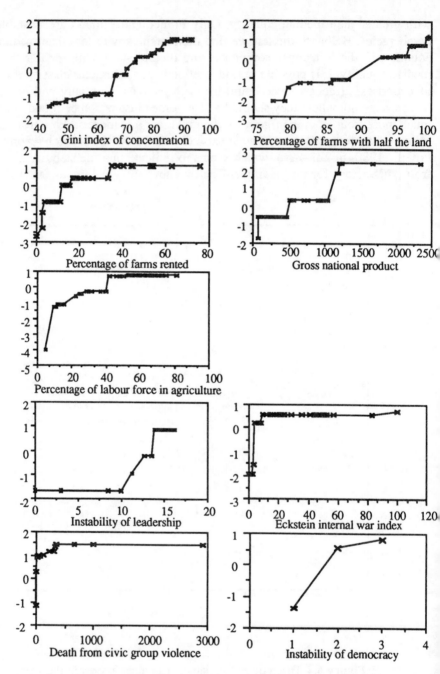

Figure 6.5 Russett's original data: transformations of the variables.

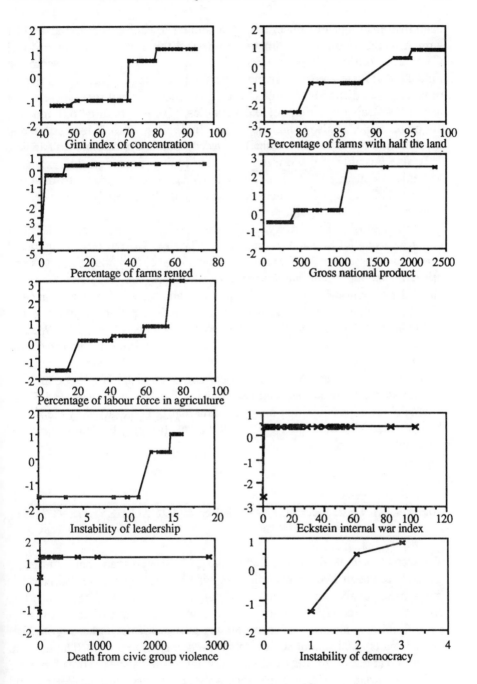

Figure 6.6 Discreticized variables: transformations of the variables.

canonical loadings are concerned, quite similar to each other, and also quite similar to the solution for dimension one and three of a three-dimensional CANALS solution given in Gifi (1980, fig. 7.7, p. 234). This seems to indicate that there are some stable effects, but they tend to occur in different dimensions in different solutions. The object scores do not seem to be stable at all. The transformations of the variables, given in Figures 6.5 and 6.6, are quite stable, as also compared with Gifi (1980, p. 236). It is possible that a solution to the instability problem is to compute more canonical variables. It is also possible that we must use more radical means in this example, for example using HOMALS or PRINCALS as a first step to quantify the variables, and perform metric canonical correlation analysis as a second step. ⇥

We could ask what CANALS adds to Russett's table 3, which classifies countries using DEMO and the median GINI index. The clusters in Figure 6.3 show that Russett's classification is a sensible one; they also suggest some refinements, but it is not yet known how stable they are. Figures 6.1 and 6.2 show the orthogonality of DEMO and INST and the extreme importance of GNPR, showing that stable democracy and equality tend to be luxury goods and indicating that possibly the fact that political scientists from stable Western democracies have constructed the indices has had some influence (India, Philippines, Uruguay are stable democracies; Brazil, Colombia, Argentina, Costa Rica, Chile are unstable democracies; Yugoslavia and Poland are dictatorships; there is greater than median equality in Taiwan and South Vietnam). It is possible that a similar data matrix constructed in the Soviet Union would look quite differently, and that this fact may explain some of the instability in the solution.

6.5 EXAMPLE 2: PREDICTION OF A SCHOOL ACHIEVEMENT TEST

The second example does not illustrate the CANALS program, but it illustrates generalizations of two-set canonical analysis and relates them to some other multivariate analysis techniques. The data are taken from CBS (1980) and consist of a three-way contingency table in which a sample of approximately 120000 children is classified according to sex (two levels), father's profession (seven levels), and school achievement test scores (five levels). The data are given in Table 6.3. In this example we want to predict test scores from the other two variables. Thus the first set consists of the two variables S (sex) and P (profession), the second set consists of the single variable T (test).

Table 6.3 Test scores as a function of sex and father's profession

		T1	T2	T3	T4	T5
S1	P1	4669	6653	5543	3380	787
S1	P2	5757	12052	9526	6003	949
S1	P3	974	2188	2714	1865	276
S1	P4	468	1395	1774	1492	334
S1	P5	883	2667	3807	2971	809
S1	P6	916	3238	5963	5458	1798
S1	P7	288	1607	2906	3594	1314
S2	P1	4240	6610	5541	3368	672
S2	P2	4641	9797	9403	5448	998
S2	P3	743	2263	2459	1681	301
S2	P4	379	1185	1670	1228	301
S2	P5	766	2669	3746	2897	650
S2	P6	858	3195	5322	5028	1736
S2	P7	214	1238	3141	3571	1286

An often-applied technique in situations like this is loglinear analysis. We have used the ordinary iterative proportional fitting technique to fit our different models, coded by the marginals that are fitted exactly. The results are in Table 6.4; we give the likelihood ratio goodness-of-fit and the corresponding number of degrees of freedom (d.f.). The interpretation of the models is facilitated by writing

$$\ln p_{ijk} = \alpha_{ik} + \beta_{jk} + \gamma_{ijk}, \qquad (6.4)$$

with p_{ijk} the proportion of individuals with father i, sex j, and test score k.

Table 6.4 Loglinear models that have been fitted to the CBS data

Interpretation	Fitted marginals	Chi-squared	d.f.
SP = 0	(12) (13) (23)	202.29	24
P = SP = 0	(12) (13)	20198.82	48
S = SP = 0	(12) (23)	295.32	28
P = S = SP = 0	(12) (3)	20294.08	52

The α_{ik} are the parameters for profession, the β_{jk} for sex, and the γ_{ijk} are the interactions. Thus SP = 0 (no interaction) means $\gamma_{ijk} = 0$ for all i, j, k, and SP = P = 0 means $\gamma_{ijk} = 0$ and $\alpha_{ik} = 0$ for all i, j, k. In Table 6.5 we show that we can find estimates for the effects of sex and profession and for the interaction by subtraction in various ways. The conclusion is clear: P is much more important than S; interaction SP is not very important. Because the sample is extremely large everything is significant from a statistical point of view no simplification of the saturated model is possible. Inferential statistics is not very useful here, because there simply *is* no law generating the data, and any simplifying model simply is *not* true. This situation is very common in the social sciences. The usual procedure in situations like this is to apply large sample statistical techniques on relatively small samples or to make very strong assumptions and then to test only *within* these untested and often untestable assumptions.

Table 6.5 Main effects and interactions by subtraction

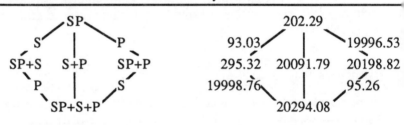

From a data analysis point of view estimates of the parameters are more interesting than global significance tests. We first study additive partitioning of chi-square, using the orthogonal function methods of Lancaster (1969). We start with the saturated model

$$p_{ijk} = q_{ij}\, r_k \left(\sum_{s=0}^{6} \sum_{t=0}^{1} \sum_{u=0}^{4} \rho_{stu} x_{is} y_{jt} z_{ku} \right), \tag{6.5}$$

where q_{ij} and r_k are the marginal proportions and where

$$\sum_{k=1}^{5} r_k z_{ku} z_{ku'} = \delta^{uu'}, \tag{6.6}$$

$$\sum_{i=1}^{7} \sum_{j=1}^{2} q_{ij} x_{is} y_{jt} x_{is'} y_{jt'} = \delta^{ss'} \delta^{tt'}. \tag{6.7}$$

Moreover $z_{k0} = x_{i0} = y_{j0} = 1$ for all i, j, k. Otherwise the x_{is}, y_{jt}, and z_{ku} are arbitrary. The saturated model can be fitted by

$$\rho_{stu} = \sum_{i=1}^{7} \sum_{j=1}^{2} \sum_{k=1}^{5} p_{ijk} x_{is} y_{jt} z_{ku}. \tag{6.8}$$

Although the choice of orthogonal functions is irrelevant from the point of view of the saturated model, it can be quite important from the data analysis point of view. In this case we have chosen orthogonal polynomials for both x_{is}, y_{jt}, and z_{ku}. Because always $\rho_{000} = 1$, $\rho_{00u} = 0$ for $u = 1,...,4$, and $\rho_{st0} = 0$ for $(s,t) \neq (0,0)$ there are 52 nontrivial ρ_{stu}. In Table 6.6 we give all $n^{1/2}\rho_{stu}$. The first row is ρ_{01u}, with $u \geq 1$; it measures the effect of S. The next six rows are ρ_{s0u}, with $s \geq 1$ and $u \geq 1$; they measure the effect of P. The final six rows are ρ_{s1u}, with $s \geq 1$ and $u \geq 1$, which measure the effect of the interaction SP. The margins in Table 6.6 contain the sum of squares of the rows and columns.

Table 6.6 Polynomial contrasts

	T				SSQ rows
S	−7.05	5.63	−0.77	−3.62	95.20
P	135.14	−4.72	0.30	6.51	18327.39
	8.90	30.49	−3.58	1.50	1023.91
	1.75	−5.69	4.46	−7.26	108.01
	8.23	−1.13	−5.80	0.17	102.60
	−9.00	−0.99	2.78	−2.31	95.04
	−4.73	−3.81	−1.33	0.39	38.87
SP	2.60	−2.67	0.63	0.81	14.98
	−0.89	0.19	2.41	−0.33	6.73
	−5.98	0.12	2.02	−6.72	85.09
	1.54	2.86	−0.64	0.98	11.96
	−1.47	2.72	5.35	−3.53	50.77
	0.74	1.25	−1.12	3.94	18.92
SSQ columns	18614.44	1057.06	117.06	190.90	

If in the population $\rho_{stu} = 0$ for all s, t, u (except for $(s,t,u) = (0,0,0)$), then the elements of Table 6.6 are independent standard normal asymptotically, and consequently the sums of squares of the rows are independent chi-squared variables with four degrees of freedom, while the sums of squares of the columns are independent chi-squares with 13 degrees of freedom. Again P is much more important than S, and again the interaction is small. We can also see from this analysis, however, that the linear component accounts for most of the variance (both for P and for T, approximately 93 per cent). The sampling theory of the statistics when the ρ_{stu} are nonzero, has been outlined recently by O'Neill (1980). The two more or less classical approaches of discrete MVA (the first one based on the 'multiplicative' and the second one on the 'additive' definition of interaction) will now be compared with canonical analysis.

Table 6.7 Four canonical analyses: components, sum, and transformations

	(a) On S + P + SP	(b) On S + P	(c) On P	(d) On S
Squared canonical	18 675.73	18 621.37	18 578.18	95.20
correlations	1 067.89	1 045.70	1 013.37	–
times n	187.51	92.76	86.57	–
	48.33	31.18	24.36	–
Sum of rows 1–4	19 979.47	19 791.02	19 702.48	95.20
Optimal scaling	1.06	1.09	1.04	–0.96
of P	0.86	0.80	0.75	–0.96
for boys	0.11	0.13	0.09	–0.96
	–0.43	–0.40	–0.45	–0.96
	–0.57	–0.50	–0.56	–0.96
	–1.22	–1.16	–1.21	–0.96
	–1.76	–1.81	–1.86	–0.96
Optimal scaling	1.01	0.99	1.04	1.04
of P	0.63	0.70	0.75	1.04
for girls	0.06	0.04	0.09	1.04
	–0.46	–0.50	–0.45	1.04
	–0.53	–0.60	–0.56	1.04
	–1.20	–1.26	–1.21	1.04
	–1.95	–1.91	–1.86	1.04

In the first form of canonical analysis we construct from S and P an interactive variable with 14 categories, and we apply multiple nominal canonical analysis to the resulting two sets of one single variable each. In this case canonical correlation analysis, homogeneity analysis, and correspondence analysis amount to the same thing. We have applied correspondence analysis to the 14 × 5 table in Table 6.3. This form of canonical analysis correspond with fitting a general nonlinear function on S × P and a general nonlinear function on T with maximal intercorrelation in the second type of canonical analysis. Canonical correlation analysis with multiple nominal variables tries to find nonlinear functions on S, P, and T *separately* in such a way that

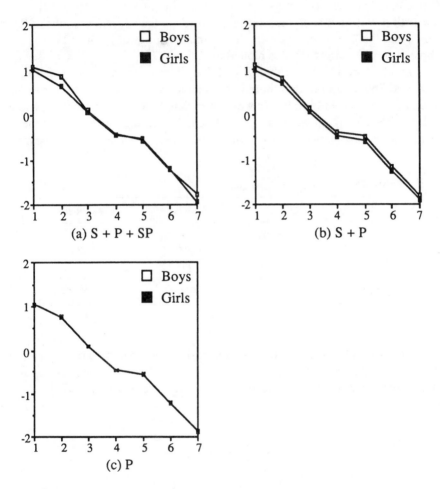

Figure 6.7 Quantifications.

$\phi(S) + \psi(P)$ correlates maximally with $\xi(T)$. Again we can compute the solution in the multiple nominal case by solving a generalized eigenvalue problem. Compute the $(7 + 2 + 5) \times (7 + 2 + 5)$ table **C** of bivariate marginals and the matrix **D**, which consists of the $(7 + 2) \times (7 + 2)$ and 5×5 principal submatrices of **C**, and solve the generalized eigenproblem corresponding with **C**, **D**. The third form of canonical analysis tries to predict T from P alone, the fourth form from S alone. These last two can be computed by correspondence analysis of the 7×5 and 2×5 tables computed from Table 6.3 by summing, respectively, over S and P. The results of the four canonical analyses are in Table 6.7. The first four rows are n times the squared canonical correlations, the fifth row is their sum, the next seven rows are the optimal scalings of P for boys, and the last seven rows are the optimal scalings of P for girls. The first three optimal scalings are also plotted in Figure 6.7. In Figure 6.7a we see that P is much more important than S and that there is not much interaction. In Figure 6.7b we require that there is no interaction, and we do not lose much predictive power. In Figure 6.7c we require that there is no interaction, and no effect of S, and again we do not lose much. The results of the three types of analyses we have performed so far are summarized in Table 6.8. It is clear that

Table 6.8 Summary of three analyses

Source	Loglinear	Additive	Canonical	d.f.
S	92.97	95.20	95.20	4
P	19 996.53	19 695.82	19 702.48	24
SP	204.58	188.45	181.79	24
Total	20 294.08	19 979.47	19 979.47	52

the conclusions that can be drawn on the basis of the analyses are very similar. Statistical inference procedures for the canonical correlations and optimal scores have been derived by Lebart (1976) and O'Neill (1978a, 1978b).

6.6 EPILOGUE

Of the asymmetric versions of the two-set problems redundancy analysis has enjoyed a lot of attention during the last decade (see the discussion of Van der Burg and De Leeuw in Van der Burg, 1988, and further references in Section 7.7).

The transfer of normalization procedure outlined in Section 6.2, as well as other theoretical and computational aspects of CANALS, are fully discussed in

Van der Burg and De Leeuw (1983), Van der Burg (1985), and Tenenhaus (1988). Van Buuren and Heiser (1989) used transfer of normalization in the context of K-means clustering with optimal scaling of variables, a situation resembling but not identical to the case where $G^1 = G_1$ is a single, *unknown* indicator matrix. Another successor of the CORALS program is TRANSREG (Kuhfeld, Young and Kent, 1987), which includes options similar to CANALS.

Multiple quantifications are not an option in CANALS, but the user can simulate the idea by introducing the same variable repeatedly in the same set. Since a quantified variable contributes to the analysis only in as much as it provides information that is independent from the other quantified variables in the same set, the various quantifications will tend to span an r-dimensional space (if the variable is repeated r times). This notion is called using *copies* (De Leeuw, 1984). The stability of the results for the Russett data was studied by De Leeuw and Van der Burg (1986) using random permutation methods.

CHAPTER 7
ASYMMETRIC TREATMENT OF SETS:
SOME SPECIAL CASES,
SOME FUTURE PROGRAMS

This chapter is somewhat different from earlier chapters. As has been mentioned in Section 6.1, there are some asymmetric cases of canonical analysis that are very important in practice. In this chapter we briefly discuss multiple regression, discriminant analysis, multivariate analysis of variance, and path analysis. For each of these special cases we have planned, but not yet written, a special computer program. Clearly we could use CANALS, but the special cases make some important simplifications possible. It is also not our intention in this chapter to review the history of each of these techniques in considerable detail, we merely discuss algorithms and planned programs. In the epilogue some more recent developments will be mentioned.

7.1 MULTIPLE REGRESSION AND MORALS

Suppose there are $K = 2$ sets, and suppose the second set contains only one variable, which is treated as a single numerical, ordinal, or nominal type. In this case we can suppose without loss of generality that $p = 1$. CANALS can be interpreted as an optimal scaling technique that maximizes the multiple correlation, and the loss function can be written quite simply as

$$\sigma_M(y^1, y_2) = SSQ(G^1 y^1 - G_2 y_2), \tag{7.1}$$

where G_2 is a complete indicator matrix and G^1 is a partitioned indicator matrix. As far as missing data are concerned, we suppose that there are no missing data *in the second set*. There are several ways in which missing data can occur in the first set and there are several of our options we can apply, but because we are not particularly interested in computational details of non-

241

existing programs we ignore missing data in this chapter and refer to Sections 5.4 and 5.5 for necessary details.

Thus, $\sigma_M(y^1, y_2)$ must be minimized over the feasible y^1 and y_2. For normalization purposes we simply require $u'D_2y_2 = 0$ and $y_2'D_2y_2 = 1$, with $D_2 = G_2'G_2$. A major difference with CANALS is that we do not have to bother about switching normalizations. The program MORALS by Young, De Leeuw, and Takane (1976) already solves a similar problem (with continuous options and without multiple nominal variables in the first set). The algorithm for adjusting y_1 is the same as in CANALS, with the obvious simplifications that result from $p = 1$.

As we have stated above, we do not intend to review the history of multiple regression in any detail. The linear theory is classical. It was started by Galton and developed further by Pearson and Yule. There have been some attempts, none of them successful, to develop a theory of partial and multiple rank correlation. Box and Cox (1964) studied parametric families of transformations to improve the fit to a linear model, Kruskal (1965) was the first to use monotonic regression in this context (the first set in this case is an ANOVA design matrix). Roskam (1968) also fitted an additive model with Kruskal-type techniques, and related it to additive conjoint measurement. An alternating least squares program ADDALS was presented for the ANOVA application by De Leeuw, Young, and Takane (1976); this was generalized to the univariate linear model by Young, De Leeuw, and Takane (1976) with the program MORALS; ADDALS was generalized to an individual differences additive model in the program WADDALS (Takane, Young, De Leeuw, 1980). An alternating least squares algorithm for additivity analysis, discriminant analysis, and multiple and polynomial regression was introduced earlier by De Leeuw (1969). French data analysts have also contributed a great deal. We mention Drouet d'Aubigny (1975), Tenenhaus (1977), and a very interesting paper by Daudin (1980).

There is, of course, much more that could be said about these generalizations of the metric linear model, but most of these additional results are in the field of gauging and stability analysis. It is also quite true that most of the questions dealing with stability and gauging have not been answered yet. In particular we can expect that the whole problem of ridge analysis, of smoothing, of cross validation, and of James–Stein estimation, have analogues in this generalized multiple regression context. ➤➤

7.2 DISCRIMINANT ANALYSIS AND CRIMINALS

In discriminant analysis we also have a single variable in the second set, but now this variable is multiple nominal. The loss function is

$$\sigma_M(Y^1, Y_2) = SSQ(G^1 Y^1 - G_2 Y_2). \tag{7.2}$$

Using $Z_1 = G^1 Y^1$ and $Z_2 = G_2 Y_2$ the normalization is $u' Z_2 = 0$ and $Z_2' Z_2 = I$. This is convenient for the algorithm, because we do not have to use transfer of normalization, but it is not the most natural way to normalize the solution. Thus if we have computed the solution, by the alternating least squares program CRIMINALS, we renormalize it in some convenient way, for example by requiring that $Z_1' Z_1 = I$ and then setting $Y_2 = D_2^{-1} G_2' G^1 Y^1$. The canonical variables for the first set Z_1 are orthonormal, while the Y_2 are the averages of the groups of individuals indicated by the second set. As far as history is concerned the same remarks apply as in Section 7.1. Discriminant analysis was invented by Fisher around 1940, it was introduced to psychologists by Rao and Slater (1949) and by Lubin (1950), and it is quite popular these days, for example in clinical psychology. A form of discriminant analysis with all variables multiple nominal was proposed by Saporta (1975); other generalizations of linear discriminant analysis are usually based on discrete MVA techniques (for example Gilbert, 1968). ➤

Interpretation of discriminant analysis is facilitated by defining the projector

$$P_2 = G_2 (G_2' G_2)^{-1} G_2', \tag{7.3}$$

and by observing that the minimum of $\sigma_M(Y^1, Y_2)$ over nonrestricted Y_2 is equal to

$$\sigma_M(Y^1, *) = \text{tr } Y^{1'} G^{1'} G^1 Y^1 - \text{tr } Y^{1'} G^{1'} P_2 G^1 Y^1. \tag{7.4}$$

The matrix $Y^{1'} G^{1'} G^1 Y^1$ is the total dispersion of the canonical variables and the matrix $Y^{1'} G^{1'} P_2 G^1 Y^1$ is the between-group dispersion. Thus we maximize the trace of the between-group dispersion matrix over all quantifications with the unit total dispersion matrix.

If all variables are single then $Y_j = y_j a_j'$ for all j in the first set. If T is the dispersion matrix of $q_j = G_j y_j$ and B is the between-group dispersion matrix, then

$$\sigma_M(Y, A, *) = \text{tr } A'TA - \text{tr } A'BA. \tag{7.5}$$

If we minimize this over $\mathbf{A}$ with the restriction $\mathbf{A'TA} = \mathbf{I}$, then clearly the result is

$$\sigma_M(\mathbf{Y},*,*) = p - \Sigma_s \lambda_s^2 (\mathbf{T}^{-1}\mathbf{B}), \tag{7.6}$$

which shows that nonlinear discriminant analysis amounts to choosing the $\mathbf{y}_j$ in such a way that the sum of the p largest eigenvalues of $\mathbf{T}^{-1}\mathbf{B}$ is maximized. A similar interpretation is possible if not all variables in the first set are single, but the notation becomes somewhat more complicated and we omit the details. It is clearly also possible in this case to choose other criteria defined in terms of the eigenvalues of $\mathbf{T}^{-1}\mathbf{B}$ which must be optimized, as in Section 5.1, but we have no experience with any other choices.

7.3 MULTIVARIATE ANALYSIS OF VARIANCE AND MANOVALS

Suppose the partitioned indicator matrix $\mathbf{G}^1$ in the first set is a design matrix of a complete orthogonal design. The requirement that the design is orthogonal is not as restrictive as it looks, because if we allow for missing data we can always make the design balanced in such a way that it becomes orthogonal. This is an old trick in the analysis of variance literature which dates back to the iterative alternating least squares method proposed by Yates (1933). The design matrix is treated as a set of orthogonal multiple nominal variables, each of the factors can be used to define a projector $\mathbf{P}_j$, and the orthogonal projectors can be used to write the loss function minimized over $\mathbf{Y}^1$ (the effects) in the form

$$\sigma_M(*,\mathbf{Y}^2) = \mathrm{tr}\ \mathbf{Y}^{2'}\mathbf{G}^{2'}\mathbf{G}^2\mathbf{Y}^2 - \Sigma_j\ \mathrm{tr}\ \mathbf{Y}^{2'}\mathbf{G}^{2'}\mathbf{P}_j\mathbf{G}^2\mathbf{Y}^2, \tag{7.7}$$

more or less as in the previous section, but with the role of $\mathbf{Y}^1$ and $\mathbf{Y}^2$ interchanged and with the added refinement due to orthogonality of the design components. The technique that is most natural from our point of view maximizes the second component of the loss function with requirements $\mathbf{Z}_2'\mathbf{Z}_2 = \mathbf{I}$, but it is a relatively small step to see that we can actually maximize any sum of the $\mathbf{Y}^{2'}\mathbf{G}^{2'}\mathbf{P}_j\mathbf{G}^2\mathbf{Y}^2$, while restricting any other sum to be unity. Thus we can maximize the sum of the main effects for fixed total, we can also maximize the sum of the first order interactions for fixed main effects, and so on. This possibility has already been indicated, in a metric context, by Abelson (1960). The first one to apply ANOVA to purely qualitative data, using optimal scaling, was Fisher (1925, example 46.2). Other references are in Section 7.1 and in De Leeuw, Young, and Takane (1976). ⇥

7.4 PATH ANALYSIS AND PATHALS

Path analysis may be unfamiliar to some of our readers, and we give a very short introduction. Suppose H_1 and H_2 are $n \times m_1$ and $n \times m_2$ data matrices. The columns of H_1 are n measurements on m_1 *exogenous* (independent) variables and the columns of H_2 are n measurements on m_2 *endogenous* (dependent) variables. The exogenous variables are assumed to influence all endogenous variables linearly; the endogenous variables are ordered in such a way that an endogenous variable is influenced linearly by all *previous* endogenous variables. These relations are written as

$$H_2 = H_1 A + H_2 B + E, \tag{7.8}$$

where A and B contain the *path coefficients*, B is upper triangular (all elements on and below the diagonal are equal to zero), and E, the residuals, are assumed to be *small*. We can require that E has other properties usually associated with residuals, for example

$$E'H_1 = 0, \tag{7.9}$$

$$E'E = \text{diag}(E'E). \tag{7.10}$$

It is of some interest that these two structural assumptions about the residuals make it possible to solve uniquely for A, B, and E. We first write the *reduced form*

$$H_2 = H_1 A (I - B)^{-1} + E (I - B)^{-1}. \tag{7.11}$$

Now define $V_{11} = H_1'H_1$, $V_{12} = H_1'H_2$, $V_{22} = H_2'H_2$, and write D_E for the diagonal matrix $E'E$. Then we find

$$V_{12} = V_{11} A (I - B)^{-1}, \tag{7.12}$$

$$V_{22} = (I - B')^{-1} A' V_{11} A (I - B)^{-1} + (I - B')^{-1} D_E (I - B)^{-1}, \tag{7.13}$$

or

$$V_{22} - V_{21} V_{11}^{-1} V_{12} = (I - B')^{-1} D_E (I - B)^{-1}. \tag{7.14}$$

Since **B** is upper triangular, this equation can be solved by Cholesky decomposition of $V_{22} - V_{21}V_{11}^{-1}V_{12}$, which is identified by setting $\text{diag}(I - B) = I$. Thus we have D_E and **B**, and we can solve **A** from (7.12) as

$$A = V_{11}^{-1}V_{12}(I - B) \tag{7.15}$$

and **E** from (7.11) as

$$E = H_2(I - B) - H_1 A. \tag{7.16}$$

We have merely shown in this section so far that, given any partitioning of the variables into endogenous and exogenous ones and given any ordering of the endogenous variables, we can perform a *complete* and *recursive* path analysis, which is merely a transformation of the data (depending on both the partitioning and the order). A path model becomes restrictive if we suppose that some of the above diagonal elements of **B** are zero or that some of the elements of **A** are zero. In this case we need a loss function again.

The loss function is simply the sum of squares of the residuals as expressed in (7.16):

$$\sigma_M(A,B) = \text{SSQ}(H_1 A - H_2(I - B)), \tag{7.17}$$

which looks like two-set canonical analysis. However, something remarkably simple is true here, due to the special structure of **B**. The term $H_2(I - B)$ has column j equal to

$$h_{(2)j} - \sum_l h_{(2)l} b_{lj} \quad (l = 1,...,j - 1), \tag{7.18}$$

and because $h_{(2)j}$ does not occur homogeneously (it has no coefficient) we do not have to normalize it in any way. Moreover, we can solve for each column separately, by using ordinary multiple regression. This remains true when some of the elements of b_j or a_j are restricted to zero.

It is trivial to generalize path analysis to variables of the single numerical, ordinal, or nominal type. We write, changing the notation into the familiar form,

$$\sigma_M(A_1,A_2,Y) = \text{SSQ}(Q_1 A_1 - Q_2 A_2), \tag{7.19}$$

with $q_j = G_j y_j$ as usual, and with A_1 and A_2 restricted to have zeros on specified places; moreover, A_2 must have ones on the diagonal. If desired, more general restrictions are possible; in fact we can require in the theory that A_1 is in a convex set Γ_1, A_2 is in a convex set Γ_2, and either $0 \notin \Gamma_1$ or else

$0 \notin \Gamma_2$. Fitting Q_1 and Q_2 for fixed A_1 and A_2 is already familiar. The elements of A_1 and A_2 can be fitted column by column, but in most cases it is quite easy to find all elements of A_1 for fixed A_2 (and Q_1 and Q_2) and all elements of A_2 for fixed A_1 (and Q_1 and Q_2). As we have seen in the linear case, it is even possible if $\text{diag}(A_2) = I$ to find all elements of A_1 and all elements of A_2 for fixed Q_1 and Q_2. ↦

It is more difficult, but also quite interesting, to define recursive path models for multiple variables. The loss function is

$$\sigma_M(Y) = SSQ(\sum_{j \in J_1} G_j Y_j - \sum_{j \in J_2} G_j Y_j). \tag{7.20}$$

Instead of requiring that certain elements of A_1 and A_2 are zero, we recall that $Y_j = y_j a_j'$ in the single case. Thus it follows that in the multiple case certain *columns* of the Y_j are zero. A little thought leads to the conclusion that this amounts to fitting path models in the usual *single equation* way, as in the linear case, but now each variable obtains a different quantification in each equation that is fitted. If we predict variable *j*, then all exogenous variables get a transformation that is different from their previous transformations. All preceding endogenous variables also get a transformation that is different from their earlier ones.

7.5 PARTIAL CANONICAL CORRELATION ANALYSIS AND PARTALS

Partial canonical correlation analysis was introduced by Roy (1957, p. 143); it was discussed more extensively by Rao (1969) and by Timm and Carlson (1976). The last authors also discussed canonical equivalents of part and bipartial correlation analysis. Of course all papers deal exclusively with the case in which all variables are numerical. ↦

We have seen that CANALS in *p* dimensions transforms or quantifies the variables in such a way that the sum of the *p* largest canonical correlations is maximized. This is one straightforward way in which we can proceed: simply define PARTALS as the technique that maximizes the sum of the first *p* partial, part, or bipartial canonical correlation coefficients. For completeness, and because the numerical techniques are not very familiar yet, we repeat some of the definitions given by Timm and Carlson (1976). In partial canonical correlation analysis there are three sets; the third set is partialled out of the first two. The equations are the same as for ordinary canonical correlation analysis, except for the fact that the dispersion matrix of the first two sets is replaced by the conditional dispersion matrix, with the third set partialled out. In part canonical correlation we only remove the third set from one of the two other sets. In bipartial canonical correlation analysis we have four sets; set three is removed from set one and set four is removed from set two. 'Removing' and 'partial' are always understood in a linear sense.

Although we have called these generalizations straightforward, it is not at all clear how to incorporate these objectives in a suitable algorithm. However, note that for partial canonical correlation there is an easy possibility, using CANALS. We add the third set to both other sets, and perform a canonical correlation analysis over the two enlarged sets. Because of the adding there will be a number of canonical correlations equal to unity (equal to the number of variables in the third set, or even more than that if some variables in the third set are multiple nominal). The remaining compounds, however, will all be orthogonal to these first 'trivial' compounds, which means that they will be orthogonal to the third set, as required. This is essentially the approach suggested by Roy (1957, p. 142), and it is easy to generalize it to both single and multiple nonnumerical variables.

7.6 SOME EXAMPLES

We again use the CBS data used in Section 6.5 for our examples in this chapter. In order to avoid possible misunderstandings we emphasize that all computations in this section are done with *ad hoc* (APL) programs. We use three variables from the CBS data: father's profession (Prof), school achievement test score (Test), and type of secondary education chosen (Educ).

The first analysis is of the multiple regression type. We want to predict Educ from Prof and Test. The model is illustrated in Figure 7.1a. We first interpret this in a general nonlinear sense, i.e. $\phi(\text{Educ}) = \phi(\text{Prof,Test})$, and then in the nonlinear but additive sense, $\phi(\text{Educ}) = \phi(\text{Prof}) + \phi(\text{Test})$. Table 7.1 contains canonical correlations and quantifications of Educ. They are extremely similar in both analyses. They illustrate the large distance between VWO/HAVO and MAVO, and the fact that LEAO/LMO is much closer to MAVO than LTS/LHNO is to MAVO. Similar results were found by De Leeuw and Stoop (1979). The quantifications of the (Prof,Test) combinations are given in Table 7.2, and plotted in Figures 7.2 and 7.3.

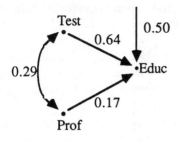

Figure 7.1a Regression path model.

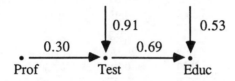

Figure 7.1b Another path model.

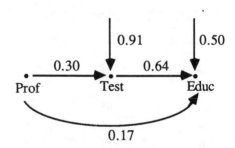

Figure 7.1c A third path model.

Table 7.1 Results of completely nonlinear and additive regression analysis on CBS data

	ϕ(Educ) = ϕ(Prof,Test)	ϕ(Educ) = ϕ(Prof) + ϕ(Test)	
Canonical loadings	0.71	0.70	
	0.37	0.35	
	0.07	0.07	
Category quantifications	1.51	1.52	VWO/HAVO
	0.24	0.23	MAVO
	-1.30	-1.29	LTS/LHNO
	-0.83	-0.83	LEAO/LMO

Table 7.2 Quantifications of (Prof,Test) combinations for completely nonlinear and additive regression analysis

	Test 1	Test 2	Test 3	Test 4	Test 5
Prof 1	−1.65	−0.97	0.03	0.93	1.69
Prof 2	−1.58	−1.01	−0.06	0.89	1.52
Prof 3	−1.52	−0.80	−0.09	0.83	1.79
Prof 4	−1.35	−0.64	0.29	1.17	1.71
Prof 5	−1.35	−0.54	0.34	1.11	1.70
Prof 6	−1.33	−0.37	0.43	1.34	1.69
Prof 7	−1.04	−0.06	0.72	1.41	1.79
Prof 1	−1.51	−0.96	−0.06	0.90	1.49
Prof 2	−1.54	−0.99	−0.09	0.87	1.46
Prof 3	−1.48	−0.93	−0.04	0.93	1.52
Prof 4	−1.27	−0.72	0.18	1.14	1.73
Prof 5	−1.23	−0.68	0.22	1.18	1.78
Prof 6	−1.08	−0.53	0.36	1.33	1.92
Prof 7	−0.89	−0.34	0.55	1.52	2.11

Table 7.3 Regression results

Model:	T→E	P→T	T+P→E
Educ	−1.52	–	−1.52
	−0.23	–	−0.23
	1.28	–	1.29
	0.88	–	0.83
Test	1.59	1.66	1.59
	0.82	0.79	0.82
	−0.18	−0.23	−0.18
	−1.15	−1.01	−1.15
	−1.75	−1.96	−1.73
Prof	–	1.04	0.69
	–	0.75	0.88
	–	0.09	0.63
	–	−0.45	−0.50
	–	−0.56	−0.62
	–	−1.21	−1.18
	–	−1.86	−1.95
Path coefficients:			
Test	0.69	–	0.64
Prof	–	0.30	0.17
Loss	0.53	0.91	0.50

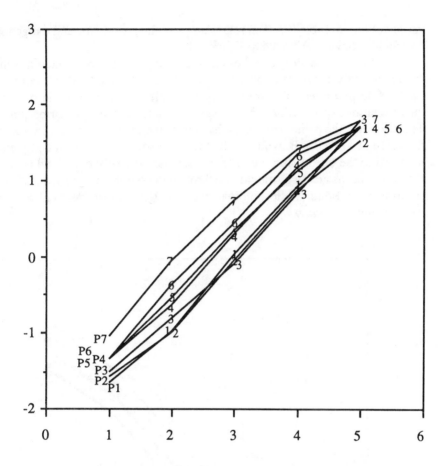

Figure 7.2 (Prof,Test) combinations after nonlinear regression analysis.

It is clear from Figures 7.2 and 7.3 that children from the lower Prof levels choose their secondary education too low and that this effect is strongest for the lower levels of Test and weakest for the higher levels of Test. Thus not too much 'talent' is 'lost' in the high Test region, but a great deal is lost in the low Test region. We must warn the reader that our conclusions are framed in a very suggestive way, and refer only to transformations and interactions found by our technique. Whether the scales constructed are stable or socially relevant remains to be seen. There are two other path models in Figure 7.1b and c. We fit them by using the multiple nominal approach, which means that we simply have to combine results of various regression analyses. The results (transformations standardized to unit variance, regression weights, and residual variance) are in Table 7.3. The regression weights, residuals, and exogenous correlations have been added to Figure 7.1.

We can also use Prof and Test as predictors (single nominal) in a discriminant analysis problem, where the four levels of Educ (multiple nominal) have to be predicted. The results are in Table 7.4. We give the multiple quantifications of the four Educ levels, the standardized transformations of Test and Prof, the path coefficients that determine the two canonical axes, and the value of $1 - \text{tr } T^{-1}B$, which is the loss. The solution is almost completely identical to the third column of Table 7.3, which is easily explained because the group means are on a straight line through the origin, and consequently the solution with Educ single nominal and $p = 1$ is almost identical to Educ multiple nominal and $p = 2$.

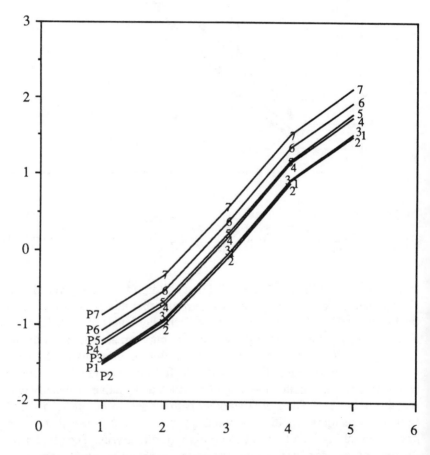

Figure 7.3 (Prof,Test) combinations after additive regression analysis.

Table 7.4 Discriminant analysis results	
T+P → E	
−0.37	−0.66
−0.05	−0.10
0.31	0.56
0.20	0.36
1.58	
0.83	
−0.18	
−1.16	
−1.73	
0.69	
0.89	
0.66	
−0.59	
−0.66	
−1.17	
−1.91	
0.11	0.13
0.30	0.56
0.50	

Table 7.5 Partial and ordinary nonlinear regression	
P|T→E|T	P→E
−2.04	−1.50
−0.31	−0.23
1.83	1.32
0.52	0.50
–	–
–	–
–	–
–	–
–	–
1.01	0.70
0.97	0.52
1.39	0.43
−0.26	−0.77
−0.27	−0.65
−0.77	−1.12
−2.07	−2.58
0.95	0.87

In Table 7.5 the results of a partial canonical correlation are given. We predict Educ|Test from Prof|Test (first column) and compare this with predicting Educ from Prof (second column). The results are quite different. Educ cannot be predicted very well from Prof (compare also the analysis in Chapter 6, where Test is predicted from Prof) and if we partial out Test, prediction becomes worse. Observe from Tables 7.3 to 7.5 that Prof accounts for 'more' variance of Educ than of Test. The difference is 0.04, which is approximately equal to the partial canonical correlation.

7.7 EPILOGUE

Multiple regression is, of course, borrowing a word from Tukey's, the workhorse of statistics, on which many innovations have been tested. The Box–Cox approach was extended to spline transformations by Winsberg and

Ramsay (1980) (also see De Leeuw, 1986, and Ramsay, 1988). The approach in Section 7.1 maximizes the multiple correlation over the class of transformations instead of maximizing the likelihood. It may be compared with ACE (Breiman and Friedman, 1985), which does not exactly optimize over a class of transformations, but is based on considerations of smoothing. Van der Lans and Heiser (1988) have considered transformations for which the category transformations are constrained to be the same across variables within subsets of the predictor set. An interesting discussion of the case in which all variables are nominal can be found in Israëls (1987a, ch. 6).

The type of discriminant analysis as described in Section 7.2 is often alternatively called *canonical variate analysis* (cf. Gittins, 1985). Robust estimation of the canonical variates is developed by Campbell (1982). Interesting special forms of discriminant analysis, which can also be viewed as special forms of correspondence analysis, have been developed by Ter Braak (1986, 1987) and Takane (1987).

It will frequently be useful to view multivariate analysis of variance as a special case of *reduced rank regression* (Izenman, 1980; Davies and Tso, 1982), which is equivalent to redundancy analysis. As remarked by Ten Berge (1985), redundancy analysis is in turn equivalent to Fortier's (1966a) procedure, called simultaneous linear prediction. The latter author builds upon work by Horst and discusses the accommodation of nonlinearity as well (Fortier, 1966b). Redundancy analysis of qualitative variables is fully covered in Israëls (1984), who also connects it with homogeneity analysis with linear restrictions.

Path analysis with optimal scaling has been further developed by Coolen and De Leeuw (1987), Israëls (1987b), and De Leeuw (1987). Partial canonical correlation analysis has been studied in the context of correspondence analysis by Israëls (1987a, ch. 5), and Ter Braak (1987, ch. 7); related work is by Yanai (1986).

CHAPTER 8
MULTIDIMENSIONAL SCALING AND
CORRESPONDENCE ANALYSIS

Multivariate analysis (MVA) and multidimensional scaling (MDS) have developed along rather separate historical traditions. In MVA the basic concepts are those of *variance, covariance,* and *correlation*; there is little emphasis on graphical representations of the objects and there is much concern about distributional assumptions for the variables. In MDS the interest is much more in (joint) plots of objects and variables, and a basic concept is that of Euclidean – or other – *distance*, often related to the concept of *similarity* or *dissimilarity* (between objects, between variables, or between an object and a variable). In this chapter we want to indicate some points of contact between the two traditions. We will show how homogeneity analysis can be viewed as a technique that minimizes the distances among homogeneous groups of objects. Next we discuss correspondence analysis and its relationship with MDS and homogeneity analysis. Finally, we give some gauging results by studying the representation of certain 'scales', i.e. particular patterns in binary data. The last subject is elaborated in Chapter 9.

8.1 HOMOGENEITY AND SEPARATION

Let us consider a binary data matrix (not yet coded as an indicator matrix). It might have the following interpretation: rows refer to individuals, columns refer to items, and the cell of the matrix is 1 if the individual *chooses* the item, 0 otherwise. What 'to choose' means depends on the context; it might mean agreement with a statement, preference for a kind of food, a vote in favour of some proposal, etc. In the sixties a variety of suggestions has been made about how to analyse such a matrix (Coombs, 1964; Lingoes, 1968; De Leeuw, 1969) with nonmetric (i.e. ordinal) scaling techniques. Typical of such proposals is that it was stressed that assumptions about the structure of the data should be weak. Typical also was that rows and columns of the matrix were treated asymmetrically: the matrix was defined either as 'row conditional', or as 'column conditional'. Row conditionality implies the general idea that in a spatial representation the columns are displayed as *points*, in such a way that for each row there is some *separating curve* or *surface* distinguishing the ones from the zeros. Under column conditionality the role of points and separating curves is reversed. Figure 8.1 gives a very simple example, in which the columns are represented as five points, and for each of the three rows there is a straight line with the ones for that row on one side and the zeros on the other.

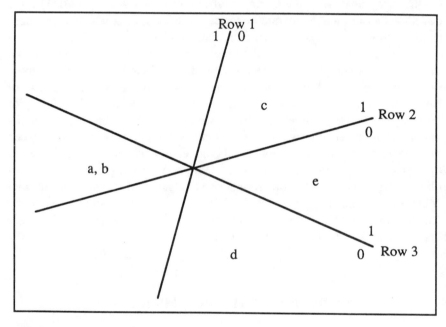

Figure 8.1 Row-conditional scaling solution.

A different representation of the same data is given in Figure 8.2, illustrating the so-called *unfolding model* (for an early account of this model, see Coombs, 1964). Now items and individuals are both represented as points in a joint space, in such a way that an individual is closer to the item(s) that is chosen and further away from the item that is not chosen. The figure illustrates this with circles around the individual points: chosen items are always within the circle and not chosen items are outside. It is obvious that the representations offered in Figures 8.1 and 8.2 are far from unique. One reason for this is that the example is a very small one. Another, more important, reason is that, even with many individuals and items, the data matrix does not impose many restrictions (Kruskal and Carroll, 1969; Heiser and De Leeuw, 1979). A consequence is that it is difficult to construct algorithms that avoid *degenerate solutions*.➥ An example of such a degenerate solution is shown in Figure 8.3 for the same data matrix as was used for Figures 8.1 and 8.2. All individuals are represented as a single point, coinciding with the items 'a', 'b', 'c', and 'e', and contrasting them with item 'd', chosen by none of the individuals. So the distances in the representation are weakly monotonic with the data, but most of the original information is lost.

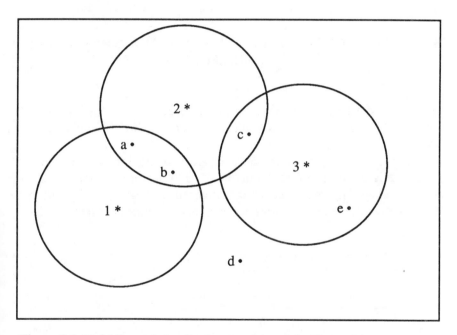

Figure 8.2 Unfolding solution for data matrix used for Figure 8.1.

Now suppose we drop the explicit goal of separation and replace it by the principle that homogeneous groups of objects should be close together. There are different ways to define homogeneous groups, and there are different ways to obtain representations with some distances smaller than others. In any case, homogeneity as used in this chapter refers to *groups of objects*, not to *groups of variables*. As we shall see, especially in Section 8.7, regular arrangements of groups of objects frequently imply the possibility of good separation, so that this goal is not completely lost.

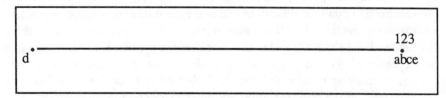

Figure 8.3 Degenerate solution for data matrix used for Figures 8.1 and 8.2.

8.2 MINIMUM DISTANCE ANALYSIS OF HOMOGENEOUS GROUPS OF OBJECTS

In this section the focus is on a binary data matrix F. Following the MDS tradition we refer to its n_r rows as the *row objects* and to its n_c columns as the *column objects*. The notation F is used rather than H, since we are not going to code the data into indicator matrices. Also, we do not use the symbol G, because F is not necessarily a (partitioned) indicator matrix. In the next section we shall allow F to contain any nonnegative values whatsoever, in which case it is called a *correspondence table* in the French literature (Benzécri, 1973).

Table 8.1 Miniature example 'pick 2' design

		Items			
		a	b	c	d
	1	1	0	1	0
Indivi-	2	1	1	0	0
duals	3	0	1	1	0
	4	0	0	1	1
	5	0	1	0	1

The goal of the analysis is to find a configuration of points in p-dimensional space, in which homogeneous groups of objects are together as closely as possible (within each group). This can be done by interpreting each column as a partitioning of row objects (column-conditional approach), by interpreting each row as a partitioning of column objects (row-conditional approach), or by interpreting the whole table as one joint partitioning of row and column objects (joint homogeneity approach). All partitionings are into one class of homogeneous objects and another class of heterogeneous objects. It will be shown that, for the specific class of loss functions used, these three approaches essentially amount to the same thing (up to a rescaling of axes). ⇥

For a miniature example we shall use the matrix given in Table 8.1, for five row objects and four column objects, corresponding to a Coombs 'pick two' design (the row individual has to tell which two out of four items are liked best). The 'pick two' design has the consequence that rows of the table add up to the same number, a property that is not essential for the present derivations.

8.2.1 Scaling the row objects associated with F

We first look upon F as column conditional, i.e. we want to scale row objects so that in their spatial representation row points are closer together to the extent that row objects are more *similar*. Let $\approx_j$ (read: 'is similar with respect to column j to') be an *equivalence relation* for pairs of row objects o_i, o_k in the sense that

$$o_i \approx_j o_k \qquad \text{if and only if} \quad f_{ij} = f_{kj} = 1. \tag{8.1}$$

The ideal representation would be a configuration in p-dimensional space where the coordinates of the point for o_i are the same as those for o_k when we have $o_i \approx_j o_k$. This objective, of course, will generally be impossible to achieve, since $o_i \approx_j o_k$ does not imply $o_i \approx_l o_k$ for $j \neq l$. We therefore have to compromise and must define a specific loss function to measure departure from the ideal representation. To this end we define for each of the n_c columns of F an $n_r \times n_r$ *dissimilarity* matrix with elements

$$\delta_{(j)ik} = \begin{cases} 0 \text{ if } o_i \approx_j o_k \\ 1 \text{ otherwise.} \end{cases} \tag{8.2}$$

In addition, we define for each column a matrix of *weights*, with elements

$$w_{(j)ik} = \begin{cases} 1/c_j & \text{if } o_i \approx_j o_k \\ 0 & \text{otherwise} \end{cases} \tag{8.3}$$

where c_j is the total of the jth column of **F**. Thus a dissimilarity of zero is associated with a weight that is inversely proportional to the size of the group of points: small groups get more weight than large groups. Points that are not to be considered as a homogeneous group get an (arbitrary) dissimilarity of one, and weight zero.

For the example, the four dissimilarity matrices are shown in Table 8.2a. The four matrices $\mathbf{W}_j$ are shown in Table 8.2b. Note that $\mathbf{W}_j$ gives the dissimilarities a weight reciprocal to the size of the jth equivalence class and that the rows and columns of each $\mathbf{W}_j$ add up to the jth column of Table 8.1.

Table 8.2a Dissimilarity matrices $\{\delta_{(j)ik}\}$ for $j = 1,...,4$

$j = 1$	$j = 2$	$j = 3$	$j = 4$
0 0 1 1 1	1 1 1 1 1	0 1 0 0 1	1 1 1 1 1
0 0 1 1 1	1 0 0 1 0	1 1 1 1 1	1 1 1 1 1
1 1 1 1 1	1 0 0 1 0	0 1 0 0 1	1 1 1 1 1
1 1 1 1 1	1 1 1 1 1	0 1 0 0 1	1 1 1 0 0
1 1 1 1 1	1 0 0 1 0	1 1 1 1 1	1 1 1 0 0

Table 8.2b Matrices of weights $\{w_{(j)ik}\}$ for $j = 1,...,4$

$j = 1$	$j = 2$	$j = 3$	$j = 4$
1/2 1/2 0 0 0	0 0 0 0 0	1/3 0 1/3 1/3 0	0 0 0 0 0
1/2 1/2 0 0 0	0 1/3 1/3 0 1/3	0 0 0 0 0	0 0 0 0 0
0 0 0 0 0	0 1/3 1/3 0 1/3	1/3 0 1/3 1/3 0	0 0 0 0 0
0 0 0 0 0	0 0 0 0 0	1/3 0 1/3 1/3 0	0 0 0 1/2 1/2
0 0 0 0 0	0 1/3 1/3 0 1/3	0 0 0 0 0	0 0 0 1/2 1/2

Table 8.2c Sum matrix **P**

5/6	3/6	2/6	2/6	0
3/6	5/6	2/6	0	2/6
2/6	2/6	4/6	2/6	2/6
2/6	0	2/6	5/6	3/6
0	2/6	2/6	3/6	5/6

The metric two-way MDS problem is to approximate all dissimilarity matrices with one matrix of distances among row points in a Euclidean space with a fixed number of dimensions (Torgerson, 1958; Shepard, 1962; Kruskal, 1964a, 1964b, 1977; Guttman, 1968; for a survey see Carroll and Arabie, 1980; a more theoretical discussion is De Leeuw and Heiser, 1982). The usual least squares loss function (called STRESS) for this approximation problem is

$$\sigma_1(X) = \sum_j \sum_{i<k} w_{(j)ik} [\delta_{(j)ik} - d_{ik}(X)]^2, \tag{8.4}$$

where $d_{ik}(X)$ denotes the Euclidean distance between the two rows i and k of the $n_r \times p$ matrix X of coordinates in the representation of the row objects. The summation is over the pairs $i < k$, since all matrices involved are symmetric. The metric MDS problem now is to find a solution for X that minimizes $\sigma_1(X)$.

This problem can be much simplified in the present case by using a special property of the dissimilarity and weight structure. For all i, j, k we have

$$w_{(j)ik}\delta_{(j)ik} = 0, \tag{8.5}$$

and therefore the two terms in which these cross products appear drop out:

$$\sigma_1(X) = \sum_j \sum_{i<k} w_{(j)ik}d_{ik}^2(X) = \sum_{i<k} d_{ik}^2(X)(\sum_j w_{(j)ik}). \tag{8.6}$$

So it turns out that we shall not really approximate the quantities $\delta_{(j)ik}$ with the corresponding distances $d_{ik}(X)$, as is the proper aim of STRESS minimization, but actually we will be *minimizing the weighted average (squared) distance* among the points – hence the title of this section.

The elements $\sum_j w_{(j)ik}$ can be collected in the sum matrix P of the n_c matrices W_j. For the miniature example P is given in Table 8.2c. It is not difficult to see that in terms of the original table $P = FD_c^{-1}F'$, with D_c the diagonal matrix of the column totals of F. We may write (using f_j to denote a column vector of F)

$$FD_c^{-1}F' = \sum_j c_j^{-1}f_jf_j', \tag{8.7}$$

and by definition $c_j^{-1}f_jf_j' = W_j$. Thus P is the sum of rank-one idempotent matrices (cf. Section 3.8.1, where an analogous sum matrix was mentioned), also called the *average similarity matrix.* ➤ It then follows that the simplified STRESS function becomes:

$$\sigma_1(X) = \sum_{i<k} p_{ik} \, d_{ik}^2(X) = \sum_{i<k} p_{ik} \left[\sum_a (x_{ia} - x_{ka})^2\right]$$

$$= \mathrm{tr}\, X'D_r X - \mathrm{tr}\, X'FD_c^{-1}F'X, \tag{8.8}$$

and we arrive at an optimization problem that is formally equivalent to HOMALS for the quantification of row objects when F is taken as a partitioned indicator matrix, not necessarily complete, and when we take the

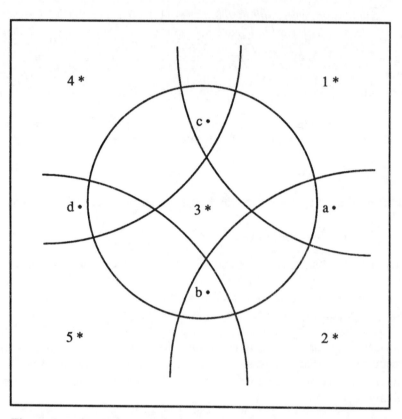

Figure 8.4 HOMALS unfolding solution for Table 8.1. Column points are the centre of gravity of row points.

normalization restriction $X'D_r X = I$ (note that in HOMALS terminology D_r is M_*, the total number of nonmissing values). This statement may be checked by comparing (8.8) with (3.33) and (3.55 to 3.59).

For the example, we find the two-dimensional solution

$$\mathbf{X} = \begin{pmatrix} +0.5 & +0.5 \\ +0.5 & -0.5 \\ 0.0 & 0.0 \\ -0.5 & +0.5 \\ -0.5 & -0.5 \end{pmatrix}$$

with loss $\sigma_1(\mathbf{X}) = 1/3$. (Loss becomes 0 if we allow a third dimension, with quantification $\mathbf{x}_3' = (1\ 1\ -4\ 1\ 1) / \sqrt{20}$.) The solution is displayed in Figure 8.4, where the item points are drawn as centres of gravity of individuals who have 'picked' them. To illustrate unfolding, circle segments are drawn around each individual point, showing that items that are 'picked' are inside such circles and items that are not 'picked' are outside.

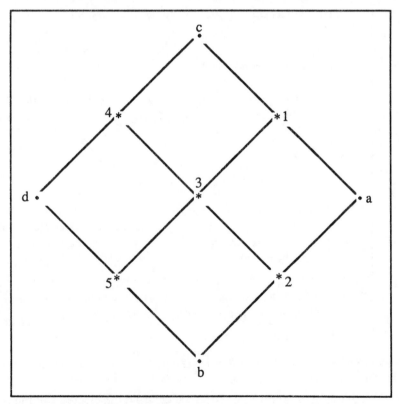

Figure 8.5 HOMALS unfolding solution for Table 8.1 with row points as centroids of column points.

8.2.2 Scaling the column objects associated with F

The dual approach is to look upon $\mathbf{F}$ as row conditional, i.e. we want to represent the column objects as points in p-dimensional space with coordinates $\mathbf{Y}$, on the basis of the way they are grouped into homogeneous groups by the rows. We shall not repeat the details of the derivations. It can be shown that the simplified STRESS function now becomes

$$\sigma_2(\mathbf{Y}) = \text{tr } \mathbf{Y}'\mathbf{D}_c\mathbf{Y} - \text{tr } \mathbf{Y}'\mathbf{F}'\mathbf{D}_r^{-1}\mathbf{F}\mathbf{Y}, \tag{8.9}$$

which is formally the same as the HOMALS category quantification problem for an (incomplete) partitioned indicator matrix. This equivalence may also be checked with (3.55) to (3.59).

For the example the column point solution is displayed in Figure 8.5, which is essentially the same as Figure 8.4 but now with individuals in the centre of gravity of their chosen items.

8.2.3 Joint homogeneity analysis of F

The third approach focuses on similarity relations within pairs of one row object and one column object. We simply define a dissimilarity matrix $\mathbf{\Delta}$ with

$$\delta_{ij} = \begin{cases} 0 \text{ if } f_{ij} = 1 \\ 1 \text{ if } f_{ij} = 0 \end{cases} \tag{8.10}$$

and a matrix of weights $\mathbf{W}$ with $w_{ij} = f_{ij}$. The STRESS function now truly becomes of the unfolding type, comparing the Euclidean distances of pairs of points from two different sets with δ_{ij}, taking into account only those deviations that correspond with $f_{ij} = 1$. In this case STRESS simplifies into

$$\sigma_3(\mathbf{X},\mathbf{Y}) = \text{tr } \mathbf{X}'\mathbf{D}_r\mathbf{X} + \text{tr } \mathbf{Y}'\mathbf{D}_c\mathbf{Y} - 2 \text{ tr } \mathbf{X}'\mathbf{F}\mathbf{Y}, \tag{8.11}$$

a function of both $\mathbf{X}$ and $\mathbf{Y}$. When we minimize (8.11) over $\mathbf{X}$ unconstrained we find that the optimal $\mathbf{X}^*$ must satisfy

$$\mathbf{D}_r\mathbf{X}^* = \mathbf{F}\mathbf{Y}, \tag{8.12}$$

i.e. the optimal row points must be in the centroid of selected column points, and substituting this solution into (8.11) we obtain (8.9). Analogously, when we minimize (8.11) over $\mathbf{Y}$ unconstrained we find that the optimal $\mathbf{Y}^*$ must satisfy $\mathbf{D_c Y}^* = \mathbf{F'X}$, i.e. the optimal column points must be in the centroid of selected row points, and substituting this solution into (8.11) we obtain (8.8). In the first solution $\mathbf{Y}$ has to be $\mathbf{D_c}$-orthonormal and $\mathbf{X}$ satisfies the centroid principle, while in the second solution $\mathbf{X}$ has to be $\mathbf{D_r}$-orthonormal and $\mathbf{Y}$ satisfies the centroid principle.

Summarizing, we may state that all three approaches lead to the same result, up to a rescaling depending upon the normalization. Typical for the joint formulation of homogeneity analysis is that it tries to obtain $d_{ij}(\mathbf{X},\mathbf{Y}) = 0$ if $f_{ij} = 1$. This is a stronger requirement than is usual in the previous, nonmetric, approaches mentioned in the introduction of this chapter, where in the row-conditional interpretation one merely tries to achieve the implication

if $f_{ij} = 0$ and $f_{il} = 1$,

then $d_{ij}(\mathbf{X},\mathbf{Y}) \geq d_{il}(\mathbf{X},\mathbf{Y})$, $\qquad$ (8.13)

or, in the column-conditional interpretation

if $f_{ij} = 0$ and $f_{kj} = 1$,

then $d_{ij}(\mathbf{X},\mathbf{Y}) \geq d_{kj}(\mathbf{X},\mathbf{Y})$. $\qquad$ (8.14)

The minimum distance solution, therefore, is obtained at the price of a strict, metric interpretation of the data, and at the price of stronger normalization conditions, with an arbitrariness in normalizing either row points or column points. Yet scaling techniques with weaker assumptions still require a choice as to the interpretation of $\mathbf{F}$ as row or column conditional; they also tend to produce degenerate solutions, precisely because of their weaker assumptions.

8.3 CORRESPONDENCE ANALYSIS

Now let $\mathbf{F}$ be a general $n_r \times n_c$ correspondence table, i.e. its entries may contain any nonnegative value whatsoever. In a contingency table, for instance, which is the most important special case, cell f_{ij} contains the frequency with which row category i co-occurs with column category j. Let $\mathbf{r} = \mathbf{Fu}$ be the vector of row totals and $\mathbf{c} = \mathbf{F'u}$ the vector of column totals, and let $N = \mathbf{u'Fu} = \mathbf{u'c} = \mathbf{u'r}$ be the grand total. Also, let $\mathbf{D_r}$ again be the diagonal matrix containing the elements of $\mathbf{r}$ and $\mathbf{D_c}$ the diagonal matrix containing the elements of $\mathbf{c}$.

In *L'Analyse des Correspondances,* as developed by Benzécri (1973), the aim is to find a representation $\mathbf{X}$ of the rows of $\mathbf{F}$ in such a way that the Euclidean distances among the row points in $\mathbf{X}$ approximate certain profile distances between the row profiles in $\mathbf{F}$. Dually, we want to find a representation $\mathbf{Y}$ of the columns of $\mathbf{F}$ in such a way that the Euclidean distances among the column points in $\mathbf{Y}$ approximate certain profile distances between column profiles in $\mathbf{F}$. These aims clearly generalize the objectives of the previous section.

More specifically, the squared distance δ_{ik}^2 between rows i and k of $\mathbf{F}$ is defined as

$$\delta_{ik}^2 = N \sum_j (f_{ij}/r_i - f_{kj}/r_k)^2 / c_j. \tag{8.15}$$

We shall call such distances χ^2 distances; this name will be explained later in this section. First note that δ_{ik}^2 is the same as the squared Euclidean distance between rows i and k of the matrix $\mathbf{B}$ with elements

$$b_{ij} = (f_{ij}/r_i)(N/c_j)^{1/2}. \tag{8.16}$$

This matrix can be written as

$$\mathbf{B} = N^{1/2} \mathbf{D_r^{-1} F D_c^{-1/2}}. \tag{8.17}$$

A simple but particularly important lemma is: a configuration $\mathbf{X}$ is a Euclidean representation of $\mathbf{B}$ if $\mathbf{BB' = XX'}$.

Proof:

$$\begin{aligned}\delta_{ik}^2 &= \mathbf{b}_i'\mathbf{b}_i + \mathbf{b}_k'\mathbf{b}_k - 2\mathbf{b}_i'\mathbf{b}_k \\ &= \mathbf{x}_i'\mathbf{x}_i + \mathbf{x}_k'\mathbf{x}_k - 2\mathbf{x}_i'\mathbf{x}_k = d_{ik}^2(\mathbf{X}).\end{aligned} \qquad \text{Q.E.D.}$$

A solution for $\mathbf{X}$ can be found as follows. Consider the SVD

$$\mathbf{D_r^{-1/2} F D_c^{-1/2}} = \mathbf{K \Lambda L'}, \tag{8.18}$$

as usual with $\mathbf{K'K = I}$, $\mathbf{L'L = I}$, and Λ the diagonal matrix of singular values.

Now define

$$X = N^{1/2} D_r^{-1/2} K \Lambda, \tag{8.19}$$

from which it follows that

$$XX' = N \ D_r^{-1/2} K \Lambda^2 K' D_r^{-1/2}. \tag{8.20}$$

Also, the SVD (8.18) yields $D_r^{-1/2} F D_c^{-1} F' D_r^{-1/2} = K \Lambda^2 K'$, so that

$$XX' = N \ D_r^{-1} F D_c^{-1} F' D_r^{-1} = BB', \tag{8.21}$$

and from the lemma we find that the Euclidean distances among the rows of X are equal to the Euclidean distances among the rows of B, and via (8.17) they are equal to the χ^2 distances among the rows of F. The representation X has one uninteresting column that can be removed. The matrix $D_r^{-1/2} F D_c^{-1/2}$ has a left singular vector $D_r^{-1/2} u$, with corresponding right singular vector $D_c^{-1/2} u$. This is seen from the following equalities:

$$D_r^{-1/2} F D_c^{-1/2} D_c^{1/2} u = D_r^{-1/2} F u = D_r^{1/2} u, \tag{8.22}$$

$$D_c^{-1/2} F' D_r^{-1/2} D_r^{1/2} u = D_c^{-1/2} F' u = D_c^{1/2} u, \tag{8.23}$$

which also shows that the corresponding singular value is equal to 1. The unit normalized version of $D_r^{1/2} u$ is $N^{-1/2} D_r^{1/2} u$, and it follows that X must have a column with elements equal to 1. This constant column obviously does not contribute to the distance between two rows of X. We may therefore drop this column from X. Dropping the constant column from X comes to the same as computing the SVD of

$$D_r^{-1/2} F D_c^{-1/2} - D_r^{1/2} uu' D_c^{1/2} / N = K \Lambda L', \tag{8.24}$$

in which the trivial singular vectors cannot occur. This adjustment in turn corresponds to a replacement of B by $B - N^{-1/2} uu' D_c^{1/2}$.

8.3.1 Joint analysis of rows and columns

Given the above solution for the row points X, coordinates for the column points can be found as $Y = N^{1/2} D_c^{-1/2} L$, with the effect that $D_r^{-1} F Y = X$. The latter equation shows that row points are the centre of gravity of the column points, weighted by their frequency in the row profile. Benzécri calls this 'le principe barycentrique'. We have met the idea before as a basic principle in

HOMALS (cf. Section 3.8.3). We will come back to this principle in Section 8.3.4.

Obviously, the correspondence analysis solution could also have been developed the other way round, with χ^2 distances between column profiles of **F** (instead of row profiles) and with the column points in the centre of gravity of row points. It turns out that this is a matter of renormalization only; using the same SVD the solution becomes

$$\mathbf{Y} = N^{1/2}\mathbf{D}_c^{-1/2}\mathbf{L}\Lambda, \tag{8.25}$$

$$\mathbf{X} = N^{1/2}\mathbf{D}_r^{-1/2}\mathbf{K}, \tag{8.26}$$

where the reversal of the centroid principle is equivalent to the choice between the two HOMALS normalizations discussed in Section 3.8.

The best p-dimensional solution is obtained by retaining the first p columns of **X** and **Y**, denoted by $\mathbf{X}^*$ and $\mathbf{Y}^*$. What remains invariant in the two alternative solutions (and therefore also in low-dimensional graphical displays of them that we examine in practice) is the approximation

$$\mathbf{X}^*\mathbf{Y}^{*'} \approx N\,\mathbf{D}_r^{-1}\mathbf{F}\mathbf{D}_c^{-1} - \mathbf{u}\mathbf{u}'. \tag{8.27}$$

With respect to graphical displays this approximation implies the following. If we draw a line through a row point and project the column points on this line, the signed magnitudes of these projections will be proportional to a row of $\mathbf{X}^*\mathbf{Y}^{*'}$. Also, if we draw a line through a column point and project the row points on this line, the projections will have signed magnitudes proportional to a column of $\mathbf{X}^*\mathbf{Y}^{*'}$. The elements of $\mathbf{X}^*\mathbf{Y}^{*'}$ approximate the elements of the matrix at the right of (8.27), i.e. the values $(f_{ij} - e_{ij})\,/\,e_{ij}$, where e_{ij} is the expected value of the cell frequency under the hypothesis of independence. ➤

8.3.2 The chi-square metric

Above it was found that χ^2 distances between rows of **F** are the same as the Euclidean distances between rows of $\mathbf{B} - N^{-1/2}\mathbf{u}\mathbf{u}'\mathbf{D}_c^{1/2}$. An element of the latter matrix can be written as

$$
\begin{aligned}
b_{ij} - (c_j\,/\,N)^{1/2} &= (f_{ij}\,/\,r_i)\,.\,(c_j\,/\,N)^{-1/2} - (c_j\,/\,N)^{1/2} \\[2mm]
&= \frac{(f_{ij}\,/\,r_i) - (c_j\,/\,N)}{(c_j\,/\,N)^{1/2}}
\end{aligned}
\tag{8.28}
$$

The numerator of the expression on the right of (8.28) gives the algebraic difference between a row proportion f_{ij} / r_i and the corresponding marginal proportion c_j / N. The denominator is the square root of the marginal proportion. Provided that row proportions can be interpreted as an estimate of the marginal proportions, we may write

$$\sum_j [b_{ij} - (c_j / N)^{1/2}]^2 \to \chi_i^2. \tag{8.29}$$

This explains why the distances δ_{ik} were called χ^2 distances. Also, the elements of

$$N^{1/2}(D_r^{-1/2}FD_c^{-1/2} - D_r^{1/2}uu'D_c^{1/2} / N) = N^{1/2}K\Lambda L', \tag{8.30}$$

in which the trivial singular vectors are excluded from the SVD, have values equal to $(f_{ij} - e_{ij}) / e_{ij}^{1/2}$, where e_{ij} again is the expected value (compare Section 3.10). It follows that the overall chi-square statistic X^2 for F, i.e. the sum of squared elements of the left-hand side of (8.30), is equal to the trace of

$$N\,K\Lambda L'L\Lambda K' = N\,K\Lambda^2 K'. \tag{8.31}$$

Obviously, this matrix has trace equal to $N \sum\lambda_s^2$, so that we have the equality

$$X^2 = N \sum\lambda_s^2, \tag{8.32}$$

which, on the assumption of independence between rows and columns of F, converges to χ^2 with $(n_r-1)(n_c-1)$ degrees of freedom.

For a miniature example, let

$$F = \begin{pmatrix} 4 & 4 & 2 & 0 \\ 1 & 2 & 8 & 9 \\ 1 & 6 & 8 & 15 \end{pmatrix}$$

with $r' = (10\ 20\ 30)$, $c' = (6\ 12\ 18\ 24)$, $N = 60$. *The expected values are*

$$E = \begin{pmatrix} 1 & 2 & 3 & 4 \\ 2 & 4 & 6 & 8 \\ 3 & 6 & 9 & 12 \end{pmatrix}.$$

The matrix with elements $(f_{ij} - e_{ij}) / e_{ij}^{1/2}$ *will be called* $N^{1/2}A$:

$$N^{1/2}\mathbf{A} = \begin{pmatrix} 3.00 & 1.41 & -0.58 & -2.00 \\ -0.71 & -1.00 & 0.82 & 0.35 \\ -1.15 & 0.00 & -0.33 & 0.87 \end{pmatrix}.$$

The sum of the squared elements of $N^{1/2}\mathbf{A}$ is $X^2 = 19.82$. $\mathbf{A}$ has the SVD solution $\mathbf{K\Lambda L'}$ with

$$\mathbf{K} = \begin{pmatrix} 0.921 & 0.029 \\ -0.282 & 0.766 \\ -0.296 & -0.642 \end{pmatrix},$$

and eigenvalues $\lambda_1^2 = 0.307$, $\lambda_2^2 = 0.023$. This confirms $X^2 = N \sum \lambda^2$. The solution for $\mathbf{X}$ becomes

$$\mathbf{X} = N^{1/2}\mathbf{D_r}^{-1/2}\mathbf{K\Lambda} = \begin{pmatrix} 1.238 & 0.011 \\ -0.271 & 0.203 \\ -0.232 & -0.139 \end{pmatrix}.$$

$\mathbf{X}$ gives the representation of the rows of $\mathbf{F}$. The Euclidean distances between rows of $\mathbf{X}$ are equal to the χ^2 distances between the rows of $\mathbf{F}$. The squared distances are

$$\{\delta_{ij}^2\} = \begin{pmatrix} 0.00 & 2.32 & 2.18 \\ 2.32 & 0.00 & 0.12 \\ 2.18 & 0.12 & 0.00 \end{pmatrix}.$$

These distances are the same as those among the rows of $\mathbf{B} - N^{-1/2}\mathbf{uu'D_c}^{1/2}$. The solution for the column representation becomes

$$\mathbf{Y} = N^{1/2}\mathbf{D_c}^{-1/2}\mathbf{L} = \begin{pmatrix} 2.416 & 0.762 \\ 0.819 & -1.370 \\ -0.280 & 1.269 \\ -0.804 & -0.457 \end{pmatrix}.$$

Figure 8.6 gives the joint plot. The distances between the row points in the plot are the χ^2 distances. Row points are the centre of gravity of column points weighted with respect to the frequencies in the row. The figure further illustrates that projections of the y points on the line through x_2 are proportional to the second row

of **XY'** *and that projections of the x points on the line through* $\mathbf{y}_3$
are proportional to the third column of **XY'**. *The elements of* **XY'**
are $(f_{ij} - e_{ij})/e_{ij}$:

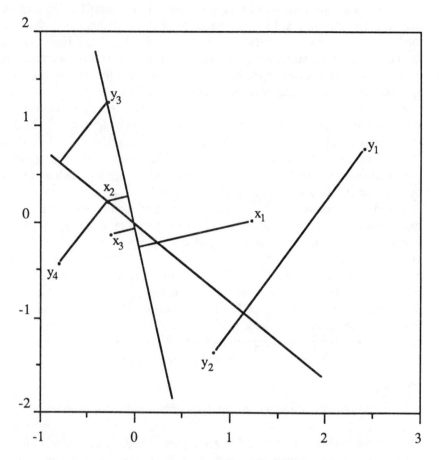

Figure 8.6 Correspondence analysis of miniature example.

$$\mathbf{XY'} = \begin{pmatrix} 3.000 & 1.000 & -0.333 & -1.000 \\ -0.500 & -0.500 & 0.333 & 0.125 \\ -0.667 & 0.000 & -0.111 & 0.250 \end{pmatrix}.$$

8.3.3 The centroid principle and reciprocal averaging

The centroid principle has been formulated independently by many authors. The oldest reference is probably Richardson (1933, quoted in Horst, 1935), who used it in 'the method of reciprocal averages'. Horst (1935) remarked that this method 'is precisely the same as that for determining the factor loadings for a group of tests' (p. 373), which shows that Horst saw the relation with PCA. Horst also acknowledged the connection with quantification of nominal categories, but he does not mention the possibility of approximating continuous nonlinear transformations in this way. Neither does Guttman in his justly famous 1941 paper, in which he makes the important step to relate nonlinear PCA with chi-square. Later contributions can be found in Johnson (1950), Lord (1958), Bock (1960), De Leeuw (1973), Nishisato (1980), and, of course, Benzécri (1973). The latter's influence probably works through in other publications of the French school, such as Dauxois and Pousse (1976) and Lafaye de Michaux (1978).

A typical example where the method is reintroduced (even under its oldest name: method of reciprocal averaging) is the paper by Hill (1974). This is a paper on the *ordination problem* in phytosociology; it is formally the same problem as the *seriation problem* in archaeology (or the Guttman *scalogram problem* in the social sciences). Hill's method is identical with what in this book is called HOMALS with an incomplete indicator matrix. Hill compares this procedure with what he calls a PCA solution (HOMALS with complete indicator matrix, option (*ii*)) and also with PCA without standardization (SVD of **G** itself). As one would expect, he finds the latter method less satisfactory (it contaminates scale values with marginal frequencies).

8.3.4 Relation with homogeneity analysis of two indicator matrices

The relation between correspondence analysis and HOMALS is very close. In Section 2.6 it was shown how to create, for a given contingency table, a corresponding indicator matrix **G**. This **G** becomes an $N \times (n_r + n_c)$ matrix. Its category quantifications will be the partitioned vector **y** that satisfies $\mathbf{Cy} = \psi^2 \mathbf{Dy}$. In terms of **F**, $\mathbf{D_r}$, and $\mathbf{D_c}$ the Burt table **C** becomes

$$\mathbf{C} = \begin{pmatrix} \mathbf{D_r} & \mathbf{F} \\ \mathbf{F}' & \mathbf{D_c} \end{pmatrix} \tag{8.33}$$

so that the SVD used in this chapter can be substituted to obtain

$$D^{-1/2}CD^{-1/2} = \begin{pmatrix} I & D_r^{-1/2}FD_c^{-1/2} \\ D_c^{-1/2}F'D_r^{-1/2} & I \end{pmatrix}$$

$$= \begin{pmatrix} I & K\Lambda L' \\ L\Lambda K' & I \end{pmatrix}. \tag{8.34}$$

It can be immediately verified that this matrix has eigenvectors $\{K'\ L'\}'$ with eigenvalues $I + \Lambda$ and eigenvectors $\{K' - L'\}'$ with eigenvalues $I - \Lambda$. It follows that the quantifications are given by $Y_r = D_r^{-1/2}K$ for row categories and $Y_c = D_c^{-1/2}L$ for column categories (where we use the nonstandard normalization $Y'DY = I$). The difference with the solution by correspondence analysis, therefore, is only in normalization.

It can also easily be shown that Y_r and Y_c yield stationary values for the correlation between rows and columns of F. Since $u'D_rY_r = 0$ and $u'D_cY_c = 0$, the columns of Y_r and Y_c are in deviations from their means. If we normalize $Y_r'D_rY_r = I$ and $Y_c'D_cY_c = I$ (instead of $Y'DY = I$), $Y_r'FY_c = \{\rho_{st}\}$ becomes a matrix of correlations. However,

$$Y_r'FY_c = K'D_r^{-1/2}FD_c^{-1/2}L = K'K\Lambda L'L = \Lambda, \tag{8.35}$$

so that the singular values are the *canonical correlations*. The relation with the earlier notation for the HOMALS eigenvalues is

$$\Psi^2 = I + \Lambda. \tag{8.36}$$

In terms of χ^2 we now have the relation that χ^2 / N must be equal to the sum of the squared canonical correlations.

8.4 CONTINGENCY AND CORRELATION

We may give a few comments on the history of the relations between chi-square and canonical correlation. Pearson (1904) defined a measure of dependence, called the *mean square contingency* $\phi^2 = \chi^2 / N$. The underlying argument is that for a continuous normal distribution $f(\underline{x})$ there is an infinite set of orthogonal transformations $\psi^{(s)}(\underline{x})$ of degree s ($s=1,...,\infty$) (called *Hermite–Chebyshev polynomials*), for which it can be shown that the correlation between $\psi^{(s)}(\underline{x})$ and $\psi^{(s)}(\underline{y})$ equals ρ^s, where $f(\underline{x}, \underline{y})$ is a binormal distribution with correlation parameter ρ. It can be shown that χ^2 / N is the sum of the squares of these correlations (in fact, this is consistent with the result of the previous section, since for the continuous binormal stochastic distribution, the series ρ^s is a series of canonical correlations). This implies

$$\chi^2 / N = \rho^2 + \rho^4 + \rho^6 + \ldots = \frac{\rho^2}{1 - \rho^2}, \qquad (8.37)$$

so that

$$\rho^2 = \frac{\chi^2}{N + \chi^2} = \frac{\phi^2}{1 + \phi^2}. \qquad (8.38)$$

Pearson also applied the latter coefficient to contingency tables. Insofar as a contingency table can be assumed to be a discretization of a binormal distribution, with many classes for $\underline{x}$ and $\underline{y}$, the correlation parameter ρ can be estimated as $\phi / (1 + \phi^2)^{1/2}$. Pearson called this the *coefficient of contingency*.

Gebelein (1941) further developed the theory. He showed that there are transformations $\xi(\underline{x})$ and $\psi(\underline{y})$ for which the correlation between $\xi(\underline{x})$ and $\psi(\underline{y})$ is maximized, and that this maximum correlation $\kappa_{\underline{xy}}$ is related to eigenvalues. He also made the point that, if both ξ and ψ are linear, the 'maximum' (there is not much to maximize in this case) measure is the correlation coefficient ρ. If either ξ or ψ is linear, the maximum becomes the correlation ratio ($\varepsilon_{\underline{xy}}$ or $\varepsilon_{\underline{yx}}$) and $\rho \leq \varepsilon \leq \kappa$. Gebelein's work was generalized by Rényi (1959) who formulated a checklist of criteria for an ideal measure of dependence $\delta_{\underline{xy}}$:

A: $\delta_{\underline{xy}}$ is defined for all nonconstant $\underline{x}$ and $\underline{y}$.

B: $\delta_{\underline{xy}} = \delta_{\underline{yx}}$.

C: $0 \leq \delta_{\underline{xy}} \leq 1$.

D: $\delta_{\underline{xy}} = 0$ if and only if $\underline{x}$ and $\underline{y}$ are independent.

E: When $\underline{x} = \xi(\underline{y})$ or $\underline{y} = \psi(\underline{x})$, then $\delta_{\underline{xy}} = 1$.

F: If ξ and ψ are one-to-one mappings, then $\delta_{\underline{xy}} = \delta_{\xi(\underline{x})\psi(\underline{y})}$.

G: If $(\underline{x}, \underline{y})$ is binormal with correlation parameter ρ, then $\delta_{\underline{xy}} = |\rho|$.

Applying this checklist to four measures of dependence – the squared correlation coefficient $\rho_{\underline{xy}}^2$, the maximum ε^2 of the correlation ratio's $\varepsilon_{\underline{xy}}^2$ – or $\varepsilon_{\underline{xy}}^2$, the coefficient of contingency $\phi / (1 + \phi^2)^{1/2}$, and Gebelein's $\kappa_{\underline{xy}}^2$ the following table can be set up:

	A	B	C	D	E	F	G
ρ_{xy}^2	0	1	1	0	0	0	1
ε^2	0	1	1	0	1	0	1
$\phi / (1+\phi^2)^{1/2}$	0	1	1	1	0	1	1
κ_{xy}	1	1	1	1	1	1	1

Lancaster (1959, 1960a, 1960b) has generalized some of these results for dependence between more than two variables.

Turning our attention to contingency tables, Hirschfeld (1935) discussed the problem of whether it is possible to quantify rows and columns of such a table, with normalization (in our notation) $u'D_rx = 0$, $u'D_cy = 0$, $x'D_rx = 1$, $y'D_cy = 1$, so that we would obtain $D_r^{-1}Fy = \rho x$, $D_c^{-1}F'x = \rho y$. In words: is there a quantification so that both regressions become linear? Hirschfeld showed that the solutions are related to the SVD of $D_r^{-1/2}FD_c^{-1/2}$ (as shown in Section 8.3); he also showed relations with the continuous case.

Fisher (1940) rediscovered this technique and inspired the first applications by Maung (1941a, 1941b). These studies also point out relations between (in the terminology of this book) maximum canonical correlation between indicator matrices, maximum product moment correlation for a bivariate frequency table, and maximum discrimination between categories for such tables. The same aspects are mentioned in Guttman's 1941 paper. (Further developments and applications can be found in Yates, 1948; Johnson, 1950; Williams, 1952; Lancaster, 1957; Bock, 1960.) Maung described the representation

$$F = D_r \{uu' + \Sigma_s \lambda_s x_s y_s'\}D_c \qquad (8.39)$$

(in our notation) with references to the comparable case of continuous variates. Earlier, Mehler (1866) had formulated for this case the representation

$$f(x, y) = f(x) f(y) [1 + \Sigma_s \rho^s \psi_s(x) \psi_s(y)], \qquad (8.40)$$

where ψ_s are the Hermite–Chebyshev polynomials. The Maung–Fisher equivalent is

$$f(x, y) = f(x).f(y) [1 + \Sigma_s \lambda_s \xi_s(x) \psi_s(y)], \qquad (8.41)$$

where λ_s (canonical correlations) replace ρ^s and where ξ_s and ψ_s are the canonical transformations. (For further developments see Lancaster, 1958;

Hannan, 1961; Venter, 1966; Cambanis and Liu, 1971; Jensen, 1971; Chesson, 1976.)

One specific question is: when are the canonical transformations polynomial functions? The question is of interest, because in many applications of nonlinear MVA polynomial, or almost polynomial, transformations are found (examples can be found in Barrett and Lampard, 1955; McFadden, 1966; McGraw and Wagner, 1968; Lee, 1971). Even more strongly, in many applications the best transformation is almost linear, the second best almost quadratic, etc. Studies by Eagleson (1964), Eagleson and Lancaster (1967), and Lancaster (1975) show that for many special bivariate distributions (not only the normal, but also the Poisson, the gamma, the hypergeometric, etc.) the canonical transformations must be classical orthogonal polynomials.

A further question then is: suppose that the canonical transformations are polynomials, what restrictions does this impose upon the canonical correlations? As indicated earlier, for a binormal distribution the canonical correlations are powers of the correlation parameter ρ. Other cases are discussed in Sarmanov and Bratoeva (1967), Eagleson (1969), Griffiths (1969, 1970), and Tyan and Thomas (1975). Generalizations to the multinormal case are treated in Appell and Kampé de Fériet (1926), Erdelyi (1953), Sarmanov and Zacharov (1960), Venter (1966), Naouri (1970), and Dauxois and Pousse (1976).

8.5 THE PROGRAM ANACOR

The program ANACOR has been developed to perform correspondence analysis. Essentially, the program solves for $\mathbf{X}$ and $\mathbf{Y}$ as defined in Section 8.3. It does not use alternating least squares, but an explicit SVD routine. The program has three options for normalization:

(a) $\mathbf{X} = N^{1/2} \mathbf{D}_r^{-1/2} \mathbf{K} \boldsymbol{\Lambda}$.

$\mathbf{Y} = N^{1/2} \mathbf{D}_c^{-1/2} \mathbf{L}$, so that $\mathbf{X}'\mathbf{D}_r\mathbf{X} = N\boldsymbol{\Lambda}^2$

$$\mathbf{Y}'\mathbf{D}_c\mathbf{Y} = N\,\mathbf{I}$$

$$\mathbf{X} = \mathbf{D}_r^{-1}\mathbf{F}\mathbf{Y}$$

where the last equality shows that this option scales row points in the centre of gravity of column points.

(b) $X = N^{1/2} D_r^{-1/2} K.$

$Y = N^{1/2} D_c^{-1/2} L\Lambda$, so that $X'D_r X = N I$

$$Y'D_c Y = N\Lambda^2$$

$$Y = D_c^{-1} F'X$$

with column points in the centre of gravity of row points.

(c) The third option drops the centroid principle and treats rows and columns symmetrically:

$X = N^{1/2} D_r^{-1/2} K\Lambda^{1/2}.$

$Y = N^{1/2} D_c^{-1/2} L\Lambda^{1/2}$, so that $X'D_r X = N\Lambda$

$$Y'D_c Y = N\Lambda.$$

The most straightforward application of ANACOR is to two-dimensional contingency tables (as discussed in Section 8.3). ANACOR then quantifies row and column categories in the same way as HOMALS would do (apart from normalizations), but ANACOR of course does not quantify the HOMALS objects. Another difference is the fact that ANACOR computes the complete set of singular values for all possible dimensions (whereas HOMALS iterates for given p), although it does not give more than p quantifications. An application of ANACOR on a contingency table is given in Section 8.5.1.

However, it should be noted that ANACOR also can handle two-dimensional tables where the entries are not frequencies but a different type of nonnegative numbers. Section 8.5.2 will give an example where the entries are derived from actual distances. Finally, ANACOR can be used to handle higher-dimensional tables, but only after reducing them to something bivariate, in the sense that the input for ANACOR is either the Burt table of bivariate marginals or a two-dimensional frequency table obtained by grouping dimensions. Both procedures will be illustrated in Section 8.5.3.

8.5.1 An example with a simple contingency table

Table 8.3 gives a 7 × 7 table of occupational status of fathers versus occupational status of their sons for a sample of 3497 British families. These data are also discussed in Glass (1954), Goodman (1965, 1969), Haberman (1974), and Bishop, Fienberg and Holland (1975). The standard ANACOR program can produce singular values λ_s and quantifications of row and column categories x_s, y_s for six dimensions. The program also can give plots of x_s and y_s for all pairs of dimensions. For the first two dimensions the singular values (canonical correlations) are $\lambda_1 = 0.526$, $\lambda_2 = 0.267$. The quantifications are given in Table 8.4, plotted in Figure 8.7; the solution is based on option (b): sons are in the centre of gravity of fathers.

Table 8.3 Occupational mobility data

Occupation of father	Occupation of son							
	PROF	EXEC	HSUP	LSUP	SKIL	SEMI	UNSK	
PROF	50	19	26	8	18	6	2	129
EXEC	16	40	34	18	31	8	3	150
HSUP	12	35	65	66	123	23	21	345
LSUP	11	20	58	110	223	64	32	518
SKIL	14	36	114	185	714	258	189	1510
SEMI	0	6	19	40	179	143	71	458
UNSK	0	3	14	32	141	91	106	387
	103	159	330	459	1429	593	424	3497

PROF:	professional and high administrative
EXEC:	managerial and executive
HSUP:	higher supervisory
LSUP:	lower supervisory
SKIL:	skilled manual and routine nonmanual
SEMI:	semi-skilled manual
UNSK:	unskilled manual

Table 8.4 Quantifications of occupational mobility data

Occupation of father			Occupation of son		
PROF	−4.021	2.870	PROF	−2.319	0.872
EXEC	−2.113	−1.639	EXEC	−1.051	−0.450
HSUP	−0.660	−1.474	HSUP	−0.491	−0.290
LSUP	−0.063	−0.865	LSUP	−0.019	−0.283
SKIL	0.330	−0.017	SKIL	0.160	−0.044
SEMI	0.668	0.865	SEMI	0.319	0.241
UNSK	0.756	1.193	UNSK	0.376	0.300

Clearly, the first dimension orders the categories of occupational status from high to low. The second dimension does the same with the notable exception for category PROF: one could also maintain, however, that the second dimension is a quadratic function of the first. As a result, in Figure 8.7 the sequence of labels for fathers (Fa....) or for sons (So....) appears as a rather smooth curve, with category PROF much separated from the other categories. Figure 8.8 (not a standard ANACOR plot) is a re-edit of Table 8.3, with distances between categories corresponding to the solution of the first dimension, again with normalization as in option (b). In addition the two lines for linear regression are drawn.

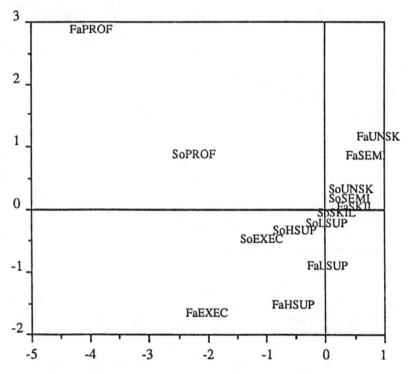

Figure 8.7 ANACOR solution for occupational mobility data.

As to their substantive interpretation, it should be noted that the above results have mainly *correlational* significance: optimal scaling of father and son categories cannot produce a correlation larger than 0.526. The analysis completely ignores any difference there might be between fathers and sons in their *average score* (the ANACOR solution gives the two variables a zero average) or any difference in *variability*. Clearly ANACOR highlights only one aspect of occupational mobility.

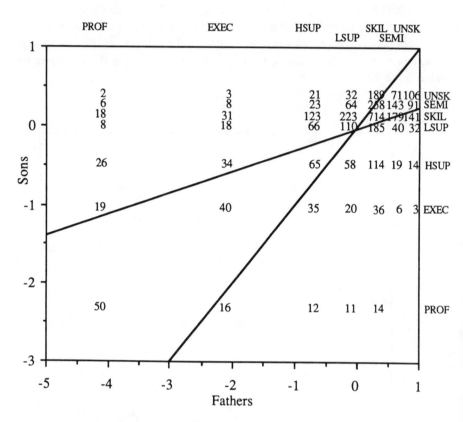

Figure 8.8 Table 8.3 graphed as a regression plot on the basis of the first ANACOR solution.

8.5.2 An example with a similarity table

To illustrate that the application of ANACOR is not restricted to frequency data and to show its potential as an MDS technique, the program is now used for analysing a table of similarities derived from physical dis tances between 23 cities in the Netherlands. Such a table is, by necessity, symmetric, and it follows that row objects and column objects will obtain an identical quantification. Note that in this case there is no HOMALS equivalent, since it is impossible to create an indicator matrix along the lines of Section 2.6.

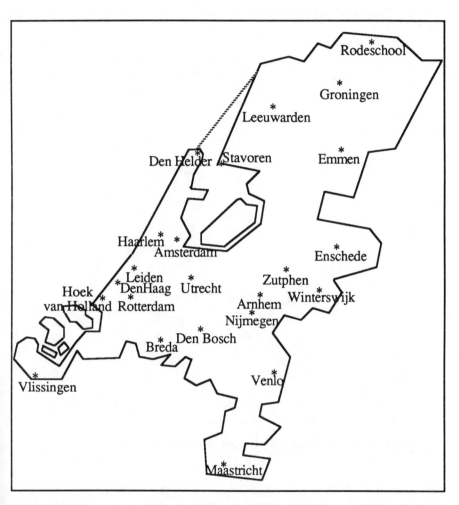

Figure 8.9a ANACOR solution for distances between 23 Dutch cities: distances as the crow flies.

The analysis has been done in three ways, all three with normalization option (c). In the first analysis, **F** contains similarities constructed by taking distances 'as the crow flies' and subtracting these from the largest distance.

In the second analysis, similarities based on shortest railway connections were taken. In the third analysis, **F** is a binary table based on the railway connections, with elements 0 if the distance is larger than the median distance and 1 if it is shorter. The results are shown in Figure 8.9. The first analysis produces a figure that is quite similar to the real map of the Netherlands.

Figure 8.9b ANACOR solution for distances between 23 Dutch cities: distances by railway.

Figure 8.9b, for railway distances, gives a typical distortion because there is no direct railway connection between Den Helder and Stavoren, whereas cities in the Western part of the country (where there are many direct connections) tend to cluster (similarly in the North-Eastern part). Figure 8.9c gives the results for the dichotomized distances. Given the severe simplification of the input, the recovery of the previous map is surprisingly good.

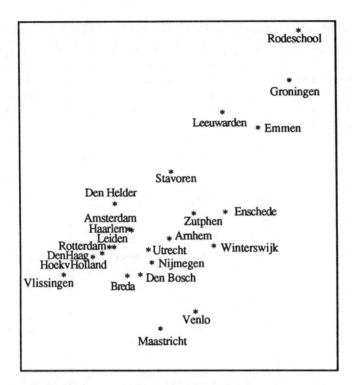

Figure 8.9c ANACOR solution for distances between 23 Dutch cities: dichotomized distances.

8.5.3 An example with a multidimensional contingency table

Table 8.5 gives data collected in 1968 for a stratified sample of 1616 students from different Dutch universities (Lammers, 1969; see also De Leeuw, 1973). The table shows for each university a bivariate subtable of frequencies for *faculties* versus *political preference*. There are 13 different faculties all together, but no university has all faculties and some universities have only one faculty. As to political preference, the five categories (from left to right in the table) are:

CDA   :   any denominational party (KVP, ARP, CHU, GPV, SGP)
VVD   :   conservative-liberal
PvdA   :   labour party
PACO   :   any of the smaller left-wing parties (PSP, CPN)
D'66   :   pragmatic-liberal

Table 8.5 Political preference (order: CDA, VVD, PvdA, PACO, D'66)

	LEIDEN					A'DAM VU					A'DAM GU					UTRECHT				
Law	13	32	9	2	18	1	0	0	1	1	3	14	12	2	10	7	16	1	0	5
Med.	3	20	7	4	10	7	1	3	1	2	4	6	4	7	18	6	11	9	2	1
Scienc.	2	15	4	5	15	6	0	1	0	1	1	4	13	4	19	10	14	16	7	21
SocSc.	5	12	4	3	14	2	0	1	0	2	1	8	28	11	19	6	4	5	10	1
Arts	5	8	6	3	5	2	0	0	1	0	7	8	16	8	16	3	10	4	4	1
Techn.																				
Pol.											1	1	5	0	4					
Vet.																3	6	1	1	5
Dent.																0	3	0	0	1
Theo.	1	1	3	2	0	5	0	0	0	0	1	0	0	0	0	5	0	0	2	1
Agri.																				
Phil.	0	0	1	2	0	0	0	0	0	1	2	2	6	5	7	0	1	0	1	4
Econ.						4	0	0	0	2	7	16	8	6	20					

	WAGENINGEN					GRONINGEN					TWENTE					NIJMEGEN				
Law						2	8	6	0	7						5	7	3	1	6
Med.						3	7	3	2	11						6	9	3	2	5
Scienc.						7	5	9	2	12						7	2	2	4	6
SocSc.	0	1	1	0	0	4	5	11	6	10						4	2	5	4	22
Arts						1	5	6	2	6						4	0	8	2	4
Techn.											0	7	3	4	5					
Pol.																				
Vet.																				
Dent.						1	2	0	1	1						1	2	0	0	2
Theo.						1	0	2	0	1						6	0	0	0	5
Agri.	11	14	7	6	14															
Phil.						0	1	1	1	0						0	0	1	0	0
Econ.	0	0	0	0	1	1	13	5	0	7										

	TILBURG					ROTTERDAM					EINDHOVEN					DELFT				
Law	0	0	1	0	2	2	2	1	0	0										
Med.						8	7	1	1	7										
Scienc.																				
SocSc.	5	1	0	3	6	1	1	3	1	1										
Arts																				
Techn.											12	7	3	0	13	24	66	22	20	50
Pol.																				
Vet.																				
Dent.																				
Theo.																				
Agri.																				
Phil.																				
Econ.	3	11	2	0	9	9	33	15	2	16										

Legend: Med. = medicine; Scienc. = natural sciences; SocSc. = social sciences; Techn. = technology; Pol. = political sicence; Vet. = veterinary science; Dent. = dentistry; Theo. = theology; Agri. = agriculture; Phil. = philosophy; Econ. = economy.

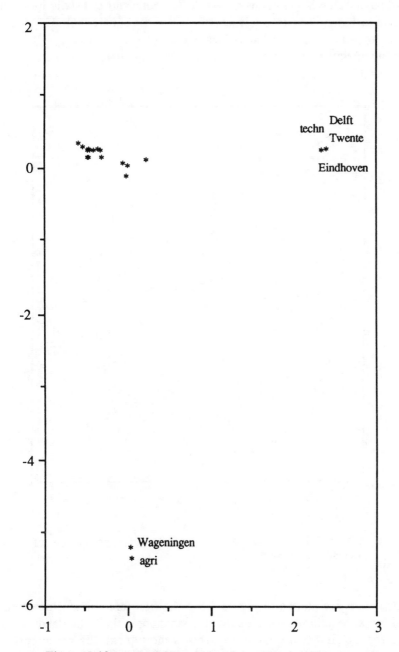

Figure 8.10a ANACOR solution for political preferences of students: dimensions 1 and 2.

Although Table 8.5 is trivariate, ANACOR forces us to handle it in some bivariate fashion. The first analysis is based on the square table C (the Burt table). For this example C becomes a $(12+13+5) \times (12+13+5)$ table of univariate (diagonal) and bivariate (off-diagonal) marginals.

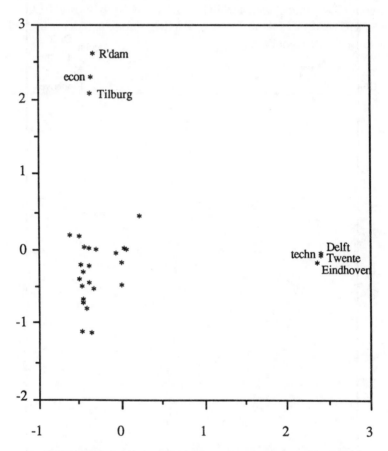

Figure 8.10b ANACOR solution for political preferences of students: dimensions 1 and 3.

The results are plotted in Figure 8.10. The first three dimensions mainly reveal trivial peculiarities of the academic structure in the Netherlands. Figure 8.10a, for the first two dimensions, shows the typical difference between *technical universities*, with only one faculty, not to be found at other universities (DELFT, EINDHOVEN, TWENTE), WAGENINGEN (where there is a faculty AGRICULTURE that is nowhere else), and the other universities. The third dimension (Figure 8.10b) capitalizes on the fact that the ECONOMY faculties in ROTTERDAM and TILBURG are dominant. Results start to become more interesting in dimensions 4 and 5 (Figure 8.10c). It shows (we mention only a few aspects) that students in THEOLOGY, at AMSTERDAM VU, favour CDA; that students in POLITICAL SCIENCE or PHILOSOPHY at AMSTERDAM GU favour PvdA or PACO; and that students of the LAW faculty in LEIDEN favour VVD.

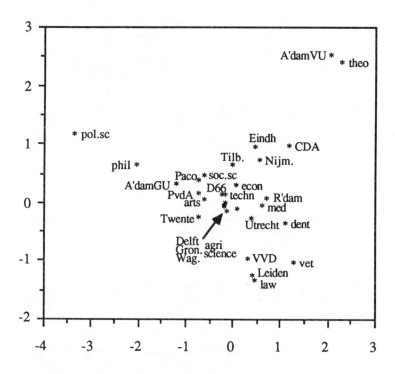

Figure 8.10c ANACOR solution for political preferences of students: dimensions 4 and 5.

The other possibility is that we combine variables into an *interactive variable* (as in Section 2.8). For the present example, UNIVERSITIES and FACULTIES were combined, in order to accommodate their interaction. Theoretically this gives $12 \times 13 = 156$ combinations, but many of them do not occur, so that 63 actual combinations are left. ANACOR will now produce the same quantification of categories as a two-variable HOMALS would do, with 5 columns for G_1 (political preference) and 63 for G_2 (combinations).

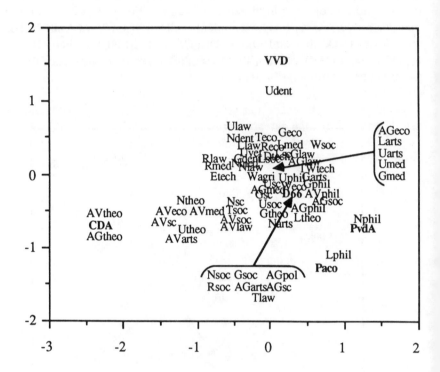

Figure 8.11 ANACOR solution for political preferences of students, with universities and faculties grouped as 63 categories of one variable.

The analysis results for the first two dimensions are shown in Figure 8.11, with canonical correlations of 0.353 and 0.328. In the figure, CDA is located towards the left, PvdA and PACO at the lower right, VVD towards the top, and D'66 in the centre. Some other aspects worth mentioning are that AMSTERDAM VU is favouring CDA for all faculties except PHILOSOPHY; PHILOSOPHY tends to favour PvdA and PACO, especially in LEIDEN and NIJMEGEN; THEOLOGY favours CDA except in LEIDEN and GRONINGEN; LAW faculties tend to favour VVD except at AMSTERDAM VU. The trivial dimensions that dominated the preceding solution have now vanished.

8.6 THE PROGRAM ANAPROF: ANALYSIS OF PROFILE FREQUENCIES

The progam ANAPROF is adapted to a *reduced profile frequency* matrix as input (cf. Table 2.3). The program is quicker, and cheaper, than HOMALS when the number of objects n is much larger than the number of response patterns (or profiles) q. In general, the total number of possible response profiles is equal to $\Pi_j k_j$, but the actual number of profiles q may be smaller.

Let $\mathbf{S}$ be a $q \times \sum k_j$ binary matrix. A row of $\mathbf{S}$ identifies a possible profile by having entry '1' for the category present in the profile and '0' for the absent categories. The rows of $\mathbf{S}$ include only those profiles that appear at least once in the HOMALS data matrix $\mathbf{H}$ (hence the term *reduced* profile frequency matrix).

Let $\mathbf{G_p}$ be an $n \times q$ *profile indicator matrix* with element $g_{(p)ih} = 1$ if the ith object has the hth profile in $\mathbf{S}$, and $g_{(p)ih} = 0$ otherwise. The ordinary partitioned indicator matrix can now be written as $\mathbf{G = G_p S}$. Furthermore, $\mathbf{G_p' G_p}$ is the diagonal matrix of marginal profile frequencies. HOMALS is equivalent to a correspondence analysis on $\mathbf{G}$, i.e. it can be based on the SVD in (8.18), substituting $\mathbf{G}$ and its marginals:

$$m^{-1/2} \mathbf{G} \mathbf{D}^{-1/2} = \mathbf{K \Lambda L'}. \tag{8.42}$$

Now, since $\mathbf{G}$ satisfies $\mathbf{G_p(G_p' G_p)^{-1} G_p' G = G}$ (the rows of $\mathbf{G}$ that correspond to the nonzero entries in a column of $\mathbf{G_p}$ are equal), it must also be true that $\mathbf{K}$ satisfies

$$\mathbf{G_p(G_p' G_p)^{-1} G_p' K = K,} \tag{8.43}$$

because the singular value decomposition is unique. Substituting the row equivalence relations $\mathbf{G = G_p S}$ into (8.42) and premultiplying by $m^{1/2} \mathbf{G_p'}$ we obtain

$$\mathbf{G_p' G_p S D^{-1/2}} = m^{1/2} \mathbf{G_p' K \Lambda L'}. \tag{8.44}$$

Let us consider a correspondence analysis on the table $\underline{\mathbf{F}} = \mathbf{G_p' G_p S}$. Its row marginals are $\underline{\mathbf{D}}_r = m \mathbf{G_p' G_p}$, and its column marginals are $\underline{\mathbf{D}}_c = \mathbf{D}$. Substituting $\underline{\mathbf{D}}_c$ and $\underline{\mathbf{F}}$ into (8.44) and premultiplying by $\underline{\mathbf{D}}_r^{-1/2}$ gives us

$$\underline{\mathbf{D}}_r^{-1/2} \underline{\mathbf{F}} \underline{\mathbf{D}}_c^{-1/2} = (\mathbf{G_p' G_p})^{-1/2} \mathbf{G_p' K \Lambda L'} = \underline{\mathbf{K}} \Lambda \mathbf{L'}, \tag{8.45}$$

with $\underline{\mathbf{K}} = (\mathbf{G_p' G_p})^{-1/2} \mathbf{G_p' K}$. Note that $\underline{\mathbf{K}}$ is orthonormal:

$$\underline{K}'\underline{K} = K'G_p(G_p'G_p)^{-1}G_p'K = I, \qquad (8.46)$$

due to (8.43). So $\underline{K}$ will be the matrix of left singular vectors used in a correspondence analysis of $\underline{F}$. We also have $K = G_p(G_p'G_p)^{-1}G_p'K = G_p(G_p'G_p)^{-1/2}\underline{K}$, so we can always recover the extended K from the reduced $\underline{K}$. This fact implies that it is sufficient to compute the correspondence analysis of $\underline{F}$, which can be solved more quickly since it only involves a $q \times \Sigma k_j$ matrix, whereas G is an $n \times \Sigma k_j$ matrix (and our assumption was $n \gg q$).

Thus ANAPROF is like HOMALS in the sense that it is based on the analysis of indicator matrices, created from the actual data, which are the profiles S and their frequencies $G_p'G_p$; it is like ANACOR in the sense that it uses an explicit SVD routine to compute object scores and category quantifications (including estimates of their stability, cf. Section 12.3.4).

What has been shown above is actually a proof of the fact that the results of a correspondence analysis remain invariant if identical profiles are added into a single one. In the French literature this fact is called the *principle of distributional equivalence* (Benzécri, 1973). A related piece of theory exists for the situation in which the profiles are not necessarily identical, but we do wish groups of object scores to be equal (see Section 10.2).

8.6.1 An example with binary survey data

To illustrate the use of ANAPROF we shall analyse the survey data of Sugiyama (1975), where $n = 4243$ Japanese individuals responded to $m = 6$ binary questions about religious practices. The questions are listed in Table 8.6. Theoretically, there are a total of $2^6 = 64$ possible profiles. Table 8.7 lists

Table 8.6 Sugiyama items

A	Do you make it a rule to practise religious conduct, such as attending religious services, religious worship, and missionary works and do you occasionally offer prayers or chant sutras?
B	Do you visit a grave once or twice a year?
C	Do you occasionally read religious books, such as the Bible or the Buddhist Scriptures?
D	Do you visit shrines and temples to pray for business prosperity, success in an entrance examination, and so forth?
E	Do you keep a talisman, such as an amulet, charm, or mascot, near you?
F	Did you draw a fortune, consult a diviner, or have your fortune told within the last years?

them, together with their frequency of occurrence. Three response profiles have zero frequency (so $q = 61$), many have very low frequency, and some have very large frequency.

Table 8.7 Sugiyama profile frequency table

1	1	1	1	1	1	042	0	1	1	1	1	1	011
1	1	1	1	1	0	033	0	1	1	1	1	0	007
1	1	1	1	0	1	006	0	1	1	1	0	1	002
1	1	1	1	0	0	017	0	1	1	1	0	0	005
1	1	1	0	1	1	012	0	1	1	0	1	1	004
1	1	1	0	1	0	029	0	1	1	0	1	0	008
1	1	1	0	0	1	008	0	1	1	0	0	1	004
1	1	1	0	0	0	082	0	1	1	0	0	0	044
1	1	0	1	1	1	051	0	1	0	1	1	1	072
1	1	0	1	1	0	069	0	1	0	1	1	0	126
1	1	0	1	0	1	020	0	1	0	1	0	1	045
1	1	0	1	0	0	054	0	1	0	1	0	0	142
1	1	0	0	1	1	034	0	1	0	0	1	1	080
1	1	0	0	1	0	124	0	1	0	0	1	0	258
1	1	0	0	0	1	027	0	1	0	0	0	1	137
1	1	0	0	0	0	317	0	1	0	0	0	0	760
1	0	1	1	1	1	001	0	0	1	1	1	1	000
1	0	1	1	1	0	002	0	0	1	1	1	0	002
1	0	1	1	0	1	000	0	0	1	1	0	1	000
1	0	1	1	0	0	009	0	0	1	1	0	0	004
1	0	1	0	1	1	001	0	0	1	0	1	1	004
1	0	1	0	1	0	011	0	0	1	0	1	0	003
1	0	1	0	0	1	007	0	0	1	0	0	1	006
1	0	1	0	0	0	059	0	0	1	0	0	0	030
1	0	0	1	1	1	008	0	0	0	1	1	1	033
1	0	0	1	1	0	023	0	0	0	1	1	0	048
1	0	0	1	0	1	007	0	0	0	1	0	1	038
1	0	0	1	0	0	035	0	0	0	1	0	0	064
1	0	0	0	1	1	010	0	0	0	0	1	1	042
1	0	0	0	1	0	055	0	0	0	0	1	0	096
1	0	0	0	0	1	013	0	0	0	0	0	1	090
1	0	0	0	0	0	194	0	0	0	0	0	0	718

For the example, S as defined above is a 61×12 matrix, with $\mathbf{G} = \mathbf{G_p S}$, the complete indicator matrix. ANAPROF computes the full SVD solution in $p_{max} = 6$ dimensions. The first two eigenvalues are $\lambda_1^2 = 0.269$ and $\lambda_2^2 = 0.204$. As a comparison of speed, the HOMALS program needs 70 time units for this case, while ANAPROF uses 3 time units. Figure 8.12 shows the plot of the response profiles in the first two dimensions of the solution. The plot contrasts profiles with a 'yes' answer to items A and C (upper left side) to those with a 'no' answer to these two items (lower right side). The plot also contrasts profiles with '000' for the last three items (upper right side) to those with '111' for these items (lower left side); profiles with only one '1' and those with two '1's form bands between the two extremes. Item B is treated in a particular way: the answers form alternating bands.

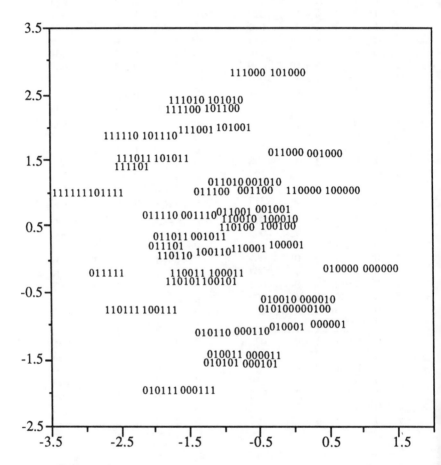

Figure 8.12 ANAPROF solution for Sugiyama data on assumption of complete indicator matrix.

The optimally scaled data matrix **Q** becomes a 4243 × 6 matrix, and its correlation matrix is given in Table 8.8, for which Figure 8.13 displays the first two dimensions of the PCA solution. The latter plot confirms the similarity between A and C, and the similarity between D, E, F, whereas item B coincides with the first axis. Actually, the example is a bit atypical, because the correlation matrix of Table 8.8 for binary variables does not depend at all

Table 8.8 Correlation matrix for Sugiyama items

1.000	0.085	0.284	0.077	0.097	−0.018
0.085	1.000	0.052	0.112	0.163	0.061
0.284	0.052	1.000	0.067	0.052	0.041
0.077	0.112	0.067	1.000	0.279	0.211
0.097	0.163	0.052	0.279	1.000	0.202
−0.018	0.061	0.041	0.211	0.202	1.000

on optimal scaling: for binary variables any scaling produces the same correlations (apart from a possible change of sign). For binary variables this soultion is identical to a PCA on the correlations among the columns of the data matrix.

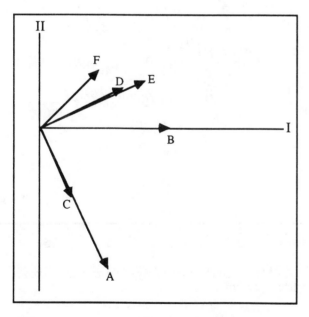

Figure 8.13 PCA solution for Sugiyama data, first two dimensions.

8.6.2 Asymmetric treatment of response categories

In the binary case a symmetric treatment of the response categories implies that
the optimal scaling approach does not contribute anything new beyond an
ordinary principal components analysis. It now becomes interesting to
compare the ANAPROF solution with a solution for the *incomplete* indicator

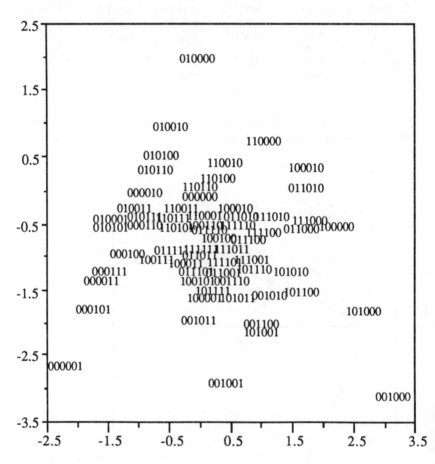

Figure 8.14 ANAPROF solution for Sugiyama data on the assumption
of an incomplete data matrix.

matrix **G** of dimension 4243 × 6 (one column only for each item, registering
'yes' as '1', but omitting the 'no' answers). Stated more generally, an

asymmetric treatment of response categories is called for when the '1' category presumably indicates homogeneous groups of objects, whereas its complement can be regarded as an indication of heterogeneity. Such a treatment will be more in the spirit of the binary unfolding methods of Section 8.2. For the present case, the ANACOR solution for the first two dimensions is displayed in Figure 8.14. The corresponding eigenvalues are $\lambda_1^2 = 0.425$, $\lambda_2^2 = 0.348$.

Indeed, Figure 8.14 looks rather different from Figure 8.12. When we take profiles with only one '1' at the positions of the items ('100000' = A, etc.), we note that the items form a *curved scale*. An item with more '1's in it is located exactly in the centroid of the profiles with only one '1' at the relevant positions. For example, '000101' is exactly midway between '000100' and '000001' and profile '111111' is the (unweighted) centre of gravity of all six item profiles with only one '1'. The unfolding structure is reflected in the fact that each profile is closer to those elementary profiles of which it is the mean than to other elementary profiles.

8.7 SOME GAUGING RESULTS FOR BINARY DATA

The considerations in the previous section bring us back to the context of multidimensional scaling. Recall that, as already indicated in Sections 2.3 and 3.12, the distinction between a *Guttman scale* and a *Coombs scale* was related to the choice between a complete or an incomplete indicator matrix. However, many more types of regular patterns for binary data can be considered, and we can study the way in which these regularities are mapped in the results of a homogeneity analysis of one sort or another. Such studies provide benchmarks for the behaviour of the technique in standardized circumstances – hence the term gauging, introduced in Section 1.5.2. Many of the results discussed in Section 8.4 can be regarded as gauging results for data from the frequency domain. In the next chapter we will study in depth various versions of one-dimensional scales for binary data; here we will consider briefly a class of structures that is defined by certain *contiguity relations* of a more complex nature than a one-dimensional ordering.

This class is characterized by *circular contiguity*. ➤ An example is given in Table 8.9, which shows for 6 items all 32 possible circular response profiles. In order to understand why these profiles are called circular, imagine Table 8.9 as being pasted on a cylinder: then the column for the last item becomes adjacent to the column for the first. A set of profiles satisfies circular contiguity if there exists a (re)ordering of the columns such that any one profile has no more than one run of '1's and no more than one run of '0's on

Table 8.9 Response profiles satisfying circular contiguity

A	B	C	D	E	F	DIA	GUT	CIR
1	1	1	1	1	1	√	√	
0	1	1	1	1	1	√		
1	0	1	1	1	1			
1	1	0	1	1	1			
1	1	1	0	1	1			
1	1	1	1	0	1			
1	1	1	1	1	0	√	√	
0	0	1	1	1	1	√		
1	0	0	1	1	1			
1	1	0	0	1	1			
1	1	1	0	0	1			
1	1	1	1	0	0	√	√	
0	1	1	1	1	0	√		
0	0	0	1	1	1	√		√
1	0	0	0	1	1			√
1	1	0	0	0	1			√
1	1	1	0	0	0	√	√	√
0	1	1	1	0	0	√		√
0	0	1	1	1	0	√		√
0	0	0	0	1	1	√		
1	0	0	0	0	1			
1	1	0	0	0	0	√	√	
0	1	1	0	0	0	√		
0	0	1	1	0	0	√		
0	0	0	1	1	0	√		
0	0	0	0	0	1	√		
1	0	0	0	0	0	√	√	
0	1	0	0	0	0	√		
0	0	1	0	0	0	√		
0	0	0	1	0	0	√		
0	0	0	0	1	0	√		
0	0	0	0	0	0	√	√	

the cylinder. Indeed, the 32 profiles of Table 8.9 all have just one run of '1' and just one run of '0' in the given order of the columns, except for the profiles '111111' and '000000', which are merely included for completeness.

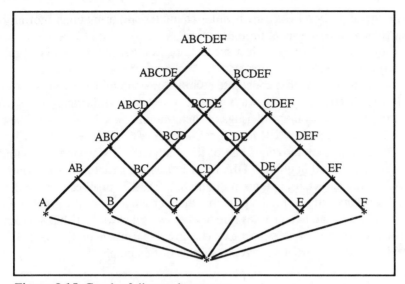

Figure 8.15 Graph of diamond structure.

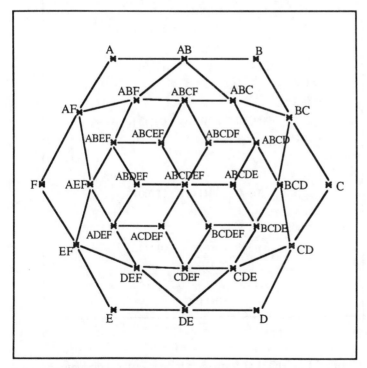

Figure 8.16 Graph of generalized circumplex structure.

Note that a property like this is not a characteristic of a single profile, but a characterization of a set of profiles.

From the table we can select subsets of rows that satisfy special regularity patterns. If we select the rows with a checkmark in the column GUT of Table 8.9 we obtain items that constitute a Guttman scale in the given order of the columns. There are many ways to select seven rows in agreement with a Guttman scale for other orderings of the columns ($m \times 2^{m-2} = 96$). The rows checked in the column CIR form a *circumplex of order three* (all profiles that can be obtained by moving a run of three adjacent '1's around the cylinder). The rows checked in column DIA form a *diamond* structure. ➤ Its contiguity relations are shown as a graph in Figure 8.15. The 21 profiles (profile 000000 is left out) are ordered on the principle that at each level of the graph a profile is the union of the two profiles directly below. All rows of Table 8.9 together form a *generalized circumplex*. The graph is given in Figure 8.16. All previous types of structure involve subsets of this graph.

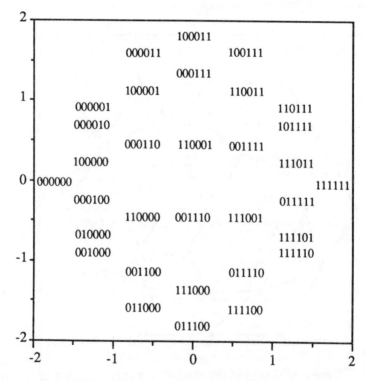

Figure 8.17a HOMALS solution for generalized circumplex (patterns of Table 8.9); dimensions 1 and 2.

The HOMALS solution for the generalized circumplex is displayed in Figure 8.17. The first dimension goes from 'north pole' (111111) to 'south pole' (000000). The graph now appears as if stretched over the surface of a sphere. Guttman scales appear as 'parallels' on the surface; circumplexes appear as 'meridians'.

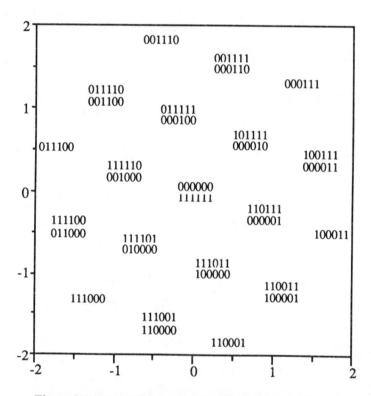

Figure 8.17b HOMALS solution for generalized circumplex (patterns of Table 8.9); dimensions 2 and 3.

It now becomes possible to recognize in the solution for the Sugiyama data of Section 8.6.2 the diamond structure. Figure 8.18 shows this for the 22 response profiles that fit a graph with 'bottom' row CABEDF (rotated 90 degrees) in agreement with the ordering in Figure 8.12 of the six elementary profiles from 000001 to 001000. This selection of profiles covers 3333 individuals (78 per cent if we include the zero profile). Figure 8.19 shows a comparable plot for the ANACOR solution (or HOMALS with all indicator matrices incomplete) that requires 111111 to become the centre of the plot and to this end folds the elementary points from 000001 to 001000 into a curve at the periphery of the plot. Note that 000000 does not appear in this plot.

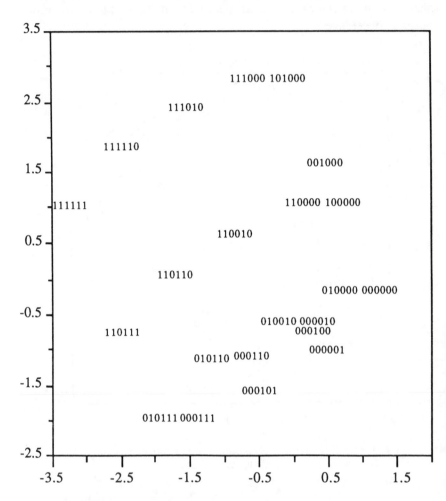

Figure 8.18 Diamond structure selected from the HOMALS configuration of the Sugiyama data, complete indicator matrix.

In conclusion, diagnosis of the particular type of contiguity is possible in both cases, although the complete indicator matrix approach requires a keen insight in three-dimensional space. The fact that in the latter approach the complement of each item is included as a category (a practice called *dédouble-*

ment in the French literature) is reflected in the appearence of mirror images on the northern and southern hemispheres that constitute the solution.

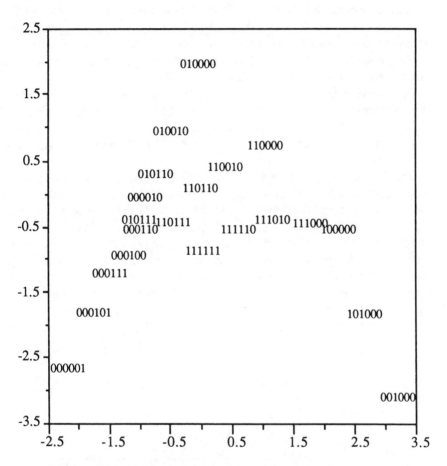

Figure 8.19 Diamond structure selected from the HOMALS configuration of the Sugiyama data, incomplete indicator matrix.

8.8 EPILOGUE

In this chapter homogeneity analysis has been discussed basically as an MDS technique that makes low-dimensional displays of data and focuses on the distances between points in these displays. Meulman (1986) took these developments a step further. She defines very general MDS techniques that can be specialized to ordinary MDS techniques on one side and to the present system of techniques on the other side. This idea is further extended in Meulman (1988, 1989) and Meulman and Heiser (1988). Degeneracies in unfolding and a remedy based on smoothing are discussed in Heiser (1989). The minimum squared distance formulation of homogeneity analysis is studied extensively in Heiser (1981, 1987a).

For a more detailed treatment of the analysis of the average similarity table, see the references mentioned in Section 3.9. The geometry of dédoublement is elegantly explained in terms of polarization by Greenacre (1984, sect. 6.1). Conditions under which dédoublement might be considered less desirable are given in Heiser (1981, 1985).

The discussion in Section 8.3 emphasizes standardized deviations from the independence model. Decomposition of deviations from other (loglinear) models has been proposed by Escofier (1983, 1984) and Van der Heijden and De Leeuw (1985). For a discussion of various allied issues in this active area of research, see Van der Heijden, De Falguerolles, and De Leeuw (1989). Anderson (1960) proposed parametrizations for the simplex, circumplex, and some similar kinds of dependencies in the plane. The book by Shye (1985) contains conceptual and algebraic studies of the diamond structure and other partial orders on profiles, as well as many examples of application.

CHAPTER 9
MODELS AS GAUGES FOR THE
ANALYSIS OF BINARY DATA

In the previous chapters we have seen many times that binary variables are somewhat special. The major reason is that quantification is unnecessary, so the distinction between numerical, ordinal, and nominal becomes irrelevant. We simply compute product moment correlations (phi coefficients) between the variables and perform linear multivariate analysis on the resulting correlation matrix. This assumes, of course, that there are no missing data, so that all $n \times 2$ indicator matrices G_j are complete. In Chapter 8 we have seen, however, that binary data are also interesting if they occur in incomplete indicator matrices.

Another reason to treat binary data separately is, of course, that they are very common. In the book by Coombs (1964), for example, this is amply illustrated. There are many 'yes–no' data matrices and many other data matrices are actually derived from this more basic form. Most data in psychological testing, archaeological seriation, ecological ordination, and political voting analysis are of this type. As a consequence many probabilistic and algebraic models have been developed which can be used as *gauges* for our techniques. This is one of the main preoccupations of this chapter. Because probabilistic gauges are quite prominent in this chapter we use the random variable notation. In Section 1.7.1 we have already indicated that the ordinary algebraic matrix notation we have used so far is a special case.

9.1 SOME GENERAL FORMULAS

Suppose $\underline{h}_j$ and $\underline{h}_l$ are binary variables. Define $\pi_j \equiv \text{AVE}(\underline{h}_j)$ and $\pi_l \equiv \text{AVE}(\underline{h}_l)$. Moreover, $\pi_{jl} \equiv \text{AVE}(\underline{h}_j\underline{h}_l)$. Then $\text{VAR}(\underline{h}_j) = \pi_j (1 - \pi_j)$ and $\text{VAR}(\underline{h}_l) = \pi_l (1 - \pi_l)$ and $\text{COV}(\underline{h}_j, \underline{h}_l) = \pi_{jl} - \pi_j\pi_l$. Because $0 \leq \pi_{jl} \leq \min(\pi_j, \pi_l)$ we find that

$$-\pi_j\pi_l \le \text{COV}(\underline{h}_j, \underline{h}_l) \le \min[\pi_j(1-\pi_l), \pi_l(1-\pi_j)]. \tag{9.1}$$

Thus if we define v_j by $v_j \equiv [\pi_j / (1-\pi_j)]^{1/2}$, then

$$-v_jv_l \le \text{COR}(\underline{h}_j, \underline{h}_l) \le \min(v_j/v_l, v_l/v_j). \tag{9.2}$$

Thus the phi coefficient $\phi_{jl} \equiv \text{COR}(\underline{h}_j, \underline{h}_l)$ is bounded by functions that depend on the marginals. We can only have $\phi_{jl} = +1$ if $\pi_j = \pi_l$; we can only have $\phi_{jl} = -1$ if $\pi_j = \pi_l = 1/2$. Because psychometricians like their correlations to be high they have suggested ϕ / ϕ_{max} as a measure of association, where ϕ_{max} is the upper bound derived above. There are other reasons to study ϕ/ϕ_{max}. They will become clear when we introduce some of the more common gauges.

9.2 MONOTONE LATENT TRAIT MODELS

Suppose $\underline{x}$ is an unobservable latent trait such that $p_j(x) \equiv \text{AVE}(\underline{h}_j \mid \underline{x} = x)$ is an increasing function of x, called the *traceline*, or *item characteristic curve*. Of course $\pi_j = \text{AVE}(p_j(\underline{x}))$, where the expectation is taken with respect to the distribution of the latent trait. We also assume *conditional* or *local independence:* for all $j \ne l$ we have

$$\text{AVE}(\underline{h}_j, \underline{h}_l \mid \underline{x} = x) = p_j(x)\, p_l(x), \tag{9.3}$$

which implies

$$\pi_{jl} = \text{AVE}(p_j(\underline{x})\, p_l(\underline{x})). \tag{9.4}$$

Because

$$\text{COV}(\underline{h}_j, \underline{h}_l) = \text{AVE}\{[p_j(\underline{x}) - p_j(\underline{y})]\,[p_l(\underline{x}) - p_l(\underline{y})]\}, \tag{9.5}$$

it is clear that $\text{COV}(\underline{h}_j, \underline{h}_l) \ge 0$, and thus $\phi_{jl} \ge 0$. In monotone latent trait models correlations are *never negative*. This improves the lower bound in Section 9.1. We know from Perron–Frobenius theory that nonnegativity of the correlation matrix implies the existence of a unique largest eigenvalue with corresponding nonnegative eigenvector. So the assumption of a monotone latent trait model imposes some structure on the observed correlation matrix, but not much. Figure 9.1 shows the item characteristic curves in a general latent trait model.

$p_j(x)$

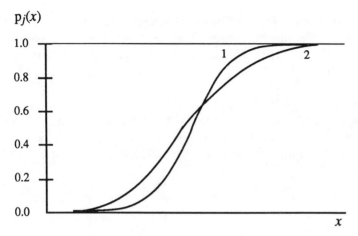

Figure 9.1 Item characteristic curves in a general latent trait model.

9.2.1 Holomorph items

Additional structure is imposed if we assume that the tracelines $p_j(x)$ do not cross. When item j is more difficult than item l for a single individual, this implies that the probability to answer item j correctly is smaller than the probability to answer item l correctly *for all individuals*, i.e. that $p_j(x) \leq p_l(x)$ for one single x means that $p_j(x) \leq p_l(x)$ for all x. Such curves $p_j(x)$ are called

$p_j(x)$

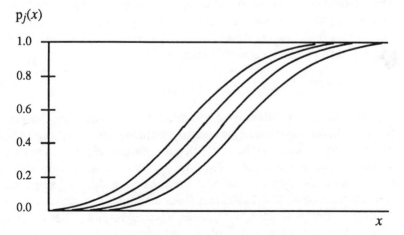

Figure 9.2 Holomorph item characteristic curves.

holomorph by Mokken (1970). The item characteristic tracelines are depicted in Figure 9.2. A measurement theory analysis of holomorphic systems, given by Levine (1970, 1972, 1975), shows that they can be written in the form $p_j(x) = p_j(x - \theta_j)$, with θ_j a location parameter. They form a family of shifted functions.

Now suppose $\theta_j \geq \theta_l$. We can make 2×2 tables of items j and l with another item g, and compare the entries. The tables are given in Figure 9.3, together with their marginals. With a '+' we indicate that this element is larger than the corresponding element in the other table, which then gets a '–'. Equal elements in both tables are denoted '='.

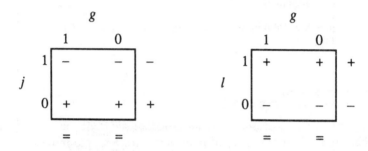

Figure 9.3 Comparing the cells of the bivariate distribution of holomorph items.

The order of the (1,1) and the (1,0) cells should be the same, and opposite to the order of the (1,0) and the (0,0) cells. The proofs of these relations are simple. If $\theta_j \geq \theta_l$ then $p_j(x) \leq p_l(x)$ for all x. The result for the (0,0) cell, for example, is then proved by observing that

$$[1-p_j(x)] \, [1-p_g(x)] \geq [1-p_l(x)] \, [1-p_g(x)] \tag{9.6}$$

and by taking expected values over both sides of this inequality. This ordering of the cells is an interesting structural property, which is explored, for example, in the book of Mokken (1970), but unfortunately it does not seem to have very clear consequences for the ϕ_{jl}, unless we make additional assumptions on the $p_j(x)$. In the next sections a number of gauges will be studied by generating correlation matrices according to specific models and inspecting their eigenvectors. Because with binary variables there is neither a distinction between numerical, ordinal, and nominal treatment, nor between single and multiple, this means that applying HOMALS comes to the same as applying linear PCA: there is only *one* correlation matrix.

9.2.2 The Guttman scale

Guttman (1944, 1950a, 1950b) introduced a model with items that constitute a breakpoint on the continuum: when an individual is located on the left-hand side of the breakpoint, the item will be answered incorrectly; when located on the right-hand side, the item will be answered correctly. The additional assumptions on $p_j(x)$ mentioned above are written as

$$p_j(x) = \begin{cases} 0 \text{ if } x < \theta_j \\ 1 \text{ if } x \geq \theta_j. \end{cases} \tag{9.7}$$

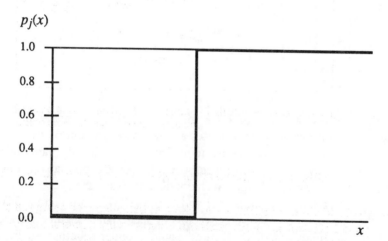

Figure 9.4 Item characteristic curve of a Guttman item.

The traceline of a Guttman item is given in Figure 9.4. The model is of breath-taking simplicity; it has a very simple interpretation in terms of physical operations and the probabilistic aspect is essentially trivialized, which makes the model algebraic. It follows from Guttman's assumption that $\pi_{jl} = \min(\pi_j, \pi_l)$, and in the 2×2 bivariate distribution at least one of the off-diagonal cells is zero. If we order the variables in such a way that $\theta_1 \geq \theta_m$, then also $\pi_1 \leq \dots \leq \pi_m$ and also $v_1 \leq \dots \leq v_m$. For $j \leq l$ this implies $\phi_{jl} = v_{jl} = v_j / v_l$, and for $j \geq l$, $\phi_{jl} = v_l / v_j$. It is thus necessary (and trivially also sufficient) for the existence of a Guttman scale (or *perfect* scale) that $\phi = \phi_{max}$ for all pairs of variables, or that $\omega = \phi / \phi_{max}$ is equal to 1 for all pairs. The use of ω in item analysis dates back to Loevinger (1947, 1948). Mokken (1970) uses ω_{jl} as a measure of the holomorphy of items, but this is somewhat risky. There are perfectly holomorphic items with arbitrarily small ω_{jl}. Examples can be constructed by using, for example, the formula for ϕ / ϕ_{max} in the Rasch model of Section 9.2.5. The programs developed by Mokken and his students must consequently be interpreted as methods to construct Guttman scales, not as tests of general holomorphy.

Table 9.1 Perfect Guttman scale

	A	B	C	D	E	F
1	0	0	0	0	0	0
2	1	0	0	0	0	0
3	1	1	0	0	0	0
4	1	1	1	0	0	0
5	1	1	1	1	0	0
6	1	1	1	1	1	0
7	1	1	1	1	1	1

Table 9.1 shows a data matrix that follows a perfect Guttman scale; it will be clear that Guttman's model imposes an enormous amount of structure on the correlation matrix. It is interesting to find out whether the same thing is true for the eigenvalues and eigenvectors of the correlation matrix. Guttman (1950a) gives a very satisfactory mathematical analysis; the necessary results were derived earlier by Gantmacher and Krein (1937). Guttman (1954) discusses the interpretation of the eigenvectors in detail, but since all eigenvectors are of course functions of the θ_j one could wonder whether such interpretations are useful.

We briefly discuss the most important results for the eigenvalues and eigenvectors. In the first place one could conjecture that if $\mathbf{b}$ is the eigenvector corresponding with the dominant eigenvalue and $v_1 \leq ... \leq v_m$, then also $b_1 \leq ... \leq b_m$. This is not true, however. In fact it is possible to prove that $b_1 \leq b_2$ and that $b_{m-1} \geq b_m$. We only show how to prove the first inequality; the second one is proved in the same way. We suppose that $\mathbf{b}$ is an eigenvector of a matrix $\mathbf{R}$ with positive elements, of order $m \geq 3$, and that all elements of $\mathbf{b}$ are also positive. Moreover, we know that $1 = r_{jj} \geq r_{j,j+1} \geq ... \geq r_{jm}$ for all j as well as $1 = r_{jj} \geq r_{j,j-1} \geq ... \geq r_{j1}$ for all j, because these are the ordinal properties of the correlation matrix of a perfect

Guttman scale, also called a *simplex*. The ordinal properties of the simplex are depicted in Figure 9.5.

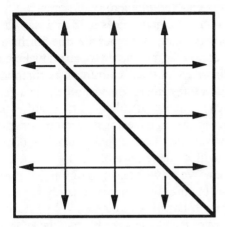

Figure 9.5 Ordinal properties of a simplex.

Because **b** is an eigenvector,

$$\sum_j (r_{1j} - r_{2j})\, b_j = \lambda^2\, (b_1 - b_2) \tag{9.8}$$

or

$$[\lambda^2 - (1 - r_{12})]\, (b_1 - b_2) = \sum_j (r_{1j} - r_{2j})b_j \quad (j = 3,...,m). \tag{9.9}$$

Because $r_{1j} \leq r_{2j}$ and $b_j > 0$ the term on the right is nonpositive. Moreover, $\lambda^2 > 1 - r_{12}$, because λ^2 is the unique largest eigenvalue of **R** (Perron–Frobenius theorem). Thus $b_1 \leq b_2$, and we have $b_1 = b_2$ if and only if $r_{1j} = r_{2j}$ for all $j = 3,...,m$. This is a negative result, but there are also some positive ones. For proofs we refer to Guttman (1950a), Gantmacher and Krein (1937, 1960), and Karlin (1964, 1968).

In the first place the inverse of the correlation matrix of a perfect scale is tridiagonal. This is a matrix with nonzero elements on its diagonal and in the entries immediately above and below the diagonal (the subdiagonals), and with zero elements elsewhere. In the second place it is useful to plot the elements of the eigenvectors b_s against the rank numbers of the v_j and connect successive points by straight line segments. This produces m polygonal functions, one for each eigenvector, with interesting properties. Function s, corresponding with eigenvector $\mathbf{b}_s$ and eigenvalue λ_s^2, has $s - 1$ zeros. Moreover, the zeros of successive functions are interwoven: between each pair of successive zeros of function s is exactly one zero of function $s - 1$. It is also true that if we choose the sign of the eigenvectors in such a way that all b_{1s} are nonnegative, then the same separation properties are true for the transpose of the matrix of eigenvectors. It is clear that these properties make it easy to recognize the correlation matrix of a Guttman scale from its eigenvalue-eigenvector properties. Another interesting property is that all eigenvalues are different if all v_j are different, which is true if and only if all π_j are different.

We have seen that the elements of the eigenvector $\mathbf{b}_1$ are not monotonic with the π_j. We must recall, however, that homogeneity analysis does not solve $\mathbf{Rb} = \lambda^2\mathbf{b}$ but $\mathbf{Cy} = \psi^2\mathbf{Dy}$, with $\mathbf{C}$ the Burt table and $\mathbf{D}$ the diagonal matrix of the univariate marginals. We have seen in Section 3.9 that the eigenvalues λ^2 and ψ^2 are the same, but the eigenvectors certainly are not. For one thing $\mathbf{y}$ has $2m$ elements, while $\mathbf{b}$ has only m elements. In fact, it was shown in Section 3.9 that the squares of the elements of $\mathbf{b}_1$ give the HOMALS discrimination measures.

Guttman (1950a) proves that the m 'positive' elements of $\mathbf{y}$ *are* monotone with π_j. We only indicate the general outline of the proof, working in the general context of minimization of

$$\sigma(\underline{x},\alpha,\beta) = m^{-1} \sum_j \text{SSQ}(\underline{x} - \alpha_j\underline{h}_j - \beta_j(1 - \underline{h}_j)). \tag{9.10}$$

The $\underline{h}_j$ and $\underline{x}$ are random variables defined on a common probability space, with elements that are supposed to be ordered. The $\underline{h}_j$ are known, and there are constants θ_j in the domain of the functions $h_j(\theta)$ such that

$$h_j(\theta) = \begin{cases} 0 \text{ if } \theta < \theta_j \\ 1 \text{ if } \theta \geq \theta_j \end{cases} \tag{9.11}$$

which implies that the $\underline{h}_j$ are all monotonic. Because $\pi_j = \text{AVE}(\underline{h}_j)$ and we suppose that $\pi_1 \leq ... \leq \pi_m$, we also have $\theta_1 \geq ... \geq \theta_m$. We now apply an alternating least squares algorithm, with the normalization conditions $\text{AVE}(\underline{x}) = 0$ and $\text{VAR}(\underline{x}) = 1$. We know that it converges from almost all starting points to the appropriate solution. Suppose we start with a monotonic $\underline{x}$. Then the least squares estimate of α_j is $\text{AVE}(\underline{x} \mid \underline{x} \geq \theta_j)$ and that of β_j is $\text{AVE}(\underline{x} \mid \underline{x} < \theta_j)$. Thus $\alpha_1 \geq ... \geq \alpha_m \geq 0$ and $0 \geq \beta_1 \geq ... \geq \beta_m$. In the second step of the alternating least squares algorithm we compute a new $\underline{x}$ for current $\boldsymbol{\alpha}$ and $\boldsymbol{\beta}$. This new $\underline{x}$ is the standardized version of

$$z = \sum_j \alpha_j h_j + \sum_j \beta_j (1 - h_j). \tag{9.12}$$

Because the elements of α are nonnegative and those of β are nonpositive, z is again monotonic. From the definition of α and β, moreover, $\text{AVE}(z) = 0$. After one cycle of the alternating least squares algorithm, starting with a monotone x, we have found weights (or category quantifications) α and β in the appropriate order, and a new x, which is still monotone. Thus subsequent iterations will not change the order properties, and we converge to a monotone x and to α and β in the appropriate order. The vector y, which has the form $y' = (\alpha', \beta')$, is consequently also in the appropriate order. The oscillation theorems about sign changes or zeros which were true for the eigenvectors b_s of R are also true for the vectors y_s. This situation accounts for the so-called *horseshoe* phenomenon in two-dimensional HOMALS solutions.

The properties of a Guttman scale are illustrated in Tables 9.2 to 9.4. The first row in Table 9.2 is π, the second row is the variance $\pi(1 - \pi)$, the third row is v, the next nine rows contain the correlation matrix R, and the last row contains the eigenvalues of $m^{-1}R$. Observe that the items in this example are rather

Table 9.2 Guttman scale statistics

π	0.01	0.04	0.08	0.14	0.20	0.26	0.33	0.39	0.45
$\pi(1-\pi)$	0.01	0.04	0.07	0.12	0.16	0.19	0.22	0.24	0.25
v	0.10	0.20	0.30	0.40	0.50	0.60	0.70	0.80	0.90
R	0.50								
	0.33	0.67							
	0.25	0.50	0.75						
	0.20	0.40	0.60	0.80					
	0.17	0.33	0.50	0.67	0.83				
	0.14	0.29	0.43	0.57	0.71	0.86			
	0.13	0.25	0.38	0.50	0.63	0.75	0.88		
	0.11	0.22	0.33	0.44	0.55	0.67	0.78	0.89	
λ^2	0.58	0.18	0.39	0.05	0.03	0.02	0.02	0.01	0.01

difficult; all π_j are less than 0.50. As a consequence, the variances are an increasing function of the variable number, which is not true in other examples with both difficult and easy items.

Table 9.3 Eigenvectors and generalized eigenvectors of Guttman scale

0.14	0.49	0.69	0.48	0.20	0.05	0.01	0.00	0.00
0.24	0.52	0.12	-0.51	-0.56	-0.28	-0.08	-0.01	-0.00
0.31	0.39	-0.29	-0.31	0.39	0.58	0.29	0.06	0.00
0.36	0.20	-0.40	0.17	0.36	-0.37	-0.57	-0.22	-0.03
0.38	0.00	-0.29	0.39	-0.16	-0.33	0.48	0.49	0.10
0.39	-0.16	-0.08	0.29	-0.40	0.27	0.16	-0.62	-0.29
0.38	-0.26	0.12	0.03	-0.22	0.38	-0.44	0.24	0.56
0.37	-0.32	0.26	-0.21	0.12	0.00	-0.16	0.38	-0.68
0.34	-0.32	0.31	-0.32	0.34	-0.34	0.34	-0.34	0.34
1.42	4.94	6.93	4.85	2.01	0.55	0.09	0.01	0.00
1.25	2.72	0.63	-2.65	-2.91	-1.45	-0.39	-0.05	-0.00
1.13	1.42	-1.05	-1.11	1.41	2.11	1.04	0.22	0.02
1.04	0.57	-1.16	0.49	1.06	-1.08	-1.65	-0.65	-0.08
0.96	0.01	-0.72	0.98	-0.40	-0.82	1.20	1.24	0.26
0.89	-0.35	-0.18	0.66	-0.92	0.61	0.35	-1.40	-0.66
0.82	-0.56	0.26	0.07	-0.47	0.82	-0.94	0.51	1.20
0.75	-0.65	0.54	-0.43	0.25	0.00	-0.34	0.78	-1.40
0.68	-0.65	0.63	-0.65	0.67	-0.68	0.68	-0.69	0.69

Table 9.4 Zeros of polygonal Guttman eigenvectors, generalized eigenvectors, and transposed eigenvectors

```
                                5.02
                         2.29        6.39
                  1.49        3.64        7.14
           1.26        2.59        4.70        7.64
    1.16        2.32        3.61        5.55        8.00
1.11      2.21        3.34        4.54        6.26        8.32
  1.08     2.14        3.22        4.31        5.44      6.72      8.53
1.05       2.10        3.15        4.20        5.26        6.34        7.45        8.67

                                3.19
                         2.57        4.44
                  2.33        3.70        5.49
           2.01        3.43        4.71        6.41
    1.72        3.22        4.42        5.60        7.20
1.59      2.68        4.13        5.36        6.47        7.65
  1.54     2.55        3.56        4.63        6.00        7.30        8.36
1.51       2.51        3.49        4.49        5.50        6.50        7.50        8.50

                                5.02
                         2.37        6.41
                  1.65        3.69        7.14
           1.41        2.67        4.73        7.65
    1.27        2.41        3.66        5.57        8.00
1.19      2.27        3.39        4.58        6.27        8.33
  1.14     2.19        3.26        4.34        5.47        6.73        8.53
1.10       2.13        3.18        4.23        5.28        6.36        7.46        8.67
```

Table 9.3 contains the eigenvectors of **R** *in the first nine rows and the generalized eigenvectors, which are the variances, in the last nine rows. The eigenvectors of* **R** *are plotted against the variable number in Figure 9.6; the first set of nine points of the first eigenvector is the first polygonal function and the horizontal axes show the property of the increasing number of zeros. The interlacing of zeros is illustrated in Table 9.4, with the location of the zeros of the polygonal functions in Figure 9.6 in its first eight rows. We also have computed the zeros of the transpose of the matrix of eigenvectors and of the generalized eigenvectors. These zeros are given, respectively, in the other two subtables of Table 9.4. Observe that all components are difficulty factors in the sense of Guilford (1941) or Ferguson (1941), because they are all functions of the item difficulty.*

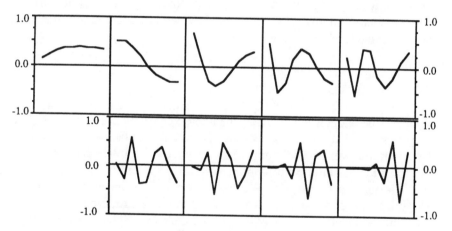

Figure 9.6 Eigenvectors of a Guttman correlation matrix.

If we do not know the distribution of the individuals on the latent continuum, then we cannot recover the θ_j, only their order. Thus we have shown that the one-dimensional HOMALS solution recovers the interesting information without distortion and that the remaining solutions show an interesting pattern which makes clear that we are dealing with something close to a perfect scale. There are some interesting special cases, e.g. $v_j = v^j$, in which the eigenvalues and eigenvectors of **R** can be determined explicitly (Guttman, 1950b; Goldberg, 1958, pp. 184–189), but we shall not go into this any further. It is also clear that computing eigenvalues and eigenvectors of **R** is not the simplest way to test if we are dealing with a perfect scale. Of course it is easier to use the relationship $-\ln \phi_{jl} = |\eta_j - \eta_l|$, with $\eta_j \equiv \ln v_j = 1/2$ logit π_j, in combination with, for example, a scaling program. Such a procedure, however, makes no sense if the data do not conform to the perfect scale pattern, while eigenvector–eigenvalue techniques also provide interesting information in nonperfect cases.

It is sometimes said in psychological scaling literature that the Guttman scale is not realistic, and consequently not interesting. In the first place this remark seems based on the wrong interpretation of the concept of a gauge (or model). The Guttman Scale, the Normal Distribution, the Spearman Hierarchy, the Republic of Plato, and the Life of Jesus are all *norms:* they show what happens if things are perfect. It is not relevant here that the real world is not perfect. In the second place, the other models for binary data are not very realistic either. People have rejected Guttman scaling partly because it became connected with nonmetric error theories and heuristic algorithms based on permutation and search. We have observed the oscillatory principal components in almost all properly constructed attitude and aptitude scales, usually in the form of the horseshoe.

9.2.3 The Spearman hierarchy

We describe another gauge under the name *Spearman hierarchy*. This name may be somewhat confusing because Spearman's one factor model, which we have in mind, was certainly not meant for binary variables. Spearman started out from the assumption of continuous variation, and constructed a simple linear latent structure model. We construct a similar linear model for binary variables and show that it produces a similar correlation matrix as the continuous model. This is why we still use the name 'Spearman hierarchy'.

We now assume that $p_j(x) = \alpha_j x + \beta_j$. Without loss of generality we can also assume that $\mathrm{AVE}(\underline{x}) = 0$ and $\mathrm{VAR}(\underline{x}) = 1$. This implies immediately that $\pi_j = \beta_j$ and $\mathrm{COV}(\underline{h}_j, \underline{h}_l) = \alpha_j \alpha_l$. Thus if $\mu_j \equiv \alpha_j / [\beta_j(1 - \beta_j)]^{1/2}$, then $\phi_{jl} = \mu_j \mu_l$. The Spearman hierarchy, which is defined by the linear item characteristic function shown in Figure 9.7, thus gives a correlation matrix of the form $\mathbf{R} = \mathbf{\mu}\mathbf{\mu}' + \Delta^2$, with Δ^2 diagonal. If δ_j^2 is the jth diagonal element of Δ^2, then $\delta_j^2 = 1 - \mu_j^2$. Observe also that if variables j and l have the same parameters α and β, and thus also the same μ, then $\phi_{jl} = \mu^2$, which shows that $\mu_j^2 = 1 - \delta_j^2$ can be interpreted as the *reliability* of variable j.

At this point the reader may lose patience, because it is somewhat outlandish to model probabilities by a linear model. The $p_j(x)$ must be between zero and one, and linear functions are unbounded on the real line; thus the variation of x is restricted by the values of all the α_j and β_j. This seems undesirable, and although the Spearman hierarchy gives a simple correlation matrix, depicted in Figure 9.8, it does not seem obvious at all that it is a useful gauge or even a powerful norm such as the Guttman scale. However, consider the following argument.

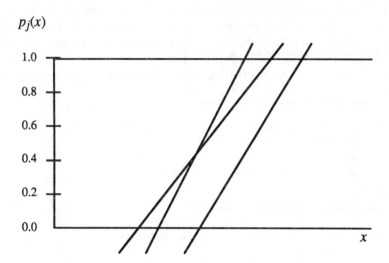

Figure 9.7 Item characteristic curves according to the Spearman hierarchy.

In a general latent trait model with item characteristic curves $p_j(x)$ we represent the latent variable x in the form $x = \tau + \varepsilon z$, with $\text{AVE}(z) = 0$ and $\text{VAR}(z) = 1$. We define $\alpha_j \equiv p_j'(\tau)$, assuming that p_j is differentiable at τ, and $\beta_j \equiv p_j(\tau)$. Moreover, μ_j is defined in terms of α_j and β_j as in the Spearman hierarchy. What happens if ε is small, i.e. if x is distributed closely around τ? It is clear that continuity of p_j at τ already implies that $\phi_{jl} \to 0$ if $\varepsilon \to 0$. Assuming that p_j is twice continuously differentiable at τ, however, makes it possible to show that

$$\lim_{\varepsilon \to 0} \phi_{jl}/\varepsilon^2 = \mu_j\mu_l, \tag{9.13}$$

or

$$\phi_{jl} = \varepsilon^2\mu_j\mu_l + o(\varepsilon^2). \tag{9.14}$$

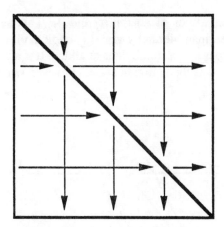

Figure 9.8 Ordinal properties of a correlation matrix of a Spearman hierarchy.

Thus for small dispersion on the latent continuum parts of a monotone traceline can very well be linearly approximated (see Figure 9.9). We find a correlation matrix (with small correlations) that approximates the correlation matrix of a Spearman hierarchy.

We have seen that in the perfect Guttman scale the successive eigenvectors of the correlation matrix have more and more peaks and zeros. Results for the

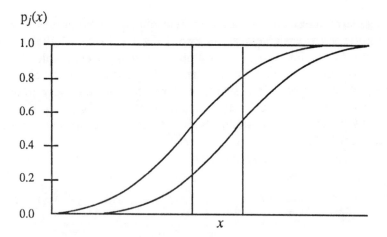

Figure 9.9 Local approximations of parts of monotone tracelines are possibly linear.

Spearman hierarchy are quite different. Suppose the diagonal elements in Δ^2 are ordered as $\delta_1^2 \geq ... \geq \delta_m^2$. The eigenvalues of $\mathbf{R} = \mathbf{\mu}\mathbf{\mu}' + \Delta^2$ then satisfy

$$\lambda_1^2 \leq \delta_1^2 + \mathbf{\mu}'\mathbf{\mu}, \tag{9.15}$$

$$\lambda_1^2 \geq \delta_1^2 \geq \lambda_2^2 \geq ... \geq \lambda_m^2 \geq \delta_m^2, \tag{9.16}$$

$$\lambda_m^2 \leq \delta_m^2 + \mathbf{\mu}'\mathbf{\mu}. \tag{9.17}$$

These results have been proved in many places; a convenient reference is Bunch, Nielsen, and Sorensen (1978). Thompson (1976) is also particularly elegant. Suppose $\mathbf{b}_s$ is the eigenvector corresponding with λ_s^2, and normalize $\mathbf{b}_s$ in such a way that $\mathbf{b}_s'\mathbf{\mu} = 1$ (thus we do not consider those eigenvectors with $\mathbf{b}_s'\mathbf{\mu} = 0$, which causes no real loss of generality). Then

$$b_{js} = \frac{\mu_j}{\lambda_s^2 - \delta_j^2)} . \tag{9.18}$$

Plotting the elements of the eigenvectors against the variable number gives interesting plots. For the first eigenvector, corresponding with the largest eigenvalue, the plot is nonnegative and monotone (if $\mathbf{\mu} \geq 0$). For the other eigenvectors there is only one sign change – the plot jumps from high positive down to low negative. The location of the jump shifts one place for each successive eigenvector; in the interval in which there is no jump the polygonal functions are increasing. Tables 9.5 and 9.6 contain

Table 9.5 Spearman hierarchy statistics

μ	0.10	0.20	0.30	0.40	0.50	0.60	0.70	0.80	0.90
R	0.02								
	0.03	0.06							
	0.04	0.08	0.12						
	0.05	0.10	0.15	0.20					
	0.06	0.12	0.18	0.24	0.30				
	0.07	0.14	0.21	0.28	0.35	0.42			
	0.08	0.16	0.24	0.32	0.40	0.48	0.56		
	0.09	0.18	0.27	0.36	0.45	0.54	0.63	0.72	
λ^2	3.33	0.99	0.95	0.90	0.82	0.72	0.59	0.44	0.26

*the Spearman hierarchy illustrations. Table 9.5 has μ in its first row, **R** in its next nine rows, and λ², the eigenvalues of **R**, in its last row. Table 9.6 has the eigenvectors of **R**, which are plotted in Figure 9.10.*

Table 9.6 Spearman hierarchy eigenvectors

0.07	0.99	0.07	0.04	0.03	0.02	0.02	0.01	0.01
0.14	−0.09	0.97	0.12	0.07	0.05	0.03	0.03	0.02
0.21	−0.05	−0.17	0.94	0.16	0.09	0.06	0.04	0.03
0.27	−0.03	−0.09	−0.25	0.90	0.18	0.10	0.07	0.04
0.33	−0.03	−0.06	−0.12	−0.34	0.84	0.20	0.11	0.07
0.38	−0.02	−0.05	−0.09	−0.16	−0.44	0.76	0.20	0.10
0.42	−0.02	−0.04	−0.07	−0.11	−0.19	−0.55	0.66	0.18
0.45	−0.02	−0.03	−0.05	−0.08	−0.13	−0.22	−0.67	0.52
0.48	−0.01	−0.03	−0.05	−0.07	−0.10	−0.14	−0.24	−0.82

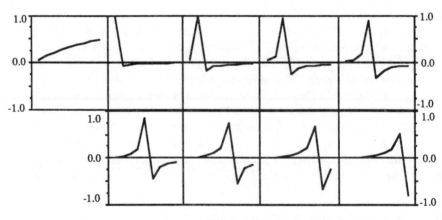

Figure 9.10 Eigenvectors of a Spearman correlation matrix.

The pattern in the eigenvector plots is again very characteristic and very different from the patterns of the Guttman scale. The patterns with the peaks have already been discussed informally in some older factor analysis literature, mainly in papers by Burt and Thomson (for example, Thomson, 1934). In the binary data context we again do not use the hierarchy as a realistic model, but as a normative gauge. Guttman shows us what happens when perfect discrimination is possible; Spearman shows us what happens when discrimination is hardly possible. It is perfectly trivial to remark that real data are usually in between these two extremes.

9.2.4 The latent distance model

There have been a number of attempts to make the Guttman scale more realistic, usually by relaxing the zero–one property that makes the Guttman model algebraic rather than probabilistic. These attempts have not been very successful, mainly because making the model more realistic means that it becomes less successful as a norm. There are many adjustable parameters, and there is a certain risk that we descend completely from the *a priori* level of norms, gauges, models, and theories to the empirical level of *fitting* models with an unlimited number of degrees of freedom. This is what happened in factor analysis, when the Spearman hierarchy was replaced by Thurstone's multiple factor analysis. Although many people realized this at the time, and Guttman has emphasized it again and again in the fifties, this actually meant that all theory was banished from factor analysis, and that factor analysis was no longer psychology but statistics. Because we are primarily interested in gauges in this chapter, and not in description of reality, we do not discuss the latent polynomial models or the latent class models of Lazarsfeld or the multidimensional models of Coombs and Kao. They generalize the Guttman scale in the direction of multiple factor analysis. We refer to Coombs (1964, chs. 10, 11, 12), to Lazarsfeld and Henry (1968), and to McDonald (1967) for some interesting discussion of these models.

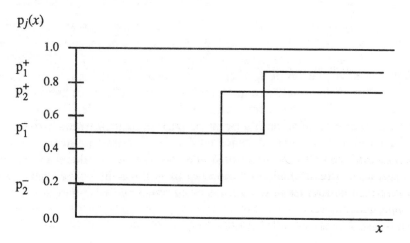

Figure 9.11 Tracelines in the latent distance model.

In this section we discuss the simplest modification of the Guttman scale, also due to Lazarsfeld (1950). It is called the latent distance model. Its tracelines are given in Figure 9.11.

The latent distance model is defined by

$$p_j(x) = \begin{cases} \pi_j^- & \text{if } x < \theta_j \\ \pi_j^+ & \text{if } x \geq \theta_j \end{cases} \tag{9.19}$$

with, of course, $0 \leq \pi_j^- \leq \pi_j^+ \leq 1$. Thus $p_j(x)$ is still a step function, but it does not step from 0 to 1 any more. Also define

$$\kappa_j \equiv \text{prob}(\underline{x} < \theta_j), \tag{9.20}$$

$$\varepsilon_j \equiv \pi_j^+ - \pi_j^-. \tag{9.21}$$

After some computation it follows that

$$\pi_{jl} - \pi_j \pi_l = \begin{cases} \kappa_j(1 - \kappa_l) \, \varepsilon_j \varepsilon_l & \text{if } \theta_j \leq \theta_l \\ \kappa_l(1 - \kappa_j) \, \varepsilon_j \varepsilon_l & \text{if } \theta_j \geq \theta_l . \end{cases} \tag{9.22}$$

Thus if

$$\xi_j \equiv \kappa_j \varepsilon_j / [\pi_j(1 - \pi_j)]^{1/2}, \tag{9.23}$$

$$\zeta_j \equiv (1 - \kappa_j)\varepsilon_j / [\pi_j(1 - \pi_j)]^{1/2}, \tag{9.24}$$

then $\phi_{jl} = \xi_j \zeta_l$ if $\theta_j \leq \theta_l$, $\phi_{jl} = \zeta_j \xi_l$ if $\theta_j \geq \theta_l$, and $\phi_{jj} = 1$ for all j. This means that we can write $\mathbf{R} = \mathbf{R}_o + \Delta^2$, with Δ^2 diagonal and nonnegative and with $\mathbf{R}_o$ a *one-pair matrix* in the sense of Gantmacher and Krein (1937). One-pair matrices have the property that their inverse is tridiagonal, in this case symmetric tridiagonal, and that their eigenvectors have the oscillation properties we discussed in connection with the perfect scale. Thus after fitting the 'uniqueness' Δ^2 we are back in the perfect scale situation. Or, to put it differently, for the latent distance model the correlation matrix is a *quasi-simplex*. If we know the order of θ_j then we can fit the model by using the logarithmic transformation of the correlations (or the covariances) again. Eigenvector properties of $\mathbf{R}$ itself are not simple in general, they are of course simple if $\Delta^2 = \delta^2 \mathbf{I}$.

9.2.5 The Rasch model

Another probabilistic generalization of the Guttman scale is the Rasch model. It has some very interesting mathematical properties which make it quite successful as a gauge, although we shall see that the correlation properties are not very simple. ⇢ In the Rasch model (Rasch, 1960, 1961, 1966; Fischer, 1974) the individuals are distributed on the positive real axis that defines the latent trait, and

$$p_j(x) = x/(x + \theta_j).$$

(9.25)

This looks simple and because it is a rational function, it is fairly easy to integrate. However, the most important simplification is the following. It is possible to prove that

$$\pi_{jl} = \frac{\pi_j \theta_j - \pi_l \theta_l}{\theta_j - \theta_l},$$

(9.26)

and this relation is true no matter what the distribution of individuals on the latent continuum is. The proof is based on the algebraic identity

$$[1 - p_j(x)]\,[1 - p_l(x)] = \frac{\theta_l[1 - p_j(x)] - \theta_j[1 - p_l(x)]}{\theta_l - \theta_j},$$

(9.27)

which assumes, of course, that $\theta_j \neq \theta_l$. If we integrate both sides of the identity and collect terms we find the desired result. The most important consequence of the result is that

$$\frac{\pi_l - \pi_{jl}}{\pi_j - \pi_{jl}} = \frac{\theta_j}{\theta_l}$$

(9.28)

which makes it possible to compute the θ_j without any assumption about the distribution of x. Rasch calls this *specific objectivity,* and shows that it is characteristic for his model. It certainly is a very nice property, analogous to measurement situations in the physical sciences and quite uncommon in psychometrics. It is this property that makes the Rasch model a suitable gauge.

Some additional computation gives the following formula for the phi coefficient

$$\phi_{jl} = \frac{1}{\theta_j - \theta_l} \left(\theta_j \frac{v_j}{v_l} - \theta_l \frac{v_l}{v_j} \right) \qquad (9.29)$$

This implies that the Loevinger–Mokken coefficient ω_{jl}, in the case that $\theta_j \geq \theta_l$, is given by

$$\omega_{jl} = \frac{1}{\theta_j - \theta_l} \left(\theta_j - \theta_l \frac{v_l^2}{v_j^2} \right) \qquad (9.30)$$

This formula illustrates our remark in Section 9.2.1 that perfectly holomorph items, in this case Rasch items, can have small ω_{jl}.

It is possible to substitute some distributions on the latent continuum into these formulas and to compute the corresponding correlation matrices. This is done, for example, in De Leeuw (1973, sect. 3.20) and in Gifi (1980, sect. 2.2.6). We shall not perform these computations here, but we mention an additional theoretical result of some interest. Gantmacher and Krein (1937, p. 457) show that the Cauchy determinant formula implies that the function $1/(x_j + y_j)$ is *totally positive*. According to the general composition formula of Karlin (1968, pp. 16–18) this means that the matrix with off-diagonal elements π_{jl} and diagonal elements $\text{AVE}(p_j^2(\underline{x}))$ is totally positive. This implies that the correlation matrix $\mathbf{R}$ of the Rasch model is of the form $\mathbf{R} = \mathbf{R}_o + (\Delta^2 - \mu\mu')$, with $\mathbf{R}_o$ totally positive, with Δ^2 diagonal, and with μ nonnegative. This representation is useful, because totally positive matrices have the same eigenvector properties as the perfect scale, and thus the Rasch correlation matrix can be thought of as the difference between $\mathbf{R}_o$, a perfect scale-type correlation matrix, and $\mu\mu' - \Delta^2$, a Spearman-type correlation matrix.

9.3 NONMONOTONIC LATENT TRAIT MODELS

Models with increasing $p_j(x)$ are inspired by psychological test theory. If individual i is 'better' than individual k, then the probability that i gives a correct response is larger than the probability that k gives a correct response. For many binary data matrices, however, the notions of correctness and 'being better' do not apply. In these cases individuals give positive responses to questions with which they agree, and the idea is that in many situations it is more realistic to suppose that the $p_j(x)$ are *unimodal*. Plants, for example, need a certain amount of moisture. If the soil is too moist, then certain types of plants will not occur; if the soil is too dry they will not occur either. There is a certain range that they can handle, different types of plants have different ranges. There are some plants, however, with monotone moisture behaviour. Plants that grow in water are a simple example. It is unwise of course to use land and water plants in the same analysis, also because most models do not take into account that moisture is bounded both from above and from below. A similar situation occurs in archaeology. Some utensils occur in graves during a particular time interval, and not outside this interval. It is possible that there are utensils that occur more and more as time progresses, but this is highly unlikely, and we never know if eventually the rate of occurrence will decrease again. Thus unimodal models are more natural in this case too, and monotone characteristic curves are best thought of as having a mode somewhere near infinity. In psychology unimodal models seem more natural in many similarity and preference contexts. To use Carroll's (1972) example: people who like cold beverages usually do not want them to be completely frozen, and people who like their coffee hot do not want it to be actually boiling.

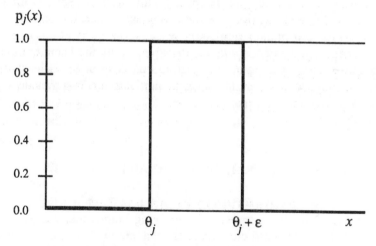

Figure 9.12 Traceline of the Coombs scale.

The model corresponding with the perfect scale in unimodal situations is the *Coombs scale* (see Figure 9.12). Here $p_j(x) = 1$ if $\theta_j \leq x \leq \theta_j + \varepsilon_j$, and $p_j(x) = 0$ otherwise. There is no special reason why the Coombs scale is less compelling or less natural than the Guttman scale, but it is considerably less important in practice for technical reasons. In fact, if items vary in both their θ_j and their ε_j, then there is no natural way to order items completely, although there are many ways to order them partially. The *restricted Coombs scale* solves this problem by letting $p_j(x) = 1$ if and only if $\theta_j \leq x \leq \theta_j + \varepsilon$, with ε the same for all j. This certainly solves the problem because now we can simply order the θ_j and apply correspondence analysis to the *incomplete* indicator matrix to recover the order (an observation due to Mosteller in 1949; cf. Torgerson, 1958, p. 338). This result is also a consequence of general results in the next section. It indicates that we should *not* analyse **R** in the case of a Coombs scale, because analysis of **R** corresponds with analysis of the complete indicator matrix. See also Chapter 8 and the epilogue to this chapter. ➤

The Coombs scale is deterministic in the same sense as the Guttman scale. There are some obvious probabilistic generalizations possible based on standard unimodal densities from the exponential family. Consider, for example, a latent variable distributed on the real line, with

$$p_j(x) = K_j \exp[-1/2\ \eta_j(x - \theta_j)^2]. \tag{9.31}$$

This combines nicely with a normal distribution on the latent continuum; it also gives totally positive π_{jl} in the same way as the Rasch model. It is not, however, a satisfactory gauge, because we can think of hundreds of models like this, choose convenient *conjugate* distributions on the latent continuum, and are left with an appalling degree of arbitrariness, as in Bayesian statistics. It seems quite likely, however, that a gauge similar to the Rasch model can be derived for nonmonotonic items as well. ➤

9.4 ORDER ANALYSIS OF BINARY MATRICES

In Section 9.2.2 we have shown that applying homogeneity analysis to a perfect scale gives category quantifications that are in the correct order. This argument applies to complete indicator matrices of order $n \times (2m)$, where in fact n can be infinite. In the previous section we have stated, but not shown, that applying correspondence analysis to the restricted Coombs scale also produces quantifications in the correct order, provided we code them in an incomplete indicator matrix, with order $n \times m$.

In this section we provide the tools to prove this and similar statements. These tools can be used for order analysis of binary matrices, in which we order the rows and columns of a binary matrix in the order of the left and right dominant eigenvectors from a correspondence analysis. In the case of the perfect scale and the incompletely coded Coombs scale this reordering produces a diagonal pattern of 1's along the diagonal of the table.

Now suppose C is a pointed, closed, convex cone that is solid (has non-empty interior) (cf. Appendix C). We use the theory of K-positive matrices; a review is in Berman (1973, pp. 51–54), who also gives the necessary references. First define the cone of $n \times n$ matrices

$$\Pi(C) = \{A \mid AC \subseteq C\}, \tag{9.32}$$

where we assume that C is a cone in $\mathfrak{R}^n$. The notation in (9.32) expresses the fact that A maps all vectors in C into C itself. Thus if $A \in \Pi(C)$ then $Ax \in C$ for all $x \in C$. The following results are very useful. If $A \in \Pi(C)$ then $\rho^2(A)$, the spectral radius of A, is an eigenvalue of A, and there is a nonzero $x \in C$ such that $Ax = \rho^2(A)x$ and a nonzero $y \in C^0$ (the polar cone, cf. Appendix C) such that $A'y = \rho^2(A)y$. If we assume K-positivity, i.e. $A(C - \{0\})$ is a subset of the interior of C, then it even follows that $\rho^2(A)$ is the unique largest eigenvalue, that x is in the interior of C, and that x is the only eigenvector in C. If P and Q are frames for C and C^0, i.e.

$$C = \{x \mid x = Pt \text{ for some } t \geq 0\}, \tag{9.33}$$

$$C^0 = \{y \mid y = Qu \text{ for some } u \geq 0\}, \tag{9.34}$$

then $A \in \Pi(C)$ if and only if $Q'AP \geq 0$ (elementwise). Furthermore, if $\rho^2(A)$ is an eigenvalue of A and A is similar to a diagonal matrix, then $A \in \Pi(C)$ for some full (= pointed, solid, closed, convex) cone C. If $\rho^2(A)$ is the unique largest eigenvalue of A, then A maps some full cone into its interior.

We can now apply these results to the linear transformations used in correspondence analysis. If F is a nonnegative matrix, and D_C and D_R are the diagonal matrices of column totals and row totals, then we are interested in the operator $D_R^{-1}FD_C^{-1}F'$. Very similar results are of course possible for $D_C^{-1}F'D_R^{-1}F$. In the first place matrices of this form are always semi-simple (similar to a diagonal matrix) and the spectral radius is always an eigenvalue of $D_R^{-1}FD_C^{-1}F'$. The same thing is true if we remove the trivial solution and work with the operator A defined as

$$A \equiv D_R^{-1}(F - N^{-1} D_R \, uu'D_C)D_C^{-1}(F - N^{-1}D_R uu'D_C)'$$

$$= D_R^{-1}FD_C^{-1}F' - N^{-1}uu'D_R. \tag{9.35}$$

In this formula the grand total $N \equiv \mathbf{u}'\mathbf{D}_C\mathbf{u} = \mathbf{u}'\mathbf{D}_R\mathbf{u}$. The general theory tells us that we can check whether a cone C is reproduced by this operator by using frames for C and C^o. We do this for the cone

$$C = \{\mathbf{x} \mid x_1 \leq \ldots \leq x_n\}. \tag{9.36}$$

A frame for C consists of the columns of the matrix $\mathbf{S}$, which has zeros above the diagonal and ones below and on the diagonal, together with the vector $(-\mathbf{u})$. A frame for C^o consists of the $n - 1$ vectors $(\mathbf{e}_{i+1} - \mathbf{e}_i.)$ Since our correspondence analysis operator $\mathbf{A}$, with the trivial solution removed, satisfies $\mathbf{A}\mathbf{u} = \mathbf{0}$, we must have $(\mathbf{e}_{i+1} - \mathbf{e}_i)'\mathbf{A}\mathbf{S} \geq 0$, which we can write as

$$\sum_k a_{i+1,k} \leq \sum_k a_{i,k} \quad (k = 1,\ldots,l) \tag{9.37}$$

for all $l = 1, \ldots, n$. This condition is fairly easy to check in many cases. ➤ It is sometimes more convenient to work with two cones C_X and C_Y, and the operators $\mathbf{D}_R^{-1}\mathbf{F}$ which maps $\mathfrak{R}^m$ into $\mathfrak{R}^n$, and $\mathbf{D}_C^{-1}\mathbf{F}'$ which maps $\mathfrak{R}^n$ into $\mathfrak{R}^m$. We want to find out if $\mathbf{D}_R^{-1}\mathbf{F}C_Y$ is a subset of C_X and $\mathbf{D}_C^{-1}\mathbf{F}'C_X$ is a subset of C_Y. Clearly this implies that $\mathbf{D}_R^{-1}\mathbf{F}\mathbf{D}_C^{-1}\mathbf{F}' \in \Pi(C_X)$ as well as $\mathbf{D}_C^{-1}\mathbf{F}'\mathbf{D}_R^{-1}\mathbf{F} \in \Pi(C_Y)$. This is actually how we proceeded with the perfect scale. The method can also be easily adapted to infinite dimensional spaces, as we have illustrated in Section 9.2.2 where x is a random variable with finite variance, and not necessarily an *n-element* vector.

9.5 DICHOTOMIZED MULTINORMAL DISTRIBUTIONS

Another popular model for binary data, not based on latent trait theory, is the dichotomized multinormal. We have seen in Section 1.6.1 that the popularity of this model is due to Pearson's philosophy that continuous variation is the norm, that correlation should replace causation, and that measures of association should be approximations of the product moment correlation of the underlying normal process. Although the fact that these techniques have been used a great deal makes them familiar and even natural, we must ask the reader to reflect a moment on the assumption that archaeological findings, answers to multiple choice items, or occurrences of vegetation in certain areas result from dichotomizing a multinormal process. It seems to us that the assumption is not very natural in most situations, and sometimes very silly indeed. It is also not clear in what sense this model is very special, what turns it into an interesting gauge. For completeness, however, we list some of the consequences of the assumptions.

We suppose that $z_1, ..., z_m$ are joint multivariate normal, with zero means and unit variances, and with correlations ρ_{jl}. We also suppose that the observed h_j are formed by the rule $h_j = 0$ if $z_j < \theta_j$ and $h_j = 1$ if $z_j \geq \theta_j$. In connection with principal components analysis and factor analysis this model was used for the first time by Lawley (1944), although he used the classical tetrachorical theory of Pearson in his derivations. We can use the Hermite–Chebyshev polynomials ψ_s to show that we can represent the correlations in the form

$$COR(h_j, h_l) = \sum_{s=1}^{\infty} \rho_{jl}^s \tau_{js} \tau_{ls} \qquad (9.38)$$

with

$$\tau_{js} \equiv (s!)^{-1/2} \phi(\theta_j) \psi_{s-1}(\theta_j) \{\Phi(\theta_j) [1 - \Phi(\theta_j)]\}^{-1/2}, \qquad (9.39)$$

where ϕ is the density and Φ is the distribution function of the standard normal. Observe that

$$\Phi(\theta_j) [1 - \Phi(\theta_j)] = VAR(h_j). \qquad (9.40)$$

For $s = 1$ and $s = 2$ we find

$$\tau_{j1} = \Phi(\theta_j) VAR^{-1/2}(h_j), \qquad (9.41)$$

$$\tau_{j2} = 1/2\sqrt{2} \; \theta_j \phi(\theta_j) VAR^{-1/2}(h_j). \qquad (9.42)$$

Lawley (1944) supposes in addition that the items have a Spearman structure, i.e. $\rho_{jl} = \alpha_j \alpha_l$ for all $j \neq l$. Then

$$COR(h_j, h_l) = \alpha_j \alpha_l \, \tau_{j1} \tau_{l1} + \alpha_j^2 \alpha_l^2 \, \tau_{j2} \tau_{l2} + O \, (\alpha_j^3 \alpha_l^3). \qquad (9.43)$$

Because $\tau_{j2} > 0$ for a difficult test and $\tau_{j2} < 0$ for an easy test, the second factor in principal components analysis depends on the difficulty of the test. A plot of the first two factors gives, as with the continuous normal distribution and the Guttman scale, a quadratic horseshoe. We also remark that our gauges give important contributions to the problem of difficulty factors which has bothered factor analysts for a very long time. The problem is discussed very nicely in McDonald (1967), who also gives the necessary references.

Alternative techniques for presumably multinormally generated binary data are usually based on tetrachoric correlations. Divgi (1979) reviews and develops efficient computational methods. Estimates of the parameters of the one-factor model are discussed by Lawley (1944) and more efficient ones by Bock and Lieberman (1970). The corresponding multiple factor models have been developed by Christofferson (1975, 1977) and Muthèn (1978). Compare also Bartholomew (1980). If the multinormal is categorized in more than two classes we need polychoric correlation. An interesting historical overview, and a new estimation method, are in Lancaster and Hamdan (1964). The definite proof that among the most important Swedish export products are partial derivatives of likelihood functions is given by Olsson (1979).

9.6 EPILOGUE

An up-to-date, general introduction to the Rasch model can be found in Andrich (1988a). The reader is referred to Van den Wollenberg (1982) and Molenaar (1983) for a discussion of diagnostics for failure of the model to hold, and to De Leeuw and Verhelst (1986) for a lucid review of maximum likelihood estimation under the Rasch model.

A Rasch model of the Coombs scale was studied in depth by Jansen (1983); also see Andrich (1988b). It should be noted that the tracelines of the Coombs scale usually refer to persons, not to items (persons are selecting, rather than being selected). Also, single peakedness does appear to give a satisfactory gauge, albeit not a unique one. This fact is illustrated in Heiser (1985, 1987a), Ter Braak (1985), and Ter Braak and Barendregt (1986), who studied correspondence analysis in connection with the Gaussian single-peaked response model.

For an example of the checking condition (9.37) in the case of an **F** with the consecutive ones property, see Heiser (1981, ch. 3). Further results on ordering in the context of bivariate probability distributions are given in Schriever (1985).

CHAPTER 10
REFLECTIONS ON RESTRICTIONS

We have seen in the previous chapters that HOMALS is a special case of PRINCALS, and PRINCALS is a special case of OVERALS. If all sets contain a single variable then OVERALS becomes PRINCALS; if all variables are multiple nominal then PRINCALS becomes HOMALS. The relationship with ANACOR is more complicated. If we are dealing with two multiple nominal variables then HOMALS becomes ANACOR; if we apply singular value decomposition to indicator matrices, complete or incomplete, or to the matrix C with bivariate cross products, then ANACOR becomes HOMALS. It is consequently better to think of ANACOR as a different technique to minimize similar loss functions, which does not fit naturally into the order HOMALS – PRINCALS – OVERALS. The techniques treated in Chapters 6 and 7 are again special cases of OVERALS in which there is an asymmetric treatment of sets of variables, with often a special structure of one of the sets. Thus the implied partial order of generality is as indicated in Figure 10.1.

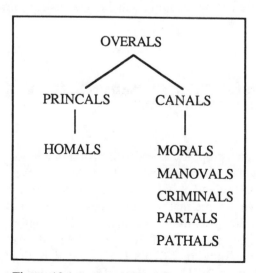

Figure 10.1 Partial order of MVA techniques.

This partial order is consistent with classification of MVA techniques in terms of partitioning of the variables into sets; it is also consistent with the classification into join techniques and meet techniques.

On the other hand, there is the purely technical fact that both join and meet techniques can be fitted by using σ_M, meet loss, and the idea that interactive variables and additivity restrictions can be used to introduce sets into homogeneity analysis. These two points are clearly illustrated in Sections 4.4 and 5.3. As a consequence we can formulate the most general problem in which we are interested in this chapter as the minimization of

$$\sigma(X,Y) = m^{-1} \sum_j \text{tr}(X - G_jY_j)'M_j(X - G_jY_j) \tag{10.1}$$

over X and Y restricted in various ways. Because of the inclusion of the diagonal binary matrices M_j we can suppose without loss of generality that the G_j are complete indicator matrices. When we are using option missing data single category or missing data multiple categories for missing data then $M_j = I$ for all j. When we use option missing data passive for missing data, then $M_j \neq I$, and the corresponding incomplete indicator matrix is M_jG_j. We can rewrite (10.1) as

$$\sigma(X,Y) = m^{-1} \text{ tr } X'(M_* - mI)X + m^{-1} \sum_j \text{SSQ}(X - M_jG_jY_j), \tag{10.2}$$

which shows the difference between the missing data treatments very clearly. The normalization conditions in the general restricted problem will not be defined explicitly, because we shall be interested primarily in classes of restrictions on X and/or Y that are general enough to include the usual normalization conditions.

10.1 THE CLASS OF RESTRICTIONS CLOSED UNDER LINEAR TRANSFORMATIONS

There is an interesting invariance property, already discussed in Section 5.2, that simplifies the restricted homogeneity analysis problem. Suppose we write the general constraints as $X \in \Omega_X$ and $Y \in \Omega_Y$, where Ω_X and Ω_Y are not necessarily cones. Suppose we assume that $Y \in \Omega_Y$ implies that $YU \in \Omega_Y$ for all square nonsingular U. Suppose, moreover, that $X \in \Omega_X$ implies that $X'M_*X = I$. Then the same argument as in Section 5.2 shows that our problem is equivalent to maximizing $\text{SSQ}(X'GY)$ over $X \in \Omega_X$ and $Y \in \Omega_Y$ that satisfy in addition $Y'DY = I$. Here G is the partitioned indicator matrix consisting of the m matrices $M_j G_j$ next to each other and D is the diagonal matrix of column totals of G. M_*, the sum of the M_j, is the diagonal matrix of row totals of G.

In the same way, if $X \in \Omega_X$ implies that $XT \in \Omega_X$ for all square nonsingular T and $Y \in \Omega_Y$ implies that $Y'DY = I$, then our problem is equivalent to maximizing $SSQ(X'GY)$ over $X \in \Omega_X$ with $X'M_*X = I$ and $Y \in \Omega_Y$. Finally, if $X \in \Omega_X$ implies that $X'M_*X = I$ and $Y \in \Omega_Y$ implies that $Y'DY = I$, then the problem we are solving is equivalent to maximizing $\text{tr}(X'GY)$ over $X \in \Omega_X$ and $Y \in \Omega_Y$.

These three results are the somewhat more abstract counterpart of the theorem familiar from Chapters 3 and 8 that we can *either* normalize over X *or* normalize over Y and find the same solution (up to a nonsingular transformation). Normalizing both over X and over Y gives the same solution in simple situations such as the ones discussed in Chapters 3 and 8, but may give different solutions in more complicated situations. The interesting thing is that choice of normalization remains irrelevant in general constrained situations, provided the constraint sets are closed under multiplication on the right with a nonsingular matrix. This condition is true for constraints of the form $Y_k = S_k Y^k$ that are used to incorporate sets of variables; it is also true for constraints like $Y_j = y_j a_j'$ that we use for single variables. If *all* variables are single then the matrix $X'GY$ transforms to $X'QA$, where column j of Q is given by $q_j = M_j G_j y_j$. Moreover $Y'DY = A'BA$, with $B = \text{diag}(Q'Q)$.

This summarizes most of our theory so far, and shows that it is comparatively easy to incorporate several types of constraints on X and Y. This does not mean that the computer programs we have discussed can routinely handle some of these constraints. It does mean, at least in some cases, that the constraints indicate that preprocessing of the data is all that is needed before we apply one of our programs.

10.2 EQUALITY CONSTRAINTS

We illustrate the use of equality constraints in some simple special cases. Our analysis shows clearly how more general cases should be handled. Suppose we require $X = G_C X_C$ for some indicator matrix G_C. This means that some of the object scores are required to be equal, for example because the objects are considered to be replications or because we wish all women to have one score and all men to have another. We also require $u'M_*X = u'M_*G_C X_C = 0$ and $X'M_*X = X_C'G_C'M_*G_C X_C = I$. Observe that indicator matrices G

have the pleasant property that $\mathbf{G'DG}$ is diagonal for any diagonal matrix $\mathbf{D}$. Thus $\mathbf{G_C'M_*G_C}$ is diagonal too. Suppose we also constrain the $\mathbf{Y}_j$ by the requirement that $\mathbf{Y}_1 = ... = \mathbf{Y}_m$ or $\mathbf{Y}_j = \mathbf{Y}_C$ for all j; this restriction only makes sense, of course, if all variables have an equal number of categories. Equality of category quantifications across variables seems appropriate, although certainly not strictly necessary, if the variables are Likert items, or rating scales, or equal appearing interval judgments, or Q-sorts, or successive category judgements, and so on (cf. Torgerson, 1958). All classical psycho-physical and attitude scaling methods are based on the assumption that the category quantifications are the same for all variables, and most of these classical scaling methods assume in addition that the quantifications are also known *a priori*.

The results of the previous section apply directly to constraints of this form. It is not too difficult to show that we compute $\mathbf{X_C}$ and $\mathbf{Y_C}$ from a correspondence analysis of the table

$$\mathbf{F_C} = \mathbf{G_C'} \Sigma_j \mathbf{M_j G_j}. \tag{10.3}$$

This is a nice and simple result. It easily generalizes to the situation where we only want some of the $\mathbf{Y}_j$ to be equal. We then must replace the indicator matrices for these variables in the partitioned indicator matrix by their sum, and apply ANACOR to the resulting table, consisting of sums of indicator matrices. We now discuss two examples in somewhat more detail.

In the method of successive intervals (Guilford, 1954, ch. 10; Torgerson, 1958, ch. 10) we also make the assumption that all $\mathbf{Y}_j$ are equal, but contrary to most other methods we do not assume them to be known *a priori*. The method scales the category boundaries and the objects simultaneously, on one-dimensional scales, so we may as well assume that $p = 1$ too. The idea is that the objects, called stimuli in this context, are compared with the boundaries and that stimulus i is less than boundary j with probability $\Phi(b_j - s_i)$, with Φ the cumulative normal. Thus in the stimuli by categories table the appropriate probabilistic model is an $n \times (m + 1)$ table, with for each stimulus a row of conditional probabilities $\Phi(b_j - s_i) - \Phi(b_{j-1} - s_i)$, with $b_0 = -\infty$ and $b_{m+1} = +\infty$. It follows from the results of Section 9.4 that correspondence analysis recovers the correct order of the s_i and b_j if the model holds exactly, and this result remains valid if Φ is replaced by any other cumulative distribution function. As in the case of the Guttman scale we cannot expect to recover more than the order without making explicit assumptions about Φ. This is one of the basic messages of Coombs (1964). We can only find metric representations if we make essentially arbitrary assumptions. In our programs the arbitrariness is hidden in a special choice of the optimality criterion.

Another application of the use of equality constraints is in the analysis of preference rankings. If n judges rank m options, then the data consist of n permutation matrices. Now permutation matrices are very special indicator matrices. So homogeneity analysis seems to be appropriate, but there are two caveats. In the first place we must pay attention to the fact that the ranked options in the experiment are the objects in the HOMALS analysis and that the judges in the experiment are the variables in the HOMALS program. We are dealing with a *reversed indicator matrix*, also discussed in Section 3.13. In the second place the ranking information is not picked up by HOMALS: because the G_j are permutation matrices they satisfy $G_j'G_j = G_j G_j' = I$, and this implies that all nonzero HOMALS eigenvalues are equal to 1. Any centered m-vector can be chosen for x, and the permutated quantifications $y_j = G_j'x$ give a perfectly homogeneous solution. It is consequently necessary to impose some sort of constraints. We have already discussed one way to deal with the problem in Chapter 4. If we suppose that all variables are single ordinal, then we can use PRINCALS and recover the familiar nonmetric approach to preference rankings using the vector model. Single nominal will not do; the same argument that showed triviality of HOMALS in this context also shows that a trivial perfect single nominal solution exists in one dimension. Single numerical is possible, of course; we shall discuss it shortly.

Another constraint is possible, without going single: we require that all Y_j are equal. The analysis now amounts to performing correspondence analysis on the sum of the permutation matrices. This matrix is known in psychophysics as the rank-order frequency matrix (Guilford, 1954, ch. 8). It can be used in combination with the normalized rank method, which uses a fixed nonlinear scale for the rank numbers and then computes scale values of the options by averaging (a first iteration of reciprocal averaging). It can also be used in combination with Thurstonian methods, and again we can show that correspondence analysis recovers the correct orders if the simplest (case V) models are true.

Equality constraints can also be used in combination with single variables. Thus we can require $Y_j = y_C a_j'$, for example, in the preference rankings situation, where the vector y_C is the same for all j. In addition, we can impose ordinal or linear constraints on y_C. If we require that y_C is linear with the rank numbers, for example, then the technique becomes equivalent to the singular value decomposition of the $n \times m$ matrix of row centered rank numbers, a technique previously invented by Slater (1960), Carroll and Chang (1964), and Benzécri (1965). This technique has the property that the first principal component for the options is often very similar to the average rank numbers across judges. This redundancy can be removed by requiring in addition that the weights a_j sum to zero over all j (judges in the ranking experiment). It turns out that we must now centre the matrix of rank numbers by rows and by columns before computing the singular value decomposition. This double centering can be interpreted in terms of the squared distance model of preferential choice, a special case of unfolding (Ross and Cliff, 1964). When y_C is not completely known, but only ordinally constrained or not constrained at all, then the situation becomes slightly more complicated. Yet it is straightforward to adapt the alternating least squares algorithm of PRINCALS to this case.

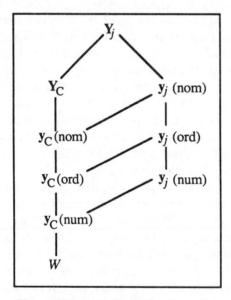

Figure 10.2 Partial order for the analysis of preference rankings.

Summarizing the possibilities for the analysis of preference rankings we find the partial order in Figure 10.2. The lower we are in the order, the more loss we incur. Methods with index j do not require equality; methods with index C do require equality. Thus Y_j and y_j(nom) have trivial perfect solutions; method y_C(num) is Carroll and Chang's MDPREF or Benzécri's 'analyse des préférences' if we actually use the rank numbers as *a priori* quantifications. The W stands for Kendall's coefficient of concordance (Kendall, 1962, ch. 6), which does not only require y_C(num) but also $a_j = 1$ for all j.

More or less the same things can be said about other situations that involve subjective judgement of stimuli. However, it is not necessarily true that $\mathbf{Y}_j$ and $y_j(\text{nom})$ are trivial. In a Q-sort technique or for partial rankings, for example, the $\mathbf{G}_j$ are not square, and triviality does not apply. ➤

Table 10.1 Guilford's spot pattern data. Rows: stimulus groups, columns: response categories

	1	2	3	4	5	6	7	8	9
15	14	18	7	1					
16	16	19	3	2					
17	7	18	11	4					
19	8	18	9	3	2				
20	3	12	14	3	6	2			
22	1	11	14	12	2				
24		3	12	14	9	2			
26		2	9	18	9	2			
28			2	20	17	1			
30				26	11	3			
32			2	10	16	9	3		
35				8	17	14	1		
37				8	18	10	4		
40				2	14	14	10		
43					12	19	9		
46				2	6	18	14		
49					2	14	23	1	
53						10	25	5	
56						12	22	6	
60						5	22	11	2
64							14	20	6
69							7	17	16
74							6	20	14

The data in our first example are Guilford's spot pattern data (Guilford, 1954, p. 203). He used 100 different cards with spot patterns. There were 25 groups of four cards, patterns in each group having the same number of spots. One single judge sorted the deck in nine ordered piles, with the beautiful old-fashioned instruction in mind, that he had to attempt to keep interpile distances psychologically equal. There were ten replications of this

same experiment. The data can be collected into ten indicator matrices of order 100 × 9, and can then be analysed by HOMALS. These matrices are not reported by Guilford; however, he reports a single 23 × 9 matrix of groups against piles. It is not clear to us where the other two groups went. Aggregating data like this can be fitted into our framework if we suppose that HOMALS must give equal scores to cards in the same groups and equal category quantifications for all ten replications. The homogeneity approach has a natural interpretation in psychophysical contexts, in which we very commonly suppose that there 'exists' a one-dimensional scale and that the different variables are merely replications.
Guilford's data matrix is reproduced as Table 10.1. The first singular value of the ANACOR analysis of this table is 0.93; the remaining singular values conform closely to the rule $\lambda_s = (0.93)^s$,

Table 10.2 Guilford's spot pattern data: quantifications

Stimulus groups		Response categories	
15	−1.35	1	−1.49
16	−1.36	2	−1.37
17	−1.24	3	−1.11
19	−1.21	4	−0.56
20	−0.96	5	−0.17
		6	0.42
22	−0.98	7	0.98
24	−0.65	8	1.53
26	−0.58	9	1.69
28	−0.39		
30	−0.38		
32	−0.09		
35	−0.01		
37	0.02		
40	0.31		
43	0.37		
46	0.48		
49	0.74		
53	0.91		
56	0.90		
60	1.10		
64	1.36		
69	1.50		
74	1.51		

which implies that we only have to pay attention to a single dimension. Table 10.2 contains optimal quantifications of spot patterns and of piles (intervals). Both transformations are plotted in Figures 10.3 and 10.4. It is clear from the transformation of the intervals that correspondence analysis does not follow the instruction to keep interpile distances equal: the intervals in the middle are larger, near the endpoints the distances are smaller. The transformation of the stimuli is fairly linear; deviations from linearity are in the direction of concavity. In this sense correspondence analysis, which produces the 'best' scale in a specific least squares sense, confirms Fechner's law, but this is obviously a very weak confirmation. It could very well be that a similar technique with a least absolute deviation loss function refutes Fechner's law.

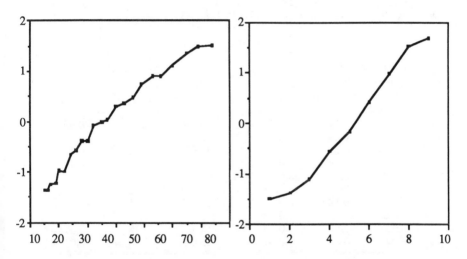

Figure 10.3 Guilford's spot pattern data: transformation of the number of spots in the stimuli.

Figure 10.4 Guilford's spot pattern data: transformation of the response categories.

Our second exampleare concerns the Roskam's journal preference data already discussed in Section 4.8. We first investigate the results with a multiple two-dimensional Y_C; so the response categories are restricted to be equal for all 39 judges. This amounts to performing correspondence analysis on the sum of the 39 *permutation matrices. This sum is given in Table 10.3, the results of the correspondence analysis are in Table 10.4, and they are plotted in Figures 10.5 and 10.6. The two-dimensional trans*

Table 10.3 Roskam's journal preference data: sum of permutation matrices

	1	2	3	4	5	6	7	8	9	10
JEXP	8	5	2	3	4	5	5	3	1	3
JAPP	3	4	5	5	6	3	9	3	1	0
JPSP	5	6	3	3	7	9	3	2	0	1
MVBR	1	4	3	5	3	3	6	4	8	2
JCLP	5	2	2	5	3	1	2	4	8	7
JEDP	1	2	2	5	2	4	4	8	6	5
PMET	4	2	5	3	3	3	3	3	5	8
HURE	1	5	1	0	2	6	4	8	8	4
BULL	6	9	12	7	1	2	0	1	0	1
HUDE	5	0	4	3	8	3	3	3	2	8

Table 10.4 Roskam's journal preference data: correspondence analysis for multiple equal Y_C

JEXP	0.17	0.26
JAPP	0.21	0.35
JPSP	0.29	0.51
MVBR	−0.19	0.01
JCLP	−0.27	−0.38
JEDP	−0.33	−0.05
PMET	−0.12	−0.30
HURE	−0.39	0.18
BULL	0.67	−0.53
HUDE	−0.04	−0.04

1	0.28	−0.06
2	0.33	0.04
3	0.46	−0.43
4	0.17	−0.25
5	0.07	0.36
6	0.03	0.48
7	−0.09	0.42
8	−0.35	0.07
9	−0.57	−0.25
10	−0.33	−0.37

λ^2	0.17	0.08

Table 10.5 Roskam's journal preference data: equal, single nominal y_C

JEXP	0.38	−0.02
JAPP	0.11	0.23
JPSP	−0.19	0.37
MVBR	0.35	−0.35
JCLP	−0.49	−0.05
JEDP	−0.22	−0.34
PMET	0.45	−0.27
HURE	−0.31	−0.22
BULL	0.18	0.67
HUDE	−0.26	−0.04

1	0.37
2	0.30
3	0.35
4	0.30
5	0.10
6	−0.11
7	−0.09
8	−0.26
9	−0.50
10	−0.46

formation of the rank numbers is horseshoe-like, with clusters (1, 2, 3, 4), (5, 6, 7), and (8, 9, 10) along the horseshoe. These three clusters also make it possible to think of the solution in terms of a triangle.

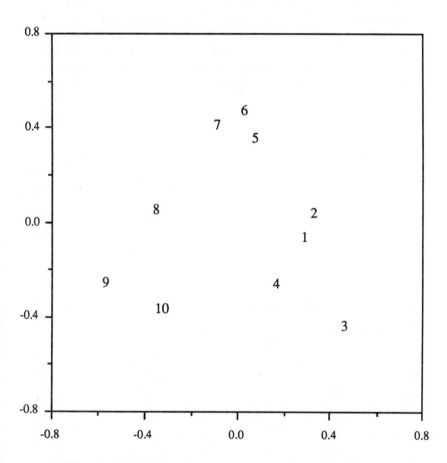

Figure 10.5 Roskam's journal preference data: category quantifications in the equal multiple analysis.

The same triangle can also be found in Figure 10.6, where the Psychological Bulletin is one corner of the triangle (associated with (1, 2, 3, 4): popular with many kinds of psychologists), the journals JEXP, JAPP, JPSP are another corner (associated with (5, 6, 7): popular with many, unpopular with some), and the

remaining journals are scattered around the third corner (associated
with (8, 9, 10): unpopular with many, popular with some). The
solution can perhaps be characterized by the fact that it tries to find
the one-dimensional scale and put it on a horseshoe. For these data
this is not very possible, and the reason is clear from a closer
inspection of Table 10.3. Most of the rows of this table are
bimodal, sometimes even U-shaped, except for BULL and for
JCLP and PMET, which are very flat and quite unpopular. In this
particular case it is not very satisfactory to treat judges as
replications; the options invoke too many systematic individual
differences.

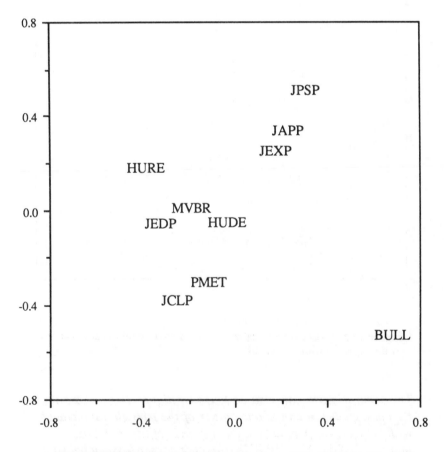

Figure 10.6 Roskam's journal preference data: object scores in the
equal multiple analysis.

A second analysis of the same example could be characterized as being in between the ordinal and the linear analysis already discussed in Chapter 4. In the ordinal analysis we require that $Y_j = y_j a_j'$, *with the* y_j *in the appropriate order; in the linear analysis we require that* $Y_j = y_C a_j'$ *with* y_C *the centered rank numbers, i.e. a fixed vector. An interesting intermediate case is the common transformation analysis with* $Y_j = y_C a_j'$ *with both* y_C *and* a_j *free (in our example we only required* y_C *not to be in the appropriate order).*

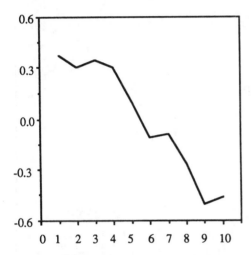

Figure 10.7 Roskam's journal preference data: quantifications plotted against the rank numbers in the common transformation analysis.

A special purpose APL program gave the transformation of the rank numbers plotted in Figure 10.7, the object scores in Figure 10.8, and the judge directions a_j *in Figure 10.9. The interesting feature of Figure 10.7 is that the transformation of the rank numbers is very flat at the lower and also rather flat at the upper end. Thus the extreme opinions are flattened out in order to achieve a better fit. The dimensions in Figure 10.8 clearly have to do with specific–general and with hard–soft; if we compare the solution with Figure 4.6 then the clinical cluster (HURE, JCLP) has merged with the developmental cluster (JEDP, HUDE). The hard cluster (PMET, JEXP, MVBR) is still there, and so is the general cluster (BULL, JAPP, JPSP), which has become somewhat clearer. In Figure 10.9 we also see that the developmental psychologists and the clinical people are not separated any more.*

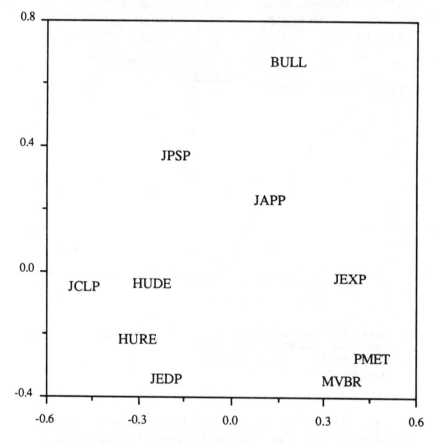

Figure 10.8 Roskam's journal preference data: object scores in the common transformation analysis.

On the whole this analysis seems less informative than the ones in Chapter 4. Numerical results for this analysis are given in Table 10.5.

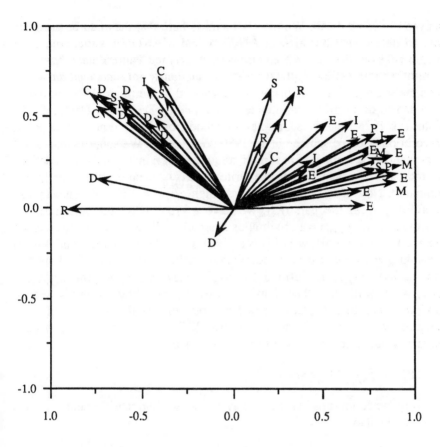

Figure 10.9 Roskam's journal preference data: loadings in the common transformation analysis.

10.3 OTHER LINEAR CONSTRAINTS

Equality constraints are a simple special case of linear constraints. Additivity constraints, which we use to define sets of variables, are a special case too. In general we can require that $X = S_0 B_0$ and $Y_j = S_j B_j$ for any *design matrices* S_j without too many complications. If we require, for example, that the category quantifications must be polynomials of degree less than or equal to s, then the S_j are sets of basis polynomials on the category numbers, presumably orthogonal with respect to D_j. Additivity constraints have already been analysed in some detail. We merely mention here that they can be used not only to define sets but also in ANOVA and MANOVA situations; their categorical extensions have been discussed briefly in Chapters 6 and 7. ↠

Another interesting application of linear constraints is to paired comparisons scaling. This was originally discussed in Guttman (1946) and was extended to other more complicated situations by De Leeuw (1973, ch. 4). Nishisato (1978) rederives Guttman's method from a different starting point; his solution is also discussed in his 1980 book (ch. 6). We suppose that all m^2 pairs that can be formed of m stimuli are judged by n individuals. This defines m^2 indicator matrices, which we write as $G_{(j,l)}$, meaning that pair (j,l) is compared. Now we use the coding in which $g_{(j,l)i1} = 1$ if individual i picks stimulus j from the pair (j,l), we have $g_{(j,l)i2} = 1$ if the individual is undecided, and $g_{(j,l)i3} = 1$ if stimulus l is picked. The linear constraints we use are $Y_{(j,l)} = S_{(j,l)}B$, with $S_{(j,l)}$ a given $3 \times m$ design matrix, and with B the unknown $m \times p$ matrix of stimulus quantifications we are looking for. The matrix $S_{(j,l)}$ is defined by $s_{(j,l)1j} = s_{(j,l)3l} = +1$ and $s_{(j,l)1l} = s_{(j,l)3j} = -1$, while all other elements of $S_{(j,l)}$, including the entire second row, are equal to zero. Constrained homogeneity analysis of paired comparisons, with the normalization condition $X'X = I$, amounts to computing a singular value decomposition of the $n \times m$ matrix

$$F_{PC} = \Sigma_{(j,l)} \, G_{(j,l)} S_{(j,l)}. \tag{10.4}$$

By using the definitions in the appropriate way we find that element f_{ij} of this matrix is equal to

$$f_{ij} = \Sigma_l \, [(g_{(j,l)i1} - g_{(j,l)i3}] - [g_{(l,j)i1} - g_{(l,j)i3})]. \tag{10.5}$$

When there are no ties, then this simplifies because $g_{(j,l)i1} + g_{(j,l)i3} = 1$, and the same thing is true for $G_{(l,j)}$. If we only ask the individual for $1/2m(m-1)$ judgements and insert the other half by mirroring, then more simplification becomes possible because $g_{(j,l)i1} = g_{(l,j)i3}$ and $g_{(j,l)i3} = g_{(l,j)i1}$. If the individual is perfectly consistent and judgements represent a strict order without ties of the m stimuli, then the matrix F_{PC} is the row-centered matrix of rank numbers again. Thus we recover the same techniques as in the previous section, but in a more general way because we allow for ties and inconsistencies, and with slight modifications even for missing data.

There is an interesting generalization of this technique which allows for response bias with respect to order of presentation. Here B is $(2m) \times p$ and $S_{(j,l)}$ is $3 \times (2m)$, with the first row equal to $(e_j - e_{m+l})$ instead of $(e_j - e_l)$ and with the third row equal to $(e_l - e_{m+j})$ instead of $(e_l - e_j)$. This gives a double quantification of all stimuli, one as the first member of a pair and one as the second member. If the individuals are consistent, and thus unbiased towards choosing the first element of the pair or the second, then this new method is no generalization, because both stimulus quantifications will be equal.

Because the coding method is a bit complicated we illustrate it in Table 10.6. A single individual has judged the nine pairs that can be formed with three stimuli a, b, and c. The judgements are in Table 10.6a. All $G_{(j,l)}S_{(j,l)}$ are 1×3 matrices we write them below each other in Table 10.6b, together with their sum. We also give the double coding, which allows for response bias in Table 10.6c.

Table 10.6a Judgements of a single individual			Table 10.6b Paired comparison codings plus sum			Table 10.6c Double paired comparison coding					
a	=	a	0	0	0	0	0	0	0	0	0
a	>	b	+1	-1	0	+1	0	0	0	-1	0
a	>	c	+1	0	-1	+1	0	0	0	0	-1
b	>	a	-1	+1	0	0	+1	0	-1	0	0
b	=	b	0	0	0	0	0	0	0	0	0
b	<	c	0	-1	+1	0	-1	0	0	0	+1
c	<	a	+1	0	-1	0	0	-1	+1	0	0
c	>	b	0	-1	+1	0	0	+1	0	-1	0
c	=	c	0	0	0	0	0	0	0	0	0
			+2	-2	0	+2	0	0	0	-2	0

It is, of course, perfectly legitimate to analyse the $G_{(j,l)}$ by HOMALS without restrictions. This means that we compute correlations between binary judgements and that we quantify, either by HOMALS or by principal components analysis of the phi coefficients, all pairs of stimuli. These pair quantifications can then be used in a subsequent analysis, for example to derive a single scale of stimuli by analysis of variance methods such as the ones outlined by Scheffé (1952) or Bechtel (1967, 1971, 1976). The scores for the individuals can be used directly, without further analysis. Observe that if the unfolding model is true, then the individuals that prefer stimulus j to stimulus l can be separated from the individuals that prefer stimulus l to stimulus j by a hyperplane perpendicular to the line connecting j and l and bisecting this line. From the general theory of Chapter 4 this means that PRINCALS with *continuous* ordinal option on the binary data matrix would give a perfect solution. However, we have seen that continuous ordinal does not always work properly. We have also seen that the discrete options can be interpreted as approximating the continuous ones, and that consequently HOMALS or PRINCALS on the $G(j,l)$ can be thought of as approximating the positions of the individual points under the unfolding model. ⇥

10.4 ZEROS AT SPECIFIC PLACES

Another type of linear constraint demands that some of the parameters are equal to fixed known values, while others are free to vary. In the single numerical case, for example, we fix the transformations and find optimal loadings, weights, and scores. In PATHALS we require that some of the regression coefficients are equal to zero and others are equal to one. We give one more interesting example of requiring zeros at specified places. Suppose we are fitting PRINCALS with n individuals, m variables, and p dimensions. Then usually $p < m < n$, because $p > m$ is just a waste of space: we can obtain a perfect fit in $p = m$ dimensions. Recall that PRINCALS with single variables and without option i missing data passive maximizes the sum of the p largest eigenvalues of the correlation matrix of the transformed variables. If $p = m$, then this sum is always equal to m, no matter how the transformations are chosen, and consequently there is nothing to maximize.

Now suppose that $m < p < n$ and that we require zeros in specified places in the matrix of weights A. The loss function is as usual. Thus

$$\sigma(X,Y,A) = m^{-1} \sum_j SSQ(X - G_j y_j a_j') \tag{10.6}$$

and the normalizations are $X'X = I$, $u'G_j y_j = 0$, and $y_j' D_j y_j = 1$. We apply alternating least squares, as usual. Fitting y_j for fixed X and A is the same as in Chapter 4. Fitting A for fixed X and Y is also quite simple. We use the $n \times m$ matrix Q with columns $q_j = G_j y_j$. We also define the unconstrained minimum of (10.6) over a_j as $\tilde{a}_j \equiv X'q_j$, which enables us to split (10.6) into two components:

$$\sigma(X,Y,A) = m^{-1} \sum_j SSQ(X - q_j \tilde{a}_j') + m^{-1} \sum_j SSQ(\tilde{a}_j - a_j), \tag{10.7}$$

where we have used the fact that $SSQ(q_j) = 1$. It follows that the optimal a_j for given X and Y is equal to $\tilde{a}_j$, except that the elements of $\tilde{a}_j$ are replaced by zeros on the required places. Computing the optimal X for fixed A and Y is a bit more complicated. It amounts to maximizing tr $X'QA$ over $X'X = I$, with both X and QA matrices of order $n \times p$, but with $\text{rank}(QA) \leq m < p$. The solution is $X = K_1L_1' + K_2TL_2'$, with K_1 the $n \times q$ matrix of left singular vectors corresponding with the nonzero singular values of QA, $q = \text{rank}(QA)$; L_1 the corresponding $p \times r$ matrix of right singular vectors; K_2 an $n \times (n - q)$ matrix of left singular vectors corresponding with the zero singular values; L_2 an $p \times (p - q)$ matrix of corresponding right singular vectors, and T any $(n - q) \times (p - q)$ matrix that satisfies $T'T = I$. Because $q < p < n$ it follows that the solution for X is not unique, which corresponds to the notorious *factor score indeterminacy* problem. Clearly the choice of T determines the further course of the iterations, but the basic convergence theorems for alternating least squares still apply, basically because the mapping of QA to the set of all X of the form $X = K_1L_1' + K_2TL_2'$ is upper semi-continuous, so that Zangwill's general convergence theorems (Zangwill, 1969, ch. 4) apply. The fact that we must choose an arbitrary T in each iteration has made some people, such as Takane, Young, and De Leeuw (1979), very nervous. They consequently prefer a loss function which is defined in the space of correlation matrices $R = Q'Q$, which makes the regression problems that adapt the y_j considerably more complicated. Of course if we define $\sigma(*,Y,A)$ as the minimum of $\sigma(X,Y,A)$ over X for fixed Y and A, then

$$\sigma(*,Y,A) = m^{-1} [mp + \text{SSQ}(A) - 2 \sum_j \lambda_j(A'RA)], \qquad (10.8)$$

with $\lambda_j^2(A'RA)$ the eigenvalues of $A'RA$; thus λ_j are the singular values of QA. This is also a loss function defined in terms of R. We can now use the fact that the last term, the sum of the singular values of QA, is convex in A for given Q and convex in Q for given A to use the majorization techniques explained in a different context by De Leeuw and Heiser (1980). ➨

The most interesting application of this algorithm is to factor analysis. This was the application studied by Takane, Young, and De Leeuw (1979), who developed the algorithm FACTALS. Our point in this section is that it is possible to perform factor analysis with PRINCALS, by allowing for the possibility that $p > m$ and by making it possible to require certain patterns of zeros in A. For unrestricted common factor analysis, for example, A is $m \times (t + m)$, and the last m columns of A must be diagonal. For restricted common factor analysis we can also require zeros in the first t columns of A. Again we are not saying that the current version of PRINCALS can actually do all this, we are merely pointing out that the necessary modifications are easy to implement, either for us or for anyone else who is interested in categorical or nonlinear common factor analysis.

10.5 NONLINEAR RESTRICTIONS

Just for completeness we also discuss some interesting nonlinear restrictions. The constraints $y_j \in \Omega_j$ we have used many times are nonlinear, and so are the constraints $Y_j = y_j a_j'$ An interesting possibility, which we merely mention, is the three mode generalization of our techniques. Suppose, for example, that the same variables have been measured on a number of different occasions or different matched groups of individuals. We can analyse the occasions separately, for example by using HOMALS. This gives category quantifications $Y_{j\tau}$, of variable j at occasion τ. However, we can also do all the HOMALS analyses simultaneously, with the restriction that $Y_{j\tau} = Y_j B_\tau$, with B_τ either a full or a diagonal $p \times p$ matrix. Alternating least squares algorithms for this class of constraints are easy to construct. Observe that if the same individuals are tested on different occasions (possibly even with different variables), then we can also require $X_\tau = X B_\tau$, and if we have both the same variables and the same individuals we can combine the two types of restrictions. Moreover, it can be combined with single or additivity constraints and so on. We have not the faintest idea of whether this is useful, but it certainly is in the spirit of modern psychometrics, and it guarantees the possibility of an unending stream of programs and publications in the appropriate journals.

In several sections of this book we have discussed the difference between HOMALS and PRINCALS in the following terms. HOMALS computes p different solutions with an approximate join rank equal to one; PRINCALS computes a single solution with an approximate join rank equal to p. This interpretation is necessary if we want to see both programs as methods for nonlinear principal components analysis, but we have also seen that alternatively we can interpret both programs as methods of generalized canonical analysis of K sets. The principal components interpretation left us with the rather mysterious possibility of computing p different solutions with an approximate join rank equal to q, which generalizes both HOMALS and PRINCALS. We have been very vague about this possibility so far, mainly because it cannot be done with the existing programs and because we have never tried it in practice. In this programmatic and speculative chapter we can suggest a fairly simple possibility. We use our ordinary loss function, with one variable in each set,

$$\sigma(X,Y) = m^{-1} \sum_j \text{SSQ}(X - G_j Y_j), \tag{10.9}$$

but we choose the dimensionality equal to the product of p and q. This implies that $\mathbf{Y}_j$ is $k_j \times pq$. We partition $\mathbf{Y}_j$ in p submatrices $\mathbf{Y}_{(j)s}$ of order $k_j \times q$, and for each of the $\mathbf{Y}_{(j)s}$ we require that $\mathbf{Y}_{(j)s} = \mathbf{y}_{(j)s}\mathbf{a}'_{(j)s}$. We still normalize by $\mathbf{X}'\mathbf{X} = \mathbf{I}$. If we partition $\mathbf{X}$ in a similar way into p matrices $\mathbf{X}$, then we can write

$$\sigma(\mathbf{X},\mathbf{Y}) = m^{-1} \sum_j \sum_s \mathrm{SSQ}(\mathbf{X}_s - \mathbf{G}_j\mathbf{y}_{(j)s}\mathbf{a}'_{(j)s}), \tag{10.10}$$

with normalization $\mathbf{X}'_s\mathbf{X}_s = \mathbf{I}$ and $\mathbf{X}'_s\mathbf{X}_t = \mathbf{0}$ if $s \neq t$. This shows that we work with p PRINCALS problems at the same time, all of dimensionality q. If $p = 1$ this is PRINCALS, if $q = 1$ this is HOMALS. Again alternating least squares programs along the lines of PRINCALS are quite simple; we must require that the submatrices $\mathbf{y}_{(j)s}$ are of rank one, and thus perform rank one approximation of submatrices. This formulation also shows that it is not essential at all that q is the same for all p.

10.6 EPILOGUE

For an example of HOMALS on Q-sort data, see Slooff and Van der Kloot (1985). Further work on transformations that are common across variables has been reported in Van der Lans and Heiser (1988). A more extensive analysis of possibilities of coding paired comparison data can be found in Heiser (1981, ch. 4). Equality constraints on the object scores can be used to make one variable (the one containing the equality constraints) dominate the analysis. A similar idea has been discussed, with a lot of auxiliary results, by Nishisato (1984, 1988) under the name *forced classification*.

Restrictions of the form $\mathbf{X} = \mathbf{G}_C\mathbf{Y}_C$ with $\mathbf{G}_C$ an unknown indicator matrix requires transfer of normalization (cf. Section 6.2), and leads to a cluster analysis technique (Van Buuren and Heiser, 1989). Alternative loss functions that optimize some other aspect of the correlation matrix of quantified variables with a majorization strategy have been discussed by Tijssen and De Leeuw (1989) and De Leeuw (1989).

CHAPTER 11
NONLINEAR MULTIVARIATE
ANALYSIS: PRINCIPLES AND
POSSIBILITIES

There are a number of unifying ideas of different levels in this book. Some of them are of a general methodological nature. We take the point of view, for example, that mathematical statistics is one of the possible methods to investigate stability of data analytic techniques and that probabilistic models are possible gauges for data analytic techniques. We do not go out of our way to practise inductive inference from sample to population; we are also not trying to make coherent and rational decisions all the time. A second unifying principle is of a data theoretical kind. In the final analysis all data are categorical or discrete and the measurement level is prior information that *can* be incorporated in the form of restrictions. Thus a measurement level is not a property of the data; it is an abuse of language to say that a variable *is* ordinal or interval.

A data analytic unifying set of tools is the use of least squares loss functions in combination with geometrical interpretations. These tools are used to illustrate the essential unity of multivariate analysis and multidimensional scaling. Of course, least squares is also used because of computational convenience and because of the historical link with classical linear multinormal analysis. The major technical tools are optimal scaling, alternating least squares, and the singular value decomposition. Other ideas that occur frequently are the distinction between the multiple and the single treatment of a variable, and the 'first-step' approach which transforms variables nonlinearly in such a way that subsequent linear multivariate analysis techniques give 'better' results. The classical distinction between the analysis of dependence and interdependence is generalized to the distinction between join and meet techniques.

In this chapter we formalize this last distinction, starting with the finite-dimensional case and then proceeding more generally. This framework makes it possible to discuss some other related general problems such as choice of

basis, discretization of continuous variables, infinite dimensional gauges, and
analysis of stochastic processes.

11.1 JOIN AND MEET

Suppose L is the set of all subspaces of $\Re^n$. We can order L partially by
letting $L_1 \leq L_2$ if L_1 is a subspace of L_2. We can make L into a complete
lattice by defining MEET(L_1, L_2) to be the *intersection* of L_1 and L_2 and
JOIN(L_1, L_2) to be the *linear sum* of L_1 and L_2 (i.e. the set of all $\mathbf{x} \in \Re^n$ of
the form $\mathbf{x} = \mathbf{y} + \mathbf{z}$ with $\mathbf{y} \in L_1$ and $\mathbf{z} \in L_2$). The meet of L_1 and L_2 is the
greatest lower bound of L_1 and L_2, i.e. the largest subspace contained in both
L_1 and L_2. The join of L_1 and L_2 is the least upper bound of L_1 and L_2, i.e.
the smallest subspace that contains both L_1 and L_2. If $L_1,...,L_m$ are subspaces
of $\Re^n$ we can also define MEET$(L_1,...,L_m)$ and JOIN$(L_1,...,L_m)$.

On the lattice of subspaces we can define the valuation DIM( ), the
dimensionality of the subspace. Using DIM( ) we define the *meet rank* of
$L_1,...,L_m$ as

$$\text{MRK}(L_1,...,L_m) \equiv \text{DIM}(\text{MEET}(L_1,...,L_m)), \tag{11.1}$$

and the join rank as

$$\text{JRK}(L_1,...,L_m) \equiv \text{DIM}(\text{JOIN}(L_1,...,L_m)). \tag{11.2}$$

We say that $L_1,...,L_m$ have a *p-meet* if MRK$(L_1,...,L_m) \geq p$ and $L_1,...,L_m$
have a *p-join* if JRK$(L_1,...,L_m) \leq p$.

A very useful feature of these definitions is that they are coordinate-free,
i.e. they do not depend on the choice of a basis in $\Re^n$. Another useful aspect
is that we can use the general results from lattice theory (Birkhoff, 1967),
which give interesting results when interpreted in the data analysis framework
implied by our definitions. We first give this interpretative framework. The
meet problem (qualitative version) is to find a subspace L of $\Re^n$ such that
DIM$(L) = p$ *and* $L \approx$ MEET$(L_1,...,L_m)$. Thus given $L_1,...,L_m$ and given an
integer p with $1 \leq p \leq n$ we try to find a subspace of dimension p that is
approximately contained in all L_j. Clearly it is possible to find a subspace that
satisfies these conditions exactly if the $L_1,...,L_m$ have a p-meet. The meet
problem is an abstract version of the K-set problems we solved in previous
chapters. An important difference is that in these chapters we have introduced

a least squares loss function and we can consequently define the meet problem *quantitatively.* The *join problem,* which is solved in principal components analysis, has a similar abstract version. We start with $L_1,...,L_m$ and an integer p. We now try to find a subspace L such that DIM(L) = p and $L \approx$ JOIN($L_1,...,L_m$). Again an exact solution is possible if and only if $L_1,...,L_m$ have a p-join, and a quantitative formulation of the problem is possible by introducing loss functions.

It is clear from the definitions that if $L_1,...,L_m$ have a p-meet, then they also have a q-meet for all $q \leq p$, and if they have a p-join, then they also have a q-join for all $q \geq p$. Thus we want to choose p as large as possible in meet problems and as small as possible in join problems. Again 'as possible' is defined qualitatively, and can be translated into quantitative terminology only by introducing loss functions. Thus in the meet problem we want to find subspaces of large dimension contained in all given subspaces; in the join problem we want to find subspaces of small dimension that contain all given subspaces.

Before we continue with the abstract discussion of these concepts, it is useful to give some concrete examples of the subspaces we usually deal with. In linear K-set canonical analysis we solve a meet problem, and the K subspaces are defined as the column spaces of K given matrices. In homogeneity analysis, interpreted as a meet problem, there are m subspaces, defined as the column spaces of the indicator matrices G_j. Thus $x \in L_j$ if $x_i = x_k$ whenever individuals i and k are in the same category of variable j and DIM(L_j) = k_j, the number of categories of variable j. In linear principal components analysis we solve a join problem. Each variable defines a one-dimensional subspace, the ray of all n-vectors proportional to the observations on that variable. In nonlinear principal components analysis with discrete variables the indicator matrices are a basis for the subspace of all nonlinear transformations of the variable. Thus nonlinear principal components analysis solves the join problem for the column spaces of the G_j.

We now state some simple results on the functions MRK and JRK. This is merely classical linear algebra, translated in our notation. We start with

$$0 \leq \text{MRK}(L_1,...,L_m) \leq \min_j (\text{DIM}(L_j)) \quad (j = 1,...,m) \tag{11.3}$$

with equality on the left if and only if the intersection of the L_j is the null vector and with equality on the right if and only if one of the L_j contains all the others. For JRK the corresponding inequality is

$$0 \leq \text{JRK}(L_1,...,L_m) \leq \sum_j \text{DIM}(L_j), \tag{11.4}$$

with equality on the left if and only if all L_j are equal to the null vector and equality on the right if and only if the pairwise intersection of the L_j is the null vector (i.e. if and only if the linear sum is equal to the direct sum). We can also relate the two concepts directly

$$\text{MRK}(L_1,...,L_m) \leq \text{JRK}(L_1,...,L_m),$$ (11.5)

with equality if and only if $L_1 = ... = L_m$. Some of the general lattice theory results we mentioned earlier are the distributive inequalities

$$\text{MEET}(L_1, \text{JOIN}(L_2, L_3))$$

$$\geq \text{JOIN}(\text{MEET}(L_1, L_2), \text{MEET}(L_1, L_3)),$$ (11.6)

$$\text{JOIN}(L_1, \text{MEET}(L_2, L_3))$$

$$\leq \text{MEET}(\text{JOIN}(L_1, L_2), \text{JOIN}(L_1, L_3)).$$ (11.7)

Because $\Re^n$ is finite dimensional the lattice of subspaces is *modular*. This means that if L_1 is a subspace of L_3 then

$$\text{JOIN}(L_1, \text{MEET}(L_2, L_3)) = \text{MEET}(\text{JOIN}(L_1, L_2), L_3).$$ (11.8)

An extremely important result, which shows why the case of two subspaces is very special indeed, is

$$\text{MRK}(L_1, L_2) + \text{JRK}(L_1, L_2) = \text{DIM}(L_1) + \text{DIM}(L_2).$$ (11.9)

Another important result can be described in words as follows. Suppose L_j is the join of k_j one-dimensional subspaces. Then $\text{JOIN}(L_1,...,L_m)$ is the join of all Σk_j one-dimensional subspaces that define the L_j. The dual result for the meet is that if L_j is the meet of k_j subspaces, then $\text{MEET}(L_1,...,L_m)$ is the meet of all Σk_j subspaces defining the L_j. This result is far less interesting, because we cannot add the description 'one-dimensional' here. The result for the join problem tells us that we can assume without loss of generality that $\text{DIM}(L_j) = 1$ for all j. No such thing is true for the meet problem.

11.1.1 Matrix formulation

Before we introduce loss functions, we first reformulate our definitions in terms of matrices. Suppose L_j is the subspace of all linear combinations of the columns of a matrix G_j which is not necessarily an indicator matrix. Then $L_1,...,L_m$ have a p-meet if and only if there exists an $n \times p$ matrix X of rank p and $k_j \times p$ matrices Y_j such that

$$X = G_1 Y_1 = ... = G_m Y_m. \tag{11.10}$$

Moreover, $L_1,...,L_m$ have a p-join if and only if there exists an $n \times p$ matrix X and $k_j \times p$ matrices Y_j such that

$$G_j = X Y_j', \tag{11.11}$$

for all $j = 1,...,m$. By joining the G_j in an $n \times \sum k_j$ partitioned matrix G and the Y_j in a $\sum k_j \times p$ partitioned matrix Y, we can also write this last condition as $G = XY'$, which is the matrix formulation of the last result from the previous section, that only the one-dimensional subspaces matter. Observe also that we do not have to require that X has rank p in this case; if $RANK(X) = q < p$, then $L_1,...,L_m$ have a p-join, although they also have a q-join.

The matrix definitions suggest how to define the least squares loss functions. *Meet loss* is defined by

$$\sigma_M(X;Y_1,...,Y_m) = m^{-1} \sum_j SSQ(X - G_j Y_j). \tag{11.12}$$

The subspaces $L_1,...,L_m$ (we can also say in this case the matrices $G_1,...,G_m$) have a p-meet if and only if the minimum of σ_M over all Y_j and over all X of rank p exists and is equal to zero. From a computational point of view, however, this condition is useless. The *infimum* of σ_M over all Y_j and over all X of rank p is always equal to zero, no matter what the G_j are. We choose the Y_j arbitrary, select an arbitrary rank p matrix X_0, and let $\alpha \to 0$ in $X = \alpha X_0$. The most convenient way out of this problem follows from the fact that the $L_1,...,L_m$ have a p-meet if and only if the minimum of σ_M over all Y_j and over all X satisfying $X'X = I$ exists, and is equal to zero. It will be shown in the sequel that this minimum always exists, but is certainly not always equal to zero.

In the same way we can also define the *join loss*. This is

$$\sigma_J(X;Y_1,...,Y_m) = m^{-1} \sum_j SSQ(G_j - X Y_j'). \tag{11.13}$$

There are some important differences compared to σ_M. In the first place the partitioning of the matrix G into submatrices G_j is irrelevant for σ_J, and certainly not for σ_M. In fact we can write

$$\sigma_J(X;Y) = m^{-1} \, \text{SSQ}(\mathbb{G} - XY'). \tag{11.14}$$

A second difference is that we do not need rank restrictions on X; the subspaces have a p-join if and only if the minimum of σ_J over all Y_j and all X is equal to zero; moreover, this minimum is always attained and not always equal to zero. Again this will be shown in the sequel. A third way to express σ_J is sometimes also useful. We can write

$$\sigma_J(X;Y) = m^{-1} \sum_l \text{SSQ}(g_l - Xy_l), \tag{11.15}$$

where the summation is over all columns of the matrix $\mathbb{G}$, i.e. we have reduced the join problem to its one-dimensional subspaces again.

11.1.2 Further analysis of meet loss

We now study σ_M more closely by using familiar results from least squares and eigenvalue theory. Define

$$\sigma_M(*;Y_1,\ldots,Y_m) \equiv \min\{\sigma_M(X;Y_1,\ldots,Y_m) \mid X\}. \tag{11.16}$$

Thus the minimum is computed over all X, not normalized or restricted in any way. Also define

$$\tilde{X} \equiv m^{-1} \sum_j \mathbb{G}_j Y_j. \tag{11.17}$$

Then

$$\sigma_M(X;Y_1,\ldots,Y_m) = \text{SSQ}(X - \tilde{X}) + m^{-1} \sum_j \text{SSQ}(\tilde{X} - \mathbb{G}_j Y_j), \tag{11.18}$$

and thus

$$\sigma_M(*;Y_1,\ldots,Y_m) = m^{-1} \sum_j \text{SSQ}(\tilde{X} - \mathbb{G}_j Y_j). \tag{11.19}$$

We can define the matrix C, of order $(\sum k_j) \times (\sum k_j)$, with submatrices $C_{jl} \equiv \mathbb{G}'_j \mathbb{G}_l$, and the diagonal matrix D of the same order, with diagonal submatrices $D_j \equiv C_{jj}$. Then

$$\sigma_M(*;Y_1,\ldots,Y_m) = m^{-1} \, (\text{tr } Y'DY - m^{-1} \, \text{tr } Y'CY). \tag{11.20}$$

These equations are familiar from the analysis of homogeneity in Chapter 3. In the analysis of variance terminology $Y'DY$ is the total dispersion of the $\mathbb{G}_j Y_j$

$m^{-1}\mathbf{Y'CY}$ is the *between-set* dispersion, and thus $\mathbf{Y'DY} - m^{-1}\mathbf{Y'CY}$ is the *within-set* dispersion.

Now let $r_j \equiv \text{RANK}(\mathbb{G}_j)$ and $r \equiv \text{RANK}(\mathbf{D}) = \Sigma r_j$, and $s = \min(p,r)$. Define

$$\sigma_M(*;*,\ldots,*) \equiv \min\{\sigma_M(*;\mathbf{Y}_1,\ldots,\mathbf{Y}_m) \mid \mathbf{Y'DY} = \mathbf{I}_s\}, \tag{11.21}$$

where $\mathbf{I}_s$ is the diagonal matrix of order p with s elements equal to one and the remaining diagonal elements equal to zero. Then

$$\sigma_M(*;*,\ldots,*) = s - \Sigma_t \lambda_t^2(m^{-1}\mathbf{D}^{-1/2}\mathbf{C}\mathbf{D}^{-1/2}), \tag{11.22}$$

where $\lambda_1^2 \geq \ldots \geq \lambda_s^2$ are the ordered eigenvalues and where $\mathbf{D}^{-1/2}$ is the symmetric square root of the Moore–Penrose inverse of $\mathbf{D}$.

We now reverse the roles of $\mathbf{X}$ and $\mathbf{Y}$ in this derivation. We first define

$$\sigma_M(\mathbf{X};*,\ldots,*) \equiv \min\{\sigma_M(\mathbf{X};\mathbf{Y}_1,\ldots,\mathbf{Y}_m) \mid \mathbf{Y}_1,\ldots,\mathbf{Y}_m\}, \tag{11.23}$$

where the minimum is computed over all $\mathbf{Y}_j$ unrestricted. Define, using the Moore–Penrose inverse $(\mathbb{G}_j'\mathbb{G}_j)^+$,

$$\tilde{\mathbf{Y}}_j \equiv (\mathbb{G}_j'\mathbb{G}_j)^+\mathbb{G}_j'\mathbf{X}. \tag{11.24}$$

Then

$$\sigma_M(\mathbf{X};\mathbf{Y}_1,\ldots,\mathbf{Y}_m) = m^{-1} \Sigma_j \text{SSQ}(\mathbf{X} - \mathbb{G}_j\tilde{\mathbf{Y}}_j)$$
$$+ m^{-1} \Sigma_j \text{tr} \, (\mathbf{Y}_j - \tilde{\mathbf{Y}}_j)'\mathbb{G}_j'\mathbb{G}_j(\mathbf{Y}_j - \tilde{\mathbf{Y}}_j), \tag{11.25}$$

a partitioning that is already familiar from Chapter 4. Thus

$$\sigma_M(\mathbf{X};*,\ldots,*) = m^{-1} \Sigma_j \text{SSQ}(\mathbf{X} - \mathbb{G}_j\tilde{\mathbf{Y}}_j). \tag{11.26}$$

By using partitioned matrices this can again be written in more compact notation as

$$\sigma_M(\mathbf{X};*,\ldots,*) = \text{tr} \, \mathbf{X'X} - m^{-1} \, \text{tr} \, \mathbf{X'}\mathbb{G}\mathbf{D}^+\mathbb{G}'\mathbf{X}. \tag{11.27}$$

If

$$\sigma_M(*;*,\ldots,*) \equiv \min\{\sigma_M(\mathbf{X};*,\ldots,*) \mid \mathbf{X'X} = \mathbf{I}_s\}, \tag{11.28}$$

then

$$\sigma_M(*;*,\ldots,*) = s - \Sigma_t \lambda_t^2(m^{-1}GD^+G').$$ (11.29)

However, the nonzero eigenvalues of GD^+G' are the same as those of the matrix $D^{-1/2}CD^{-1/2} = D^{-1/2}G'GD^{-1/2}$. Consequently, the fact that we have defined $\sigma_M(*;*,\ldots,*)$ in two different ways does not matter; it is indeed true that

$$\min\{\sigma_M(X;Y) \mid X, \; Y'DY = I_s\} = \min\{\sigma_M(X;Y) \mid X'X = I_s, \; Y\}.$$ (11.30)

We have already used this result in many places, both for theoretical purposes and for algorithm construction. We now see that it is true for meet loss in general, even if some of the G_j are singular.

Thus the subspaces L_j (or the corresponding matrices G_j) have a p-meet if and only if the matrices $m^{-1}GD^+G'$ or $m^{-1}D^{-1/2}CD^{-1/2}$ have p eigenvalues equal to 1. We have included the possibility that $s < p$. In this case

$$MRK(L_1,\ldots,L_m) \le JRK(L_1,\ldots,L_m) \le r = s < p,$$ (11.31)

which implies that the L_j have no p-meet. These results can also be used to derive some simple lower bounds for $\sigma_M(*;*,\ldots,*)$. Because

$$GD^+G' = \Sigma_j G_j(G_j'G_j)^+G_j',$$ (11.32)

we see that

$$\Sigma_t \lambda_t^2(GD^+G') \le \Sigma_j \Sigma_t \lambda_t^2[G_j(G_j'G_j)^+G_j'] = \Sigma_j \min(s,r_j)$$ (11.33)

with equality if and only if $MRK(L_1,\ldots,L_m) \ge s$. It follows from this inequality that

$$\sigma_M(*;*,\ldots,*) \ge m^{-1} \Sigma_j \max(0,s-r_j).$$ (11.34)

Thus $\sigma_M(*;*,\ldots,*) = 0$ is possible only if $s \le r_j$ for all j. In most situations we have $s = p$ and $r_j = k_j$, with k_j the number of columns of G_j as usual. In that case meet loss can vanish only if $p \le k_j$ for all j. If $p \ge r$, then $s = r \ge r_j$, and thus

$$\sigma_M(*;*,\ldots,*) \ge (m-1)r / m.$$ (11.35)

If $p \le r$ and $p \ge r_j$ for all j, then

$$\sigma_M(*;*,\ldots,*) \ge p - r / m.$$ (11.36)

11.1.3 Further analysis of join loss

For join loss the situation is much simpler. We use the compact formulation $\sigma_J(X;Y)$. Define

$$\sigma_J(*;Y) \equiv \min\{\sigma_J(X;Y) \mid X\}. \tag{11.37}$$

If

$$\tilde{X} \equiv \mathbb{G}Y(Y'Y)^+, \tag{11.38}$$

then

$$\sigma_J(X;Y) = m^{-1}\{SSQ(\mathbb{G} - \tilde{X}Y') + tr(X - \tilde{X})Y'Y(X - \tilde{X})'\}, \tag{11.39}$$

and consequently

$$\sigma_J(*;Y) = m^{-1} SSQ(\mathbb{G} - \tilde{X}Y') = m^{-1} tr(I - Y(Y'Y)^+Y')C, \tag{11.40}$$

with $C = \mathbb{G}'\mathbb{G}$ as usual. If

$$\sigma_J(*;*) \equiv \min\{\sigma_J(*;Y) \mid Y\}, \tag{11.41}$$

then

$$\sigma_J(*;*) = \sum_s \lambda_s^2(m^{-1}\mathbb{G}'\mathbb{G}) \quad (s = p+1,\ldots,\sum k_j). \tag{11.42}$$

If

$$\sigma_J(X;*) \equiv \min\{\sigma_J(X;Y) \mid Y\}, \tag{11.43}$$

then because

$$\sigma_J(X;Y) = m^{-1}\{SSQ(\mathbb{G} - X\tilde{Y}') + tr(Y - \tilde{Y})X'X(Y-\tilde{Y})'\}, \tag{11.44}$$

with

$$\tilde{Y} \equiv X(X'X)^+X'\mathbb{G}, \tag{11.45}$$

we have

$$\sigma_J(X;*) = m^{-1} SSQ(\mathbb{G} - X\tilde{Y}') = m^{-1} tr(I - X(X'X)^+X')\mathbb{G}\mathbb{G}'. \tag{11.46}$$

Thus

$$\sigma_J(*;*) = \min\{\sigma_J(X;*) \mid X\} = \sum_s \lambda_s^2(m^{-1}GG'), \tag{11.47}$$

where the index s runs as $s = p + 1,...,n$. Home.

This derivation shows, again, that there is no need to normalize if we minimize join loss. It also shows that the partitioning of G into submatrices is irrelevant for join loss.

11.1.4 Solving meet and join problems

We have been concerned with definitions and with perfect fit so far. In the case of meet loss we found that the subspaces have a p-meet if $m^{-1}GD^+G$ has at least p eigenvalues equal to 1 and they have a p-join if G has at most p nonzero singular values. It is now easy to define meet and join solutions in an approximate sense: we simply carry out the minimizations in the previous sections to find solutions for X and Y for *given p*. These are the best meet and join approximations for that particular value of p. We can use the value of $\sigma_M(*;*,...,*)$ and of $\sigma_J(*;*)$ to find out whether join rank or meet rank are actually equal to p, and of course, they probably are not. Indeed, the concepts of meet and join are *gauges*: it is usually not true that there is a 'true' rank to discover. The only reasonable model in most cases is that the meet rank is zero and the join rank is $\sum k_j$ (if there are no build-in singularities). We are explicitly interested in the best approximation of the p-meet and p-join for fixed p and this is the technique that should be gauged.

We have shown in the previous sections that join and meet are independent concepts, defined first without using coordinates in the lattice of subspaces and later by using coordinates in terms of the eigenvalues of certain matrices. There are, however, some interesting relationships between the two concepts, which are most easily explained in terms of the loss functions. We have already commented on these relationships in previous chapters; we formulate them more generally here.

Suppose $G_j'G_j = I$ for all j. Then

$$\sigma_M(X;Y) = SSQ(X) - 2m^{-1} \operatorname{tr} X'GY + m^{-1} SSQ(Y). \tag{11.48}$$

For join loss we have the expression

$$\sigma_J(X;Y) = m^{-1} \sum_j k_j - 2m^{-1} \operatorname{tr} X'GY + m^{-1} SSQ(XY'). \tag{11.49}$$

The term $\sum_j k_j$ follows from the assumption $G_j'G_j = I$, where I is of order $k_j \times k_j$. If we assume in addition that $X'X = I$, which can be done without loss of generality, then

$$\sigma_M(X;Y) = \sigma_J(X;Y) + (p - m^{-1} \sum_j k_j). \tag{11.50}$$

Thus if $G_j'G_j = I$ the problems only differ by a constant, and optimizing join loss and meet loss obviously has the same solution. In both cases the optimal X and Y can be computed from the eigenanalysis of GG' or $G'G$, or from the singular value decomposition of G. The relationship between the loss functions also implies

$$\sigma_M(X;Y) \geq p - m^{-1} \sum_j k_j \tag{11.51}$$

and

$$\sigma_J(X;Y) \geq m^{-1} \sum_j k_j - p. \tag{11.52}$$

The condition $G_j'G_j = I$ for all j may look very special, but if we recall that the G_j were defined as any matrix for which L_j is the column space, then we see that the assumption can actually be made without loss of generality. To put it differently, if we choose the G_j as an orthonormal basis for L_j, then both the join and the meet solution can be computed from the singular value decomposition of the matrix G. Moreover, it is clear that the definitions of join rank and meet rank are certainly independent from the choice of spanning set G_j.

There is another interesting special case, which seems a bit weird at first sight. Suppose $k_j = 1$ for all j. Clearly the meet rank of $L_1, \ldots, L_m$ in this case is either one or zero, but we can still use the meet loss as defined. We then find

$$\sigma_M(X;Y) = \sigma_J(X;Y) + (p - 1), \tag{11.53}$$

which shows that a general algorithm to minimize σ_M can also solve problems involving σ_J. We have used this fact extensively in Section 4.5, and actually PRINCALS minimizes meet loss in order to solve the join problem.

11.1.5 Some extensions of the join and meet framework

Our formulation of meet and join so far starts with subspaces, then introduces bases or spanning sets which are essentially arbitrary, and then uses these spanning sets to reduce the meet and join problems to eigenvalue–eigenvector problems. This is only possible if the variables we deal with are treated either as single numerical or as multiple nominal. We have seen in the previous chapters that in that case our problems reduce to eigenproblems for known matrices. The situation is considerably less simple if we decide to treat some variables as single nominal or single ordinal. In this case some of the columns of some of the G_j are unknown, and the definition of join and meet becomes less natural and more complicated.

We first define L_j, which is the set of subspaces of L generated by all permissible choices of the columns of G_j. Thus we choose columns of G_j that are unknown or partially unknown in their respective feasible regions (usually convex cones). For each choice of these columns that we can compute the subspace spanned by these columns, the set of all such subspaces is L_j. We then define

$$\mathrm{MRK}(L_1,\ldots,L_m) \equiv \max\{\mathrm{MRK}(L_1,\ldots,L_m) \mid L_1 \in L_1,\ldots,L_m \in L_m\},$$

$$(11.54)$$

$$\mathrm{JRK}(L_1,\ldots,L_m) \equiv \min\{\mathrm{JRK}(L_1,\ldots,L_m) \mid L_1 \in L_1,\ldots,L_m \in L_m\}.$$

$$(11.55)$$

These definitions make it possible to repeat most of our earlier results in this chapter, with the additional complication that in the loss functions we also have to choose the columns of the G_j in the appropriate way. Our previous theorems show directly that the meet problem quantifies the variables in such a way that the largest eigenvalues of GD^+G' are maximized and the join problem quantifies variables in such a way that the smallest eigenvalues of $G'G$ are minimized. More precisely, we either maximize the sum of the p largest eigenvalues or we minimize the sum of the p smallest. If each G_j has only one column, then the eigenvalues of GD^+G are those of $D^{-1/2}G'GD^{-1/2}$, which is the correlation matrix of the variables. In this general formulation there are no problems with missing data; we simply adapt the definition of the feasible columns of G_j in the appropriate way. As we have seen in previous chapters it is sometimes desirable to normalize object scores alternatively, for example by $X'M_*X = I$ instead of $X'X = I$. Clearly using such an alternative normalization does not change the definition of perfect fit, i.e. of having a p-meet or a p-join, but it can change the approximate solutions considerably.

11.1.6 Extensions to infinite-dimensional space

In Chapter 1 we have indicated that our basic approach does not only apply to vectors of n observations but also to general random variables with finite variances, defined on the same probability space. The set of all such random variables is a separable Hilbert space, in which the covariance is the inner product and the variance is the square of the norm. The nonlinear transformations of a given random variable define a subspace of this same space, and in the general case also a subspace of infinite dimension. Now there is no reason to think that infinite-dimensional spaces are very difficult to understand, or have very complicated properties. In our case we restrict ourselves to the most simple and natural generalization of $\Re^n$, with the consequence that our coordinate-free definition of join rank and meet rank still applies, although some of the other results in the beginning of Section 11.1 are no longer true. The biggest nuisance, however, is computational. We get into trouble with infinite dimensionality as soon as we want to apply our formulas to do some actual computing; the formulas themselves do not change very much. We simply interpret $G_j Y_j$, for example, as involving an infinite summation; G_j possibly has a countable infinite number of columns. Moreover, SSQ must be replaced by VAR. Later in this chapter we shall illustrate some of the formulas in more detail, when we discuss some infinite-dimensional gauges.

11.2 CHOICE OF BASIS

In the previous sections we have seen that the choice of basis or spanning set is not important from a theoretical point of view; the minimum value of the loss functions is independent from the choice of basis. On the other hand, the choice can be important for computational and/or interpretational reasons. It clearly simplifies the computations if the G_j are chosen in such a way that $G_j'G_j$ is diagonal; this happens, for example, if G_j is a single complete or incomplete indicator matrix. For categorical variables the choice of basis is usually no problem, because the indicator matrices are both general enough and have all the desirable properties. A possible exception is the case in which the variables have a very large number of categories and in which some of the categories have very few observations. In those cases we often merge categories, which means that we choose to work in a lower-dimensional subspace of nonlinear transformations, because we expect the results to be more stable in that subspace.

An extreme case is the continuous variable. We do not mean the infinite-dimensional case; this will be treated in the next section. The number of observations is still finite and equal to n, but we assume that k_j is approximately equal to n. In this case indicator matrices for the separate categories will be approximate permutation matrices, and the situation is close to the 'm rankings' problem we discussed in Chapters 4 and 10. We have seen there that one possible way out of the problem that a perfect trivial solution exists is to use single variables and to impose ordinal or even numerical constraints. Examples such as the Roskam journal data show that this works satisfactory in these situations. On the other hand, we also have to recall the result of Section 4.3, which tells us that single ordinal will usually not work if the number of categories is close to the number of observations. The difference between the two types of situations is perhaps still worth emphasizing. In the situation discussed in Section 4.3 individuals are a sample from a population, while in the preference rankings situation 'individuals' in the program or technique are the objects that are ranked and 'variables' in the program are the individuals. Thus in Section 4.3 the rows of the matrix $\mathbb{G}$ are independent replications, whereas in the rankings situation the $\mathbb{G}_j$ are independent $n \times n$ matrices, and the argument in Section 4.3 breaks down because now we have to let $m \to \infty$. In a situation such as Section 4.3 with as many categories as individuals any arbitrary transformation of the variables will tend to give perfect homogeneity, and consequently the only thing we can do is to restrict the admissible transformations. Requiring ordinality will usually not be enough, requiring linearity defeats our purpose, which is to generalize linear techniques. Thus we continue to work with linear subspaces, but we restrict their dimensionality.

The simplest way to do this, which also works for nominal variables, is to decrease the number of categories by merging them. In the numerical case this is the familiar problem of the choice of category boundaries; in this case it is often possible to give some rough guidelines on how to categorize in a satisfactory way. This will be treated in more detail in the next chapter. If the variables are treated nominally (or ordinally) there are no clear-cut rules, although it is probably always good practice to merge (adjacent) categories with a very small number of observations.

For numerical variables there are some alternative possibilities that are also interesting. We can fit nonlinear transformations which are restricted to be polynomials of low order. This defines a low-dimensional subspace which we can parametricize in such a way that $\mathbb{G}_j' \mathbf{D}_j \mathbb{G}_j = \mathbf{I}$ (thus using $\mathbf{D}_j$ orthogonal polynomials). Polynomials have the somewhat unfortunate property that they are too rigid for flexible approximation of general functions. Sometimes we need a very high degree to get a good approximation, and of course choosing a polynomial of degree k_j-1 is equivalent to choosing a completely general nonlinear transformation, i.e. it is equivalent to using the $n \times k_j$ indicator matrix. This fact shows another advantage of considering different bases; it is sometimes nicer to interpret the Y_j in $G_j Y_j$ when G_j is a basis of D_j orthogonal polynomials, for in this case the category quantifications and their squares can be interpreted directly as the 'contributions' of the various degrees of the polynomial to the nonlinear transformation. A second disadvantage of polynomials is that additional constraints on them – for example that they are monotone, or nonnegative, or convex – are rather difficult to incorporate. Regression problems with these constraints are of course quadratic programming problems, just as monotone regression is a quadratic programming problem, but they are not *simple* quadratic programming problems.

11.2.1 Splines

Another possibility seems more promising than polynomials, although we have not compared the two systematically. Using indicator matrices in combination with continuous variables can be interpreted as approximating smooth transformations by step functions. When k_j is small, step functions are not very satisfactory, not only because they do not have enough parameters but also because it is very difficult to see from the best approximate step function what the approximated smooth function looked like. It is, of course, true that when we take a sufficiently large number of categories (jump points, knots) then arbitrary precise approximation is possible. Thus, as with polynomials, part of our criticism is that we need too many dimensions in the approximating subspace. Splines are intermediate between step functions and polynomials, and their current popularity seems to be due to the fact that they are both flexible and parsimonious. We first give a very brief introduction to splines.

Suppose $(..., a_{-1}, a_0, a_1, ...)$ is a doubly infinite increasing sequence of real numbers, called *knots*. They play the same role as category boundaries with step functions. A *spline* of *degree s* is a function that is a polynomial of degree s in each of the intervals (a_{k-1}, a_k): generally it is a different polynomial in each interval, with the additional property that the function is $s-1$ times continuously differentiable on the line. Because polynomials are infinitely many times differentiable, this last condition is a restriction only at the knots, where the $s-1$ derivatives from the right must be the same as the $s-1$ derivatives from the left. If $s = 0$ then our definition merely says that a spline should be constant in each interval between two consecutive knots, which implies that splines with $s = 0$ are just step functions. If $s = 1$ then a spline consists of broken line segments that are joined at the knots, which makes the resulting function continuous. Throughout this short introduction we suppose that the knots are fixed and known numbers and that there are no multiple knots (all a_k are different).

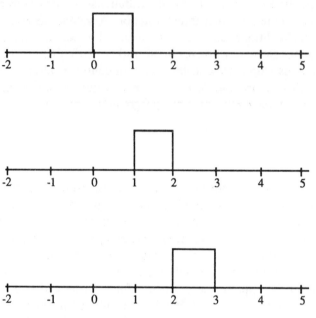

Figure 11.1 Zero degree B-splines.

It follows directly from the definition that splines of a given degree s on a fixed knot sequence a_k form a linear subspace. Because there is an infinite number of knots, the subspace is also of infinite dimension, with a finite number of observations, however. The knots that are outside the range of the observations do not matter, and so the space becomes finite dimensional. It can be shown that in our situation the dimensionality is equal to the number of interior knots plus the degree of the spline plus one. We illustrate this with a small example, in which all data points are between 0 and 3, and the knot sequence consists of the positive integers, the negative integers, and zero (if the knots are equally spaced, the corresponding splines are called *cardinal splines*). We construct the B-spline basis for the splines of degree zero, one, and two. The B-splines, or *basic splines*, were introduced by Curry and Schoenberg (1966); their properties are discussed extensively in De Boor (1978), where we can also find a proof of the theorem that the B-splines are indeed a basis of the subspace of all splines. B-splines are interesting for numerical purposes because they have *local support,* by which we mean that they are nonzero on $s + 1$ consecutive knot intervals only. For $s = 0$ this immediately gives the step (or 'block') functions in Figure 11.1. Clearly their normalization is immaterial, but we have normalized them in such a way that the $n \times 3$ matrix with the values of the n observations on the three functions is an indicator matrix, with rows that add up to 1. In Figure 11.2 we have drawn

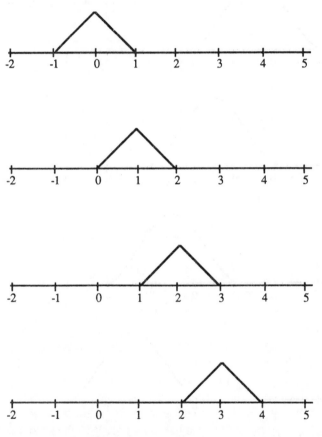

Figure 11.2 First-degree B-splines.

the four first-degree splines and in Figure 11.3 the five second-degree splines. The corresponding object $\times$ function matrices are $n \times 4$ and $n \times 5$; we have normalized the functions in such a way that the rows add up to 1 again.

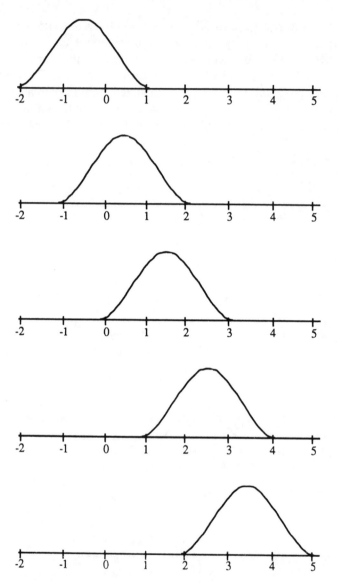

Figure 11.3 Second-degree B-splines.

Thus we now have obtained straightforward generalizations of indicator matrices through using B-splines. Increasing the degree of the spline gives a basis with one more column; local support says that at most $s + 1$ consecutive elements are nonzero. The zero-degree B-splines, starting at knot k, satisfy

$$g_k(h) = \begin{cases} 1 & \text{if } k \le h \le k + 1 \\ 0 & \text{otherwise} \end{cases} \tag{11.56}$$

The first-degree B-splines are

$$g_k(h) = \begin{cases} h - k & \text{if } k \le h \le k + 1 \\ (k + 2) - h & \text{if } k + 1 \le h \le k + 2 \\ 0 & \text{otherwise} \end{cases} \tag{11.57}$$

The second-degree B-splines are

$$g_k(h) = \begin{cases} 1/2(h - k)^2 & \text{if } k \le h \le k + 1 \\ 3/4 - [h - (k + 3/2)]^2 & \text{if } k + 1 \le h \le k + 2 \\ 1/2[h - (k + 3)]^2 & \text{if } k + 2 \le h \le k + 3 \\ 0 & \text{otherwise.} \end{cases} \tag{11.58}$$

Remember that these formulas are only for *cardinal* B-splines; for different knot sequences we have other functions. De Boor (1978) discusses efficient and stable recursive algorithms to compute B-spline values for any value of h.

Table 11.1 B-splines of degree zero, one, and two for four values of h

h	$s = 0$			$s = 1$				$s = 2$				
0.50	1	0	0	0.50	0.50	0.00	0.00	0.125	0.750	0.125	0.000	0.000
0.75	1	0	0	0.25	0.75	0.00	0.00	0.031	0.688	0.281	0.000	0.000
0.90	1	0	0	0.10	0.90	0.00	0.00	0.005	0.590	0.405	0.000	0.000
2.30	0	0	1	0.00	0.00	0.70	0.30	0.000	0.000	0.245	0.710	0.045

In Table 11.1 we illustrate how the B-splines generate matrices such as the indicator matrices, but with more nonzero entries in each row. It is clear that if $\mathbb{G}_j$ is defined using B-splines of degree s, then $\mathbb{G}_j'\mathbb{G}_j$ also has $s + 1$ consecutive nonzero elements in each row and column. This can be used to simplify the linear regression problem considerably. We also remark that Winsberg and Ramsay (1980, 1981) fit *monotonic* splines; as a basis for the monotonic splines they use integrals of B-splines. In fact, if we normalize the B-splines such that their integral equals one, then the integrated B-splines are distribution functions, and they can be used very nicely to fit latent trace models to binary variables. B-splines and their integrals have many fascinating properties, for which we refer to the book by De Boor (1978) and its references. ⇝

11.2.2 An empirical gauge for recovery of rational functions

We now give an example comparing step functions and (first-degree) splines. This example is in the nonlinear regression area. Thus there are two subspaces L_1 and L_2. As we already know, the case of two subspaces is somewhat special, but we shall approach the problem as a meet problem for these two subspaces. The example is from physics; it was used by Wilson (1926) and by Wilson and Worcester (1939) to illustrate the failure of statistical data analysis techniques, such as regression and components analysis, to give results that conform to physical theory. Willard Gibbs discovered a theoretical formula connecting the density z, the pressure y, and the absolute temperature x of a mixture of gases with convertible components. The formula is

$$\log \frac{A\,(z-A)}{(2A-z)^2} = \frac{B}{x} + \log y + C. \tag{11.59}$$

The constant A is the density of the rarer component, and it can be computed from the molecular formula. The constants B and C must be estimated. Gibbs computed them from data of experiments of Cahours and Bineau. Wilson does not state, unfortunately, how Gibbs computed these constants. Gibbs then applied his formula and the estimated constants to 65 experiments of Neumann, and he discusses the systematic and accidental divergences (residuals). Wilson gives an 'empiricist' instead of a 'rationalist' analysis of the data of Neumann. He fits three regression equations, computes partial correlations and so on, and concludes that it is very difficult to relate the results with the theory of Gibbs, and that it is impossible to deduce the rational formula of Gibbs from the regressions. Thus Wilson chooses, basically, $DIM(L_1) = 2$, with L_1 spanned by temperature and pressure, and $DIM(L_2) = 1$, with L_2 spanned by density. The regression statistics (only one canonical correlation can be computed) are given in the first row of Table 11.2. For comparison purposes we have also computed the same linear regression after applying the transformations suggested by the formula of Gibbs: they are given in the second row of Table 11.2. Of course this second analysis is quite useless; if we want to show that our techniques can recover functional types, then this clearly cannot be done by explicitly building them into the regression. The transformations dictated by the formula of Gibbs are plotted in Figure 11.4. Observe that they are monotonic, and can actually be approximated quite well by linear functions.

Table 11.2 Regression statistics for six different analyses of Gibbs' data

	r_{xy}	r_{xz}	r_{yz}	b_x	b_y	R_2
Linear	−0.380	0.816	0.153	1.022	0.542	0.917
Rational	−0.401	0.829	0.166	1.067	0.594	0.983
Step 1	−0.230	0.725	0.219	0.819	0.407	0.683
Step 2	−0.385	0.848	0.141	1.058	0.548	0.974
Spline 1	−0.389	0.839	0.166	1.065	0.581	0.991
Spline 2	−0.388	0.841	0.166	1.066	0.580	0.993

To recover functional types we need higher-dimensional subspaces. The Gibbs formula suggests that the model is additive (i.e. linear in the transformed variables). What happens if we do not constrain the transformations, and only require additivity? Then

$$L_1 = \{ \mathbf{u} \in \Re^n \mid u_i = \psi(x_i) + \phi(y_i) \},$$

$$L_2 = \{ \mathbf{v} \in \Re^n \mid v_i = \eta(z_i) \}$$

(11.60)

In this case, however, the subspaces have too many dimensions. Temperature has only nine different values, but pressure has 65 different values, and so has density. Thus L_1 has dimension $(9 - 1) + (65 - 1) = 72$ and L_2 has dimension 64. As the number of experiments is only 65, less than 72, there are 64 canonical correlations which are all equal to unity.

The first compromise is to group the observations on pressure and density in a fairly small number of classes, which means that we have to restrict our transformations to be step functions. For purposes of comparison we have used a crude discretization and a fine discretization. The number of intervals in the crude discretization, with their marginal frequencies, is given in Table 11.3a. Here $\mathrm{DIM}(L_1) = 5$ and $\mathrm{DIM}(L_2) = 2$. The transformations corresponding with the largest canonical correlation are plotted in Figure 11.5, the regression statistics are in the third row of Table 11.2.

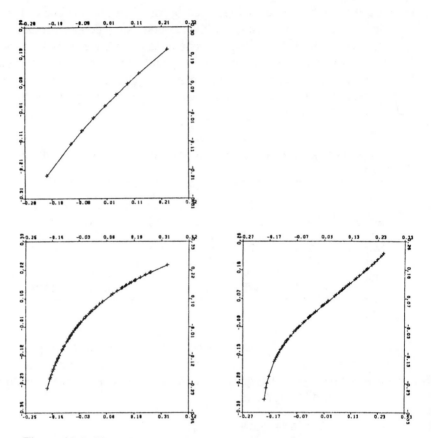

Figure 11.4 Gibbs–Wilson example: rational transformations.

The results are not satisfactory at all: the multiple correlation is very small and it is impossible to get an impression of the 'true' functional types from Figure 11.5. In the fine discretization $DIM(L_1) = 17$ and $DIM(L_2) = 8$; the marginals are given in Table 11.3b, the regression statistics are in the fourth row of Table 11.2, and the transformations in Figure 11.6. Because we know the form of the Gibbs transformations, it is possible to see that the step functions are beginning to bend in the 'correct' way; the regression statistics also become quite close to the 'rational' ones, but it still seems impossible to recover functional types from these results. In other words, in situations like these step functions do not seem to work very well. It is true, of course, that we would get a smoother plot if we used the midpoint on the interval against the transformed value, but our purpose here is to recover the transformation over the whole range.

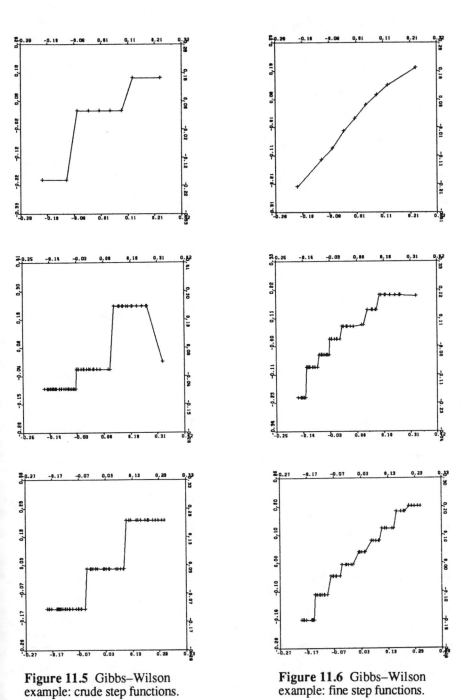

Figure 11.5 Gibbs–Wilson example: crude step functions.

Figure 11.6 Gibbs–Wilson example: fine step functions.

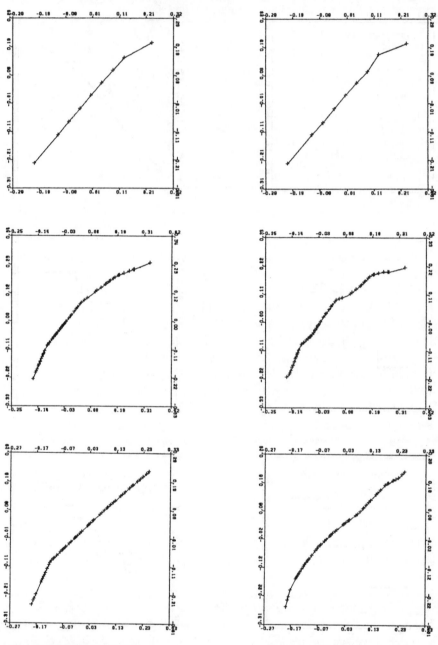

Figure 11.7 Gibbs–Wilson
example: crude splines.

Figure 11.8 Gibbs–Wilson
example: fine splines.

Table 11.3 Marginals for (a) step 1: crude step functions; (b) step 2: fine step functions; (c) spline 1: crude splines; (d) spline 2: fine splines

(a)												
Temperature	16	39	10									
Pressure	34	17	13	1								
Density	28	21	16									

(b)												
Temperature	6	10	7	8	9	8	7	2	8			
Pressure	9	12	13	8	7	2	5	4	4	1		
Density	9	12	7	9	6	6	7	4	5			

(c)												
Temperature	0.2	16.2	35.0	12.8	0.8							
Pressure	2.9	29.4	19.2	11.0	2.6							
Density	3.4	24.7	21.1	14.0	1.8							

(d)												
Temperature	6.0	10.0	7.0	8.0	9.0	8.0	7.0	2.0	8.0			
Pressure	0.2	8.3	12.2	12.6	9.3	5.6	2.5	4.6	4.6	3.1	1.1	0.8
Density	1.5	7.2	11.0	8.4	8.4	6.3	6.3	6.6	4.6	4.0	0.7	

We now compare these results with a crude and a fine basis of first-degree B-splines, differing in the number of knots. The column sums of the crude spline basis are in Table 11.3c, the regression statistics are in the fifth row of Table 11.2, and the transformations are in Figure 11.7. The transformations follow the rational curves quite closely, although $DIM(L_1)$ is only 8 and $DIM(L_2)$ is only 4. An even better recovery is obtained from the fine splines, with marginals in Table 11.3d, regression statistics in the sixth row of Table 11.2, and transformations in Figure 11.8. Here $DIM(L_1) = 19$ and $DIM(L_2) = 10$, still a considerable saving compared with the maximum dimensionality 64. However, the regression statistics indicate that the crude spline approximation has not improved considerably. The transformations in Figure 11.8 are quite smooth, but they do show some interesting dips, which could be studied in more detail (analysis of residuals).

The results of our spline analysis show that Wilson is too pessimistic about statistical (or data analysis) methods. It is indeed possible to recover some of the important features of the rational transformations; the statistical results make it possible to exclude a number of functional types from further rational search and the residuals point to outlying experiments that could be studied in more detail. If there were no theory at all, then these results could probably help in reaching 'the proper induction'. We must emphasize, however, that using social science standards implies that this example is too good to be true. In the social sciences there are no comparable rational theories that can be used as gauges, and the amount of error is usually much higher. This is also illustrated if we analyse the Gibbs–Wilson data with discrete ordinal options. Then $DIM(L_1) = 72$ and $DIM(L_2) = 64$, but we are saved from triviality by the monotonicity constraints. We do not give the transformations, but the solution monotonicity constraints. We do not give the transformations, but the solution is very similar to the fine spline solution we discussed. When we use continuous ordinal we get into trouble. Temperature is measured at only nine levels, and the continuous ordinal option uses the possibility to untie ties. It produces a perfect monotone solution, and other perfect monotone solutions when the program is started from other starting points. Some of the transformations are quite like the ones found with the spline options, but others are quite different from the spline transformations and thus also from the rational ones. The fact that $DIM(L_1) = 123$ in this case puts too heavy a burden on the monotonicity constraints.

11.2.3 Interactive variables

We have already used interactive variables at a number of different places in this book. The examples in Sections 6.5, 7.6, and 8.9 clearly show that interactive variables can be very useful for data analytic purposes. In this section we discuss a general technique to construct bases for interactive variables, given bases for the original variables. The technique is quite general, and in fact it is the correct way of interpreting most of discrete multivariate analysis in a linear algebraic context. This idea is explained in Good (1963), Benzécri (1967), and Haberman (1974).

The technique uses the *tensor product* (also: direct product, Kronecker product) of linear spaces. A nice discussion of this product is given by Halmos (1948, app. II), but many other books in linear and multilinear algebra have more extensive and more technical discussions. We first give a coordinate-free definition. Suppose L_1 and L_2 are subspaces of $\Re^n$. Define $B(L_1,L_2)$ as the set of all bilinear functions on $L_1 \times L_2$ (here the symbol $\times$ is used for the Cartesian product, the set of all pairs). Then define $L_1 \otimes L_2$, the tensor product of L_1 and L_2, as the algebraic dual of $B(L_1,L_2)$, i.e. the set of all linear functionals on $B(L_1,L_2)$.

For each $\mathbf{x} \in L_1$ and $\mathbf{y} \in L_2$ there is a $\mathbf{z} = \mathbf{x} \otimes \mathbf{y}$ in $L_1 \otimes L_2$, defined by $\mathbf{z}(\mathbf{u}) = \mathbf{u}(\mathbf{x},\mathbf{y})$ for all $\mathbf{u}$ in $B(L_1,L_2)$. Alternatively, we can start with a basis $\mathbf{x}_1,...,\mathbf{x}_r$ for L_1 and a basis $\mathbf{y}_1,...,\mathbf{y}_s$ for L_2, and define $L_1 \otimes L_2$ as the set of matrices that are linear combinations of the $n \times n$ matrices $\mathbf{x}_i\mathbf{y}_j'$. It is better to see this as an interpretation; the problem with matrices, as always, is that the same array of numbers can mean many things in many different contexts. From the definition, and from the interpretation, it is clear that the dimension of the direct product is the product of the dimension of the subspaces and that the technique can be extended quite easily to more than two subspaces (the interpretation then is in terms of linear combinations of multiway arrays of rank one). Another basic result is that the $r \times s$ tensor products $\mathbf{x}_i \otimes \mathbf{y}_j$ form a basis for $L_1 \otimes L_2$. Moreover we can define an inner product in $L_1 \otimes L_2$ by the rule

$$< \mathbf{x}_1 \otimes \mathbf{y}_1, \mathbf{x}_2 \otimes \mathbf{y}_2 > \; = \; < \mathbf{x}_1, \mathbf{x}_2> < \mathbf{y}_1, \mathbf{y}_2>, \tag{11.61}$$

where $<\mathbf{x}_1, \mathbf{x}_2>$ and $<\mathbf{y}_1, \mathbf{y}_2>$ are the original inner products in L_1 and L_2. This shows that orthogonal bases in L_1 and L_2 give an orthogonal basis in $L_1 \otimes L_2$.

The tensor product is the tool; we can now apply it in some contexts. In the simplest examples we have already replaced an $n \times k_1$ indicator matrix $\mathbf{G}_1$ and another $n \times k_2$ indicator matrix $\mathbf{G}_2$ by an $n \times (k_1 \times k_2)$ indicator matrix, which we can now write as $\mathbf{G}_1 \otimes \mathbf{G}_2$. Interesting generalizations of homogeneity analysis are possible on this basis. Rather than requiring $\mathbf{x} = \mathbf{G}_j \mathbf{y}_j$ for all j, we can now require

$$\mathbf{x} = \mathbf{G}_j \otimes \mathbf{G}_l \mathbf{y}_{jl}, \tag{11.62}$$

for all j,l. Instead of performing HOMALS with m variables we perform HOMALS with $1/2m(m-1)$ product variables. If we have a basis of polynomials or splines for each j, then the direct product of these bases can be used as a basis for the direct product. We have done this, with orthogonal polynomials, in the example in Section 6.5. In this context we have to remember that if the polynomials are orthogonal with respect to the marginals of the variables, then the direct product of the polynomials is orthogonal with respect to the direct product of the marginals, which is generally not the same thing as the multivariate distribution of the variables.

11.2.4 Infinite dimensionality

Now suppose we replace $\mathfrak{R}^n$ by an infinite-dimensional separable Hilbert space, for example the set of all random variables with finite variance on a given probability space. If the subspaces $L_1,...,L_m$ are finite dimensional, then nothing much changes. The bases for the subspaces still consist of a finite number of elements and the inner products of the elements in the bases can still be used to define finite matrices $\mathbf{C}$ and $\mathbf{D}$, which define a finite eigenvalue–eigenvector problem. The definitions of indicator functions arising from discretization and of spline functions with a given knot sequence also do not change, and of course polynomials are also easy to define.

We outline one special case, more or less as an exercise in notation. If $\underline{h}_1,...,\underline{h}_m$ are the random variables on the probability space, then one-dimensional HOMALS wants to find transformations $\phi_1,\ldots,\phi_m$ and a random variable $\underline{x}$ such that

$$\sigma_M(\underline{x};\phi_1,\ldots,\phi_m) = m^{-1} \sum_j VAR(\underline{x} - \phi_j(\underline{h}_j)) \tag{11.63}$$

is minimized, under the normalization condition $AVE(\underline{x}) = 0$ and $VAR(\underline{x}) = 1$. It is important to realize that this is a straightforward generalization of ordinary one-dimensional HOMALS, in which the random variables $\underline{h}_j$ are discrete and have k_j different possible values and in which probability is the counting measure. Just as in the previous chapters we can define

$$\sigma_M(\underline{x};*,\ldots,*) \equiv \min\{\sigma_M(\underline{x};\phi_1,\ldots,\phi_m) \mid \phi_1,\ldots,\phi_m\}. \tag{11.64}$$

with $\phi_1,\ldots,\phi_m$ not constrained in any way. The formal solution to the problem of minimizing σ_M for fixed $\underline{x}$ is given by substituting the conditional expectations

$$\phi(\underline{h}_j) = AVE(\underline{x} \mid \underline{h}_j). \tag{11.65}$$

Since conditional expectation is projection on a subspace we have (Appendix C) that

$$VAR(\underline{x}) = VAR(\underline{x} - AVE(\underline{x} \mid \underline{h}_j)) + VAR(AVE(\underline{x} \mid \underline{h}_j)) \tag{11.66}$$

for all j, which implies that

$$\sigma_M(\underline{x};*,\ldots,*) = 1 - m^{-1} \sum_j VAR(AVE(\underline{x} \mid \underline{h}_j)). \tag{11.67}$$

This result shows that $\sigma_M(\underline{x};*,\ldots,*)$ is 1 minus the average correlation ratio of $\underline{x}$ and the $\underline{h}_j$ (recall that in the linear case principal components analysis maximizes the average correlation of $\underline{x}$ and the $\underline{h}_j$). A somewhat different notation is also convenient. If we write $AVE(\underline{x} \mid \underline{h}_j)$ as $P_j(\underline{x})$ to indicate that we project on the subspace of all functions of $\underline{h}_j$ with finite variance, then idempotency of the projector shows that

$$\sigma_M(\underline{x};*,\ldots,*) = 1 - m^{-1} \sum_j COV(\underline{x}, P_j(\underline{x})). \tag{11.68}$$

Thus maximizing over $\underline{x}$ under the condition $VAR(\underline{x}) = 1$ amounts to solving the eigenproblem for the operator $m^{-1} \sum_j P_j$. Again this is a simple generalization of the finite case, in which we had $\mathbf{P}_j = \mathbf{G}_j(\mathbf{G}_j'\mathbf{G}_j)^{-1}\mathbf{G}_j'$. This notational exercise can be generalized in the familiar directions. If we project on a cone instead of on a subspace, for example, the optimal $\phi(\underline{h}_j)$ is a generalized conditional expectation and $P_j(\underline{x})$ is a nonlinear projector. When we compute more dimensions we have to choose again between single and multiple treatment of subspaces. When we have sets of variables we can use interactive coding and linear restrictions again.

An important practical problem is, of course, that it is impossible to carry out actual computations in infinite-dimensional spaces. The matrices are too big. This is not a very serious problem from a theoretical point of view. Truly infinite-dimensional problems are always gauges, because they do not only imply continuous variables but they also imply an infinite number of observations. In addition, if we have well-defined gauges the computations can often be carried out in the form of formulas. Suppose, for example, that the h_j are multinormal with mean zero, variance one, and correlations ρ_{jl}. We now apply the alternating least squares algorithm for one-dimensional HOMALS theoretically. We start with a normally distributed x with $COR(x, h_j) = \theta_j$. The first step of the ALS algorithm computes the optimum ϕ_j for given x. It gives $\phi_j(h_j) = \theta_j h_j$, because regressions in the multinormal case are linear. The second step computes a new optimal x for given ϕ_j; it makes the new x proportional to the average of the $\phi_j(h_j)$. This implies that the correlation of the new x with the h_j is proportional to $\mathbf{R}\theta$, where $\mathbf{R}$ is the correlation matrix of the h_j. Thus the correlation between x and the h_j converges to the dominant eigenvector of $\mathbf{R}$, and in the multinormal case the optimum nonlinear solution is the same as the optimum linear solution. We shall study gauges such as these in more detail in the next section, we use them here to illustrate that ALS can be carried out theoretically and can be used to prove theorems. As we illustrated in Chapter 9 it is also possible to prove ordinal properties of the eigenvector corresponding with the largest eigenvalue in this way.

For some of these theoretical computations orthogonal polynomials are quite useful. For a given probability distribution for which moments of all orders exist, the orthogonal polynomials $\psi_0(x)$, $\psi_1(x)$, ... are defined uniquely by the conditions that $\psi_s(x)$ is a polynomial of degree less than or equal to s and $AVE(\psi_s(x)\psi_t(x)) = 0$ for $t < s$, while $AVE(\psi_s^2(x)) = 1$. The classical exponential distributions such as the normal, Poisson, gamma, and binomial all have a classical set of orthogonal polynomials associated with them. We refer the reader to Tricomi (1955) for a nice treatment of orthogonal polynomials; there is also a fairly complete coverage of their statistical applications in Lancaster (1969). We illustrate how to construct $\psi_2(x)$ for the normal distribution directly. First observe that $\psi_0(x) = 1$ and $\psi_1(x) = x$ if we assume that x is standard normal. So if $\psi_2(x) = \alpha x^2 + \beta x + \gamma$, then $AVE(\psi_0(x)\psi_2(x)) = AVE(\psi_2(x)) = 0$ gives $\alpha + \gamma = 0$. Next, the condition $AVE(\psi_1(x)\psi_2(x)) = 0$ gives $\beta = 0$. Finally, the condition $AVE(\psi_2^2(x)) = 1$ yields $\alpha = 1/2\sqrt{2}$. It is clear that this process of explicit orthogonalization can rapidly become tedious, but fortunately simple recursive formulas are available to compute $\psi_s(\underline{x})$ for all values of $\underline{x}$ and s. The same thing is true for the other classical orthogonal polynomials. If ϕ_1 and ϕ_2 are two functions with representation

$$\phi_1(\underline{x}) = \Sigma_s \; \alpha_s \psi_s(\underline{x}), \tag{11.69}$$

$$\phi_2(\underline{x}) = \Sigma_s \; \beta_s \psi_s(\underline{x}), \tag{11.70}$$

for $s = 0,\ldots,\infty$, then

$$\mathrm{AVE}(\phi_1(\underline{x})\phi_2(\underline{x})) = \Sigma_s \; \alpha_s \beta_s \quad (s = 0,\ldots,\infty) \tag{11.71}$$

and thus

$$\mathrm{COV}(\phi_1(\underline{x}),\phi_2(\underline{x})) = \Sigma_s \; \alpha_s \beta_s \quad (s = 1,\ldots,\infty). \tag{11.72}$$

This result is useful because such a polynomial development is possible for all functions with finite variance. It also reduces problems with functions as arguments to problems involving infinite sums, which are easier to handle in many applications. Observe that the expansion of the covariance is not really what we are interested in, because we do not compute covariances between functions of the same variable, but covariances between functions of different variables. In such a case the polynomial expansion merely gives

$$\mathrm{COV}(\phi_1(x_1),\phi_2(x_2)) = \Sigma_s \; \Sigma_t \; \alpha_s \beta_t \; \mathrm{COV}(\psi_{s_1}(x_1),\psi_{s_2}(x_2)), \tag{11.73}$$

for $s, t = 1,\ldots,\infty$, which is clearly far less simple.

It is also possible by basically the same methods to construct orthogonal polynomials on multivariate distributions. Thus we construct polynomials, say in m variables, such that

$$\mathrm{AVE}(\phi_{s_1 \ldots s_m}(x_1,\ldots,\underline{x}_m) \; \phi_{t_1 \ldots t_m}(x_1,\ldots,\underline{x}_m)) = \delta(s_1,t_1)\ldots\delta(s_m,t_m),$$

$$\tag{11.74}$$

where $\delta(s,t)$ is the Kronecker delta. If such orthogonal polynomials are available it is possible to use them in the treatment of interactive variables, in which case it is not necessary to use the tensor product to construct bases. Dahmen (1979, 1980a, 1980b) has developed a system of multivariate B-splines which can be used in a similar way.

11.3 SOME INFINITE-DIMENSIONAL GAUGES

Some of the results in this section have already been discussed briefly, in a bivariate context, in Section 8.5. There we mentioned Mehler's formula for the bivariate normal distribution. Bivariate normality implies, in the notation of the previous section, that

$$\text{COV}(\psi_{s1}(\underline{x}_1), \psi_{t2}(\underline{x}_2)) = \delta(s,t)\rho_{12}^s. \tag{11.75}$$

Thus the covariance of two orthogonal polynomials on the standard bivariate normal distribution vanishes if the polynomials have different degree, and is equal to the ordinary correlation to the power s if they have the same degree s. Consequently, the covariance between arbitrary square integrable functions becomes

$$\text{COV}(\phi_1(\underline{x}_1), \phi_2(\underline{x}_2)) = \sum_s \alpha_s \beta_s \rho_{12}^s \quad (s = 1,\dots,\infty). \tag{11.76}$$

When we have m functions of m variables, with joint multinormal distribution, then

$$\phi_j(\underline{x}_j) = \sum_s \alpha_{js} \psi_s(\underline{x}_j) \quad (s = 0,\dots,\infty) \tag{11.77}$$

can be used to derive an interesting result on the sum of the covariances. If $\mathbf{R}^{(s)}$ is the matrix with correlations to the power s, and $\boldsymbol{\alpha}_s$ is the m-vector with the α_{js} for fixed s, then

$$\sum_j \sum_l \text{COV}(\phi_j(\underline{x}_j), \phi_l(\underline{x}_l)) = \sum_s \boldsymbol{\alpha}_s' \mathbf{R}^{(s)} \boldsymbol{\alpha}_s \quad (s = 1,\dots,\infty). \tag{11.78}$$

The sum of the variances is simply

$$\sum_j \text{VAR}(\phi_j(\underline{x}_j)) = \sum_s \boldsymbol{\alpha}_s' \boldsymbol{\alpha}_s \quad (s = 1,\dots,\infty) \tag{11.79}$$

This shows what the eigenvalue problem that must be solved in homogeneity analysis amounts to in the case of the multinormal distribution. The eigenvalues are the m eigenvalues of $\mathbf{R}^{(1)}$, the m eigenvalues of $\mathbf{R}^{(2)}$, and so on. It follows from general results on Hadamard products (Styan, 1973) that the largest eigenvalue of all these eigenvalues is always the largest eigenvalue of $\mathbf{R}^{(1)}$, which is the ordinary correlation matrix. Thus the largest eigenvalue has all transformations $\phi_j(\underline{x}_j)$ linear, a result proved by other techniques in the previous section. The second largest eigenvalue is more of a problem, however. It can be the second largest eigenvalue of $\mathbf{R}^{(1)}$ or the largest eigenvalue of $R^{(2)}$. In the first case the transformations on the second dimension are all linear too; in the second case they are all quadratic. The second case typically occurs if the largest eigenvalue of $R^{(1)}$ is much larger than the second one, because in that case the largest eigenvalue of $R^{(2)}$ is often larger than the second eigenvalue of $R^{(1)}$. In this case the plot of the first two dimensions shows a horseshoe. Thus multinormal data in the infinite-dimensional generalization of HOMALS tend to give a horseshoe. The implication is that approximately multinormal data, discreticized, and from a finite sample, will give an approximate horseshoe.

The horseshoe does not always occur in the analysis of multinormal data. We illustrate this with a small example. Suppose we have four normal variates, with correlation matrix

$$R = \begin{pmatrix} 1 & A & B & C \\ A & 1 & C & B \\ B & C & 1 & A \\ C & B & A & 1 \end{pmatrix}$$

with A, B, and C nonnegative numbers. The eigenvectors of R are, columnwise,

$$\begin{pmatrix} +1 & +1 & +1 & +1 \\ +1 & +1 & -1 & -1 \\ +1 & -1 & +1 & -1 \\ +1 & -1 & -1 & +1 \end{pmatrix}$$

and the corresponding eigenvalues are

$$\lambda_1^2 = 1 + A + B + C,$$

$$\lambda_2^2 = 1 + A - B - C,$$

$$\lambda_3^2 = 1 - A + B - C,$$

$$\lambda_4^2 = 1 - A - B + C.$$

This example is illustrative because $R^{(s)}$ has the same form as R, and consequently the eigenvalues are also of the same form. The order of the eigenvalues depends, of course, on the size of A, B, and C. If we also look at the eigenvalues of $R^{(s)}$ then the order problem becomes quite complicated, even for this small example.

> *We want to use this example to show that multinormal data do not necessarily give horseshoes. Thus we want the second eigenvalue of $R^{(1)}$ to be larger than the dominant eigenvalue of $R^{(2)}$. We choose A, B, C in such a way that their difference is as large as possible. After some problem solving we find that this occurs for $A = 1/2$ and $B = C = 0$ for which the first two eigenvalues of $R^{(s)}$ are $1 + (1/2)^s$ and the second two are $1 - (1/2)^s$. Thus the second eigenvalue of $R^{(1)}$ is 1.5 and the first eigenvalue of $R^{(2)}$ is 1.25. We now draw a sample of size 1000 from this multinormal distribution and discreticize the continuous variables in four categories. If ψ_1 and ψ_2 are the normal orthogonal polynomials (the Hermite–Chebyshev polynomials), then we expect the first four eigenvectors of HOMALS to look row-wise like*
>
> $(+\psi_1 \; +\psi_1 \; +\psi_1 \; +\psi_1)$,
>
> $(+\psi_1 \; +\psi_1 \; -\psi_1 \; -\psi_1)$,
>
> $(+\psi_2 \; +\psi_2 \; +\psi_2 \; +\psi_2)$,
>
> $(+\psi_2 \; +\psi_2 \; -\psi_2 \; -\psi_2)$,
>
> *with the first two eigenvalues equal to 1.50 and the second two to 1.25. The actual eigenvalues in the discreticized sample are 1.53, 1.43, 1.18, 1.15. The eigenvectors are plotted against category number in Figure 11.9, with each row is an eigenvector and each column a variable. It is clear that the predicted pattern is there.*

The statement that we find orthogonal polynomials when the data are approximately multinormal is much too weak. If the multivariate distribution is a *mixture* of multivariate normals, then the eigenvectors are again the Hermite–Chebyshev polynomials, but the eigenvalues are more complicated functions of the various correlation matrices. If the data are multivariate lognormal (or more generally one-to-one nonlinear transformations of normal variates), then our technique finds the inverse transformation to normality and consequently finds the Hermite–Chebyshev polynomials again. The marginals of the transformed variables can be anything, even rectangular. The work of Lancaster (1969), Eagleson (1969), Griffiths (1969, 1970), and Tyan and Thomas (1975) shows that there are many other classical bivariate distributions with orthogonal polynomials as eigenvectors. They often have the same solution structure as the multinormal, although the relation with the matrices $R^{(s)}$ is not necessarily true. The work of Karlin (1964, 1968) shows that oscillatory eigenvectors, which cannot be distinguished from orthogonal polynomials in discrete situations, often occur because of the general condition of *total positivity*. As a consequence we can make the somewhat stronger statement that HOMALS tends to find orthogonal polynomial transformations whenever it is used as a variable transformation technique on data that are strongly homogeneous. Thus the horseshoe is the HOMALS equivalent of the general factor. In many cases, however, HOMALS is used as a multidimensional scaling technique on data that are not a random sample on a set of *m* variables, or on data in which some of the variables are purely nominal, or on data in which there are positive and negative and large and small correlations. In that case we do not find the familiar quadratic pattern.

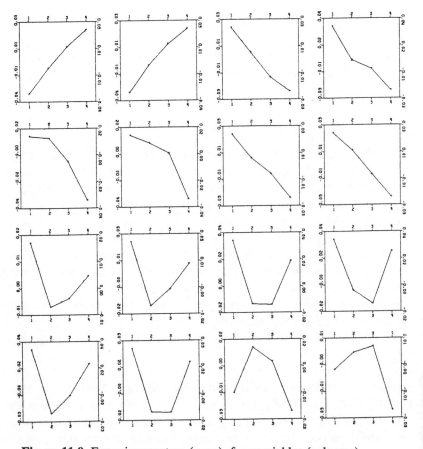

Figure 11.9 Four eigenvectors (rows), four variables (columns).

The multinormal distribution can gauge HOMALS, but can it also be used to gauge CANALS and PRINCALS? We start with OVERALS and use the multivariate version of the Hermite–Chebyshev polynomials discussed by Appell and Kampé de Fériet (1926), Erdelyi (1953), and used in this context by Venter (1966) and Dauxois and Pousse (1976). For these polynomials it can be proved that

$$COV(\psi_{s_1...s_m}(x_1,...,x_m), \psi_{t_1...t_m}(y_1,...,y_m))$$

$$= \delta(s_1,t_1)...\delta(s_m,t_m)\rho_1^{s_1}...\rho_m^{s_m}, \qquad (11.80)$$

where $\rho_1 \geq ... \geq \rho_m$ are the canonical correlations (we assume for the moment that both sets contain an equal number of variables, merely to simplify notation). This implies

$$COV(\phi_j(x_j), \phi_l(x_l)) = \Sigma_s \alpha_s' R_s \alpha_s. \qquad (11.81)$$

In this equation the x_j are vectors of random variables, the index s is a vector $(s_1,...,s_m)$, and for each value of this index there is a 2×2 matrix R_s with off-diagonal element

$$(R_s)_{12} = \rho_1^{s_1}...\rho_m^{s_m} \qquad (11.82)$$

where the ρ_s are the canonical correlations. This also shows that CANALS with all variables multiple nominal amounts to solving eigenvalue problems for the matrices R_s separately, in the same way as HOMALS. Ordering the eigenvalues of the various matrices is even more complicated in this case, although it is easy to see that the largest eigenvalue is the largest eigenvalue of $R_{(1,0,...,0)}$, the matrix with the off-diagonal element equal to the largest canonical correlation, which again corresponds to linear transformations of all variables. If there are more than two sets of variables, the situation becomes much more complicated again.

For single variables additional complications are introduced too. We have seen that HOMALS on the multivariate normal can select its successive solutions from the various $R^{(s)}$, depending on the relative size of the eigenvalues of these matrices. PRINCALS computes a single solution (transformation), and the single restriction forces PRINCALS to select its p dimensions from a single $R^{(s)}$. It may be possible to construct a correlation matrix for which the sum of the p largest eigenvalues of $R^{(2)}$ is larger than the sum of the p largest eigenvalues of $R = R^{(1)}$, but we have not found one yet. Generally PRINCALS selects its p dimensions from $R^{(1)}$, which means that using PRINCALS in the multinormal (and similar) cases should give the same solution as using HOMALS as a first step, followed by linear principal components analysis 'as a second step'. A similar result is true for CANALS with single variables. We emphasize, however, that we have not proved that the linear solution gives the maximum; we have merely proved that the linear solution (and all other solutions based on the orthogonal polynomials) give stationary values of the loss function. The proof is simply by substitution in the stationary equations and is not given here. ⇥

11.4 ANALYSIS OF STOCHASTIC PROCESSES

Nonlinear components analysis techniques can also be used for the analysis of stochastic process data. This does not add anything new for discrete time; the time points t_1, t_2,... define variables. In practical data analysis problems there is only a finite number of time points, and consequently a finite number of variables. For continuous time the situation is considerably more interesting. For details we refer the reader to the admirable paper by Deville and Saporta (1980). We merely emphasize the formal similarity with HOMALS. The eigenequation we have to solve is

$$\int_T \mathbf{D}_t^{-1} \, \mathbf{C}_{ts} \, \mathbf{y}_s \, \mathrm{d}s = \lambda^2 \mathbf{y}_t, \tag{11.83}$$

with $\mathbf{D}_t = \mathbf{G}_t' \mathbf{G}_t$ and $\mathbf{C}_{ts} = \mathbf{G}_t' \mathbf{G}_s$, and with $\mathbf{G}_t$ the indicator matrix (or the basis of indicator functions, polynomials, or splines) at time t. Thus we now have an infinite number of variables, for which our basic join and meet philosophy does not necessarily apply without further regularity conditions on the subspaces and operators. In full generality these conditions are treated by Dauxois and Pousse (1976). For the particular application we discuss here they simplify considerably, and are discussed by Deville and Saporta (1980). The simplifications are due to the fact that we deal with a discrete and finite state space here.

Deville and Saporta (1980) also discuss data analysis aspects of continuous time. Again we have to discreticize time in some way or another to get finite matrices. The first idea is to select a point in each of a finite number of time intervals and use the indicator matrix of this time point, weighted with the length of the time interval, in a correspondence analysis. A better alternative is to use generalized indicator matrices, in which element $g_{(j)ir}$ indicates the percentage of time spend by individual i in state r in time interval j. Thus the $\mathbf{G}_j$ are not binary, but their rows still add up to 1. Again the generalized indicators are weighted by the length of the time interval and subjected to correspondence analysis.

Deville and Saporta (1980) do not mention any gauges for their techniques, but quite a number of interesting results are available. In the same way as in the previous section gauges can be derived from the canonical decomposition of bivariate distributions, in this case from the transition probabilities of symmetric stationary Markov processes with a continuous time parameter. For continuous state space this was done by Wong and Thomas (1962) and by Sarmanov (1963), and for birth and death processes with a discrete state space by Eagleson (1969). These authors investigate in detail the conditions that guarantee that the canonical variables are orthogonal polynomials, with results similar to those obtained in the static case. In the very interesting paper of Cooper, Hoare, and Rahman (1977) the transition probabilities of a class of Markov chains are analysed. The canonical variables in this case are discrete orthogonal polynomials. When various parameters approach their limiting values, continuous state processes are obtained with continuous orthogonal polynomials as canonical components.

11.5 ALTERNATIVE CRITERIA

In Chapter 5 we have discussed a number of different criteria that could be optimized in the case of linear K-set canonical analysis. We can now formulate the same problem more generally. Suppose $x_1,\ldots,x_m$ are m-vector valued random variables, which can be discrete or continuous, and which can be defined on a finite or infinite probability space. We study the transformations $\phi_j(x_j)$, where all kinds of restrictions can be imposed on the ϕ_j: they could be additive, additive univariate splines, additive and monotonic, and so on. We agree to choose the ϕ_j from feasible sets Φ_j in such a way that a function of the correlations $COR(\phi_j(x_j), \phi_l(x_l))$ is maximized or minimized. As explained in previous chapters this formulation covers HOMALS (maximize the largest eigenvalue of the correlation matrix, compute more than one solution), it covers PRINCALS (maximize the sum of the p largest eigenvalues), it covers CANALS (maximize, for a given partitioning into two sets, the sum of the first p canonical correlations), and it covers OVERALS (maximize, for a given partitioning into K sets, the sum of the first p generalized canonical correlations). It also covers the alternative criteria mentioned in Chapter 5.

It is possible to define a more general class of criteria in terms of the correlation. This shall be outlined in the next section. It is also possible to develop an alternative loss of homogeneity function, to be discussed in the subsequent section.

11.5.1 Unitarily invariant norms

It is interesting to find a more general class of criteria for which some results can be shown to be true without restricting ourselves to some particular criterion in the class. This is possible if we specify that the function of the correlation matrix $\mathbf{R}$ we are trying to maximize is a *unitarily invariant norm*. This is a matrix norm with the additional property that $\mathbf{K'RK}$ has the same norm as $\mathbf{R}$ for all orthogonal $\mathbf{K}$. It does not necessarily have the property that the norm of a product is less than or equal to the product of the norms of the matrices in the product. Thus if $\omega(.)$ is our criterion, then

$$\mathbf{R} \neq \mathbf{0} \rightarrow \omega(\mathbf{R}) > 0, \tag{11.84}$$

$$\omega(\alpha\mathbf{R}) = |\alpha|\, \omega(\mathbf{R}), \tag{11.85}$$

$$\omega(\mathbf{R}_1 + \mathbf{R}_2) \leq \omega(\mathbf{R}_1) + \omega(\mathbf{R}_2), \tag{11.86}$$

$$\omega(\mathbf{K'RK}) = \omega(\mathbf{R}). \tag{11.87}$$

The sum of the p largest eigenvalues of $\mathbf{R}$ is a particular case of a unitarily invariant norm; in fact it is quite a special case. A theorem by Fan (1951) says that $\omega(\mathbf{R}_1) \geq \omega(\mathbf{R}_2)$ for *all* unitarily invariant norms if and only if

$$\lambda_1^2(\mathbf{R}_1) \geq \lambda_1^2(\mathbf{R}_2) \tag{11.88}$$

and

$$\lambda_1^2(\mathbf{R}_1) + \lambda_2^2(\mathbf{R}_1) \geq \lambda_1^2(\mathbf{R}_2) + \lambda_2^2(\mathbf{R}_2) \tag{11.89}$$

and ... and

$$\lambda_1^2(\mathbf{R}_1) + ... + \lambda_{m-1}^2(\mathbf{R}_1) \geq \lambda_1^2(\mathbf{R}_2) + ... + \lambda_{m-1}^2(\mathbf{R}_2). \tag{11.90}$$

Thus a set of transformations is optimal for all unitarily invariant norms if and only if it is optimal for PRINCALS in $1, 2, ..., m-1$ dimensions.

Now suppose that $\mathbf{R}$ has a representation

$$r_{jl} = \sum_s \alpha_{js}\alpha_{ls}c_{jls}, \tag{11.91}$$

or, in matrix notation,

$$\mathbf{R} = \Sigma_s \, \mathbf{A}_s \mathbf{C}_s \mathbf{A}_s, \tag{11.92}$$

with $\mathbf{A}_s$ an $m \times m$ diagonal matrix, with α_{js} on the diagonal, and $\Sigma_s \, \mathbf{A}_s^2 = \mathbf{I}$. For unitarily invariant norms it is possible to prove that

$$\omega(\mathbf{R}) \le \max_{(s=1,\ldots,\infty)} \omega(\mathbf{C}_s), \tag{11.93}$$

and equality is possible by choosing $\mathbf{A}_s = \mathbf{I}$ for one s and $\mathbf{A}_s = \mathbf{0}$ for all other s. The same type of inequality can be proved for other matrix norms that do not necessarily satisfy $\omega(\mathbf{K'RK}) = \omega(\mathbf{R})$, but that do satisfy $\omega(\mathbf{R}_1\mathbf{R}_2) \le \omega(\mathbf{R}_1)$ $\omega(\mathbf{R}_2)$. The inequality implies that we can investigate all $\mathbf{C}_s$ separately and keep the one with the largest $\omega(\mathbf{C}_s)$. In the special case of the multinormal distribution $\mathbf{C}_s = \mathbf{R}^{(s)}$. We know that if we select $\omega(\mathbf{C}_s) = \lambda_1^2(\mathbf{C}_s)$, then the maximum of $\omega(\mathbf{R})$ is $\lambda_1^2(\mathbf{R}^{(1)})$. For PRINCALS we conjecture that the sum of the p largest eigenvalues of $\mathbf{R}$ is also maximized for $\mathbf{R}^{(1)}$, but we have not proved this. Oppenheim's inequality (Styan, 1973; Beckenbach and Bellman, 1965, p. 71) can be used to show that for $\omega(\mathbf{R}) = [\det(\mathbf{R})]^{1/m}$ we also only have to look at the $\det(\mathbf{C}_s)$ individually, and that if $\mathbf{C}_s = \mathbf{R}^{(s)}$, then $\det(\mathbf{R}^{(1)})$ is the *smallest* of the determinants $\det(\mathbf{R}^{(s)})$. Thus the determinant criterion in Chapter 5, which had to be minimized, gives linear transformations in the multinormal case too.

11.5.2 Least squares or other

In many places we have already given our reasons to use least squares, either explicitly or implicitly, by emphasizing pictures, inner products, or multinormal gauges. Currently popular alternatives to least squares are motivated statistically, often by robustness considerations. As explained earlier we do not think that statistical gauges are necessarily the best way to select data analysis techniques; the situation is often the other way around. On the other hand, if we have definite reasons to take some gauge seriously, then other methods than least squares certainly have to be taken into account. We do not discuss them in detail; we merely indicate how some of our techniques can be generalized. Suppose, for example, that we want to minimize

$$\sigma(\mathbf{x};\mathbf{y}_1,\ldots,\mathbf{y}_m) = m^{-1} \Sigma_j \, \xi(\mathbf{x} - \mathbf{G}_j\mathbf{y}_j), \tag{11.94}$$

where $\mathbf{x}$ must satisfy the normalization condition $\psi(\mathbf{x}) = 1$. In this formulation the functions $\xi(\,)$ and $\psi(\,)$ are now assumed to be positively homogeneous and convex. Algorithms to minimize functions of this sort are derived easily from the work of Robert (1967) or Boyd (1974); they have been used in a somewhat different context by De Leeuw (1977). The extension to random variables is easy enough; the extension to multidimensional solutions is somewhat more complicated because a natural definition of orthogonality is not necessarily available. We do believe, however, that no new complications arise when translating most of our theory into normed linear space terminology. ➤ Certainly the concept of join and meet are independent of any norm, and consequently they can be translated into loss function terminology by using any norm.

11.6 EPILOGUE

Additional applications of monotonic splines can be found in Winsberg and Ramsay (1982, 1983). Recent reviews of the area are Van Rijckevorsel (1987), Winsberg (1988), and Ramsay (1988). A version of PRINCALS that incorporates splines, called SPLINALS, is documented in De Leeuw, Van Rijckevorsel, and Van der Wouden (1981) and Coolen, Van Rijckevorsel, and De Leeuw (1982).

De Leeuw (1988) discusses HOMALS as a first step extensively, emphasizing the consequences of the data satisfying a special condition called *simultaneous linearizability*. For further work on functional data in time see Saporta (1985), Besse (1988), and Ramsay (1989). Switching to other loss functions than least squares has been tried by Heiser (1987b) for the case of correspondence analysis, using the absolute value function in formula (11.94). The consequences are much more radical than expected, leading to a clustering of the object points into $p + 1$ groups.

CHAPTER 12
THE STUDY OF STABILITY

In this chapter we give an overview of most of the techniques that can be used to analyse stability, we give some applications to nonlinear multivariate analysis, and we discuss some possible research projects in connection with stability. It is a somewhat unfortunate situation that all results are collected in one chapter; we have made it quite clear in Chapter 1 that data analysis techniques are *incomplete* without an extensive analysis of stability. Consequently it would be more appropriate to discuss the stability results in the chapters in which the various techniques are introduced. Our results, however, are still rather fragmentary, and more easily described as applications of general techniques for the investigation of stability than as substantial contributions to the analysis of stability of particular data analysis techniques.

Notice that it is true that we have already used some general techniques to assess stability in earlier chapters, in an informal way. Some examples have been analysed with different methods, for example the Roskam journal preference data in Chapters 4 and 10. More results about 'stability under method selection' will be discussed in the next chapter. Choice of measurement level of the variables, both between single and multiple, and between numerical, ordinal, and nominal, has also been investigated in various places in the previous chapters. It is a particular case of 'stability under model selection', which has been investigated more closely in Chapters 6 and 7. The general idea here is that if we analyse data by various methods, 'assuming' various models and using different techniques, then it will be possible to separate properties of the data from those of the models and techniques. This is an old idea, which also underlies the idea of experimental design. If we manipulate the factors in the design (data, model, technique, and possibly others such as algorithm, loss function, gauge), and we do this systematically and with as little bias as possible, then we can isolate the effects of all factors and their possible interactions. This is obviously an idealized statement; the selection from possible data, techniques, and models is largely subjective without any clear-cut rules. The approach is, however, considerably more realistic than the one which assumes that one particular model is 'true', that one particular principle to construct loss functions is always optimal or very close to optimal, and that consequently there also is a unique technique that should be used. It seems to us that this approach ignores the data. Data analysis is not a mechanical process, although processes such as LISREL, ALSCAL, MULTISCALE, and HOMALS are often used in such a way. Almost by definition this is not the fault of the users; at least sometimes the people who construct, produce, and sell these processes promote or do not discourage these practices.

12.1 ANALYTICAL STABILITY

12.1.1 Eigenvalue problems

We first study the problem of solving

$$AX = BX\Lambda^2, \tag{12.1}$$

$$X'BX = I. \tag{12.2}$$

The matrices A and B are given real symmetric matrices of order m; we assume that B is positive definite. The system must be solved for the matrix X, also of dimension $m \times m$, and for the $m \times m$ diagonal matrix Λ^2. The problem solved by OVERALS with all variables either numerical or multiple nominal is of this type, and consequently (for the same types of variables) CANALS, PRINCALS, HOMALS, and ANACOR are special cases.

Analytical stability in this context concerns the following problem. Suppose the matrices A and B depend on a number of parameters, thus they are not fixed matrices, but matrix valued functions defined on a parameter space. Since the solution to the generalized eigenvalue problem is usually unique (up to permutation of dimensions) these problems also define X and Λ^2 as matrix valued and diagonal-matrix valued functions of the parameters. We now want to know what the effect of a small change in the parameters is on the value of X and Λ^2. We proceed formally in this section, assuming that A and B are differentiable functions of the parameters and differentiating these functions in the neighborhood of a parameter point where all elements of Λ are different. The justification of this formal procedure is due to Rellich, who published a series of five papers on the subject in the *Mathematische Annalen* between

1937 and 1942. All the necessary results, and much more than that, are reviewed in the book of Kato (1976).

For the moment we assume that there is just one parameter, and we write the partial derivatives of a matrix with respect to that parameter by using the symbol for the matrix with a dot above it. Thus differentiating the two equations defining X and Λ gives

$$\dot{A}X + A\dot{X} = BX\dot{\Lambda}^2 + B\dot{X}\Lambda^2 + \dot{B}X\Lambda^2, \tag{12.3}$$

$$X'B\dot{X} + \dot{X}'BX + X'\dot{B}X = 0. \tag{12.4}$$

If we premultiply the first of these by X' and simplify, we find

$$X'\dot{A}X - X'\dot{B}X\Lambda^2 = \dot{\Lambda}^2 + (X'B\dot{X}\Lambda^2 - \Lambda^2 X'B\dot{X}). \tag{12.5}$$

The matrix in parentheses on the right has a zero diagonal. If we consider diagonal terms only we find

$$\dot{\Lambda}^2 = \mathrm{diag}\,(X'\dot{A}X - X'\dot{B}X\Lambda^2). \tag{12.6}$$

If we define S implicitly by $\dot{X} = XS$, then it follows from (12.4) that

$$S + S' = -X'\dot{B}X, \tag{12.7}$$

or

$$\mathrm{diag}(S) = -1/2\,\mathrm{diag}(X'\dot{B}X). \tag{12.8}$$

It follows from (12.5) and (12.2) that we may write

$$X'\dot{A}X - X'\dot{B}X\Lambda^2 = \dot{\Lambda}^2 + (S\Lambda^2 - \Lambda^2 S), \tag{12.9}$$

or

$$\mathrm{nondiag}(X'\dot{A}X - X'\dot{B}X\Lambda^2) = S\Lambda^2 - \Lambda^2 S, \tag{12.10}$$

where the nondiag( ) of a matrix is the matrix minus its diagonal. In more familiar notation the result for $\dot{\Lambda}^2$ shows that for the sth eigenvalue differentiated with respect to the kth parameter

$$\frac{\partial \lambda_s^2}{\partial \theta_k} = \Sigma_i \Sigma_j\, x_{is}\, x_{js} \left(\frac{\partial a_{ij}}{\partial \theta_k} - \lambda_s^2\, \frac{\partial b_{ij}}{\partial \theta_k} \right). \tag{12.11}$$

The corresponding results for $\dot{\mathbf{X}}$ are considerably more complicated. We first define

$$q_{st}^k \equiv \Sigma_i \Sigma_j \, x_{is} \, x_{jt} \, \frac{\partial a_{ij}}{\partial \theta_k} \, , \tag{12.12}$$

$$r_{st}^k \equiv \Sigma_i \Sigma_j \, x_{is} \, x_{jt} \, \frac{\partial b_{ij}}{\partial \theta_k} \, . \tag{12.13}$$

With this notation we find for the partials

$$\frac{\partial x_{js}}{\partial \theta_k} = \Sigma_{t \neq s} \, x_{jt} \, \frac{q_{ts}^k - \lambda_t^2 \, r_{ts}^k}{\lambda_s^2 - \lambda_t^2} - 1/2 \, x_{js} \, r_{ss}^k \tag{12.14}$$

and

$$\frac{\partial \lambda_s^2}{\partial \theta_k} = q_{ss}^k - \lambda_s^2 \, r_{ss}^k. \tag{12.15}$$

These formulas are not very interesting in themselves, but it is nice to have them around. By differentiating them again we can find formulas for the second partials, but these formulas look horrible, and we do not reproduce them here. In some interesting special cases the formulas simplify a lot. A familiar special case, for example, arises when $\mathbf{B}$ is a constant matrix. Then $\dot{\mathbf{B}} = \mathbf{0}$, and all formulas simplify. Another interesting special case is the one in which both $\mathbf{A}$ and $\mathbf{B}$ are linear combinations of matrices $\mathbf{A}_k$ and $\mathbf{B}_k$. We shall use some of the resulting simplifications in our examples below. Computationally it is interesting that the formula for the partials of eigenvalue s only involves eigenvalue and eigenvector s, while the partials for eigenvector s involve all eigenvalues and eigenvectors. This last result seems inconvenient in data analysis techniques which compute only a few (often one or two) of the largest eigenvalues and their corresponding eigenvectors. It is, however, possible to repair this by using the formulas

$$\dot{\mathbf{x}}_s = -(\mathbf{A} - \lambda_s^2 \mathbf{B})^+ \, (\dot{\mathbf{A}} - \lambda_s^2 \dot{\mathbf{B}} - \dot{\lambda}_s^2 \mathbf{B}) \mathbf{x}_s - \eta_s \mathbf{x}_s, \tag{12.16}$$

with

$$\lambda_s^2 = \mathbf{x}_s'(\mathbf{A} - \lambda_s^2 \mathbf{B})\mathbf{x}_s, \tag{12.17}$$

$$\eta_s \equiv \frac{1}{2}\mathbf{x}_s'\mathbf{B}\mathbf{x}_s. \tag{12.18}$$

In (12.16) we have used the fact that the vector $\eta_s \mathbf{x}_s$ is in the null space of $\mathbf{A} - \lambda_s^2\mathbf{B}$ (since it is an eigenvector satisfying 12.1) in order to identify the generalized inverse. It is possible to use these formulas in combination with various iterative schemes. In general it seems true, however, that the best approach to computing the partials depends on the particular application we have in mind. ⇥

12.1.2 Eliminating a variable in HOMALS

One way to formulate the problem of homogeneity analysis is solving the eigenproblem

$$\mathbf{P}_*\mathbf{X} = \mathbf{M}_*\mathbf{X}\Lambda^2, \tag{12.19}$$

$$\mathbf{X}'\mathbf{M}_*\mathbf{X} = \mathbf{I}, \tag{12.20}$$

with

$$\mathbf{P}_* = m^{-1}\sum_j \mathbf{P}_j, \tag{12.21}$$

$$\mathbf{M}_* = m^{-1}\sum_j \mathbf{M}_j, \tag{12.22}$$

$$\mathbf{P}_j = \mathbf{M}_j\mathbf{G}_j\,(\mathbf{G}_j'\mathbf{M}_j\mathbf{G}_j)^+\mathbf{G}_j'\mathbf{M}_j. \tag{12.23}$$

Now suppose we do another homogeneity analysis, on the same data but with the first variable eliminated. The matrix $\mathbf{P}_*$ changes into

$$\mathbf{P}_* + \frac{1}{m-1}\,(\mathbf{P}_* - \mathbf{P}_1), \tag{12.24}$$

and $\mathbf{M}_*$ changes into

$$\mathbf{M}_* + \frac{1}{m-1}\,(\mathbf{M}_* - \mathbf{M}_1). \tag{12.25}$$

For general perturbations of the form $\mathbf{P}_* + \varepsilon(\mathbf{P}_* - \mathbf{P}_1)$ and $\mathbf{M}_* + \varepsilon(\mathbf{M}_* - \mathbf{M}_1)$ we find, differentiating with respect to ε,

$$\lambda_s^2(\varepsilon) = \lambda_s^2(0) - \varepsilon \, \mathbf{x}_s'(\mathbf{P}_1 - \lambda_s^2 \, \mathbf{M}_1)\mathbf{x}_s + o(\varepsilon). \tag{12.26}$$

Thus, for large m, the eigenvalue s of the HOMALS solution without variable one will be approximately equal to

$$\lambda_s^2 - \frac{1}{m-1} \, \mathbf{x}_s' (\mathbf{P}_1 - \lambda_s^2 \, \mathbf{M}_1)\mathbf{x}_s. \tag{12.27}$$

This suggests that the importance of a variable for a dimension can be defined (or approximated) by the quantity $\mathbf{x}_s'(\mathbf{P}_j - \lambda_s^2 \, \mathbf{M}_j)\mathbf{x}_s$. Observe that the average over j of these quantities is zero. If $\mathbf{M}_j = \mathbf{I}$ for all j, then they are equal to $\mathbf{x}_s'\mathbf{P}_j\mathbf{x}_s - \lambda_s^2$, which is the discrimination measure of variable j on dimension s minus the average of the discrimination measures of all variables on dimension s (which is equal to the eigenvalue λ_s^2). If we eliminate a variable with a small discrimination measure the eigenvalue will not change much.

Results for eigenvectors are far less satisfactory in many ways. In the first place they are necessarily more complicated; in the second place the terms $(\lambda_s^2 - \lambda_t^2)^{-1}$ in (12.14) suggest that if eigenvalues are close, then small perturbations can have large effects on eigenvectors. Another important component in eigenvector perturbation are the off-diagonal elements q_{st}^k and r_{st}^k. If these are approximately zero, then small perturbations of the matrices will have little effect on the eigenvectors. In our example with a variable deleted from HOMALS, with in addition $\mathbf{M}_j = \mathbf{I}$ for all j, eigenvector s does not change much if the off-diagonal elements in row and column s of $\mathbf{X}'\mathbf{P}_j\mathbf{X}$ are small. This indicates that in homogeneity analysis the discrimination measures on the diagonal of $\mathbf{X}'\mathbf{P}_j\mathbf{X}$ give interesting information on the influence of the variables of the eigenvalues and the off-diagonal elements give information about the influence of the variables on the eigenvectors.

It is not difficult to see that if we add a variable, with corresponding projector $\mathbf{P}_0$, then $\mathbf{P}_*$ changes into

$$\mathbf{P}_* - \frac{1}{m+1} (\mathbf{P}_* - \mathbf{P}_0), \tag{12.28}$$

and we can apply essentially the same results in the same way.

12.1.3 Merging categories in HOMALS

Merging categories can be formalized algebraically as using indicator matrices G_{C_j}, of dimension $k_j \times l_j$, with $l_j \leq k_j$, to replace G_j by $G_j G_{C_j}$. If $k_j = l_j$, then it follows that $G_{C_j} = I$, and nothing changes for this variable. We assume that $M_j = I$ for all j to avoid unnecessary and inessential complications. P_j is replaced by

$$\tilde{P}_j \equiv G_j G_{C_j} (G'_{C_j} D_j G_{C_j})^+ G'_{C_j} G'_j, \tag{12.29}$$

and the first-order approximation to the new eigenvalues is

$$\tilde{\Lambda}^2 = \Lambda^2 + \text{diag}(X'(\tilde{P}_* - P_*)X) + o(\| \tilde{P}_* - P_* \|). \tag{12.30}$$

Again many possible applications can be studied; the simplest one is merging the first two categories of the first variable. Suppose the category frequencies of these categories are d_1 and d_2, and suppose the category quantifications are y_{1s} and y_{2s}. Then

$$\tilde{\lambda}_s^2 = \lambda_s^2 - m^{-1} \frac{d_1 d_2}{d_1 + d_2} (y_{1s} - y_{2s})^2 + o(m^{-1}) \tag{12.31}$$

Merging two categories with approximately the same quantifications hardly changes the eigenvalues when the number of variables is not too small.

12.1.4 Eliminating a subject in HOMALS

In this case it is better to start from the generalized eigenvalue problem

$$CY = mDY\Lambda^2, \tag{12.32}$$

where C contains bivariate marginals, the Burt table, and D contains only the univariate marginals. C is of the form

$$C = n^{-1} \sum_i t_i t'_i, \tag{12.33}$$

where the t_i are vectors with $\sum k_j$ elements, composed of the rows of m indicator matrices. Moreover $D = \text{diag}(C)$. If we eliminate the first subject, we find a new $\tilde{C}$, related to the old C by

$$\tilde{C} = C + \frac{1}{n-1} \ (C - t_1 t_1') \ .$$

(12.34)

If we let $T_1 = \mathrm{diag}(t_1 t_1')$, then

$$\tilde{D} = D + \frac{1}{n-1} \ (D - T_1).$$

(12.35)

It follows from our general results that, with normalization $y_s' D y_t = \delta^{st}$,

$$m\tilde{\lambda}_s^2 = m\lambda_s^2 - \frac{1}{n-1)} \ [(y_s' t_1)^2 - m\lambda_s^2 \ y_s' T_1 y_s] + o((n-1)^{-1}).$$

(12.36)

If we use the stationary equation $G y_s = m\lambda_s^2 x_s$, which is satisfied in HOMALS with the usual normalization, then $y_s' t_1 = m\lambda_s^2 \ x_{1s}$, and thus

$$\tilde{\lambda}_s^2 / \lambda_s^2 = 1 - \frac{1}{n-1} \ (m\lambda_s^2 \ x_{1s}^2 - y_s' T_1 y_s) + o((n-1)^{-1}).$$

(12.37)

Thus the quantities $m\lambda_s^2 \ x_{is}^2 - y_s' T_i y_s$ can be interpreted as the importance of subject i for dimension s, analogous to the discrimination measures. For each s these quantities sum to zero. Our general theory applies directly if we use the normalizations $y_s' D y_s = 1$, which implies together with $G y_s = m\lambda_s^2 x_s$ and $G' x_s = D y_s$ that $x_s' x_s = (m\lambda_s^2)^{-1}$. Of course we can renormalize x_s and y_s in other ways, and adjust the equations correspondingly. Again the same approach can be used if we add a subject. More generally we can define C as

$$C = \Sigma_i \ p_i \ t_i \ t_i',$$

(12.38)

in which each profile t_i has a relative frequency p_i. We may then study

$$C(\varepsilon) = \Sigma_i \ [(1 - \varepsilon)p_i + \varepsilon \delta^{i1}] \ t_i t_i' = C - \varepsilon(C - t_1 t_1')$$

(12.39)

with the same methods as before. Perturbations of this form are studied in classical asymptotic statistics and in jackknife-type methods. We shall return to them later in the chapter.

12.1.5 Problems that are not eigenvalue problems

The use of restrictions, notably the use of cone restrictions and the use of single variables, often leads to problems that are not eigenvalue problems for a given fixed matrix, but eigenvalue problems for a matrix that depends on the quantifications. For applications of this sort we need more general approaches to compute partial derivatives. We do not study problems of this kind in any detail, we simply indicate some possible approaches and we list some of the possible mathematical tools.

Suppose a loss function $\sigma(\gamma, \theta)$ is a function of two sets of parameters. The first set are the parameters we are interested in; the second set are the perturbation parameters. Define

$$\sigma(*,\theta) \equiv \min\{\sigma(\gamma,\theta) \mid \gamma \in \Gamma\}, \tag{12.40}$$

where we assume that the problem is defined in such a way that the minimum is attained for all $\theta \in \Theta$, the set of perturbations we are interested in. We also define

$$\Omega(\theta) \equiv \{\gamma \in \Gamma \mid \sigma(\gamma,\theta) = \sigma(*,\theta)\}. \tag{12.41}$$

Thus $\Omega(\theta)$ is the set of minimizers for perturbation θ; this is a point-to-set map because the minimizer is not necessarily unique. In recent years there has been an enormous amount of research in the mathematical programming and convex analysis literature on theorems that guarantee that the functions $\sigma(*; \theta)$ and $\Omega(\theta)$ are continuous or even differentiable. We do not intend to review this literature here, we merely want to point out that it is extremely useful in many problems of psychometrics, data analysis, and statistics, and that some of the key references are Demjanov and Malozemov (1974), Hogan (1973a, 1973b), Hiriart-Urruty (1978), Springarn (1980). The simplest and in many respects most useful result is that if $g(\gamma, \theta)$ is the vector of partial derivatives of σ with respect to the perturbation parameters, evaluated at $\gamma \in \Gamma$ and $\theta \in \Theta$, then

$$\sigma(*, \theta + \varepsilon\delta) = \sigma(*,\theta) + \varepsilon \min\{\delta'g(\gamma, \theta) \mid \gamma \in \Omega(\theta)\} + o(\varepsilon) \tag{12.42}$$

$(\varepsilon > 0)$, uniformly in the directions δ. This can be used in most of our problems. A simple application, for example, generalizes some of the earlier results in this chapter. If $P_*(\varepsilon) = P_* + \varepsilon(P_* - P_i)$ and if P_* has a largest eigenvalue with multiplicity r, with a corresponding $n \times r$ matrix of eigenvectors X, then

$$\lambda_{\max}^2(\mathbf{P}_*(\varepsilon)) = \lambda_{\max}^2(\mathbf{P}_*) + \varepsilon\,\lambda_{\max}^2(\mathbf{X}'(\mathbf{P}_* - \mathbf{P}_i)\mathbf{X}) + o(\varepsilon) \qquad (12.43)$$

($\varepsilon > 0$), which generalizes Section 12.1.2 where we have to assume that $r = 1$. On the other hand, the result for $r > 1$ is definitely less useful, because the perturbation of the eigenvalue is no longer asymptotically linear with the perturbation, i.e. not differentiable in the usual sense.

12.2 ALGEBRAIC STABILITY

12.2.1 Eigenvalue problems with restrictions

Consider the problem of computing

$$\sigma_1 \equiv \sup\{\mathrm{tr}(\mathbf{X}'\mathbf{B}\mathbf{X})^+\,\mathbf{X}'\mathbf{A}\mathbf{X} \mid \mathbf{X} \in \mathbf{S}\}, \qquad (12.44)$$

where $\mathbf{S}$ is a subset of $\Re^{n\times p}$, the set of all $n \times p$ matrices. We also define

$$\sigma_0 \equiv \sup\{\mathrm{tr}(\mathbf{X}'\mathbf{B}\mathbf{X})^+\,\mathbf{X}'\mathbf{A}\mathbf{X} \mid \mathbf{X} \in \Re^{n\times p}\}. \qquad (12.45)$$

We assume that $\mathbf{B}$ is positive semidefinite, that $\mathbf{A}$ is positive semidefinite, that rank$(\mathbf{B}) \geq p$, and that $\mathbf{A} \leq \mathbf{B}$ in the sense that $\mathbf{x}'\mathbf{A}\mathbf{x} \leq \mathbf{x}'\mathbf{B}\mathbf{x}$ for all $\mathbf{x} \in \Re^n$. These assumptions are satisfied in most of the problems studied in the previous chapters. They guarantee that

$$\sigma_0 = \Sigma_s\,\lambda_s^2 \qquad (12.46)$$

and that the sup is attained for $\mathbf{X} = \mathbf{Z}$, where $\mathbf{A}\mathbf{Z} = \mathbf{B}\mathbf{Z}\Lambda^2$, $\mathbf{Z}'\mathbf{B}\mathbf{Z} = \mathbf{I}$, and $\lambda_1^2 \geq\ldots\geq \lambda_p^2$ are the p largest generalized eigenvalues of this problem. Clearly $0 \leq \sigma_1 \leq \sigma_0$. In this section we first try to improve the lower bound.

For this purpose we suppose that $\mathbf{X} \in \mathbf{S}$ implies that $\mathbf{X}\mathbf{Q} \in \mathbf{S}$ for all $p \times p$ matrices $\mathbf{Q}$. We have already seen in the previous chapters that this assumption is usually satisfied in nonlinear multivariate analysis problems, even if they involve cone restrictions and single variables. Simple special cases are the restriction that each column of $\mathbf{X}$ must be in a subspace L of $\Re^n$, or the restriction that $\mathbf{X}$ must have rank less than or equal to some given q. We also define

$$\varepsilon^2 \equiv \inf\{\mathrm{tr}(\mathbf{Z} - \mathbf{Y})'\mathbf{B}(\mathbf{Z} - \mathbf{Y}) \mid \mathbf{Y} \in \mathbf{S}\}, \qquad (12.47)$$

and we suppose that the inf is attained in some $U \in S$. Remember that Z is the solution of the unrestricted generalized eigenvalue problem corresponding with A and B. By the definition of U and the fact that S is closed under multiplication on the right we see that $\text{tr}(Z - UQ)'B(Z - UQ)$ is minimized over Q by $Q = I$. This means that $U'BZ = U'BU$. Now define $W \equiv U'BU$ and $r = \text{rank}(W)$. Clearly

$$(Z - U)'B(Z - U) = I - U'BU = I - W, \tag{12.48}$$

which means that all eigenvalues of W are less than or equal to one, and

$$\varepsilon^2 = \text{tr}(I - W), \tag{12.49}$$

which shows that W is positive definite, and thus $r = p$ if $\varepsilon^2 < 1$. We now define $V \equiv UW^+$. Thus $V \in S$ and $(V'BV)^+ = W$. Moreover

$$Z'B(V - Z) = Z'BUW^+ - I = WW^+ - I. \tag{12.50}$$

This implies

$$\text{tr}(V'BV)^+V'AV = \text{tr}\, W\{Z + (V - Z)\}'A\{Z + (V - Z)\}$$

$$= \text{tr}\, W\Lambda^2 + \text{tr}\, W(V - Z)'A(V - Z) + 2\, \text{tr}\, WZ'A(V - Z). \tag{12.51}$$

The last term on the right vanishes, because

$$\text{tr}\, WZ'A(V - Z) = \text{tr}\, W\Lambda^2 Z'B(V - Z) = \text{tr}\, W\Lambda^2 (WW^+ - I) = 0. \tag{12.52}$$

Thus

$$\text{tr}\,(V'BV)^+ V'AV = \text{tr}\, W\Lambda^2 + \text{tr}\, W(V - Z)'A(V - Z). \tag{12.53}$$

Now suppose τ is such that $A - \tau B \geq 0$, i.e. $x'Ax \geq \tau x'Bx$ for all $x \in \mathfrak{R}^n$. Clearly $\tau = 0$ satisfies this condition, but a better choice is to take τ equal to the smallest generalized eigenvalue of A and B. Now

$$\text{tr}\,(V'BV)^+V'AV \geq \text{tr}\, W\Lambda^2 + \tau\, \text{tr}\, W(V - Z)'B(V - Z). \tag{12.54}$$

The second term on the right can be simplified by using

$$(V - Z)'B(V - Z) = V'BV - Z'BV - V'BZ + Z'BZ$$

$$= W^+ - 2WW^+ + I, \tag{12.55}$$

so that

$$\text{tr } (\mathbf{V'BV})^+ \mathbf{V'AV} \geq \text{tr } \mathbf{W}\Lambda^2 + \tau \text{ tr } (\mathbf{WW}^+ - \mathbf{W}). \tag{12.56}$$

Making the necessary substitutions proves our final result, which is

$$\sigma_1 \geq \sigma_0 - \Sigma_s (\lambda_s^2 - \tau) \varepsilon_s^2 - \tau(p - r), \tag{12.57}$$

where ε_s^2 is the diagonal element s of $\mathbf{I} - \mathbf{W}$. This lower bound is valid for all τ not larger than the smallest generalized eigenvalue of $\mathbf{A}$ and $\mathbf{B}$. This result generalizes a theorem by Weinberger (1974, p. 68) who supposes that $\mathbf{S}$ is defined by a *subspace* of $\mathfrak{R}^n$, and who assumes that $r = p$, in fact even that $\varepsilon^2 < 1$. An easy generalization of our result can be obtained if we replace $\mathfrak{R}^n$ by a separable Hilbert space, $\mathbf{A}$ and $\mathbf{B}$ by bounded self-adjoint positive semidefinite linear operators, and $\mathbf{S}$ by a set in a finite dimensional subspace of the Hilbert space.

We have indicated how we can bound σ_1 by computing the solution to the generalized eigenproblem without restrictions and by projecting this solution on $\mathbf{S}$ in the metric defined by $\mathbf{B}$. If $\mathbf{Z}$ is close to $\mathbf{S}$, then σ_1 is close to σ_0, and our result gives precise quantitative bounds that tell us how close. A similar result for the *solutions* of the two problems is considerably more complicated. We derive a result that is at least easy to apply.

Suppose $\mathbf{Z}$ solves the unrestricted problem and $\mathbf{Y}$ solves the restricted problem, both $\mathbf{Z'BZ} = \mathbf{I}$ and $\mathbf{Y'BY} = \mathbf{I}$. We write $\mathbf{Y}$ in the form $\mathbf{Y} = \mathbf{ZT} + \mathbf{Q}$, with $\mathbf{Q'BZ} = 0$ and $\mathbf{T} = \mathbf{Z'BY}$. Then

$$\mathbf{Y'AY} = \mathbf{T'}\Lambda^2\mathbf{T} + \mathbf{Q'AQ}, \tag{12.58}$$

$$\mathbf{Y'BY} = \mathbf{T'T} + \mathbf{Q'BQ}. \tag{12.59}$$

Multiply the second equation by λ_{p+1}^2 and subtract it from the first. This gives

$$\mathbf{Y'AY} - \lambda_{p+1}^2\mathbf{I} = \mathbf{T'}(\Lambda^2 - \lambda_{p+1}^2\mathbf{I})\mathbf{T} + \mathbf{Q'}(\mathbf{A} - \lambda_{p+1}^2\mathbf{B})\mathbf{Q}. \tag{12.60}$$

We now use the result that

$$\lambda_{p+1}^2 = \max \left\{ \frac{\mathbf{x'Ax}}{\mathbf{x'Bx}} \mid \mathbf{Z'Bx} = 0 \right\}. \tag{12.61}$$

This shows that $\mathbf{Q'}(\mathbf{A} - \lambda_{p+1}^2\mathbf{B})\mathbf{Q} \leq 0$. From this point we can proceed in many different ways; we choose what seems to be the simplest one. By taking traces we find

$$\sigma_1 - p\lambda_{p+1}^2 \le \Sigma_s (\lambda_s^2 - \lambda_{p+1}^2)\kappa_s^2, \tag{12.62}$$

where $\kappa_1 \ge \dots \ge \kappa_p$ are the singular values of $\mathbf{T}$. This implies

$$\kappa_1^2 \ge \frac{\sigma_1 - p\lambda_{p+1}^2}{\sigma_0 - p\lambda_{p+1}^2}, \tag{12.63}$$

which is a convenient, although possibly not very sharp, estimate of the size of $\mathbf{T}$. If we combine this with our upper bound on σ_1, then we find

$$\kappa_1^2 \ge 1 - \frac{\Sigma_s (\lambda_s^2 - \tau)\varepsilon_s^2}{\Sigma_s (\lambda_s^2 - \lambda_{p+1}^2)}, \tag{12.64}$$

where we have assumed that $p = r$. Thus we see that $\mathbf{Y}$ and $\mathbf{Z}$ are close if ε^2 is small, but also if λ_{p+1}^2 is much smaller than the larger λ_s^2.

12.2.2 Applications of the previous results

The applications in this section are different from those in Sections 12.1.2 to 12.1.4. There we dealt with asymptotic approximations; here we deal with inequalities that are valid without any assumption on the size of the perturbation, although the results will only be nontrivial if the perturbations are relatively small. Both in the previous sections and here we assume that an eigenvalue problem has been solved, and we want to say something about the solutions of a related problem without actually computing them. In the previous sections the related problems are eigenvalue problems of the same order; here we have considerably more freedom. The results of Section 12.2.1 can be applied as soon as we can solve the problem of minimizing $\text{tr}(\mathbf{Z} - \mathbf{Y})'\mathbf{B}(\mathbf{Z} - \mathbf{Y})$ over $\mathbf{Y}$ in $\mathbf{S}$ fairly easily. This gives us the ε_s^2 and the required upper bounds.

Suppose, for example, that we eliminate a variable from HOMALS. We use the eigenvalue problem $\mathbf{CY} = m\mathbf{DY\Lambda}^2$ and we require that $\mathbf{Y}_j = 0$. Then ε_s^2 is the discrimination measure $\mathbf{z}_{js}' \mathbf{D}_j \mathbf{z}_{js}$ of variable j on dimension s. Merging categories is equally simple. In the situation of Section 12.1.3 we require that $y_{1s} = y_{2s}$ for all s. The projection problem is very easy to solve, and we find that

$$\varepsilon_s^2 = \frac{d_1 d_2}{d_1 + d_2} (z_{1s} - z_{2s})^2, \tag{12.65}$$

as expected. If we want to eliminate a subject we start with $\mathbf{P}_*\mathbf{X} = \mathbf{M}_*\mathbf{X}\Lambda$, and require that $x_{is} = 0$ for all s. Clearly $\varepsilon_s^2 = m_{*i}x_{is}^2$. Thus the earlier examples are fairly trivial in this context; the only remarkable thing is that we always start with the 'dual' eigenproblem in this new approach. An additive perturbation in the approach of Section 12.1.1 is a subspace restriction in the approach of Section 12.2.1 and a subspace restriction is an additive perturbation in the dual eigenproblem, and vice versa.

We have already indicated that it is also possible to treat nonlinear restrictions. If we change all HOMALS (i.e. multiple nominal) variables to PRINCALS single variables, then

$$\varepsilon^2 = \sum_j \text{tr} \, (\mathbf{Z}_j - \mathbf{y}_j\mathbf{a}_j')'\mathbf{D}_j \, (\mathbf{Z}_j - \mathbf{y}_j\mathbf{a}_j') , \tag{12.66}$$

where $\mathbf{y}_j$ and $\mathbf{a}_j$ are chosen to minimize this quantity, i.e. $\mathbf{y}_j$ and $\mathbf{a}_j$ are the optimal rank-one approximation to $\mathbf{Z}_j$ in the metric $\mathbf{D}_j$. Our experience indicates that in many examples of this particular kind ε^2 will *not* be small if $p > 1$, i.e. often ε_1^2 will be small, but the other ε_s^2 will be large. We have explained this behaviour in the previous chapter; in many common gauges HOMALS creates horseshoes and PRINCALS does not.

The procedure in the previous example suggests an alternative proof of the inequalities relating σ_0 and σ_1. It turns out that they can be proved, at least as quickly but somewhat less generally, by starting from the meet loss σ_M and applying the same partitioning of meet loss as in Chapters 4 to 7. In these chapters we also used the closedness of the constraint sets under matrix multiplication on the right to show equivalence with certain singular value and eigenvalue problems. Both in this section and in Section 12.1 it is possible to generalize the treatment to, for example, dropping both subjects and variables at the same time. It is clear that in this case we have to study perturbations of the singular value problem directly, not as one of the derived dual eigenproblems. However, the singular value problem is, of course, equivalent to an augmented eigenproblem, which means that our general theory still covers it.

12.2.3 Discretization

We have already pointed out that the results of Section 12.2.1 also apply to approximation of an infinite dimensional eigenproblem by using finite-dimensional subspaces. In particular we can approximate the general homogeneity analysis problem outlined in the previous chapter by using step functions. Suppose $\phi_j(\underline{h}_j)$ is the optimal transformation without restrictions and suppose $-\infty = a_0 < \ldots < a_{k_j} = +\infty$ are the discretization points. Moreover, we suppose that $\underline{h}_j$ has marginal probability density $p_j(x)$. Then the contribution to ε^2 of variable j is

$$\Sigma_l \int_{a_{l-1}}^{a_l} [y_l - \phi_j(x)]^2 p_j(x) \, dx \quad (l = 1,\ldots,k_j), \tag{12.67}$$

where the y_l are chosen to minimize this quantity, i.e.

$$y_l = \text{AVE} \, (\phi_j(\underline{h}_j) \mid a_{l-1} < \underline{h}_j < a_l). \tag{12.68}$$

In many important gauges the optimal function we are approximating is $\phi_j(x) = x$. In this case it also makes sense, and it is computationally feasible, to define optimal discretization points a_l. These points are *not* optimal in the sense that they maximize the largest eigenvalue over all step functions and over all sets of discretization points. They are optimal in the 'marginal' sense that they minimize ε^2, and consequently minimize the upper bound to the largest eigenvalue. These marginal optimal discretization points, together with the optimal step function, can be found by using an alternating least squares algorithm to minimize ε^2. We have already indicated how to minimize the function over the y_l for fixed a_l; if we are approximating $\phi_j(x) = x$ then the optimum a_l for fixed y_l is given by

$$a_l = 1/2 \, (y_l + y_{l+1}). \tag{12.69}$$

The problem of discretization has been treated in the statistical literature by Sheppard (1898) and Cox (1957), but much more completely in the communications analysis literature under the name 'quantization'. We do not discuss this very extensive literature in detail, but some of the key references are Max (1960), Roe (1964), Wood (1969), Elias (1970), Gish and Pierce (1968), Sharma (1978), and Gersho (1979). This literature is largely unknown to statisticians and data analysts, but it contains some very interesting results. There are some interesting asymptotic expressions for ε^2 if the number of discretization points is large, and there is considerable evidence that suggests that these asymptotic approximations are already very good for small k_j. A general result of this type is that if $k_j \to \infty$ then $k_j^2 \varepsilon^2 \to c$, where c depends on $p(x)$. This suggests that discretization error in, for example, HOMALS will always be much smaller than sampling error. Results such as these are investigated using Monte Carlo methods in Van Rijckevorsel, Bettonvil, and De Leeuw (1980) for the normal distribution in HOMALS with various discretization strategies, and by Van der Burg and De Leeuw (reported in Gifi, 1980, pp. 266–277) for CANALS with normal distributions, various numbers of discretization points, and various types of intercorrelations. The conclusion is that discretization type or discretization fineness is of small influence in the cases studied; sample size is much more important, and in the case of CANALS the intercorrelation structure (degree of multicollinearity) is very important too. The discretization results in this section are much more generally valid than for normal distributions and for step functions; we can also use them to locate the knots of spline functions for any distribution. This creates the problem of a 'robust' choice of knots, which is good for most distributions in a particular family of gauges. The empirical results so far suggest that with both a uniform spacing of knots or discretization points (which is optimal for the rectangular distribution) and a choice of knots so that the marginals become rectangular the system is fairly robust.

12.2.4 Interactive variables

We have seen in Chapters 5 and 11 that the general OVERALS problem can be interpreted as HOMALS with restrictions on the category quantifications. These restrictions often enforce additivity, which is used to create groups of variables. Additivity is a linear restriction, and it is easy to estimate ε^2 for additivity. In fact computing ε^2 amounts to performing an analysis of variance on the unrestricted quantifications, although the design is often not orthogonal because we have to weight by the marginals in $\mathbf{D}$. As the examples in Chapters 6 and 11 using the CBS data show, it is often possible to predict quite precisely what CANALS does from what HOMALS on the interactive variables does, because there typically seem to be only small interactions in examples like these.

12.2.5 Other techniques for algebraic stability

The method discussed in Section 12.2.1 can be extended in various directions. A very common one is to extend it to other linear spaces of infinite dimensionality. This work is often of limited practical value for the data analysis problems we are interested in, but on the other hand it does show clearly in which direction some of our results could possibly be improved. Reviews of work in this area are the book by Vainikko (1976) and the paper by Chatelin (1979). More direct generalizations of these results to not necessarily bounded and not necessarily self-adjoint generalized eigenproblems in not necessarily separable Hilbert spaces are contained in Kolata (1978); bounds for the approximation of eigenvectors are improved by

Weinberger (1960). Our approach is based on investigating how well a solution of the eigenproblem satisfies the restrictions. It is also possible to proceed the other way around. We can start with a solution of the restricted problem and check how well it satisfies the eigenequation. Bounds can also be derived from this. Some pertinent references are Wilkinson (1961), Yamamoto (1980), Symm and Wilkinson (1980), and, in an infinite-dimensional context, Werner (1974).

Many additional results can be derived from the max–min characterization of eigenvalues. These results have two forms again, one for additive perturbations and one for eigenvalues of submatrices. The two forms are 'dual' and can often be translated into each other by using the singular value decomposition. Some of the sharpest results have been proved by Robert Thompson and his collaborators. We mention Thompson and Freede (1970, 1971), Thompson and Therianos (1972a, 1972b), Thompson (1975, 1976), but this list is not even approximately complete. The min–max results, which only refer to eigenvalues and not to eigenvectors, have been used in connection with principal components analysis and correspondence analysis by Escofier and Le Roux (1972). In that paper the results are used in their simplest form; the results of Thompson and others make it possible to improve them a great deal.

The study of eigenvectors is more complicated. There are two basic papers, one by Davis and Kahan (1970) and another one by Stewart (1973a). Both papers review previous work; they are essentially concerned with converting first order analytic perturbation results into rigorous bounds in the form of inequalities (using algebraic methods). Stewart has carried on his research since 1973, additional results being reported in Stewart (1975, 1979) and in the review paper (1978). The results of Davis and Kahan (1970) have been used in components analysis by Escofier and Le Roux (1972), who study the effect of adding and dropping a variable. The results of Davis and Kahan are limited to ordinary eigenvalue problems, while the results of Stewart make it possible to extend them to generalized eigenvalue problems and singular value problems.

It is clear that this section is merely a list of references, together with some hints on how these results could be applied in nonlinear multivariate analysis. It is clearly an enormous research project to apply them to specific techniques using various types of perturbations.

12.3 REPLICATION STABILITY

12.3.1 The delta method

The delta method is one of the classical methods of asymptotic statistics. In its simplest form it assumes that a sequence x_n of random variables is given, that $n^{1/2}(x_n - \mu)$ converges in distribution to a multinormal variable with mean zero and dispersion Σ, and that $y_n = \Phi(x_n)$ defines a new sequence of random variables, where the mapping Φ is assumed to be differentiable at μ. The basic result of the delta method then tells us that the asymptotic distribution of $n^{1/2}[y_n - \Phi(\mu)]$ is the same as the asymptotic distribution of $n^{1/2}G(\mu) (x_n - \mu)$, with $G(\mu)$ the derivative of Φ evaluated at μ. This last asymptotic distribution is multinormal with mean zero and dispersion $G(\mu)\Sigma G(\mu)'$.

This basic result can be generalized in several directions; all these generalizations are useful for our purposes. In the first place we can relax the assumption that $n^{1/2}(x_n - \mu)$ is asymptotically normal, it is sufficient for some purposes to assume that $K_n(x_n - a_n)$ has some asymptotic distribution for some sequence a_n and for some sequence of norming constants $K_n \to \infty$. It remains true that $K_n[\Phi(x_n) - \Phi(a_n)]$ has the same asymptotic distribution as $K_nG(a_n)(x_n - a_n)$, which may be considerably easier to compute. In the second place, for some purposes it suffices to assume differentiability in all directions. In that case the asymptotic distribution of $n^{1/2}[\Phi(x_n) - \Phi(\mu)]$ is the same as that if $n^{1/2}G(\mu, x_n - \mu)$, with $G(\mu, z)$ the directional derivative. The last asymptotic distribution is the distribution of $G(\mu, z)$, where z is multinormal with mean zero and dispersion Σ, provided $G(\mu, z)$ is continuous in its second argument. Another generalization is that we do not insist that Φ is the same for all n, but we study $y_n = \Phi_n(x_n)$. Finally, we can be interested in the distribution of statistics such as $n[\Phi(x_n) - \Phi(\mu) - G(\mu) (x_n - \mu)]$, which give us higher order information.

How do we apply the delta method to nonlinear multivariate analysis? Our techniques compute statistics that are all functions of the number of observations of the cells in the multidimensional contingency table, or equivalently of the frequencies of the profiles. We can apply the delta method if we show that these functions are differentiable and if we embed our observations in a sequence of random variables in a reasonable way. Presumably it is obvious how a classical statistician would perform this embedding. He would suppose that the individuals are a simple random sample from an infinite population; this makes the cell frequencies multinomially distributed and consequently asymptotically normal. Some methodological comments, which repeat some of the things said in Chapter 1, are in order here. If we want to assess the variability (the replication stability) of a statistic, then the appropriate thing to do is to repeat the experiment a couple of times under the same conditions, and to recompute the statistic after each experiment. This is what is done in the experimental sciences to find out what the 'measurement error' is, to find out to what extent we do indeed control the relevant conditions. In the social sciences it is often impossible to repeat experiments; if we do repeat them we find different results, and we do not know if these results are due to measurement error or to uncontrolled variables, because we do not know what the relevant variables are. Consequently, actual replication is replaced by the statistical model of independent identically distributed random variables, and actual control is replaced by preexperimental randomization. Certainly randomization is a very useful device, which simply cannot be dispensed with. However, the statistical model, which actually tells us that replications are unnecessary because probability theory and statistics show us what replications would give *if* we performed them, carries too heavy a burden in many situations. It is nice to be able to predict what replication would give, but it is not nice that we have no conceivable way to falsify these predictions. If we try to falsify them, we usually succeed; yet we never know whether this happens because the statistical model is not true or because our attempt to falsify the predictions was faulty. We can only verify the predictions of statistical models in Monte Carlo experiments; but of course this is more or less tautological, or it is a method to find out whether we made an error in our proof of a theorem, or it is a method to find out whether a particular approximation holds 'sufficiently well'. Thus preexperimental randomization does not replace control; it is merely a passive way to dodge the criticism that some choices are arbitrary. The solution is to make all choices completely arbitrary by definition, except one or two for which the criticism does not apply because they are the experimental conditions. This is the only possible way to proceed in many situations. On the other hand, assuming a simple statistical model does not replace actual replication, in fact it does not even make replication less desirable, and there is a simple alternative way to proceed. The alternative way is *not* to assume a model and *not* to proceed as if the model was true, and to face all kinds of instabilities. All this has very little to do with inference from sample to population (which is simply a misleading way to formulate the problem of replication stability in many situations). It also has very little to do with prescriptions on how to behave rationally, which belong to ethics or normative psychology.

We first give a simple example of the delta method to show how it works. Suppose A_k are K fixed matrices of order m and A_k has probability π_k of occurring in the population from which we are sampling. A multinomial simple random sample of size n gives relative frequencies p_k; our technique is to compute eigenvalues and eigenvectors of $\mathbf{A}_* = \sum p_k \mathbf{A}_k$. In the preference rankings techniques of Chapter 10, for example, the $\mathbf{A}_k$ can be permutation matrices $\mathbf{P}_k$ or rank-one matrices of the form $\mathbf{P}_k \mathbf{yy}'\mathbf{P}_k$, with $\mathbf{y}$ the vector of centered rank numbers. From the results of Section 12.1.1 we obtain

$$\frac{\partial \Lambda^2}{\partial p_k} = \text{diag } (\mathbf{X}'\mathbf{A}_k \mathbf{X}), \tag{12.70}$$

where $\mathbf{X}$ is the matrix of normalized eigenvectors of $\mathbf{A}_*$, evaluated at π. We suppose that all elements of Λ^2, evaluated at π, are different. Clearly

$$\sum_k \pi_k \frac{\partial \Lambda^2}{\partial p_k} = \text{diag}(\mathbf{X}'\mathbf{A}_* \mathbf{X}) = \Lambda^2. \tag{12.71}$$

The p_k are jointly asymptotically normal; more precisely, the vector $n^{1/2}(\mathbf{p} - \boldsymbol{\pi})$ is asymptotically multinormal with mean zero and dispersion $\Pi - \boldsymbol{\pi}\boldsymbol{\pi}'$, where Π is the diagonal matrix with the elements of $\boldsymbol{\pi}$ on the diagonal. Thus $n^{1/2} [\Lambda^2(\mathbf{p}) - \Lambda^2(\boldsymbol{\pi})]$ is asymptotically normal with mean zero, and

$$n \,\text{COV}(\lambda_s^2(\mathbf{p}), \lambda_t^2(\mathbf{p}))$$

$$\to \sum_k \pi_k \left(\sum_i \sum_j x_{is} x_{js} a_{ijk}\right)\left(\sum_i\sum_j x_{it} x_{jt} a_{ijk}\right) - \lambda_s^2\lambda_t^2. \tag{12.72}$$

By substituting the sample x_{is}, λ_s^2 and p_k on the right-hand side we find a consistent estimate of the asymptotic variances and covariances. In the preference rankings situation we have $\mathbf{A}_k = \mathbf{P}_k \mathbf{yy}'\mathbf{P}_k$. If we define $\rho_{ks} = \mathbf{x}_s'\mathbf{P}_k\mathbf{y}$, the correlation between the eigenvector and the vector of centered rank numbers, then

$$n \,\text{COV}(\lambda_s^2(\mathbf{p}), \lambda_t^2(\mathbf{p})) \to \sum_k \pi_k \rho_{ks}^2 \rho_{kt}^2 - \lambda_s^2 \lambda_t^2. \tag{12.73}$$

Because

$$\sum_k \pi_k \rho_{ks}^2 = \lambda_s^2, \tag{12.74}$$

it follows that the asymptotic covariance of the eigenvalues is equal to the covariance of the ρ_{ks}^2. These formulas still look fairly simple and they can be interpreted to some extent. If in generalized eigenproblems the matrices $\mathbf{A}$ and $\mathbf{B}$ both depend on the parameters, possibly nonlinearly, then the situation becomes more unpleasant. If we compute derivatives of eigenvectors or higher derivatives of eigenvalues then the formulas become even more forbidding. It does not help very much to give these formulas here for specific techniques. They can be derived by using the results of Section 12.1.1.

The delta method can also be used for *bias correction*, while in the previous development we used it mainly for *variance estimation*. For the eigenvalues, for example, we can use the development

$$\lambda_s^2(\underline{p}) = \lambda_s^2(\boldsymbol{\pi}) + \Sigma_k \frac{\partial \lambda_s^2}{\partial p_k} (\underline{p}_k - \pi_k)$$

$$+ 1/2 \Sigma_k \Sigma_l \frac{\partial^2 \lambda_s^2}{\partial p_k \partial p_l} (\underline{p}_k - \pi_k)(\underline{p}_l - \pi_l) + o_p(n^{-1}), \qquad (12.75)$$

where the notation $o_p(n^{-1})$ means that n times the residual converges to zero in probability. If we take expectations on both sides we find

$$\text{AVE} (\lambda_s^2(\underline{p})) = \lambda_s^2(\boldsymbol{\pi}) + 1/2 \Sigma_k \Sigma_l \frac{\partial^2 \lambda_s^2}{\partial p_k \partial p_l} (\delta^{kl}\pi_k - \pi_k\pi_l) + o(n^{-1}).$$

$$(12.76)$$

By using the appropriate formula for second derivatives and by substituting sample quantities, we can use this formula for bias correction. Of course this practice more or less implies that we are interested in the population eigenvalues, and it certainly implies that we take the multinomial problem embedding seriously. The same thing is true if we use asymptotic normality and consistent estimates of dispersions to test the hypothesis that the smallest eigenvalues are zero. It may be of some comfort to some that such a test is fairly easy to compute, and is consistent, although it is not asymptotically most powerful against close alternatives. However, from our point of view the statistical model that assumes that some population eigenvalues are zero is useless. The model has no rational interpretation in terms of the familiar gauges, indeed most gauges predict that the eigenvalues are all nonzero. The only model that does make sense, for example in HOMALS or ANACOR, is that all eigenvalues are zero, i.e. that all variables are independent in pairs. This model will, of course, be rejected in all interesting data sets.

When we assume that all eigenvalues are zero, the theory of Section 12.1.1 does not apply any more, because there we assumed that all eigenvalues are different. The results of Section 12.1.5 can still be used, however. In Lebart (1976) this is done for correspondence analysis; we use our simpler example of preference rank orders. Suppose that the model is that $\mathbf{A}_* = \sum \pi_k \mathbf{A}_k$ is of rank one. For the first eigenvalue we still have asymptotic normality, but for the remaining eigenvalues the theory of Section 12.1.5 tells us that

$$\lambda_s^2(\underline{p}) = \lambda_s^2[\sum_k (\underline{p}_k - \pi_k) (\mathbf{I} - \mathbf{x}\mathbf{x}') \mathbf{A}_k (\mathbf{I} - \mathbf{x}\mathbf{x}')] + o_p(n^{-1/2}), \qquad (12.77)$$

with $\mathbf{x}$ the eigenvector corresponding with the largest eigenvalue. Thus directional derivatives are eigenvalues of a matrix with asymptotically normal elements. Under some additional symmetry conditions the elements of this matrix are asymptotically independent, and thus the eigenvalues are the square roots of the eigenvalues of a standard Wishart matrix. The distribution of the eigenvalues of a standard Wishart matrix has been tabulated extensively, and has been used, for example, by Lebart (1976).

We end this section with some references and some comments. The delta method is classical. It is discussed in all important textbooks, particularly neatly, for example, in Rao (1965, sect. 6a). A proper treatment of the delta method is possible if we realize that there are three types of convergence involved. In the first place there is weak convergence, or *convergence in distribution*. The key theorem here is the Mann–Wald–Rubin theorem on the preservation of weak convergence by mappings, treated, for example, by Billingsley (1968, sect. 1.5) and more generally by Topsøe (1967). The second type of convergence is *convergence of moments* (including expected values and variances), which is guaranteed by weak convergence together with uniform integrability (Billingsley, 1968, pp. 31–33). The third type of convergence is *convergence of probability measures*, which is studied for delta method expansions in Bhattacharya and Ghosh (1978). It seems to us that convergence of moments is most useful for our problems. Hurt (1976) gives some useful general theorems; first-order asymptotic distribution theory for principal components analysis is given in the normal case by Anderson (1963) and in the nonnormal case by Davis (1977) and Waternaux (1976). First-order asymptotic theory for correspondence analysis has been given by O'Neill (1978a, 1978b). O'Neill (1980) applies the delta method in the related problem of decomposition of a discrete multivariate distribution by using tensor products of orthogonal functions on the marginals.

We have incorporated the delta method in the programs ANACOR and ANAPROF. This means that these programs do not only give eigenvalues and eigenvectors, but also an estimate of the dispersion matrix of the eigenvalues, and for each row and column point an estimate of the dispersion matrix of the p coordinates. This dispersion matrix estimate can be used to draw 95 per cent confidence ellipsoids around each of the points.

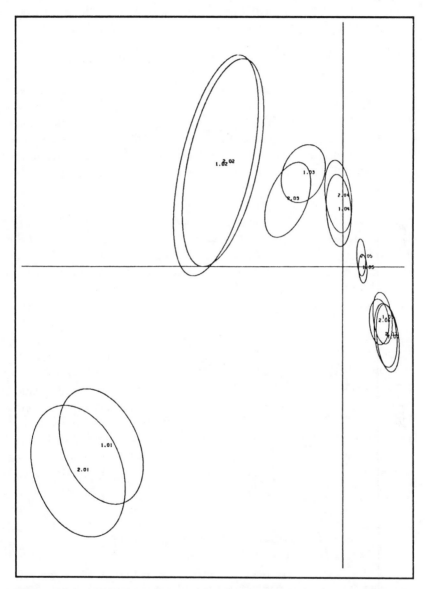

Figure 12.1 ANACOR solution of the social mobility data with 95 per cent confidence ellipses.

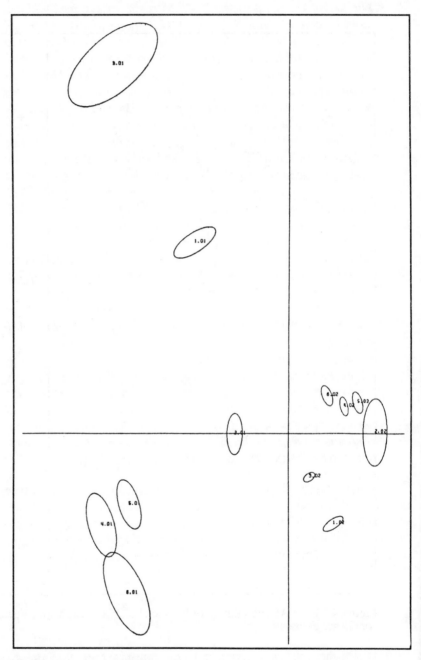

Figure 12.2 ANAPROF solution of the Sugiyama data with 95 per cent confidence ellipses.

The use of the delta method is illustrated for the occupational mobility data of Table 8.3. Figure 12.1 shows that the ANACOR structure is fairly stable, that the ellipses of row and column points have considerable overlaps, and that the categories 6 and 7 of both sons and fathers are virtually indistinguishable. The ANAPROF solution for the Sugiyama data on religious practice from Table 8.7 is plotted with 95 per cent ellipses in Figure 12.2. Because of the large sample size the ellipses are much smaller in this example, and in fact they are, with one tiny exception, all disjoint. This indicates considerable replication stability under the multinomial model. In general the ellipsoids are quite useful, even if the multinomial assumptions are not instantaneously apprehended. This is because the critical quantities in stability analysis of this type are the partial derivatives. Analytical stability gives us first-order approximations, which can in some cases be expressed as inequalities. Now inequalities are nice and rigorous, but they tend to be rather pessimistic. The delta method provides us with another method to assess the size of the derivatives, because basically the estimates of the sampling variances of the statistics are the sampling variances of the partial derivatives of the statistics. Thus we do not have to buy the statistical model in order to compute ellipsoids. An alternative way to compute ellipsoids, for example, is simply to use the second derivatives of the loss function evaluated at the solution. For some loss functions this provides asymptotically the same ellipsoids; for others they will perhaps differ from the delta method ellipsoids. The 95 per cent confidence interpretation may be natural in some situations, but it does refer to imaginary replications.

A disadvantage of the delta method in some situations is that it is sometimes difficult and/or expensive to compute partial derivatives. We have already seen that the partials of the eigenvectors are quite complicated. If we want to correct eigenvectors for bias we need the second partials, which are very complicated. The methods described in the next section have been developed partly to avoid these very complicated computations.

12.3.2 General theory of randomization methods

We describe this class of methods in the discrete case only. Suppose there are K possible profiles: the data consist of the frequencies $n_1,...,n_k$ with which these profiles occur. Let $n \equiv \sum n_k$ and $p_k \equiv n_k/n$. Observe that up to now we have not introduced any probabilistic structure. Our technique consists of the computation of a function $\phi(\mathbf{p})$, which we shall assume to be real valued for the time being. Now suppose we have a rule that associates with each pair $(\mathbf{p}, n)$ of vectors $\mathbf{p}_1(\mathbf{p}, n), ..., \mathbf{p}_m(\mathbf{p}, n)$. The vectors $\mathbf{p}_j(\mathbf{p}, n)$ are nonnegative, and their elements add up to one. With each $\mathbf{p}_j(\mathbf{p}, n)$ our rule also associates a probability $\pi_j(\mathbf{p}, n)$. Now compute

$$\mu_n(\mathbf{p}) \equiv \sum_j \pi_j(\mathbf{p}, n) \, \phi(\mathbf{p}_j(\mathbf{p}, n)), \tag{12.78}$$

$$\sigma_n^2(\mathbf{p}) \equiv \sum_j \pi_j(\mathbf{p}, n) \, [\phi(\mathbf{p}_j(\mathbf{p}, n)) - \mu_n(\mathbf{p}))]^2. \tag{12.79}$$

The vectors $\mathbf{p}_j(\mathbf{p}, n)$ are *perturbation vectors* and the $\pi_j(\mathbf{p}, n)$ are *perturbation probabilities;* the number of perturbation vectors m may depend on n. The basic idea behind the method is that we *have* observed $\mathbf{p}$, but we *might as well have* observed the $\mathbf{p}_j(\mathbf{p}, n)$.

This 'might as well have' sounds intuitive and difficult to define precisely. Before we discuss it in more detail we first discuss some general techniques to construct the $\mathbf{p}_j(\mathbf{p}, n)$. The first one, which is the most familiar one, is the *jackknife.* Miller (1974) reviews theory and applications of the jackknife. It is easiest to explain the construction of the perturbations if we work with the $n \times K$ indicator matrix of the profiles $\mathbf{G}_P$; thus $\mathbf{p} = n^{-1}\mathbf{G}_P'\mathbf{u}$. In the jackknife we suppose that we *might as well have* observed any one of the $(n-1) \times K$ submatrices of $\mathbf{G}_P$ obtained by deleting a single row. There are obviously n of these submatrices, but only K of them are different, depending on which profile we delete. The corresponding perturbation vectors are (note that m in this case is equal to K)

$$\mathbf{p}_k(\mathbf{p}, n) = \mathbf{p} + \frac{1}{n-1} (\mathbf{p} - \mathbf{e}_k), \tag{12.80}$$

with probability

$$\pi_k(\mathbf{p}, n) = p_k. \tag{12.81}$$

A second method to construct the perturbations has been suggested by Hartigan (1969, 1971). It is called *subsampling.* The basic idea is that we might as well have observed any of the $2^n - 1$ subsamples, i.e. submatrices of $\mathbf{G}_P$, excluding only the empty sample, all with equal probability. In other versions of the method we might as well have observed a particular set of submatrices, which has a particular balanced group structure, again all with equal probability (cf. Hartigan, 1969; or Gordon, 1974a, 1974b). It is clear that the marginals of the submatrices can have all possible values $v_1,...,v_k$ provided that $v_k \leq n_k$ for all k. The probability of a perturbation vector found by norming $\mathbf{v}$ is equal to

$$\binom{n_1}{v_1}\binom{n_2}{v_2}...\binom{n_k}{v_k} / (2^n - 1). \qquad (12.82)$$

The third and last method we discuss is due to Efron (1979). It is called the *bootstrap*. The basic idea is that we *might as well have* observed any matrix $\mathbf{G}_P$ of dimension $n \times K$ consisting of the same rows, but in different frequencies. The probability of another $\mathbf{G}_P$ is n^{-n} times the multinomial coefficient of $v_1,...,v_n$, where v_i indicates the number of times row i occurs in the new $\mathbf{G}_P$. Many of these new $\mathbf{G}_P$ have the same marginals; the probability of a set of marginals $v_1,...,v_K$ is the probability that we observe $v_1,...,v_K$ if we take a sample of size n from the multinomial with parameters $p_1,...,p_K$. Jackknifing, subsampling, and bootstrapping are the main randomization methods. In discrete situations it seems that the jackknife is the simplest method and the bootstrap the most natural one. Efron emphasizes that if the data are a sample from a multinomial then the perturbation vector is related to the sample proportions in the same way as the sample proportions are related to the population parameters. Interesting theoretical and practical work on subsampling has been done by Hartigan (1975), Forsythe and Hartigan (1970), and Gordon (1974a, 1974b). In the next sections we concentrate on the bootstrap and the jackknife. ➤

The idea that we might as well have observed something else is related to the idea of a random sample. If the data are a random sample from a population, then we might indeed have observed any of the perturbation vectors as well, with the indicated probabilities, which estimate corresponding population probabilities. All three approaches assume that the n individuals are equally important and interchangeable; conversely, if we assume this then the randomization methods make sense. On the other hand, it is not strictly necessary to commit oneself to any probability model concerning the original observations. We shall see that the randomization methods can be used as Monte Carlo methods or as discretization methods to approximate first and/or second derivatives of ϕ. We already know that first and second derivatives of loss functions and solutions give interesting information on stability in general.

12.3.3 Approximation properties of the jackknife

In case of the jackknife we have seen that

$$\mu_n(\mathbf{p}) = \Sigma_k p_k \, \phi \left(\mathbf{p} + \frac{1}{n-1} \, (\mathbf{p} - \mathbf{e}_k) \right). \tag{12.83}$$

When we assume that ϕ has a bounded third derivative on the unit simplex in $\Re^K$, then

$$\mu_n(\mathbf{p}) = \phi(\mathbf{p}) + 1/2 \, \frac{1}{(n-1)^2} \, \text{tr} \, \mathbf{H}(\mathbf{p}) \, (\mathbf{P} - \mathbf{pp}') + O((n-1)^{-3}), \tag{12.84}$$

with $\mathbf{H}(\mathbf{p})$ the matrix of second derivatives of ϕ, evaluated at $\mathbf{p}$, and with $\mathbf{P}$ the diagonal matrix with the p_k. We can use this relationship directly for bias correction. It follows that

$$n \, \phi(\mathbf{p}) - (n-1) \, \mu_n \, (\mathbf{p}) = \phi(\mathbf{p}) - 1/2 \frac{1}{n-1} \text{tr} \, \mathbf{H}(\mathbf{p}) \, (\mathbf{P} - \mathbf{pp}') + O((n-1)^{-2})$$

$$\tag{12.85}$$

If $\mathbf{p}$ is based on a random sample with multinomial probability $\boldsymbol{\pi}$, then we know that the delta method gives

$$\text{AVE}(\phi(\underline{\mathbf{p}})) = \phi(\boldsymbol{\pi}) + 1/2 \, n^{-1} \, \text{tr} \, \mathbf{H}(\boldsymbol{\pi})(\boldsymbol{\Pi} - \boldsymbol{\pi\pi}') + O(n^{-2}). \tag{12.86}$$

Combining the last two results gives

$$\text{AVE} \, (n \, \phi(\underline{\mathbf{p}}) - (n-1) \, \mu_n(\underline{\mathbf{p}})) = \phi(\boldsymbol{\pi}) + O(n^{-2}). \tag{12.87}$$

Thus computing $n \, \phi(\underline{\mathbf{p}}) - (n-1)\mu_n(\underline{\mathbf{p}})$ corrects for bias, without making it necessary to compute second, or even first, derivatives.

In the same way it is possible to prove that

$$\sigma_n^2(\mathbf{p}) = (n-1)^{-2} \, \mathbf{g}(\mathbf{p})'(\mathbf{P} - \mathbf{pp}')\mathbf{g}(\mathbf{p}) + O((n-1)^{-3}), \tag{12.88}$$

which means that the variance of the *pseudo-values*

$$n \, \phi(\mathbf{p}) - (n-1) \, \phi(\mathbf{p} + \left(\frac{1}{n-1} \, (\mathbf{p} - \mathbf{e}_k) \right), \tag{12.89}$$

each with probability p_k, is $\mathbf{g(p)'(P - pp')g(p)} + O((n-1)^{-1})$, where $\mathbf{g(p)}$ is the vector of partials at $\mathbf{p}$. It follows that if $\underline{\mathbf{p}}$ is multinomial, then the variance of the pseudo-values can be used as a consistent estimate of the asymptotic variance of $\phi(\underline{\mathbf{p}})$, again computable without formulas for derivatives.

We have just explained the *theoretical jackknife*, which computes all K different pseudo-values, as well as their average and variance, exactly. There are some familiar variations of the jackknife: in one variation we omit s observations at the same time, in another variation we add one observation, a third variation omits observations continuously, and so on. For these variations we refer to the literature. There is another variation that interests us more. Suppose $\varepsilon_1,...,\varepsilon_r,...,\varepsilon_R$ are independent and identically distributed random variables, which assume the values $e_1,...,e_r,...,e_K$ with probabilities $p_1,...,p_K$. Define

$$\mu_n(\mathbf{p}) = R^{-1} \sum_r \phi \left(\mathbf{p} + \frac{1}{n-1} (\mathbf{p} - \varepsilon_r) \right). \tag{12.90}$$

Now $\mu_n(\mathbf{p})$ is a random variable that converges in probability if $R \rightarrow \infty$ to the constant $\mu_n(\mathbf{p})$. In fact $R^{1/2}(\mu_n(\mathbf{p}) - \mu_n(\mathbf{p}))$ converges to a normal distribution with mean zero and variance $(n-1)^{-2}\mathbf{g(p)'(P - pp')g(p)} + O((n-1)^{-3})$. These results do not depend on any probabilistic assumptions on the distribution of $\mathbf{p}$; in fact $\mathbf{p}$ is assumed to be a fixed set of constants. We can also compute the R pseudo-values:

$$n \phi(\mathbf{p}) - (n-1) \phi \left(\mathbf{p} + \frac{1}{n-1} (\mathbf{p} - \varepsilon_r) \right). \tag{12.91}$$

The expected value of their sample mean is equal to

$$\phi(\mathbf{p}) - 1/2 \frac{1}{n-1} \operatorname{tr} \mathbf{H(p)} (\mathbf{P - pp'}) + O((n-1)^{-2}) \tag{12.92}$$

and the expected value of their sample variance is equal to

$$\frac{R-1}{R} \mathbf{g(p)'(P - pp')g(p)} + O((n-1)^{-1}). \tag{12.93}$$

If both $\underline{\mathbf{p}}$ and ε_r are random, the ε_r are defined conditionally on $\underline{\mathbf{p}}$, and $\underline{\mathbf{p}}$ is marginally multinomial with parameter $\boldsymbol{\pi}$, then the expected value of the sample mean of the R pseudo-values

$$n \, \phi(\mathbf{p}) - (n-1) \, \phi \left(\mathbf{p} + \frac{1}{n-1} \, (\mathbf{p} - \boldsymbol{\varepsilon}_r) \right) \tag{12.94}$$

is equal to $\phi(\mathbf{x}) + O(n^{-2})$, and the expected value of the sample variance is

$$\frac{R-1}{R} \, \mathbf{g}(\mathbf{x})'(\mathbf{II} - \mathbf{x}\mathbf{x}')\mathbf{g}(\mathbf{x}) + O((n-1)^{-1}). \tag{12.95}$$

This could be called the *Monte Carlo version of the jackknife*. It is easy to apply if K is very large, in which case the theoretical jackknife can easily become computationally rather demanding.

12.3.4 Approximation properties of the bootstrap

For the bootstrap we can write

$$\mu_n(\mathbf{p}) = \sum_{\mathbf{v}} \pi(\mathbf{v}|\mathbf{p}) \, \phi(\mathbf{v}/n), \tag{12.96}$$

where $\pi(\mathbf{v}|\mathbf{p})$ is the probability of integer vector $\mathbf{v}$ if we draw a random sample of size n from a multinomial with parameters $\mathbf{p}$. A first remark that is in order here is that $\mu_n(\mathbf{p})$ is the nth Bernstein polynomial of ϕ, evaluated in $\mathbf{p}$. There is an enormous amount of literature dealing with Bernstein polynomials; the older literature is reviewed in Lorenz (1953), but many more results have been derived since that book appeared. The literature is valuable, because any result about Bernstein polynomials is a result about the bootstrap. We proceed in the same way as with the jackknife. In the first place

$$\mu_n(\mathbf{p}) = \phi(\mathbf{p}) + 1/2 \, n^{-1} \, \mathrm{tr} \, \mathbf{H}(\mathbf{p})(\mathbf{P} - \mathbf{p}\mathbf{p}') + O(n^{-2}). \tag{12.97}$$

Thus

$$2 \, \phi(\mathbf{p}) - \mu_n(\mathbf{p}) = \phi(\mathbf{p}) - 1/2 \, n^{-1} \, \mathrm{tr} \, \mathbf{H}(\mathbf{p})(\mathbf{P} - \mathbf{p}\mathbf{p}') + O(n^{-2}), \tag{12.98}$$

and in the multinomial case

$$\mathrm{AVE}(2 \, \phi(\mathbf{p}) - \mu_n(\mathbf{p})) = \phi(\mathbf{x}) + O(n^{-2}). \tag{12.99}$$

Thus the theoretical bootstrap can also be used to correct for bias, in the same way as the jackknife, although generally with more computational effort. For the bootstrap we also find

$$\sigma_n^2(\mathbf{p}) = n^{-1} \, \mathbf{g(p)'(P - pp')g(p)} + O(n^{-2}). \tag{12.100}$$

Bootstrap pseudo-values are defined by $2 \, \phi(\mathbf{p}) - \phi(\mathbf{v}/n)$, which has probability $\pi(\mathbf{v}|\mathbf{p})$. Their variance is $\sigma_n^2(\mathbf{p})$. In the theoretical bootstrap summation is over all possible $\mathbf{v}$. This is expensive, even in small examples, and it is simply impossible in larger ones. Consequently we need a Monte Carlo version of the

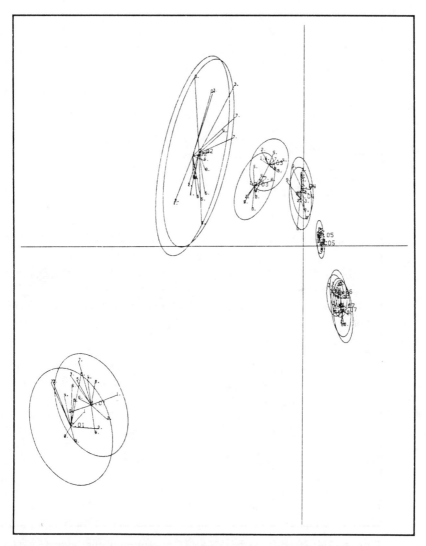

Figure 12.3 Stability of the ANACOR solution of the social mobility data: 95 per cent ellipses and ten bootstraps.

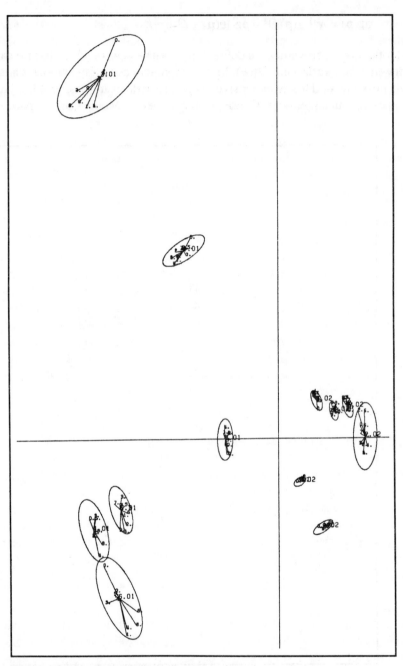

Figure 12.4 Stability of the ANAPROF solution of the Sugiyama data: 95 per cent ellipses and ten bootstraps.

bootstrap too. We define it in terms of pseudo-values as $2\ \phi(\mathbf{p}) - \phi(\underline{v}_r/n)$, where the $\underline{v}_r$ are independent samples from the multinomial with probabilities $\mathbf{p}$. If the original observations or indicator matrices are still available, we can sample the $\underline{v}_r$ by drawing simple random samples with replacement from the n objects or from the n rows of the indicator matrices, until we have a sample size n too. We repeat this procedure R times, and use these R *bootstrap samples* to compute pseudo-values. In the same way as before the mean of the R pseudo-values can be used to estimate $\phi(\mathbf{\varkappa})$ in the multinomial case; the variance of the pseudo-values can be used to estimate the standard error of $\phi(\underline{p})$. The formulas are very much like those for the jackknife, and we do not give them here.

Figures 12.3 and 12.4 give ANACOR/ANAPROF solutions for ten bootstrap samples from the occupational mobility and religious practice data analysed in Chapter 8. The ellipses are computed with the delta method; they are the same as in Figures 12.1 and 12.2. It is clear that the bootstrap gives roughly the same information as the delta method in these examples, although in some cases the bootstrap vectors emanating from the solution point tend to point in one direction only. This implies that the average of the bootstrap pseudo-values will give an appreciable bias correction.

12.3.5 Comparison of jackknife and bootstrap

We have seen that both methods lead to roughly the same formulas, and any choice between them should be based on comparison of second order terms in the expansions. Our experience so far suggests that it is difficult to recommend one of the two techniques on asymptotic grounds. Moreover, we must also compare the techniques with subsampling and with the delta method. Theoretical work in mathematical statistics seems to suggest that the delta method corrected for bias will give the smallest mean square error, but we should not only compare the methods in terms of loss functions. In fact it is clear from the examples in this chapter and the applications in the next chapter that the jackknife and the bootstrap are very powerful data analysis techniques. ➵ We prefer the Monte Carlo version of the bootstrap for the time being, because it is very easy to explain, very easy to compute, and very natural in multinomial situations. If it is feasible to apply the delta method, the two techniques give similar information on stability. The bootstrap values

$\phi(\underline{v}_r/n)$ have a probabilistic interpretation and an asymptotic distribution even if **p** is nonrandom; the jackknife fails in some situations in which the bootstrap does not fail. This is because the bootstrap smooths more: in spline terminology the bootstrap is variation diminishing and can approximate even discontinuous functions. The theoretical jackknife is easier to compute than the theoretical bootstrap, because there are fewer interpolation points. For the bootstrap $\mu_n(\mathbf{p})$ is a polynomial in **p**, which is also convenient in theoretical investigations.

12.4 EPILOGUE

When applying the theoretical results in Sections 12.1 and 12.2 it is important to pay attention to a number of details. For instance, in correspondence analysis we have to deal with a matrix **B** that is a function of both **A** and *N*, the grand total. Estimates for the variances and covariances of all parameters are provided as an option in the program ANACOR (Groenen and Gifi, 1989). The delta method and its application to correspondence analysis and qualitative redundancy analysis is also discussed in Israëls (1987a, app. B).

Jackknife and bootstrap methods have gained much popularity over the past ten years. A lot of general, theoretical work is reviewed in Hinkley (1988) and DiCiccio and Romano (1988). Beneficial effects are found from smoothing the bootstrap distribution, balancing the resampling scheme, as well as from mechanisms for bias correction. Applications of the bootstrap in nonlinear multivariate analysis are reported in Meulman (1982), Greenacre (1984), De Leeuw (1985), Saporta and Hatabian (1986), De Leeuw and Meulman (1986), and Van der Burg (1988).

CHAPTER 13
THE PROOF OF THE PUDDING

In this chapter various applications of nonlinear data analysis are presented. The purpose of this chapter is to show what kind of results the various techniques do produce, what kind of plots can be made, how such plots should be interpreted, etc. Another purpose is to discuss some of the typical problems often encountered in practice. For all examples the discussion remains focused on aspects of the data analysis, and not on details of substantive interpretation. It might be emphasized that we do not claim that the analyses presented are better than any other analysis that could have been applied to the same data.

13.1 MULTIPLE CHOICE EXAMINATION

A multiple choice examination consists of items, where for each item the individual can make a choice from a number of alternative answers, one of which is 'correct'. Every examiner knows that it is difficult to construct 'wrong' answers that nevertheless have some plausibility (so that the student really has to know the subject-matter in order to reject the 'wrong' alternatives). Every examiner also knows that it often happens that an alternative supposed to be wrong, nevertheless is chosen by many of the 'best' students. This indicates a conflict between *a priori* and *a posteriori* evaluation of items.

The present example refers to a multiple choice examination in 'Introduction to Psychology', taken by 190 undergraduate psychology students in Leiden, January 1980. There were 30 items, each of them with four response alternatives. The indicator matrix therefore becomes a 190×120 matrix, with 30 submatrices G_j, each of them having four categories.

Table 13.1 gives the basic data. This table lists for each item the marginal frequencies of the four response alternatives, with an additional column for missing values. The correct alternative is underlined. A '+' before the number of the item means that the item had been used in earlier examinations (such items are supposed to be more discriminative than new items that have not been validated earlier). The right half of the table summarizes results in terms of correct/incorrect answers, with missing counted as incorrect. The table also gives columns for discrimination measures; they will be discussed in the next section.

Table 13.1 Marginal frequencies and missing entries, discrimination measures, and eigenvalues for the multiple choice examination

	four categories						two categories (correct/wrong)			
No.	Marginal frequencies				Discrimination measures	No.	Marginal frequencies			Discrimination measures
	M	a	b	c	d			M	Correct	Wrong	
+ 1	0	16	24	10	_140_	0.103	+ 1	0	140	50	0.105
+ 2	0	11	_147_	25	7	0.285	+ 2	0	147	43	0.264
3	0	17	34	32	_107_	0.067	3	0	107	83	0.088
+ 4	0	22	9	17	_142_	0.068	+ 4	0	142	48	0.041
5	0	_102_	0	81	7	0.001	5	0	102	88	0.004
6	0	1	29	68	_92_	0.284	6	0	92	98	0.155
7	0	4	44	52	_90_	0.264	7	0	90	100	0.182
8	0	49	11	_100_	30	0.182	8	0	100	90	0.214
9	0	1	_153_	31	5	0.062	9	0	153	37	0.102
+10	0	_109_	26	2	53	0.136	+ 10	0	109	81	0.113
11	0	26	_121_	39	4	0.260	11	0	121	69	0.195
+12	0	11	36	5	_138_	0.201	+ 12	0	138	52	0.183
13	0	90	4	_92_	4	0.081	13	0	92	98	0.018
14	0	19	9	36	_126_	0.026	14	0	126	64	0.021
15	0	40	14	28	_108_	0.134	15	0	108	82	0.128
+16	0	5	5	_144_	36	0.302	+ 16	0	144	46	0.211
17	0	_106_	48	27	9	0.074	17	0	106	84	0.110
18	0	5	17	15	_153_	0.262	18	0	153	37	0.144
+19	0	66	1	_102_	21	0.058	+ 19	0	102	88	0.000
20	1	3	35	_138_	13	0.252	20	0	138	52	0.228
+21	0	12	47	_103_	28	0.093	+ 21	0	103	87	0.102
22	0	22	38	10	_120_	0.102	22	0	120	70	0.105
23	0	8	11	_148_	23	0.150	23	0	148	42	0.142
24	0	44	_126_	6	14	0.154	24	0	126	64	0.116
25	0	_154_	11	10	15	0.380	25	0	154	36	0.213
+26	1	_103_	29	35	22	0.058	+ 26	0	103	87	0.040
27	0	13	_135_	37	5	0.131	27	0	135	55	0.102
28	0	_117_	28	39	6	0.163	28	0	117	73	0.198
29	0	5	_167_	9	9	0.377	29	0	167	23	0.290
+30	0	27	_149_	11	3	0.309	+ 30	0	149	41	0.293
Eigenvalue						0.167	Eigenvalue				0.137

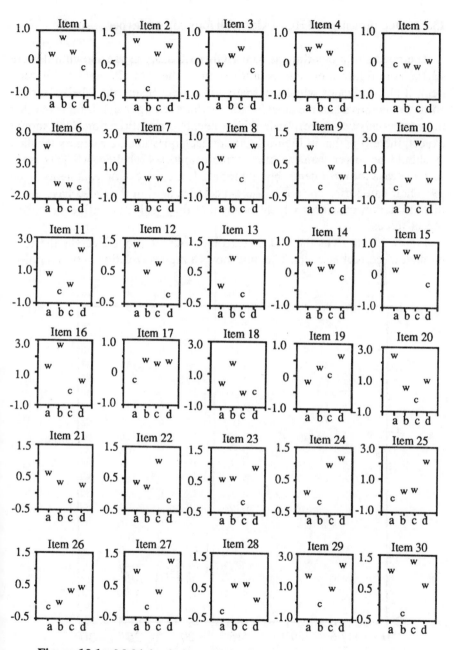

Figure 13.1a Multiple choice examination. Transformation plots of items: c = correct answer; w = wrong answer.

13.1.1 One-dimensional HOMALS with four categories per variable

The results of a one-dimensional HOMALS analysis are shown in Figure 13.1a. This figure plots the quantification of the four alternatives for each item; it also indicates which alternative is correct (using the label 'c' for the correct alternative and the label 'w' for the three wrong ones). It turns out that the correct answer corresponds with a negative quantification. One should expect, therefore, that in Figure 13.1a for each question the category labelled 'c' should be lower than the other three categories labelled 'w'. This is, in fact, the case for most items, but not for all. Look, for example, at items 5 and 19. Their categories are not in the expected order but they also have small discrimination measures which means that they had little influence on the object scores.

Figure 13.1b gives the HOMALS object scores plotted versus the total number of correct answers. The plot shows a high correlation between object

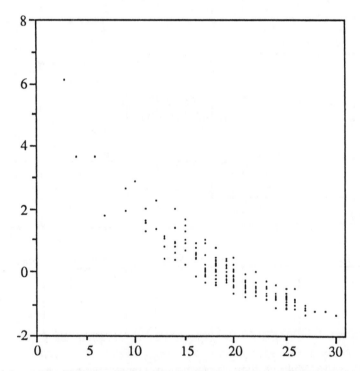

Figure 13.1b Multiple choice examination. Scatter diagram of object scores (vertical axis) versus the number of correct answers (horizontal axis). Four categories per variable.

scores and number of correct answers. Nevertheless, there are some dis-
crepancies. For example, the individual with the smallest object score,
suggesting the best examination result, has 27 correct responses, whereas
there are three individuals with a larger number of correct responses. The
explanation is simple: the individual with the smallest object score made three
errors, on items 5, 13, and 19. These are items with a small discrimination

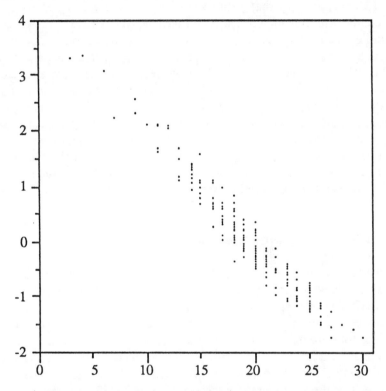

Figure 13.1c Multiple choice examination. Scatter diagram of
object scores (vertical axis) versus the number of correct answers
(horizontal axis). Two categories per variable.

measure, and the individual's wrong answers are precisely those answers
selected by most individuals with a larger number of correct responses. These
items obviously do not discriminate between individuals with many correct
answers on other items, versus individuals with few correct answers; these
items could easily have been skipped from the examination.

13.1.2 One-dimensional HOMALS with merged wrong categories

An alternative analysis is that wrong answers are merged into one category, as in the right part of Table 13.1. This implies that the HOMALS indicator matrix becomes a 190 x 60 matrix, with two columns for each item. Figure 13.1c displays the scatter diagram of the object scores versus the total number of correct answers. The discrimination measures, listed in the right-hand side of Table 13.1, again show that items 5 and 19 are among the poor ones. As expected, the discrimination measures for the analysis based on two categories are smaller than those for the analysis with four categories. Clearly, the HOMALS analysis with two categories per variable gives less information than the HOMALS analysis with four categories. Once 'incorrect' answers are merged, we shall never be able to discriminate between incorrect answers that are blatantly incorrect and incorrect answers that are only slightly incorrect (in the sense that they are often chosen by individuals with many correct answers on other items). In the example, item 13 has the correct answer in category (c), but this answer competes with the incorrect alternative (a) which is far 'less incorrect' than the other two categories (b) and (d). Such a result indicates, in fact, that the examiner should take a close look at item 13.

13.1.3 HOMALS on the transposed data matrix

A third analysis has been performed on the transpose of the data matrix, as if the individuals were sorting the items into four categories (cf. Section 3.13). Now the indicator matrix is a 30 x 760 matrix. Results are trivial. Since individuals tend to select the correct answer in the first place, they sort the items into four groups: one group where category (a) is the correct answer, a second group with category (b) as the correct answer, a third group where (c) is correct, and a fourth group where (d) is correct. Figure 13.1d displays the two-dimensional object scores for the items, identified by the category labels. The figure shows that items with (b) or (d) as the correct answer form clusters (in fact, category (d) is more often correct than any other category). One expects that category (a) and category (c) will form similar clusters in subsequent HOMALS dimensions. The analysis of the reversed indicator matrix gives discrimination measures for individuals. Such discrimination measures will be correlated with the total number of correct answers, to the extent that the latter depends on items with category (b) or (d) as the correct answer. In fact, it turns out that the third dimension indeed reveals a discrimination between items with (a) or (c) as the correct category, with items 5 and 19 (the items that discriminate poorly) excepted.

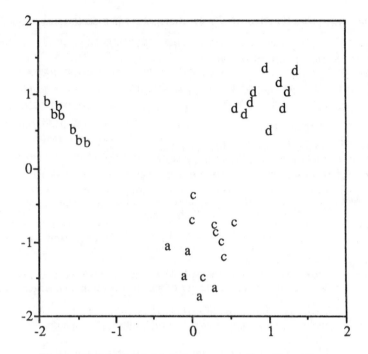

Figure 13.1d Multiple choice examination. Two-dimensional object points of the items obtained on the basis of the reversed indicator matrix.

In conclusion, the analysis of the reversed indicator matrix turns out to be a rather naive idea for this example. Although the results of the analysis tell us something about individuals (in terms of discrimination measures), they tell us more about how the examiner allocated correct answers to the categories (a) up to (d).

13.2 CONTROVERSIAL ISSUES

The data to be analysed in this section have resulted from a survey among 575 respondents in 1974 (Veenhoven and Hentenaar, 1975). The description of the 37 questions in this survey can be found in Table 13.2a. The marginal frequencies and number of missing values are given in Table 13.2b.

Table 13.2a Description of the 37 questions asked in the survey

3	questions with respect to capital punishment (CP1–CP3),
16	questions about abortion (A01–A16),
5	questions about euthanasia (EU1–EU5),
5	questions about sexual freedom (SF1–SF5),
8	questions on background.

CP   Response from 1 = 'agree completely' to 5 = 'disagree completely':

CP1   Taking hostages should be punishable by death.
CP2   Murder should be punished by death.
CP3   Killing people in time of war is justifiable.

A01–A08   Response 1 = 'agree', 2 = 'disagree'. All eight statements start with 'I find abortion justifiable if'; they continue with:

A01   prolonged pregnancy is a danger to the mother's life or health;
A02   the woman wants it for whichever reason, and if there are no medical counterindications;
A03   there is a high chance that the child will be deformed or handicapped;
A04   the expectant mother is unmarried and does not want to marry the father of the child;
A05   the pregnancy is the result of rape or indecent attack;
A06   the woman has a large family already and it is undesirable to have more children;
A07   the expectant mother is unmarried and not able to marry the father of the child;
A08   there are chances that the child will have an unhappy childhood because its parents do not really love it.

A09–A10   Responses: 1 = 'justifiable until 3 months', 2 = 'until 4 months', 3 = 'until 5 months', 4 = 'until 6 months', 5 = 'after six months', 6 = 'not justifiable'.

A09   A woman, 45 years of age, when menstruation fails to come, thinks menopause has started, and does not worry. Later she appears to be pregnant. She has a family with grown-up children. Until which month of pregnancy do you feel that abortion in this special case is still justified? Or in your judgement is abortion in this case not justified?
A10   A girl of 15 -unmarried- suspects she is pregnant. She is scared to talk about it with the family doctor or with her parents. As a result it takes much longer for her than necessary to enlist for medical aid. Until which month of pregnancy do you feel that abortion in this special case is still justified? Or in your judgement is abortion in this case not justified?

A11–A14   Response from 1 = 'agree completely' to 5 = 'disagree completely'.

A11   It is the woman's right to have abortion when she wants it .
A12   Medical practitioners who perform abortion are no better than murderers.
A13   People who agree with abortion have little respect for life.
A14   Abortion is justifiable under no circumstances.

A15   One statement with three response categories: 1 = 'abortion law, only special cases', 2 = 'law should make abortion difficult', 3 = 'no law, doctor decides'.

A15   During the last years, up to now, politicians of different parties have been working on proposals with respect to abortion. In your judgement, should there be a law that allows for abortion in special cases only, or should there be a law that makes it difficult to have an abortion, or do you think there is no need for a law and that it is up to the doctor to decide whether or not he will help the woman?

A16   One statement with three response categories: 1 = 'after 12 weeks absolutely forbidden', 2 = 'after 12 weeks in special cases only', 3 = 'no time limit'.

A16   Discussions on abortion during the last half of this year focused on whether abortion should be permitted or not after the 12th week of pregnancy. In your opinion, should there be a law that rules out abortion absolutely after 12 weeks, or a law that limits abortion after 12 weeks to special cases only, or do you say that the law should not specify a time limit for abortion?

EU   Responses: 1 = 'justifiable', 2 = 'unjustifiable'. All statements begin with 'I find euthanasia justifiable if'; they continue with

EU1   the ill person asks for it because he or she knows that the illness is terminal;

EU2   close relatives ask for it, and the ill person is unconscious, where there is no hope for recovery;

EU3   at the birth of a child it becomes evident that the child can be kept alive in a strictly technical medical sense but never will be able to have human contact;

EU4   in this way dying persons suffering from unbelievable pain can be relieved from their misery;

EU5   elderly people no longer are able to take care of themselves and express the wish they prefer to die.

SF   Five statements on sexual freedom. Response from 1 = 'agree completely', to 5 = 'disagree completely'.

SF1   I don't object to children below the age of ten walking around on the beach naked.

SF2   If sexual intercourse was separated from procreation it would soon become pure egoism.

SF3   Parents should forbid children to have sexual play.

SF4   Young people who have sexual intercourse before marriage do not have respect for each other.

SF5   Parents should impress upon their children that it is better to have control over yourself and not to indulge in masturbation.

Eight background questions

SEX   1 = 'male', 2 = 'female'.
AGE   1 = 'below 20', 2 = 'between 20 and 30', etc., until 6 = 'above 60'.

SOC	social class, ranging from 1 = 'high' to 8 = 'low'.
REL	religion: 1 = 'protestant', 2 = 'reformed', 3 = 'roman catholic', 4 = 'none'.
POL	political preference: 1 = 'left', 2 ='denominational', 3 = 'liberal', 4 = 'right', 5 = 'none'.
EDU	education level: 1 = 'LO, VGLO', 2 = 'ULO', 3 = 'VHMO', 4 = 'professional training or university'.
FUN	present profession or job: 1 = 'managerial, more than ten employees', 2 = 'ibid., less than ten employees', 3 = 'free profession', 4 = 'independent farmer', 5 = 'higher employees and civil servants', 6 = 'ibid., middle level', 7 = 'ibid., lower level', 8 = 'schooled workers', 9 = 'unskilled labour', 10 = 'students', 11 = 'housewives'.
URB	degree of urbanization: 1 = 'Amsterdam', 2 = 'Rotterdam', 3 = 'The Hague', 4 = 'medium size cities', 5 = 'small cities', 6 = 'industrialized rural', 7 = 'agricultural rural'.

13.2.1 One-dimensional HOMALS with all 37 variables

Table 13.2b also gives the discrimination measures in the first HOMALS dimension. The results show that variables CP 1, 2, and 3 do not discriminate very well, and that of the eight background variables only REL and POL are related to the first dimension. Figure 13.2a gives the transformation plots of the remaining 28 variables. The figure shows that the HOMALS solution is related to such basic variables as A11–14 and that negative category quantifications are in the direction of 'in favour of legal abortion' whereas positive quantifications are in the 'anti-abortion' direction. The quantification of POL shows that 'left' and 'liberal' are on the 'pro-abortion' side, CDA (the combination of denominational parties) is 'anti'.

13.2.2 Likert scales

A Likert scale consists of a number of statements to each of which the respondent can answer with a choice out of a number of response categories (usually five) ranging from 'strongly approve' to 'strongly disapprove'. It has become current practice to evaluate such items on the basis of item-total correlations and to calculate a respondent's score on the scale only on the basis of the items that attain the criterion of sufficiently high item-total correlation, whereas items that fail to attain the criterion are ignored.

In the data on controversial issues eleven items are candidates for a Likert scale: A9–14, and SF1–5. Table 13.2c shows their intercorrelations and the item-total correlations, after recoding the categories of items A11 and SF1.

Table 13.2b Controversial issues. Missing entries, marginal frequencies, and discrimination measures

Name	No.	M	1	2	3	4	5	6	7	8	9	10	11	Discrimination measures
CP1	1	1	188	129	61	113	83							0.026
CP2	2	1	167	112	77	108	110							0.019
CP3	3	3	86	131	89	108	158							0.013
A01	4	1	528	46										0.086
A02	5	7	256	312										0.508
A03	6	8	460	107										0.375
A04	7	6	217	352										0.470
A05	8	6	485	84										0.312
A06	9	4	275	296										0.519
A07	10	8	244	323										0.543
A08	11	10	259	306										0.463
A09	12	4	217	48	18	8	21	259						0.547
A10	13	8	205	62	31	12	33	224						0.566
A11	14	0	178	115	36	93	153							0.541
A12	15	0	41	32	77	111	314							0.567
A13	16	0	114	60	69	117	215							0.645
A14	17	1	43	54	62	110	305							0.594
A15	18	26	249	50	250									0.224
A16	19	33	158	266	118									0.220
EU1	20	6	396	173										0.317
EU2	21	4	299	272										0.248
EU3	22	14	405	156										0.348
EU4	23	4	435	136										0.335
EU5	24	7	104	464										0.081
SF1	25	1	130	85	56	98	205							0.170
SF2	26	2	84	67	85	115	222							0.163
SF3	27	1	124	109	100	114	127							0.159
SF4	28	1	49	42	56	126	301							0.292
SF5	29	4	124	97	88	88	174							0.225
SEX	30	1	248	326										0.002
AGE	31	1	39	100	130	95	98	112						0.077
SOC	32	0	17	23	65	120	201	75	70	4				0.012
REL	33	19	103	62	168	223								0.259
POL	34	0	176	128	92	20	159							0.354
EDU	35	20	311	134	43	67								0.041
FUN	36	3	3	27	8	13	12	70	84	56	33	21	245	0.077
URB	37	1	44	44	40	166	65	64	151					0.062
											Eigenvalue			0.283

Also, variables A9 and A10 needed *a priori* recoding: response category 6 rejects abortion under all circumstances, but categories 1 to 5 (in this order) become more 'tolerant' towards abortion; their order must be reversed in order to become consistent with category 6. The average squared item-total correlation in Table 13.2c is 0.41.

Table 13.2c Intercorrelations and item-total correlations of the original variables

	A09	A10	A11	A12	A13	A14	SF1	SF2	SF3	SF4	SF5	I-T
A09		0.55	0.39	0.35	0.41	0.37	0.27	0.24	0.25	0.26	0.20	0.61
A10			0.46	0.37	0.43	0.41	0.23	0.19	0.22	0.27	0.20	0.64
A11				0.50	0.57	0.49	0.24	0.14	0.18	0.26	0.21	0.65
A12					0.71	0.68	0.26	0.27	0.28	0.36	0.30	0.71
A13						0.70	0.28	0.27	0.28	0.39	0.34	0.77
A14							0.23	0.26	0.25	0.38	0.38	0.72
SF1								0.18	0.30	0.25	0.30	0.52
SF2									0.37	0.39	0.39	0.52
SF3										0.37	0.54	0.58
SF4											0.50	0.63
SF5												0.63

Average squared item-total correlation   0.41

Table 13.2d Intercorrelations and item-total correlations after HOMALS quantification

	A09	A10	A11	A12	A13	A14	SF1	SF2	SF3	SF4	SF5	I-T
A09		0.65	0.52	0.49	0.57	0.51	0.32	0.31	0.28	0.36	0.30	0.73
A10			0.59	0.53	0.59	0.58	0.27	0.24	0.25	0.36	0.30	0.75
A11				0.55	0.58	0.51	0.27	0.15	0.20	0.29	0.25	0.69
A12					0.73	0.68	0.30	0.28	0.28	0.40	0.33	0.78
A13						0.71	0.30	0.29	0.30	0.41	0.37	0.83
A14							0.27	0.28	0.25	0.40	0.41	0.79
SF1								0.16	0.30	0.28	0.30	0.47
SF2									0.37	0.43	0.39	0.49
SF3										0.38	0.55	0.52
SF4											0.50	0.63
SF5												0.60

Eigenvalue   0.45

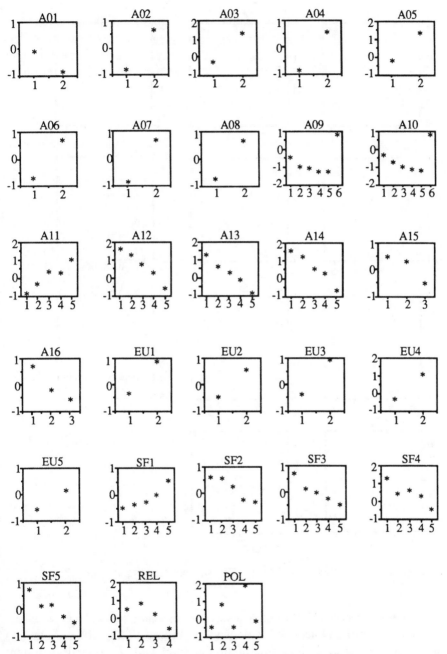

Figure 13.2a Controversial issues. Transformation plots of questions.

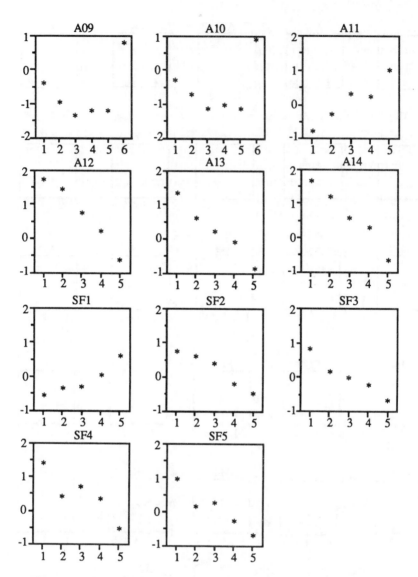

Figure 13.2b Controversial issues. Transformation plots of 'Likert' items.

A one-dimensional HOMALS analysis over the same set of eleven variables gives the quantifications that are depicted in Figure 13.2b, with correlations shown in Table 13.2d. Note, first of all, that the HOMALS quantification by itself takes care of the reversal of items A11 and SF1. Also, HOMALS very nicely takes care of the recoding of the variables A9 and A10, with the first five categories going downwards but category 6 obtaining the highest quantification. Secondly, for the HOMALS solution the average squared item total correlation is 0.45, slightly better than for the Likert scale solution. In fact, it is true that HOMALS maximizes this index (it is the HOMALS eigenvalue ψ^2).

Thirdly, the Likert scale solution would calculate individual scores as the sum of the item scores. In HOMALS, however, the individual score becomes a weighted sum of item scores: items with higher discrimination measures contribute more than items with small discrimination measures.

Fourthly, the Likert procedure assumes that variables are measured on an interval scale level. The HOMALS solution can be interpreted as a test of that assumption: if the assumption is correct, the plots in Figure 13.2b (after reordering, when necessary, such as in A9 and A10) should be linear. We find that this is roughly true for a number of items, but not for all of them. In SF1, for example, the difference between categories 2 and 3 is negligible (they might as well have been merged).

Fifthly, in the Likert scale solution the A items have larger item-total correlations than the SF items. This tendency is even stronger in HOMALS,

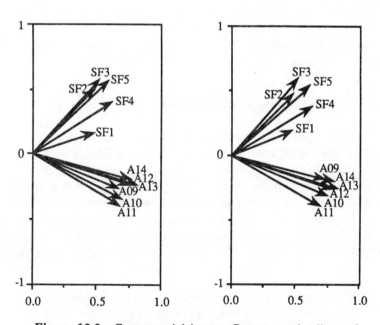

Figure 13.2c Controversial issues. Component loadings of the Abortion and Sexual Freedom items, on the basis of the original data on the left; on the basis of the optimally transformed data on the right.

where the item-total correlations for SF become smaller than in the Likert scale solution. This indicates that the eleven items are not truly one dimensional: there must be a second dimension for which the SF items are better indicators. To explore this further, a principal component analysis was applied to the correlation matrices in Tables 13.2b and 13.2c, with the component loadings shown in Figure 13.2c. The two solutions are very similar, so optimal scaling has no major effect.

We conclude that the HOMALS approach has a number of advantages over the Likert approach. It takes care of reversals and possible reordering of item categories, it does not assume interval scale level of measurement, and it produces optimal scale values for individuals. On the other hand, the results in this practical example demonstrate that the net gain of HOMALS over the Likert scale solution is small.

13.2.3 Guttman scales

A Guttman scale consists of binary items and assumes that only a limited number of response patterns can occur (see Sections 2.3 and 9.2.2). Very often items will not fit a Guttman scale perfectly, in the sense that there are response patterns that 'should not have been there'. Guttman (1946) proposes the *coefficient of reproducibility* (REP) as an index of how well empirical data approximate a perfect Guttman scale. The index is defined as

$$REP = 1 - \frac{\text{number of errors}}{\text{total number of responses}},$$

where an error is defined as a reversal one has to make in order to change an anomalous response pattern into an acceptable one. If REP is large (larger than 0.85 is the rule of thumb), one may assume that anomalous patterns can be ignored and that the hypothesis of a one-dimensional underlying structure can be maintained. An objection against REP is that the Guttman scale will order the items in such a way that the number of errors is minimized, so that REP will always be larger than zero. The minimum of REP will often be larger than 0.5 (Mokken, 1971, pp. 50-54). Another index is the MMR (coefficient of minimum marginal reproducibility), defined as

$$MMR = \frac{\text{total number of responses in modal categories}}{\text{total number of responses}},$$

where 'a response in a modal category' is a response in the category most often used. Edwards (1957) proposes that REP not only should be larger than 0.85, but also substantially larger than MMR.

Starting from the other end, one might define a coefficient of scalability

$$S = 1 - E/E_0$$

where E is the total number of errors and E_0 the expected number of errors if responses were completely random (Niemöller, 1976). This approach implies a probabilistic rather than a deterministic model. The idea has been further developed by Mokken (1971). The Mokken scale orders items according to

Table 13.2e Controversial issues. Guttman scale solution

Item	A01		A02		A03		A04		A05		A06		A07		A08		Total
Resp	A	D	A	D	A	D	A	D	A	D	A	D	A	D	A	D	
	—ERR—		—ERR—		—ERR—		—ERR—		—ERR—		—ERR—		—ERR—		—ERR—		
D 8	0	17	0	17	0	17	0	17	0	17	0	17	0	17	0	17	17
I	—ERR—																
S 7	40	4	3	41	1	43	0	44	0	44	0	44	0	44	0	44	44
A			—ERR—														
G 6	38	6	36	8	14	30	0	44	0	44	0	44	0	44	0	44	44
R					—ERR—												
E 5	97	4	91	10	98	3	4	97	6	95	4	97	1	100	2	99	101
E							—ERR—										
4	56	4	57	3	57	3	17	43	21	39	20	40	8	52	4	56	60
									—ERR—								
3	47	1	47	1	45	3	31	17	23	25	21	27	16	32	10	38	48
											—ERR—						
2	45	5	49	1	46	4	36	14	38	12	30	20	34	16	22	28	50
													—ERR—				
1	33	3	36	0	34	2	32	4	23	13	29	7	35	1	30	6	36
															—ERR—		
0	143	0	143	0	143	0	143	0	143	0	143	0	143	0	143	0	143
Sums	499	44	462	81	438	105	263	280	254	289	247	296	237	306	211	332	543
Pcts	92	8	85	15	81	19	48	52	47	53	45	55	44	56	39	61	
Error	0	27	3	23	15	15	4	78	27	50	45	27	59	1	68	1	442
B.C.	0.3556		0.7143		0.7083		0.9049		0.8438		0.8668		0.9568		0.9284		

A = agree
D = disagree

Coefficient of reproducibility = 0.8983
Minimum marginal reproducibility = 0.6680
Percentage improvement = 0.2302
Coefficient of scalability = 0.6935

the proportion p_i of positive answers to item i. However, the scale accepts items only if they obey *double monotonicity*: if p_i is increasing with i, then for each fixed item h the probability of a positive answer to both h and i must be increasing with i, and the probability of a negative response to both h and i must be decreasing with i (cf. Section 9.2). An index for scalability is H_i, essentially the ratio between the sum of the covariances between i and all other items and the sum of the maximum covariances allowed by the marginal proportions.

Table 13.2f HOMALS solution

	agree	disagree	discr.
A01	-0.07	0.82	0.058
A02	-0.88	0.72	0.623
A03	-0.29	1.22	0.342
A04	-1.05	0.65	0.673
A05	-0.24	1.33	0.308
A06	-0.84	0.78	0.657
A07	-0.98	0.74	0.718
A08	-0.84	0.72	0.598
eigenvalue			0.497

Table 13.2g Mokken scale: the resulting Mokken scale depends on the order in which the items are selected. This order is: (A05, A07), A04, A03, A06, A02, A08, A01

variable	difficulty	H_i
A01	.9183	.3928
A02	.4452	.6450
A03	.8000	.6943
A04	.3774	.7434
A05	.8435	.7203
A06	.4783	.6824
A07	.4243	.7152
A08	.4504	.6312
scalability		.6730

In the controversial issues, variables A1–8 qualify as binary items. The Guttman scale solution is given in Table 13.2e. Biserial item-total correlations order the items as

$$7 - 4 - 6 - 2 - 8 - 5 - 3 - 1$$

The HOMALS solution is given in Table 13.2f; the discrimination measure (squared item–total correlation) orders the items as

$$7 - 4 - 6 - 2 - 8 - 3 - 5 - 1$$

so that in this respect results are quite comparable (the more so since in both solutions items 3 and 5 are close together).

Results for the Mokken scale analysis are given in Table 13.2g. As to
'degree of difficulty' results are blockwise comparable to the Guttman and
HOMALS ordering. However, as to scalability, results are very much
different.

13.2.4 Two-dimensional HOMALS with 28 variables

A new analysis has been performed, this time in two dimensions, deleting the
CP variables and the non-discriminating background variables. The two-
dimensional category points are plotted in Figure 13.2d. We notice the rather
typical horseshoe, where the second dimension is a quadratic function of the

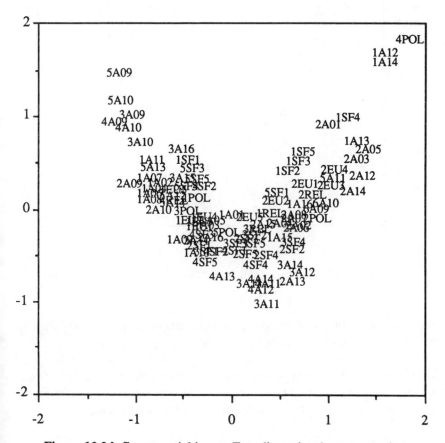

Figure 13.2d Controversial issues. Two-dimensional category points.

first, and as such has no other substantive interpretation than that it contrasts categories in the middle with those at the extremes. Going from the upper right corner downwards to the middle, and from there to the upper left corner, one finds categories ordered from 'anti-abortion' and political 'right', towards

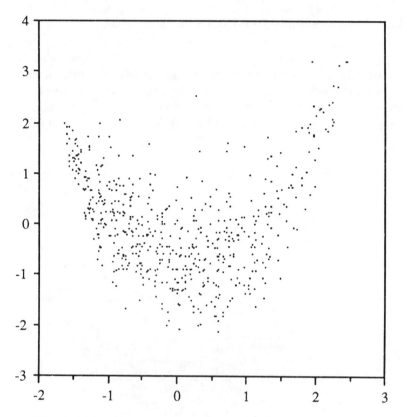

Figure 13.2e Controversial issues. Two-dimensional object points.

'pro legalization of abortion' and political 'left' or 'liberal'. Categories for variables SF (sexual freedom) follow roughly the same pattern, but because of their smaller discrimination measures they are located more towards the centre. Variable EU5 (very small discrimination measure) has all its categories near the centre of the plot. Figure 13.2e gives the corresponding points for the individuals, with a similar, but more blurred, horseshoe pattern.

13.2.5 PRINCALS

PRINCALS has been applied to a selection of 26 variables (those of Section 13.2.4, but without A1 and EU5 because of their small discrimination measure). In this analysis the variables REL and POL were treated as multiple nominal. The typical Likert items (A11–16, SF1–5) were treated as single ordinal. For binary variables (A2–8, EU2–4) the choice of measurement level is irrelevant. For the special variables A9 and A10 a single nominal transformation was required. Figure 13.2f gives a plot of the weights (correlations) of the variables with the two PRINCALS dimensions. Note that A9

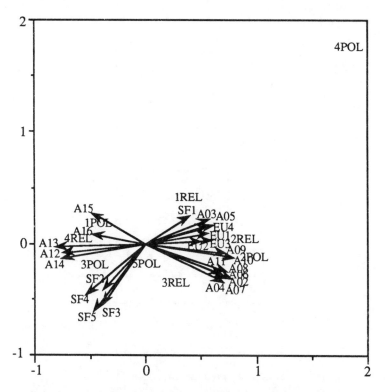

Figure 13.2f Two-dimensional PRINCALS solution abortion data.

is mainly related to the first PRINCALS dimension, whereas SF5 is also related to the second. This confirms results of the Likert scale analysis of Section 13.2.2 above. Figure 13.2g gives the corresponding plot for the object points, with the component loadings of the variables A9 and SF5. The plot also shows the boundary hyperplanes (lines) for the categories of A9 and SF5. Note also that PRINCALS, too, reorders the categories of item A9.

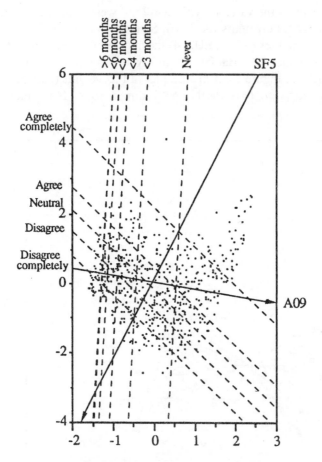

Figure 13.2g Controversial issues. Two-dimensional configuration of object points. Solid lines give the directions for A9 and SF5; the dotted lines give the optimal separation between the objects, based on the responses on these variables.

In Figure 13.2g, one finds at the upper right corner individuals who are anti-abortion, politically at the right, and religious versus in the upper left corner individuals who are tolerant with respect to abortion, are politically left or liberal, and non-religious. In the lower middle one finds individuals who favour sexual freedom to some extent, but oppose abortion.

13.3 AS YEARS GO BY

The data in this section come from a longitudinal survey on the determinants of the education level attained by children after they had left elementary school. A description of the variables is given in Table 13.3a. Details about the survey can be found in the original survey proposal (ITS, 1968), and in reports by Kropman and Collaris (1974), Collaris and Kropman (1978), Dronkers (1978), and Dronkers and Jungbluth (1979). The crucial variable is EIN. A number of variables (BVA, BIL, INT, URB, BIM, DLO, ADV, LL6, KGS, PRE, SEX) originate from a nationwide survey held in 1965 over more than 11000 children. Later it was decided to do a follow-up study for a subgroup of some 2000 children. This study was done in 1970, and data were collected for the variables OPV, OPM, AKG, ASO, ASL, OOA, DWO, BMB, INS, KLS, TON, AOS, LLS, EXT, and also EIN for those children who by that time had left secondary school. For the other children EIN was measured in 1974. Ultimately, the complete data set contained data for 1845 children. It should be added, perhaps, that the categories of EIN (final level) exclusively refer to the level of secondary full-time education: part-time education and tertiary education (MBO) have not been taken into account.

Table 13.3a Description of the 'as years go by' variables

BVA	occupational level of father: 1 = 'agricultural or unskilled labour', 2 = 'skilled labour', 3 = 'clerical', 4 = 'small business', 5 = 'farmers', 6 = 'managerial', 7 = 'higher managerial and free professions'.
OPV	education level of father.
OPM	education level of mother.
	For both variables categories ranging from 1 = 'primary school only' to 7 = 'university training'.
AKG	number of children in family: 1 = 1, etc., until 9 = 'more than 8'.
URB	degree of urbanization of residence: 1 = 'rural', 2 = 'villages', 3 = 'small cities', 4 = 'large cities'.
ASO	aspiration level of parents (Reissman, 1953): parents were asked whether they find their child later should accept a 'very good job' even if this would imply a specific disadvantage such as 'no time for hobbies'.

Eleven such disadvantages were mentioned, for each of which parents could respond with 1 = 'unacceptable' to 4 = 'acceptable'. The score is the sum of the category numbers chosen, later condensed into six classes.

ASL	aspiration level of child: same as ASO, answered by child.
BIL	scale for 'interest' in various professional activities, at lower level.
BIM	scales for 'interest' in various professional activities, at higher level.
INT	parents' interest in child's school performance, as rated by teacher in last form of primary school: 1 = 'much interest', 2 = 'not much interest'.
OOA	a five-point scale indicating agreement with the statement that a child's career should be based on advice from teachers and results of mental tests: 1 = 'agree', to 5 = 'disagree'.
DWO	a four point scale, indicating agreement with the statement that parents decide about educational and professional career of their child irrespective of what child wants: 1 = 'agree' to 4 = 'disagree'.
BMB	a four-point scale indicating agreement with the statement that professional training is unimportant for girls: 1 = 'agree' to 4 = 'disagree'.
INS	do parents agree with present educational choice? 1 = 'yes', 2 = 'no'.
KLS	has child been at nursery? 1 = 'yes', 2 = 'no'.
DLO	did child repeat a form in primary school? 1 = 'no', 2 = 'yes'.
LL6	number of children in last form of primary school. 1 = 'less than 10', 2 = '10 to 19', 3 = '20 to 30', 4 = 'more than 30'.
ADV	teacher's advice as to education after primary school, on four levels increasing from 1 to 4.
KGS	average score on achievement test in last form of primary school, from 1 = 'low' to 4 = 'high'.
PRE	result on test supposed to predict achievement in secondary school. Stanine scoring.
TON	level of secondary school first selected immediately after primary school. Categories 1 to 5 indicate increasing level.
AOS	whether first-selected secondary school had a differentiated curriculum: 1 = 'no', 2 = 'yes'.
LLS	number of pupils at first secondary school: 1 = 'less than 100', 2 = '100 to 200', etc., until 8 = 'more than 700'.
EXT	number of extra-curricular activities at secondary school: library, school-paper, excursions, clubs, school council. Categories 1 to 5 are a count of number of activities mentioned, 6 = 'child cannot mention any such activity'.
EIN	final level of secondary education: 1 = 'LO only', 2 = 'VGLO without certificate', 3 = 'VGLO with certificate, or LBO 1 year, or "brugklas" ', 4 = 'LBO or ULO/MAVO/VHMO unfinished after 2 years', 5 = 'LBO or ULO/MAVO/HAVO/MMS unfinished after 3 years', 6 = 'LBO finished', 7 = 'ULO/MAVO/HAVO/MMS unfinished after 4 years, or VHMO unfinished after 3 years', 8 = 'HAVO/MMS unfinished after 5 years, or VHMO unfinished after more than 4 years', 10 = 'HAVO/MMS finished', 11 = 'HBS finished', 12 = 'ATHENEUM/GYMNASIUM finished'.
SEX	1 = 'male', 2 = 'female'.

13.3.1 One-dimensional HOMALS with all variables

The transformation plots of all variables are given in Figure 13.3a; the corresponding discrimination measures can be found in Table 13.3b. Looking

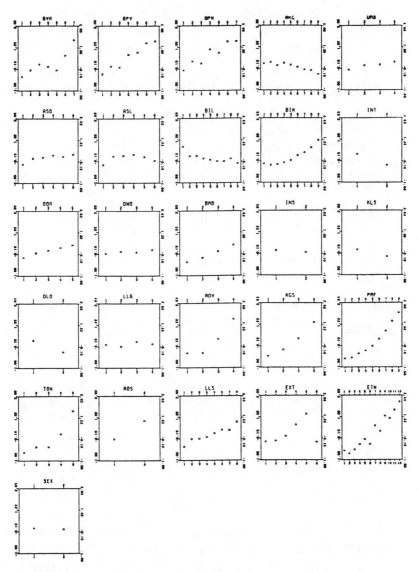

Figure 13.3a As years go by. Transformation plots of all variables.

at the variables with large discrimination measures, it becomes evident that the
first HOMALS dimension is related to school success (EIN, TON, PRE,
ADV). We shall give a few comments on the quantifications shown in Figure
13.3a.

Table 13.3b As years go by. HOMALS discrimination
measures for three different analyses

	All subjects	Girls	Boys
BVA	0.384	0.388	0.376
OPV	0.411	0.422	0.384
OPM	0.259	0.244	0.274
AKG	0.055	0.065	0.055
URB	0.034	0.044	0.026
ASO	0.049	0.058	0.041
ASL	0.067	0.088	0.052
BIL	0.059	0.038	0.112
BIM	0.186	0.134	0.273
INT	0.109	0.125	0.087
OOA	0.053	0.063	0.055
DWO	0.008	0.011	0.004
BMB	0.165	0.184	0.153
INS	0.003	0.000	0.009
KLS	0.014	0.021	0.028
DLO	0.125	0.123	0.129
LL6	0.015	0.016	0.021
ADV	0.651	0.652	0.670
KGS	0.412	0.386	0.434
PRE	0.627	0.606	0.645
TON	0.774	0.783	0.776
AOS	0.218	0.137	0.364
LLS	0.145	0.245	0.107
EXT	0.359	0.440	0.301
EIN	0.711	0.737	0.709
SEX	0.001	–	–
Eigenvalue	0.227	0.241	0.243

One interpretation of the quantification of categories is that HOMALS
transforms variables in such a way that their distribution becomes more similar
to a normal distribution. This implies that for variables that already have a
normal distribution, the HOMALS transformation will tend to be linear. Such

is the case for the stanine variables BIL, BIM, and PRE. On the other hand, HOMALS will correct for skewness (an example is ASO). Variable URB has an almost rectangular distribution before quantification; HOMALS corrects for this by making the difference between categories 1 and 2 and that between categories 3 and 4 larger than the difference between categories 2 and 3.

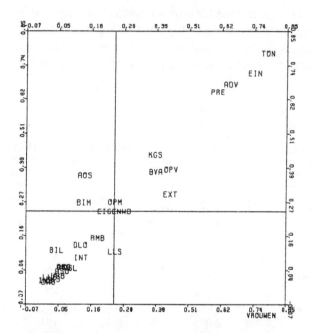

Figure 13.3b As years go by. HOMALS discrimination measures for boys (vertical axis) versus girls (horizontal axis).

A second comment concerns many of the nominal variables (BVA, OPV, OPM, ADV, TON, EIN): their quantifications are close to a monotonic function of the original category numbers. However, there are typical exceptions. One example is BVA, where categories 3 and 5 are interchanged, which, in hindsight, looks plausible enough. OPV and OPM also have some (interpretable) interchanges. For EIN we find some irregularities that cannot be understood so easily. Probably they are related (among other things) to the large differences in marginal frequency for the categories of EIN (for a more detailed discussion, see Stoop, 1980).

Finally, the quantification of EXT nicely puts category 6 (no extracurricular activities) below category 1 (only one extracurricular activity), which demonstrates that HOMALS often automatically corrects for a somewhat awkward *a priori* labelling of categories.

13.3.2 One-dimensional HOMALS for subgroups of individuals

HOMALS analyses were performed for boys and girls separately. Table 13.3b gives the discrimination measures and Figure 13.3b gives the plot of the discrimination measures for boys versus girls. The figure indicates that AOS and BIM have larger discrimination measures for boys, whereas LLS and

Table 13.3c As years go by. HOMALS discrimination measures for the categories of occupation levels (BVA)

	Agricultural + unskilled labour	Skilled labour	Clerical	Small business	Farmers	Mana-gerial	High mana-gerial +free professions
OPV	0.174	0.064	0.150	0.185	0.188	0.165	0.202
OPM	0.098	0.105	0.127	0.080	0.151	0.116	0.056
AKG	0.078	0.142	0.045	0.071	0.083	0.036	0.088
URB	0.033	0.028	0.000	0.030	0.068	0.006	0.028
ASO	0.064	0.011	0.043	0.030	0.095	0.087	0.295
ASL	0.137	0.027	0.072	0.020	0.058	0.069	0.063
BIL	0.077	0.074	0.127	0.106	0.133	0.081	0.086
BIM	0.090	0.171	0.339	0.288	0.095	0.116	0.194
INT	0.130	0.056	0.091	0.113	0.129	0.056	0.004
OOA	0.110	0.027	0.009	0.006	0.015	0.040	0.110
DWO	0.035	0.031	0.078	0.018	0.011	0.006	0.004
BMB	0.163	0.099	0.015	0.093	0.156	0.100	0.007
INS	0.008	0.013	0.033	0.016	0.006	0.001	0.005
KLS	0.028	0.028	0.001	0.035	0.000	0.023	0.006
DLO	0.173	0.090	0.171	0.162	0.167	0.191	0.117
LL6	0.019	0.027	0.088	0.025	0.006	0.010	0.108
ADV	0.613	0.683	0.646	0.633	0.711	0.684	0.580
KGS	0.308	0.293	0.389	0.292	0.317	0.417	0.447
PRE	0.576	0.635	0.723	0.632	0.684	0.611	0.691
TON	0.666	0.758	0.783	0.768	0.824	0.799	0.727
AOS	0.093	0.084	0.185	0.253	0.149	0.295	0.278
LLS	0.062	0.165	0.127	0.189	0.328	0.299	0.309
EXT	0.118	0.340	0.392	0.375	0.412	0.479	0.445
EIN	0.617	0.668	0.779	0.616	0.766	0.713	0.762
SEX	0.026	0.011	0.000	0.001	0.002	0.001	0.009
Eigenvalue	0.180	0.185	0.217	0.202	0.222	0.216	0.225

EXT have larger discrimination measures for girls; the differences, however, are far from dramatic.

A number of HOMALS solutions were computed for each of the categories of BVA (occupational level of father) separately. Table 13.3c gives the discrimination measures. On the whole, we note that variables characterizing

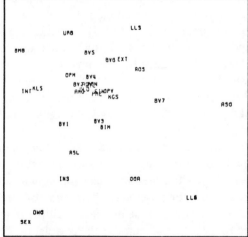

Figure 13.3c As years go by. Results of correspondence analysis on Table 13.3c. Upper plot: dimensions 1 and 2; lower plot: dimensions 1 and 3.

secondary education (AOS, LLS, EXT) have increasing discrimination measures with increasing levels of BVA. Rather large differences are found also in OPV, ASO, INT and BMB. In order to capture Table 13.3c in a graphical display, correspondence analysis was applied to the table (see Figure 13.3c). The figure shows that variables with large discrimination measures for all professional categories (EIN, TON, PRE, and ADV) are located in the centre of the plot, whereas variables for which the discrimination measure varies a lot have eccentric location. Note that the professional categories in the plot move away from the centre towards the variable with the largest discrimination for that category: BV7 has moved towards ASO and LL6, BV3 towards BIM and INS, BV1 towards ASL, etc.

Figure 13.3c illustrates how information in a large table (Table 13.3c) can be summarized in a picture. One should be cautious, however, about the interpretation. Discrimination measures are squared correlations between variables and the first HOMALS dimension, and changes in correlation may turn up for a variety of reasons. One of them is *restriction of range*. Division of data into subgroups always entails the danger that restriction of range makes correlations smaller.

13.3.3 Various applications of MORALS

As an illustration of MORALS, Table 13.3d shows results of eight analyses, all of them with EIN as the variable to be predicted, but with increasing numbers of variables in the predictor set. The first analysis makes use only of BVA, OPV, OPM, AKG, and URB as predictors, in the second analysis the variables ASO, ASL, BIL, and BIM are added to the predictor set, etc. The idea is that the selection of variables follows the 'time factor'. Variables BVA, OPV, OPM, AKG and URB can be assessed, as it were, even before the child is born, whereas variables AOS, LLS, and EXT are available only a short time before EIN itself can be assessed.

Table 13.3d shows that the multiple correlation increases to the extent that more predictor variables are added. This, of course, is a trivial result: it could not be otherwise. More interesting is that the table also shows that for each application of MORALS the correlation between EIN and the predictor variables changes (this, of course, is not the case in linear multiple regression analysis). For example, in the first MORALS analysis the correlation between EIN and BVA equals 0.468; in the last analysis this correlation goes down to 0.195. The changes in correlations are due to the fact that each MORALS analysis gives a different quantification to all variables included in the analysis.

The table also shows that, especially in the last three analyses, predictor variables are added that correlate more highly with EIN than earlier variables. For instance, once TON enters the predictor set (and TON is highly correlated with EIN), all earlier variables become more or less irrelevant. In the table of

Table 13.3d (to be continued) As years go by. Correlations for eight MORALS analyses with EIN as the criterion variable

	Correlations							
BVA	0.468	0.455	0.450	0.450	0.436	0.411	0.216	0.195
OPV	0.451	0.439	0.416	0.416	0.376	0.376	0.201	0.088
OPM	0.329	0.326	0.307	0.304	0.297	0.306	0.168	0.070
AKG	0.177	0.178	0.175	0.175	0.174	0.165	0.014	0.064
URB	0.065	0.056	0.009	0.018	0.001	0.054	0.004	0.001
ASO		0.187	0.064	0.073	0.159	0.145	0.060	0.194
ASL		0.172	0.170	0.168	0.151	0.149	0.121	0.062
BIL		0.083	0.088	0.092	0.091	0.091	0.098	0.226
BIM		0.345	0.330	0.330	0.316	0.276	0.216	0.222
INT			0.335	0.334	0.327	0.234	0.163	0.183
OOA			0.210	0.185	0.248	0.195	0.034	0.168
DWO			0.178	0.105	0.212	0.123	0.023	0.109
BMB			0.350	0.338	0.347	0.267	0.023	0.016
INS			0.238	0.220	0.232	0.161	0.163	0.237
KLS				0.083	0.080	0.087	0.090	0.070
DLO					0.431	0.398	0.395	0.272
LL6					0.072	0.094	0.062	0.033
ADV						0.744	0.485	0.456
KGS						0.366	0.207	0.145
PRE						0.674	0.395	0.170
TON							0.880	0.709
AOS								0.750
LLS								0.806
EXT								0.850
Multiple correlation	0.557	0.630	0.680	0.682	0.739	0.853	0.917	0.983

regression weights TON then becomes the only variable with substantive weight; for most other variables the regression weight goes down. Such a result can also be expected in linear regression analysis (contrary to Gresham's law we could say: a good predictor drives out bad predictors).

Typical for MORALS is that the correlations between other variables and EIN go down once TON is added to the predictor set (in linear analysis such correlations do not change, of course).

Table 13.3d (continued) As years go by. Regression weights for eight MORALS analyses with EIN as the criterion variable

	Regression weights							
BVA	0.285	0.236	0.228	0.229	0.222	0.122	0.107	0.069
OPV	0.235	0.218	0.163	0.161	0.135	0.082	0.038	0.015
OPM	0.152	0.133	0.112	0.110	0.102	0.049	0.033	0.019
AKG	0.127	0.123	0.099	0.100	0.087	0.072	0.038	0.013
URB	0.021	0.028	0.027	0.031	0.028	0.020	0.009	0.009
ASO		0.178	0.036	0.035	0.035	0.030	0.041	0.048
ASL		0.092	0.079	0.078	0.076	0.034	0.028	0.017
BIL		0.053	0.052	0.056	0.056	0.061	0.043	0.063
BIM		0.219	0.190	0.190	0.160	0.066	0.031	0.077
INT			0.210	0.210	0.201	0.113	0.105	0.092
OOA			0.049	0.048	0.055	0.033	0.041	0.034
DWO			0.067	0.038	0.115	0.042	0.031	0.076
BMB			0.097	0.090	0.084	0.024	0.013	0.004
INS			0.233	0.203	0.205	0.166	0.134	0.220
KLS				0.044	0.043	0.015	0.026	0.003
DLO					0.323	0.132	0.071	0.021
LL6					0.073	0.023	0.027	0.020
ADV						0.419	0.105	0.371
KGS						0.086	0.033	0.026
PRE						0.189	0.069	0.018
TON							0.726	0.066
AOS								0.254
LLS								0.291
EXT								0.208
Multiple correlation	0.557	0.630	0.680	0.682	0.739	0.853	0.917	0.983

The table of regression weights shows in addition that TON's dominant position is immediately undermined when the even better predictors AOS, LLS, and EXT are added to the predictor set: TON's regression weight goes down from 0.726 to 0.066 (TON's correlation does not go down that drastically). Again, such a result is not typical for MORALS. We might find the same thing in linear regression analysis, where regression weights will also become very unstable if we add new predictors correlated with old ones.

Table 13.3e (to be continued) As years go by. Correlations for nine MORALS analyses with SEX as the criterion variable

	Correlations								
BVA	0.143	0.134	0.134	0.134	0.133	0.133	0.131	0.114	0.114
OPV	0.106	0.108	0.106	0.106	0.105	0.104	0.102	0.102	0.099
OPM	0.090	0.085	0.085	0.085	0.085	0.081	0.082	0.080	0.081
AKG	0.061	0.055	0.052	0.052	0.050	0.050	0.051	0.054	0.052
URB	0.042	0.038	0.031	0.028	0.037	0.041	0.041	0.042	0.042
ASO		0.256	0.264	0.264	0.263	0.263	0.263	0.263	0.262
ASL		0.056	0.061	0.060	0.061	0.060	0.061	0.062	0.061
BIL		0.218	0.222	0.222	0.223	0.223	0.219	0.200	0.199
BIM		0.211	0.212	0.212	0.211	0.212	0.213	0.210	0.209
INT			0.192	0.193	0.192	0.190	0.191	0.183	0.182
OOA			0.135	0.134	0.100	0.123	0.067	0.080	0.150
DWO			0.190	0.190	0.190	0.186	0.190	0.189	0.184
BMB			0.220	0.220	0.220	0.215	0.220	0.216	0.213
INS			0.161	0.163	0.151	0.134	0.150	0.086	0.072
KLS				0.008	0.008	0.008	0.008	0.008	0.008
DLO					0.088	0.089	0.089	0.088	0.088
LL6					0.075	0.072	0.074	0.072	0.070
AD						0.144	0.140	0.114	0.111
KGS						0.135	0.134	0.134	0.134
PRE						0.035	0.031	0.010	0.001
TON							0.099	0.056	0.055
AOS								0.148	0.151
LLS								0.376	0.375
EXT								0.200	0.202
EIN									0.176
Multiple correlation	0.213	0.426	0.483	0.484	0.492	0.521	0.532	0.634	0.650

As a further example, the same type of successive MORALS analyses has been applied to SEX as the criterion variable (and EIN as a ninth addition to the predictor set). Results are shown in Table 13.3e. Now results are much more stable. The reason is that the new variables added in each successive analysis do not have typically larger correlations with SEX.

Table 13.3e (continued) As years go by. Regression weights for nine MORALS analyses with SEX as the criterion variable

	Regression weights								
BVA	0.141	0.129	0.130	0.130	0.132	0.123	0.118	0.115	0.126
OPV	0.108	0.103	0.104	0.103	0.102	0.102	0.103	0.101	0.102
OPM	0.089	0.073	0.070	0.072	0.065	0.052	0.055	0.060	0.060
AKG	0.061	0.061	0.062	0.062	0.067	0.071	0.066	0.057	0.058
URB	0.038	0.042	0.031	0.031	0.037	0.042	0.041	0.054	0.058
ASO		0.258	0.282	0.283	0.278	0.242	0.241	0.226	0.209
ASL		0.085	0.087	0.089	0.091	0.084	0.075	0.087	0.080
BIL		0.174	0.168	0.169	0.169	0.161	0.160	0.145	0.143
BIM		0.160	0.155	0.157	0.150	0.148	0.152	0.129	0.120
INT			0.193	0.194	0.191	0.184	0.182	0.170	0.168
OOA			0.083	0.083	0.063	0.046	0.056	0.049	0.054
DWO			0.275	0.279	0.272	0.244	0.284	0.306	0.282
BMB			0.242	0.242	0.226	0.196	0.252	0.262	0.231
INS			0.110	0.112	0.106	0.088	0.105	0.065	0.066
KLS				0.021	0.020	0.031	0.030	0.043	0.042
DLO					0.069	0.079	0.073	0.084	0.086
LL6					0.069	0.046	0.047	0.039	0.037
AD						0.131	0.156	0.109	0.114
KGS						0.138	0.126	0.118	0.112
PRE						0.052	0.026	0.008	0.012
TON							0.137	0.054	0.042
AOS								0.153	0.165
LLS								0.331	0.326
EXT								0.096	0.086
EIN									0.189
Multiple correlation	0.213	0.426	0.483	0.484	0.492	0.521	0.532	0.634	0.650

Finally, MORALS has been applied to girls and boys separately. Results are summarized in Table 13.3f (with EIN as the criterion variable, and all other variables in the predictor set). Clearly, for girls, variables BVA, OPV, INT, OOA, DLO, and KGS are more important than for boys. Interpretation of such results is again ambiguous, mainly because changes in correlations can depend upon many different things (such as restriction of range). Table 13.3f, therefore, invites us to go back to the data with new hypotheses as to what the data might reveal. Why are BVA, OPV, INT, OOA, DLO and KGS more important for girls? Table 13.3f does not give an answer to this question;

it sends us back to scrutinize the data. In fact, that is what usually happens in exploratory MVA: we do not get ready-made answers to questions we never posed, but we are made aware of peculiarities of the data that need further investigation.

Table 13.3f As years go by. MORALS on boys and girls separately, with EIN as the criterion variable

| | Correlations | | Weights | |
	Girls	Boys	Girls	Boys
BVA	0.14	0.02	0.08	0.08
OPV	0.34	0.24	0.10	0.03
OPM	0.20	0.26	0.04	0.04
AKG	0.16	0.10	0.05	0.05
URB	0.01	0.00	0.03	0.01
ASO	0.17	0.06	0.03	0.04
ASL	0.09	0.12	0.05	0.04
BIL	0.14	0.17	0.07	0.05
BIM	0.21	0.28	0.03	0.08
INT	0.26	0.08	0.10	0.07
OOA	0.14	0.02	0.03	0.02
DWO	0.13	0.21	0.05	0.06
BMB	0.09	0.10	0.00	0.01
KLS	0.10	0.09	0.00	0.01
DLO	0.46	0.37	0.08	0.04
LL6	0.08	0.02	0.02	0.03
ADV	0.65	0.62	0.19	0.25
KGS	0.30	0.13	0.05	0.05
PRE	0.48	0.40	0.09	0.09
TON	0.89	0.89	0.38	0.28
AOS	0.57	0.77	0.02	0.15
LLS	0.68	0.74	0.25	0.20
EXT	0.77	0.82	0.10	0.14
INS	0.17	0.25	0.11	0.16
Multiple correlation	0.96	0.96		

13.3.4 Comparing HOMALS and PRINCALS analyses

A one-dimensional PRINCALS analysis on 25 variables (SEX excluded) has been performed under both ordinal and linear restrictions. Table 13.3g gives the results in terms of squared correlations between rescaled variables and the first dimension of PRINCALS. The linear restriction implies that the results are the same as in classical PCA. The corresponding correlations then are classical component loadings. Ordinal PRINCALS allows for an ordinal rescaling of the categories, and under this restriction the first eigenvalue of the correlation matrix is maximized. The HOMALS solution in Table 13.3g allows for any optimal scaling of the categories; so the same first eigenvalue is now maximized without restrictions. Obviously, the first eigenvalue will increase to the extent that fewer restrictions are imposed and we find for PRINCALS linear an eigenvalue of 5.36, for PRINCALS ordinal an eigenvalue of 5.85, and for HOMALS an eigenvalue of 6.05.

Table 13.3g As years go by. PRINCALS squared component loadings (single fit) and HOMALS discrimination measure

	PRINCALS		HOMALS
	Numerical	Ordinal	multiple
BVA	0.56	0.63	0.63
OPV	0.63	0.65	0.65
OPM	0.49	0.51	0.51
AKG	0.23	0.24	0.24
URB	0.17	0.18	0.18
ASO	0.19	0.22	0.22
ASL	0.14	0.25	0.27
BIL	0.20	0.25	0.24
BIM	0.40	0.44	0.43
INT	0.39	0.35	0.35
OOA	0.22	0.22	0.22
DWO	0.09	0.09	0.09
BMB	0.44	0.42	0.41
INS	0.07	0.05	0.05
KLS	0.14	0.15	0.15
DLO	0.42	0.39	0.38
LL6	0.05	0.10	0.12
ADV	0.77	0.81	0.81
KGS	0.65	0.65	0.65
PRE	0.80	0.81	0.80
TON	0.86	0.89	0.89
AOS	0.42	0.45	0.46
LLS	0.32	0.35	0.37
EXT	0.30	0.42	0.59
EIN	0.84	0.86	0.86

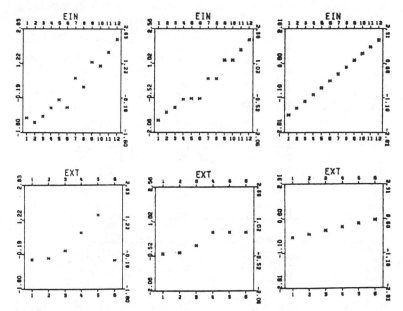

Figure 13.3d As years go by. Three different transformations for variables EIN and EXT.

Figure 13.3d compares the transformations of two selected variables, EIN and EXT, for the three different analyses. Obviously, HOMALS recognizes the special position of category 6 in EXT; PRINCALS ordinal has to compromise, and gives ties for the categories 4, 5, and 6; in PRINCALS linear, on the other hand, the fact that category 6 is special is not recognized at all. As to EIN, HOMALS shows the typical reversals, PRINCALS ordinal flattens them, and PRINCALS numerical is not allowed to act on them. For EXT the HOMALS solution is better, for the categories of EXT were a bit muddled up to start with. Therefore PRINCALS results for EXT are unduly at a disadvantage. As to EIN, we remarked in Section 13.3.1 that the HOMALS quantification may be affected by the unequal marginal frequencies; PRINCALS ordinal corrects for that.

13.3.5 Bootstrapping one-dimensional HOMALS

In the bootstrap a number of data matrices are created, which have the same dimensions as the original data matrix, by taking random samples of size n,

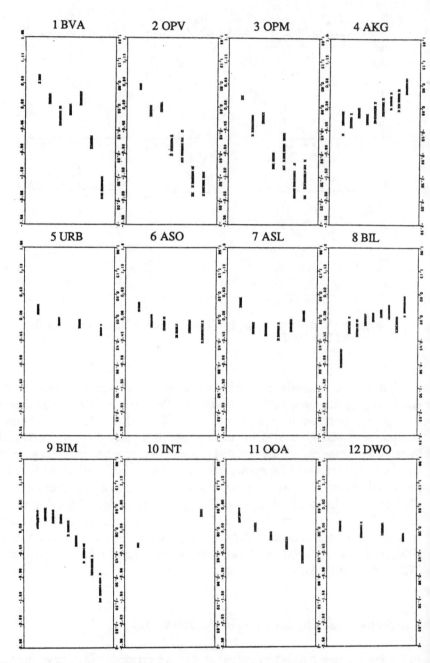

Figure 13.3e As years go by. HOMALS bootstrap for categories.

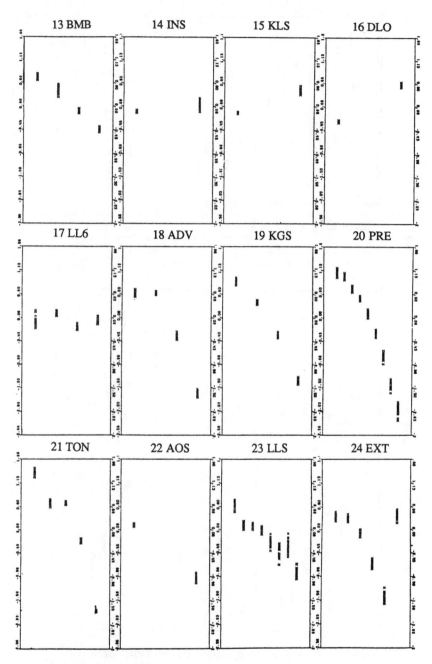

Figure 13.3e continued.

25 EIN

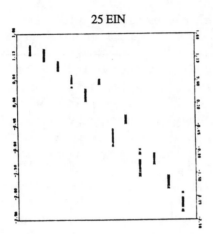

Figure 13.3e continued.

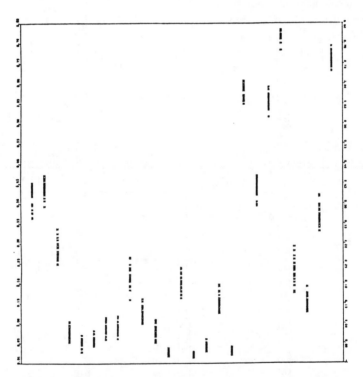

Figure 13.3f As years go by. HOMALS bootstrap for discrimination measures.

with replacement, from the rows. In the present study $n = 1845$ and 25 new data matrices were analyzed with exactly the same options as in Section 13.3.4. Figure 13.3e gives the category quantifications of 25 bootstrap samples plotted together. Except for SEX and reflection of the vertical axis, this figure can be compared with Figure 13.3a. We see, on the whole, that the bootstrap quantifications have a smaller dispersion for variables with a large discrimination measure (e.g. TON). However, there is also a connection with marginal frequency: OPV-1 with a marginal frequency of 761 has a small dispersion, whereas OPV-6 with a marginal frequency of 56 has a large dispersion.

Bootstrap results confirm the reversal of categories 3 and 5 in BVA. They also show that in OPV the categories 2 and 3, or 4 and 5, or 6 and 7 might as well be merged. Furthermore, we notice that the deviating quantifications for categories 1 and 8 of BIL are stable.

Figure 13.3f shows the bootstrap results for the discrimination measures. The figure suggests an association between the size of the discrimination measure and its stability. When a discrimination measure is small, it is very stable. Large discrimination measures tend to be more stable than intermediate discrimination measures, such as for OPV or AOS.

13.4 PARLIAMENT SURVEY

In 1968 and in 1972 members of the Second Chamber of the Dutch Parliament were asked to cooperate in an extensive questionnaire study. The data were

Table 13.4a Party allegiance of respondents in 1968

Party	Number of respondents	Description of party
CPN	0	Communist
PSP	4	Pacifist-socialist
PvdA	37	Labour
D'66	7	Pragmatic-liberal
PPR	0	Radical
KVP	42	Catholic
ARP	15	Protestant
CHU	12	Protestant
VVD	17	Conservative-liberal
BP	4	Farmers
SGP	2	Reformed
GPV	1	Reformed

kindly made available by the Department of Political Science of the University of Leiden. A selection of these data will be analysed in the present section. Previous analyses of the data can be found in Daalder and Rusk (1972), De Leeuw (1973), and Daalder and Van de Geer (1977). The data that have been selected concern the preferences of the Members of Parliament for the political parties residing in parliament at that time, diverging from extreme left-wing parties to extreme right-wing parties (Table 13.4a).

In addition, the respondents expressed their position with respect to seven political issues on a nine-point scale. The issues are listed in Table 13.4b.

13.4.1 Preference rank orders, 1968

Table 13.4b The seven issues and the meaning of lowest and highest category

	DEVELOPMENT AID	
The government should spend *more money* on aid to developing countries	1...................9	The government should spend *less money* on aid to developing countries
	ABORTION	
The government should *prohibit* abortion completely	1...................9	A *woman has the right* to decide for herself about abortion
	LAW AND ORDER	
The government takes *too strong* action against public disturbances	1...................9	The government should take *stronger action* against public disturbances
	INCOME DIFFERENCES	
Income differences should *remain* as they are	1...................9	Income differences should become *much less*
	PARTICIPATION	
Only management should decide important matters in industry	1...................9	*Workers too* must participate in decisions that are important for industry
	TAXES	
Taxes should be *increased* for general welfare	1...................9	Taxes should be *decreased* so that people decide for themselves how to spend their money
	DEFENCE	
The government should insist on *shrinking* the Western armies	1...................9	The government should insist on *maintaining* strong Western armies

Preference rankings for political parties were available from 141 Members of Parliament in 1968. Their party allegiance is given in Table 13.4a. All respondents returned a complete rank-order, coded from 1 (own party) up to 12 (least preferred party).

For the present illustration it was decided to analyse the data with PRINCALS, both with numerical and with ordinal option, applied to the *transposed* data matrix.

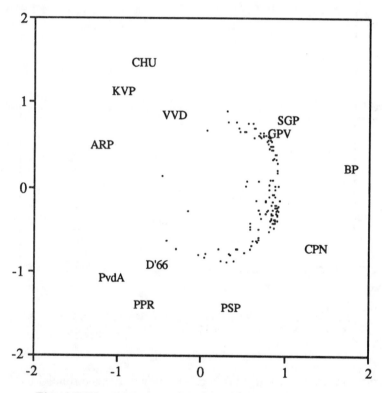

Figure 13.4a Parliament survey. Preference rank orders of 141 Members of Parliament for 12 parties. Numerical analysis.

The numerical analysis treats the rank numbers, now the columns of the data matrix, at interval scale level. Figure 13.4a shows the 12 points for the parties and the 141 points for respondents jointly. If we draw a vector through the point of a respondent and project party points on this vector, such projections will be approximately proportional to the entries of a column of the transposed data matrix. Conversely, if we draw a vector through a party point and project all 141 respondent points on it, such projections will be approximately proportional to a row of the transposed data matrix.

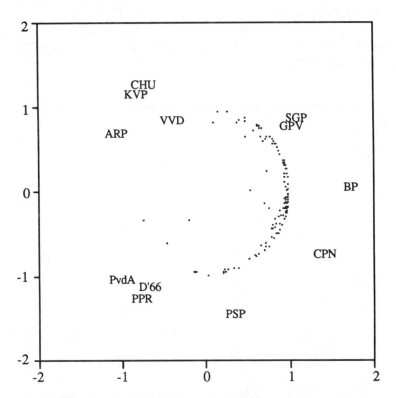

Figure 13.4b Parliament survey. Preference rank orders of 141 Members of Parliament for 12 parties. Ordinal analysis.

Figure 13.4b gives the joint configuration for the ordinal PRINCALS analysis. Since the ordinal option is less restrictive than the numerical one, the eigenvalues cannot be smaller than those for the numerical solution. For the numerical solution, the eigenvalues are $\lambda^2_1 = 0.56$ and $\lambda^2_2 = 0.24$; the corresponding eigenvalues for the ordinal solution are $\lambda^2_1 = 0.65$ and $\lambda^2_2 = 0.27$. The improvement is not impressive and, in fact, Figure 13.4b is very similar to Figure 13.4a.

Note that Figure 13.4b shows that the party points are almost located on a closed circle, whereas the respondent points show a horseshoe. This type of horseshoe should not be confused with the ones we tend to find in HOMALS. Note, for instance, that the endpoints are bending backwards. If we give parties an *a priori* ordering from left to right (as was done in Table 13.4a with the CPN at the extreme left), then the extremes on this scale are close together in the plot.

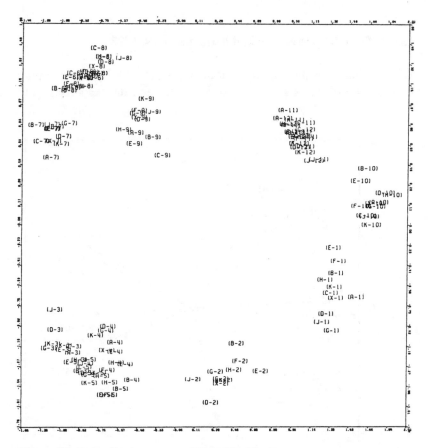

Figure 13.4c Parliament survey. PRINCALS bootstrap on party preferences.

Figure 13.4c gives the results of ten bootstrap samples analysed with the ordinal option. Note that since we transposed the data matrix we sample here from the columns. In the plot, the points are labelled from A to K (symbol I is not used), indicating the bootstrap sample. The plot is a mere juxtaposition of points, without any attempt to make the ten configurations more similar by means of rotations or scale corrections. The bootstrap clouds show some overlap between KVP and CHU, between SGP and GPV, and to some extent also between PvdA, PPR, and D'66. The two least preferred parties, CPN and BP, tend to come close together. The distinction between the denominational parties KVP, ARP and CHU on the one hand and the progressive parties PvdA, D'66 and PPR is stable.

13.4.2 Political preference and issue positions, 1972

For the data from the 1972 questionnaire, CANALS has been applied to the preference orderings as one set of variables and the responses to the issue

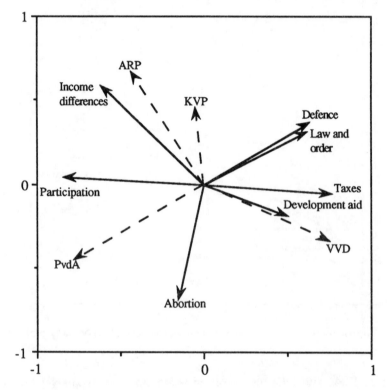

Figure 13.4d Parliament survey. Plot A. Correlations with canonical axes for issues.

statements as the second. Preference rankings were reduced to rankings for the four largest parties only (in terms of representation in the Second

Chamber: PvdA (39 members), KVP (35), VVD (16), and ARP (13)). A higher score on the preference rankings means more sympathy for that party. Although data were available from 141 Members of Parliament, three members had too many missing data and were excluded from the analysis. In this application of CANALS the preference data, although collected as row-conditional, are treated as column conditional; i.e. each of the four parties becomes a 'variable', with as many categories as the number of different rank numbers given to the party.

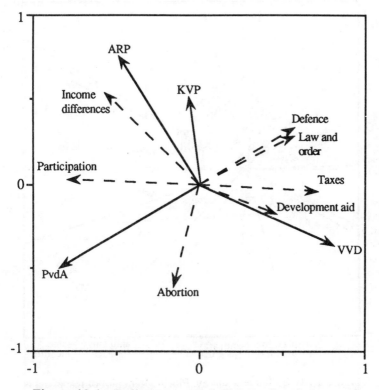

Figure 13.4e Parliament survey. Plot B. Correlations with canonical axes for preferences.

A two-dimensional CANALS with ordinal treatment of the variables produced a canonical correlation of 1.00 for the first dimension. Such a result (not uncommon in CANALS) usually means that some object has an atypical response pattern in both sets, which will create a separate dimension with perfect canonical correlation. This was also found to be the case in the present example: one member had the unique response pattern of combining very low sympathy for KVP with a missing score on LAW AND ORDER. It was therefore decided to repeat CANALS with this member left out. The two-dimensional solution now produced canonical correlations of 0.92 and 0.90, respectively. Figure 13.4d gives one possible plot of the results, with the variables as vectors, projected onto the plane of the canonical variables for

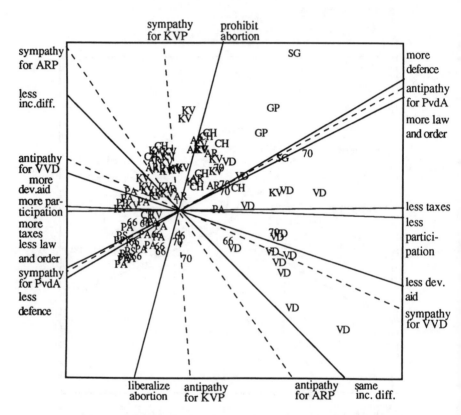

Figure 13.4f Parliament survey. Plot A: Joint plot of respondents and issue variables in the issue space. Legend: AR = ARP, CH = CHU, GP = GPV, KV = KVP, PA = PvdA, PP = PPR, PS = PSP, SG = SGP, VD = VVD, 66 = D'66, 70 = DS70.

issues in plot A, while Figure 13.4e gives the plot projected onto the plane of the canonical variables for preferences in plot B. Since the canonical correlations are high, the two plots are very similar. As to substantive interpretation of the plots, it is useful to remember that vectors may be mirrored; e.g.

TAXES and PARTICIPATION are almost opposite vectors. Both vectors point in the direction of a high score, which for TAXES means that taxes should be decreased and for PARTICIPATION that workers must have more influence. If we reverse the direction of the vector TAXES, it will point in the direction 'taxation must be increased', and this then goes along with 'more participation of workers'.

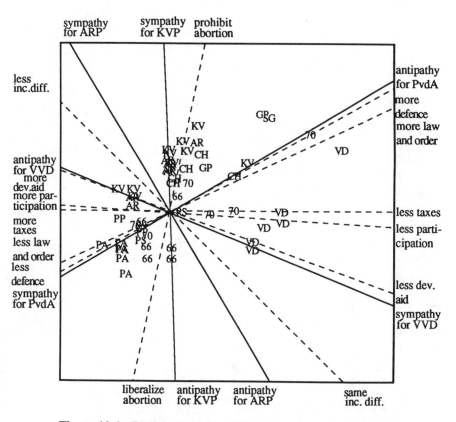

Figure 13.4g Parliament survey. Plot B: joint plot of respondents and issue-variables in the preference space. Legend: AR = ARP, CH = CHU, GP = GPV, KV = KVP, PA = PvdA, PP = PPR, PS = PSP, SG = SGP, VD = VVD, 66 = D'66, 70 = DS70.

Figures 13.4f and 13.4g give two joint plots of Members of Parliament and variables: one in the canonical plane for issues (A) and the other in the canonical plane for preferences (B). Variables are indicated as directions in the plane, labelled at the positive and the negative sides. Again, if we project the

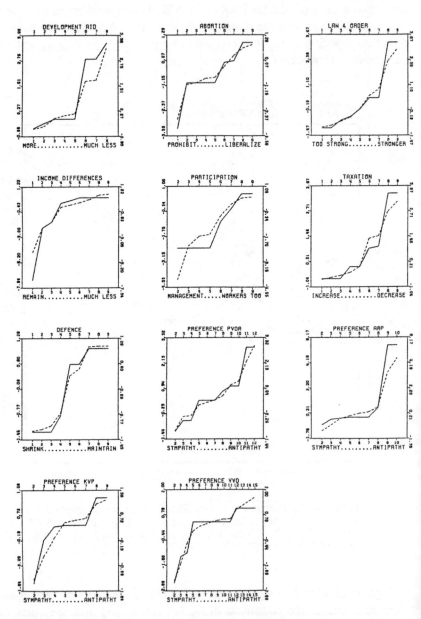

Figure 13.4h Parliament survey. Optimal scaling based on ordinal CANALS (solid lines) and averages of bootstrap results (dotted lines).

Figure 13.4i Parliament survey. Bootstrap results for ordinal CANALS.

points for members on such a direction, these projections will be approximately proportional to the quantification of the responses to that variable. The figures show, for example, that TAXES contrasts members from VVD and DS70 with members from PvdA and many from KVP; ABORTION separates members of denominational parties from the others. One should, however, remain careful with the interpretation of the figures, and realize that they always remain an approximate summary of the data; e.g. the figures strongly suggest that TAXES (decrease) and PARTICIPATION (less worker's influence) are very closely related, and indicate a characteristic position of VVD and DS70. From an inspection of the raw data, however, it could be seen that in fact VVD and DS70 have rather similar outspoken views on TAXES, whereas for PARTICIPATION VVD is on the middle of the scale (the average is close to the initial category (5)) and DS70 has a much higher average (to the direction of increased worker's influence).

For completeness it must be said that in plot A of Figure 13.4f two points for Members of Parliament have been omitted, and in plot B of Figure 13.4.g four. These are points for VVD members; their scores are so extreme (in the south-east direction) that they should be plotted far outside the figures, as shown.

Figure 13.4h gives a plot of the ordinal transformation of the category numbers of the variables. It shows that the first five categories of DEVELOPMENT AID might as well be merged, as well as categories 4 to 9 of INCOME DIFFERENCES and 2 to 5 of PARTICIPATION. As to the ordinal transformation of the preferences, the plot for ARP shows that this becomes an almost binary variable. The plots for KVP and VVD reveal degrees in sympathy more than in antipathy. Figure 13.4h also shows the average quantification for 30 bootstrap samples. These dotted curves do not deviate much from the solid curves, except for the lowest category value of PARTICIPATION. This category is infrequently chosen, and always by VVD members. The bootstrap analyses are sensitive to whether or not such response categories are sampled.

Figure 13.4i gives the dispersion of the category quantifications for the same 30 bootstrap analyses. It it striking that the dispersion is often large at one extreme of the response scales, such as category 8 of DEVELOPMENT AID, or category 1 of ABORTION, or category 10 of ARP. In all these cases marginal frequencies of the categories are low, so that the bootstrap result depends on whether or not such a category is sampled.

Figure 13.4j illustrates this further: the figure shows bootstrap variance as a function of marginal frequency, and also shows that all cases of extreme bootstrap variance are related to categories with small marginal frequency.

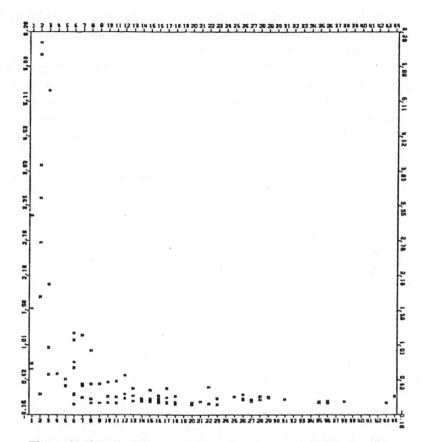

Figure 13.4j Parliament survey. Bootstrap variance plotted against marginal frequencies.

13.4.3 Relation between position on issues and party allegiance, 1972

To illustrate CRIMINALS, an analysis was performed on 119 members representing the seven largest parties in Parliament, who had no missing values. These parties are PvdA, KVP, VVD, ARP, D'66, CHU, and DS70. CRIMINALS then comes to the same as CANALS with, on the one hand, a set of seven binary variables that indicate whether or not a respondent is a member of a party, and on the other hand a set of seven variables for the issues, each with nine categories. CRIMINALS was used in two dimensions of the solution, with four different options:

(a)   *Multiple nominal*. All categories are treated as nominal; in addition, the quantification of categories for the second dimension is different from that obtained in the first dimension.

(b)   *Single nominal*. Category quantifications are the same for both dimensions. In other words, categories are optimally scaled, and the analysis gives the same results as a linear discriminant analysis on this optimally scaled data matrix, where the quantification is optimal in the sense that the sum of the first two eigenvalues is maximized.

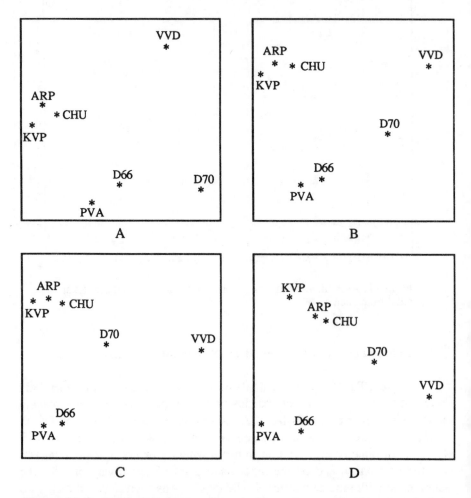

Figure 13.4k Parliament survey. Canonical group means for the seven largest parties in two-dimensional CRIMINALS.
A=multiple nominal; B=single nominal; C=single ordinal; D=numerical.

(c)   *Single ordinal.* The transformation of the variables is restricted to be
      monotonic. The same remarks as under (b) apply.
(d)   *Numerical.* The results will be identical to those of a linear discriminant
      analysis.

   Clearly, the four types of analysis, in the given order, are increasing in the
degree of constraint. It follows that the loss will also increase. The loss values
are: 0.115 for multiple nominal, 0.169 for single nominal, 0.223 for single
ordinal, and 0.304 for numerical.

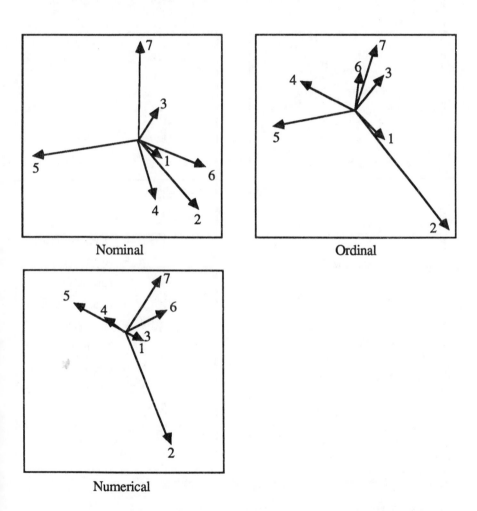

Nominal

Ordinal

Numerical

Figure 13.4l Parliament survey. Correlations between issue variables and the
canonical variables for three single CRIMINALS solutions.

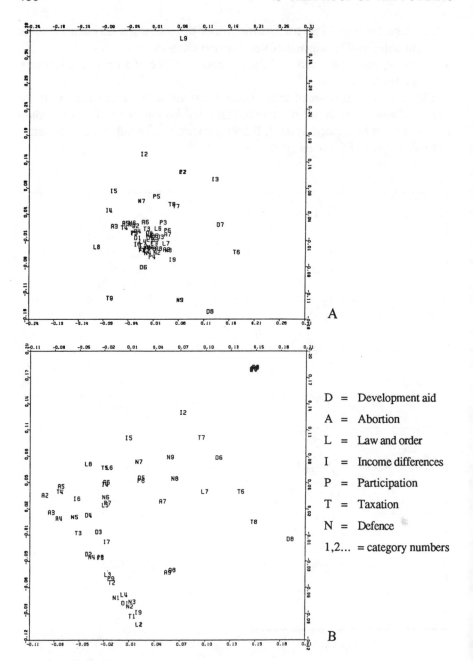

Figure 13.4m Parliament survey. Category points for
the two-dimensional multiple nominal CRIMINALS.
Plot A: category weights. Plot B: average object score
per category.

The four plots in Figure 13.4k show the canonical group means for the different solutions. The positions of the parties in the four plots are roughly comparable, except for DS70: this party moves from an eccentric position opposite to the denominational parties (multiple nominal) in a direction towards the denominational parties (single nominal), and finally comes to rest about midway between the denominational parties and the VVD (ordinal and numerical). This variability should be interpreted in relation to the fact that there is also a change in how important the different issues are for the canonical variables. Figure 13.4l shows the correlations between issues and canonical variables for the three single solutions. In the numerical analysis ABORTION is dominant, in the ordinal solution PARTICIPATION gains in influence, and in the nominal solution INCOME DIFFERENCES and TAXES play a different role.

Plot A of Figure 13.4m gives the category points for the multiple nominal CRIMINALS. Actually, these quantifications are weights for the separate columns of the indicator matrices. As usual, especially for highly collinear variables, such weights tend to behave erratically, and their interpretation is difficult. Note that, unlike in HOMALS, category quantifications are not the averages of scores of individuals within the category. Such averages are plotted in plot B of Figure 13.4m which is a plot much easier to interpret.

One should realize that multiple nominal CRIMINALS has a strong tendency to capitalize on idiosyncratic peculiarities of the data. Whenever some categories are used exclusively by individuals in the same group, CRIMINALS will select this feature for creating separate dimensions, even if the number of individuals involved is very small. In the present example there are in fact a number of categories exclusively used by VVD members. They are DEVELOPMENT AID (category 7, marginal frequency: $d=1$); LAW AND ORDER (category 9, $d=1$); INCOME DIFFERENCES (category 1, $d=2$ and category 3, $d=5$); PARTICIPATION (category 2, $d=1$, category 3, $d=1$, category 4, $d=2$, and category 5, $d=5$); TAXES (category 9, $d=1$). Also, TAXES (category 6, $d=9$ and category 8, $d=5$) and DEVELOPMENT AID (category 8, $d=3$) are used almost exclusively by VVD and DS70 members. These phenomena are the major determinants of the CRIMINALS solution. It follows that an issue that is salient in the numerical solution (such as ABORTION) becomes masked in the multiple nominal solution, because its categories do not have such idiosyncratic features. One might say that to the extent the analysis is less restrictive, it will more and more tend to highlight details that are, so to speak, at the periphery of the data.

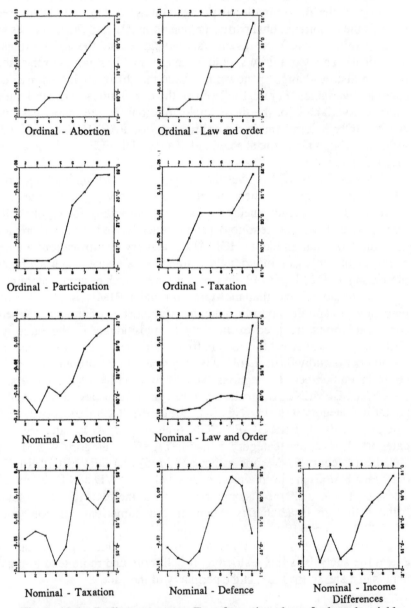

Figure 13.4n Parliament survey. Transformation plots of selected variables in two-dimensional CRIMINALS.

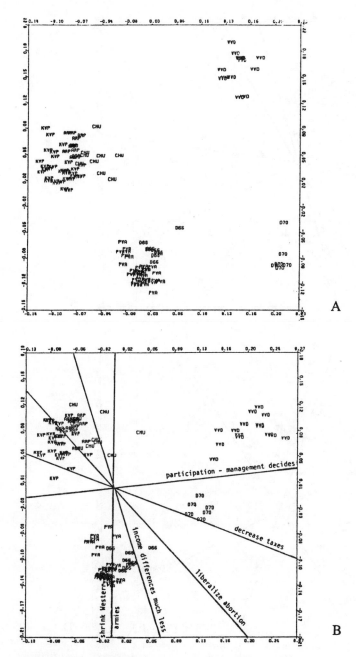

Figure 13.4o Parliament survey. Object points in two-dimensional CRIMINALS.
Plot A: multiple nominal. Plot B: single nominal.

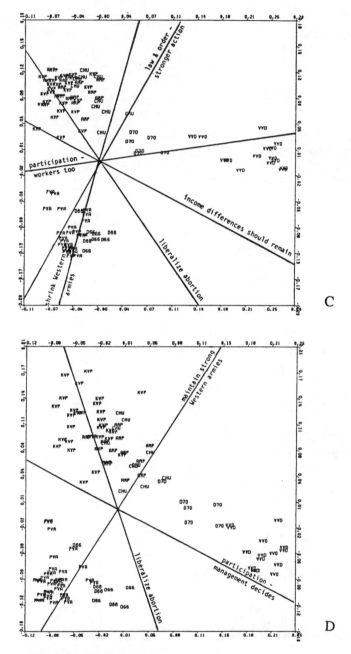

Figure 13.4o Parliament survey. Object points in two-dimensional CRIMINALS.
Plot C: single ordinal. Plot D: numerical.

Figure 13.4n gives transformation plots of selected variables for the single options. Note that the single nominal solution gives special values to category 9 of LAW AND ORDER, categories 1 and 3 of INCOME DIFFERENCES, and category 6 of TAXES – precisely those categories that also dominate in the multiple nominal solution.

Figure 13.4o shows the object points for all four CRIMINALS solutions, with for the single solutions the directions indicating for some of the more salient issues. Obviously, with an increasing amount of constraint the dispersion around the canonical group means increases. Nevertheless, in all solutions respondents from PvdA and D'66, from the denominational parties, and from VVD form strikingly homogeneous clusters.

13.5 CRIME AND FEAR

In the Department of Justice a special committee has been installed to investigate the prevention of criminality. They asked the Scientific Research and Documentation Center of the department (WODC) to make a number of surveys. We use some of the results of one of those, on judgements and feelings about the criminality question. These judgements were supposed to be relevant for understanding the priority of criminality prevention and for determining margins to the humanization of the administration of criminal law (Cozijn and Van Dijk, 1976).

The survey involved 1219 respondents, and 48 questions were asked. We used the following selection of variables (see Table 13.5a):
A. Six questions concerning judgements about the effectiveness of different methods to fight crime;
B. Four questions about feelings of helplessness and unrest;
C. Four variables with background information about the respondents.
Three individuals were left out of the analyses because they had systematic missing values and application of option (b) or (c) for missing data did not give satisfactory solutions: they gave perfect discrimination for variables with the missing values.

13.5.1 Overview of analyses

Several ways to explore relations between the variables 1–6 (CRIME), 7–10 (FEAR), and 11–14 (background variables) have been tried. Two main

Table 13.5a Description of variables

A.   Give your judgement about the effectiveness to fight crime of the following
     methods (we shall call these variables CRIME henceforth):
     1. Reeducation of criminals   (EDUC)
     2. Social work, rehabilitation   (SOWO)
     3. Better employment for criminals   (EMPL)
     4. Locking up of criminals   (LOCK)
     5. More severe punishment   (PUNI)
     6. Labour-camps   (CAMP)
     Methods 1, 2, and 3 were supposed to be 'social preventive' methods (we shall
     label them SOCIAL); methods 4, 5, and 6 were supposed to be the 'penal law'
     approach (PENAL). Answer categories are 1 = 'very ineffective', 2 =
     'ineffective', 3 = 'neither ineffective nor effective' and 'don't know', 4 =
     'effective', 5 = 'very effective'.

B.   Feelings of helplessness and unrest (these variables are henceforth called
     FEAR):
     1. You have to watch out when you walk in the city   (CITY)
     2. It is unwise nowadays to go outdoors at night   (DARK)
     3. You cannot even rely on the police any more   (POLI)
     4. When something happens to you in the street
         you cannot expect aid from someone   (AID)
     Answer categories are 1 = 'agree (completely)', 2 = 'neither agree nor disagree'
     and 'don't know', 3 = 'disagree (completely)'.

C.   Background information.

C1. Religion (RELI). Categories were 1 = 'Calvinist' (CALV), 2 = 'Protestant'
     (PROT), 3 = 'Catholic' (CATHO), 4 = 'other religions' (OTHER), 5 = 'no
     religion' (NOREL).

C2. Voting behaviour (VOTE). Categories were 1 = 'Labour Party' (LABOR),
     2 = 'Conservative-liberal Party' (CONSE), 3 = 'Catholic Denominational
     Party' (CATHP), 4 = 'Protestant Party' (PROTP), 5 = 'Protestant Party'
     (PROTP), 6 = 'Radical Party' (RADIC), 7 = 'abstention' (ABST), 8 = 'do
     not know' (DOKN), 9 = 'do not vote' (DOVO), 10 = 'other parties'
     (OTHEP).

C3. Occupational status and sex were combined into an interactive variable (see
     Section 2.8). The first seven categories apply to males, the last seven to
     females:
     1. Higher ranked employees (and comparable) (HIGHM)   8. (HIGHF)
     2. Middle ranked employees   (MIDDM)   9. (MIDDF)
     3. Small trades people   (TRADM)   10. (TRADF)
     4. Lower ranked employees   (LOWM)   11. (LOWF)
     5. Skilled workers   (SKILM)   12. (SKILF)
     6. Unskilled workers   (UNSKM)   13. (UNSKF)
     7. No profession   (NOPRM)   14. (NOPRF)

C4. Age: 1 = '16–17 years', 2 = '18–24 years', 3 = '25–34 years', 4 = '35–49 years',
     5 = '50–64 years', 6 = '65–70 years'.

approaches can be distinguished: non-linear analyses with HOMALS, PRINCALS, and CANALS, and linear analyses on the basis of one-dimensional HOMALS solutions (applications of HOMALS as a first step) – principal components analysis (PCA) and canonical correlation analysis (CCA). Within these approaches several ways to treat the background variables are discussed. They are taken as multiple nominal variables in HOMALS and PRINCALS, as single nominal variables in PRINCALS and CANALS, and a comparison has been made between them having an active or a passive role in the analysis. By active we mean that they have the same status in the analysis as the CRIME and FEAR variables. To say they have been treated as passive variables amounts to quantifying their categories afterwards on the basis of the object scores obtained from the analysis of the CRIME and FEAR variables only. Optimal ('active') and non-optimal ('passive') quantifications are used to obtain transformed variables in order to apply linear MVA afterwards.

13.5.2 Multiple join solutions with all variables

HOMALS has been applied in the first place to all variables. Transformations for CRIME and FEAR categories are monotone increasing in the first dimension. For ease of interpretation category quantifications have been split up to make four plots. Plot A of Figure 13.5a gives category points for CRIME variables and RELI plus VOTE, plot B of Figure 13.5.b gives the same points for RELI and VOTE, but now plotted with category points for FEAR variables. By inspecting positions for RELI and VOTE points, we can insert FEAR points in the plot with CRIME points. The same thing is true for Figures 13.5c and 13.5d, but now for the background variables OCCU and AGE, with the same category points for CRIME and FEAR as in Figures 13.5a and 13.5b. Starting in the centre of plot A of Figure 13.5a, we see a cluster of category 4 points. Many people think that all methods to fight crime are very effective (points labelled with 5's), with extreme opinions about PENAL as very ineffective when we arrive at the upper right corner. Going back to the centre and making the same movements but now to the upper left corner, we encounter PENAL as very effective, with at the extreme people regarding SOCIAL as very ineffective. We could interpret this configuration as representing a scale from SOCIAL '(very) ineffective' $\overset{*1}{\rightarrow}$ PENAL 'very effective' $\overset{*2}{\rightarrow}$ PENAL and SOCIAL 'effective' $\overset{*3}{\rightarrow}$ SOCIAL 'very effective' $\rightarrow$ PENAL '(very) ineffective'. The asterisks have been used to indicate the positions of the category points for FEAR variables in the plot. It should be noticed that these variables play a far less important role in the analysis.

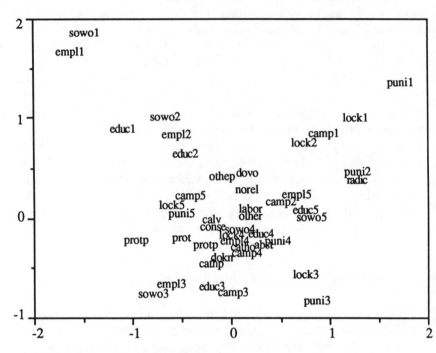

Figure 13.5a Crime and fear.
Plot A. HOMALS over all variables. RELI and VOTE plotted with CRIME.

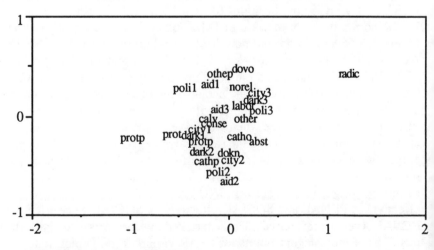

Figure 13.5b Crime and fear.
Plot B. HOMALS over all variables. RELI and VOTE plotted with FEAR.

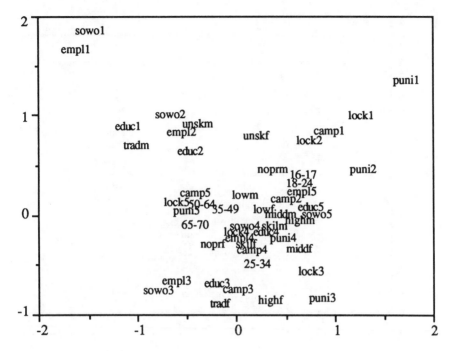

Figure 13.5c Crime and fear.
Plot A. HOMALS over all variables. OCCU and AGE plotted with CRIME.

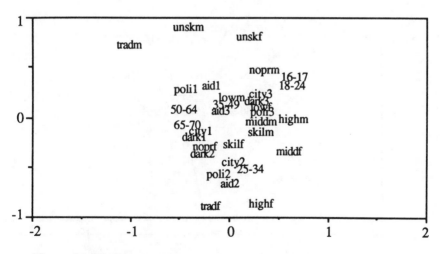

Figure 13.5d Crime and fear.
Plot B. HOMALS over all variables. OCCU and AGE plotted with FEAR.

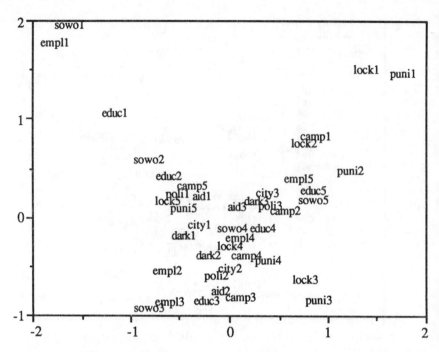

Figure 13.5e Crime and fear.
Plot A. Optimal HOMALS category quantifications for CRIME and FEAR.

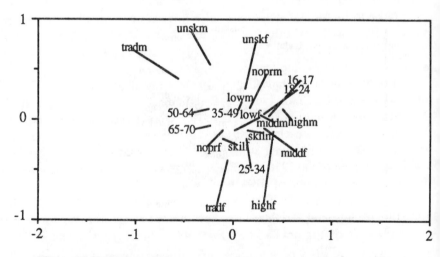

Figure 13.5f Crime and fear.
Plot B. Optimal and non-optimal HOMALS category quantifications for
OCCU and AGE.

Table 13.5b Marginal frequencies for CRIME, FEAR, and background variables

	1	2	3	4	5	6	7	8	9	10	11	12	13	14
							Categories							
EDUC	86	234	232	491	173									
SOWO	44	107	189	538	338									
EMPL	46	102	193	525	350									
LOCK	19	140	174	436	447									
PUNI	35	104	121	221	735									
CAMP	78	201	134	318	485									
CITY	505	164	547											
DARK	366	144	706											
POLI	359	266	591											
AID	292	213	711											
RELI	117	200	409	29	461									
VOTE	316	127	127	73	49	51	4	207	172	90				
OCCU	22	106	41	99	110	43	160	2	16	9	76	23	13	496
AGE	77	188	253	309	271	118								

Some remarks should be made about the background variables. Inspecting the category points for VOTE we find a very prominent position for the radical party (RADIC), close to PENAL '(very) ineffective'. Conservative-liberals (CONSE) are more to the PENAL '(very) effective' side, as are the denominational parties (PROTP) and (CATHP). The Labour Party seems to take an intermediate position, slightly closer to SOCIAL '(very) effective' and very close to FEAR '1' categories ('agree'). Category points for religion also indicate PENAL '(very) effective' for Protestants and Calvinists. For AGE we see 35–70 years PENAL '(very) effective', 16–24 years SOCIAL 'very effective' and PENAL 'ineffective'; 25–34 years are intermediate.

For occupation small tradesmen (TRADM) and unskilled males (UNSKM) are very close to SOCIAL '(very) ineffective', all kinds of employees (HIGH, MIDD, LOW) close to SOCIAL 'effective', females with no profession (most likely housewives) are close to PENAL 'effective'. The position of higher ranked females seems rather strange. Inspecting marginal frequencies gives only two women of the 1216 respondents in this category, which may account for the marginal position of HIGHF.

Plot A in Figure 13.5e gives the category points for CRIME and FEAR only; the background variables have been treated as passive. We recover the main features of Figure 13.5a (SOWO and EMPL, LOCK and PUNI '1' categories are at the extremes), but the category points for SOCIAL 'very effective' and PENAL 'ineffective' on the one side are closer together. The same applies, to an even larger extent, to their counterparts on the other side. Plot B of Figure 13.5f gives non-optimal category points, computed afterwards, for AGE and OCCU. The category points are connected with the optimal points from the previous solution. The difference between optimal and non-optimal is clear. We see points for TRADM and UNSKM move towards the centre. Those were the categories most connected with SOCIAL 'ineffective' categories. An interesting move is made by HIGHF; higher ranked females are very close now to EDUC '4', middle ranked females and corresponding categories for males.

13.5.3 Single join solutions with all variables

We continue our data analysis by looking at the linear PCA solutions for the first dimensions of the two preceding analyses. Figure 13.5g gives the component loadings on the basis of the optimal quantifications for CRIME, FEAR and background categories (eigenvalues 2.67 and 1.52). Figure 13.5h gives the loadings on the basis of non-optimal quantifications for AGE, OCCU, RELI, and VOTE (eigenvalues 2.65 and 1.47). The vectors are

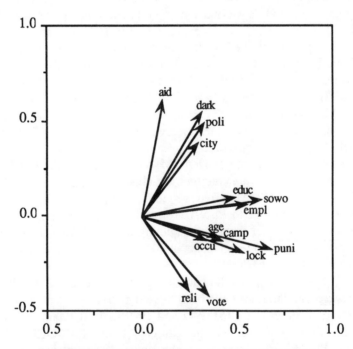

Figure 13.5g Crime and fear. PCA after HOMALS, background variables active.

plotted in one direction only (but can of course be reflected) and have the following interpretation: FEAR variables 'disagree', SOCIAL 'very effective', PENAL 'very ineffective'. The transformations of the background variables yield the following interpretations: RELI 'no religion' (versus 'Protestant'), VOTE 'radical party' (versus 'Protestant'), AGE 'young' (versus 'old'), and OCCU 'higher' and 'middle', 'male' and 'female' (versus 'no profession–female', 'unskilled–male' and 'small tradespeople–male').

In both figures we see a clear distinction between FEAR and CRIME variables. The most interesting difference between Figures 13.5g and 13.5h is the differentiation between SOCIAL and PENAL when background variables are active. With background variables passive this differentiation is more or less lost and CAMP, AGE and OCCU are more connected with SOCIAL variables.

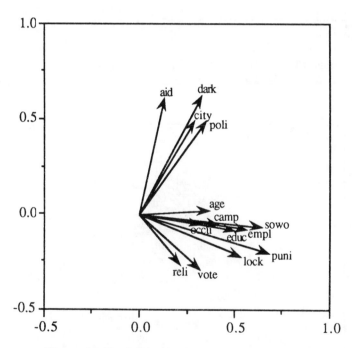

Figure 13.5h Crime and fear. PCA after HOMALS, background variables passive.

A PRINCALS analysis has been performed with single ordinal treatment of CRIME and FEAR variables and single nominal treatment of the background variables. Figure 13.5i gives the component loadings.

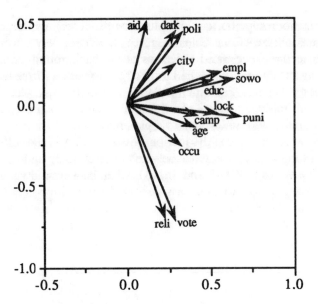

Figure 13.5i Crime and fear. PRINCALS over all variables, background variables single nominal.

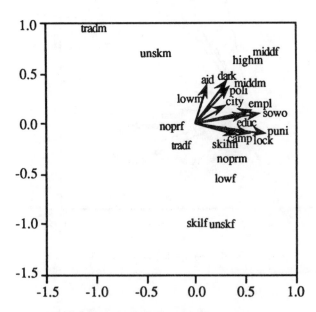

Figure 13.5j Crime and fear. PRINCALS over all variables, background variables multiple nominal.

At first sight this plot resembles the PCA plot in Figure 13.5g: a distinction between FEAR and CRIME, and a differentiation between SOCIAL and PENAL. Looking at the background variables however, some things have changed. RELI and VOTE are less related to CRIME and more to AID (and DARK, POLI, and CITY). According to the direction in which the vectors are plotted, 'no religion' and 'catholic' and the radical party by far have the highest category quantifications (see Table 13.5c in Section 13.5.6).

Another PRINCALS analysis has been performed, but now with *multiple* nominal treatment of the background variables. The component loadings are shown in Figure 13.5j, with multiple category points for OCCU. We have recovered the distinction between SOCIAL and PENAL. The rank order for OCCU categories on the first dimension is almost identical to the ordering found by HOMALS. (For completeness it should be mentioned that category point HIGHF is omitted because of its extreme position on the second dimension.)

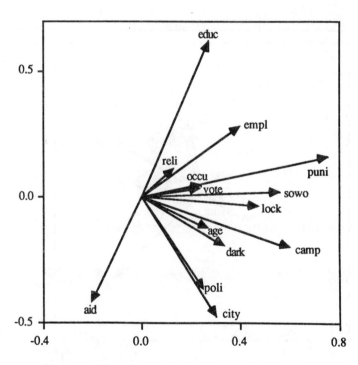

Figure 13.5k Crime and fear. CCA after HOMALS over all variables.

13.5.4 Single meet solutions over all variables

The transformed data matrices, which have been used for PCA in Section 13.5.3, have also been used for a linear canonical correlation analysis, with CRIME and FEAR in one set and the background variables (active or passive) in the other. Results are presented for CCA with non-optimal quantifications for RELI, OCCU, VOTE and AGE only, because the figures would be almost identical and canonical correlations are only slightly larger (0.383/0.159). The vectors are plotted in the plane of the canonical variables for CRIME and FEAR variables (Figure 13.5k). Very striking is the almost opposite direction of EDUC 'very effective' and AID 'disagree'. These variables have the closest connection with RELI 'Catholics' and RELI 'Protestants', respectively. OCCU (HIGHM, HIGHF, MIDDM, MIDDF) is closely connected with PUNI and LOCK 'ineffective' and SOWO 'effective'; the same applies to VOTE 'radical party'. AGE 'young' is closely related with CAMP 'ineffective' and DARK 'disagree'.

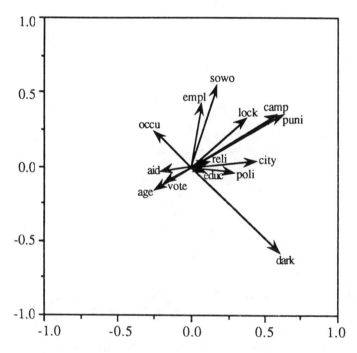

Figure 13.5l Crime and fear. CANALS over all variables.

CANALS has been applied with single ordinal treatment of CRIME and FEAR variables and single nominal treatment of the background variables. The first two dimensions of the four-dimensional solution are presented; the canonical correlations are 0.375 and 0.288 (Figure 13.5l). The most interesting feature is the long vector for DARK, perfectly correlated with OCCU. The main distinction in category quantifications is between males and females (except HIGHF and MIDDF); MIDDM, HIGHM and TRADM are at the other extreme. RELI and EDUC are opposite to AID (as in Figure 13.5k), which is pointing to the other direction compared to the other FEAR variables. AGE and VOTE have the closest connection with PENAL 'ineffective'. SOWO represents *very* effective only. All other category quantifications have a negative sign and must be thought of as lying in the opposite direction.

13.5.5 Single meet solutions with the CRIME and FEAR variables

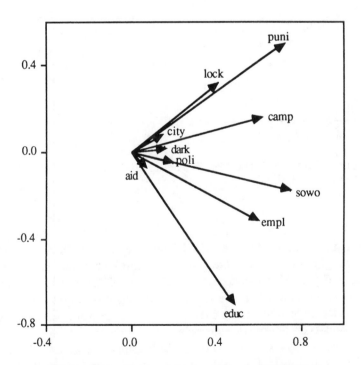

Figure 13.5m Crime and fear. CCA after HOMALS over CRIME and FEAR.

The optimally transformed CRIME and FEAR variables from the HOMALS analysis with the background variables passive have been used for a linear CCA with the CRIME variables in the first set and the FEAR variables in the second set. Results are rather disappointing for projections of the FEAR variables in the plane of the canonical variables of the CRIME set (Figure 13.5m). The canonical correlation on the second dimension is very small (0.235/0.099). CITY and DARK are pointing in the same direction as PENAL 'ineffective'; AID and POLI are more connected with SOCIAL 'effective'.

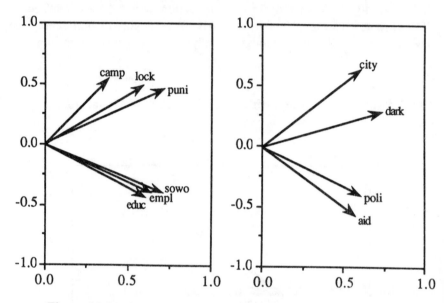

Figure 13.5n Crime and fear. PCA after HOMALS over CRIME and HOMALS over FEAR.

It seemed interesting to compare results for CCA when one-dimensional HOMALS had been applied to the CRIME variables and the FEAR variables *separately*. Single join solutions are given in Figure 13.5n; the CCA solution is presented in Figure 13.5o (canonical correlations 0.225/0.099). Comparing Figure 13.5o with Figure 13.5m we see almost identical solutions. In this case it makes no difference whether we use transformed data from *one* HOMALS analysis or use data transformed *separately*.

Finally, these analyses of two sets can be compared with the CANALS single ordinal solution in Figure 13.5p (canonical correlations 0.256 and 0.128). The relation between the CRIME and the FEAR variables shows the same pattern: AID and POLI are related with SOCIAL, DARK and CITY with PENAL. Relations between the FEAR variables no longer give perfect correlation between LOCK and PUNI; between CRIME variables themselves SOWO and EMPL are closer related. The rank order of the projections on the first dimension is the same as the one resulting from the linear canonical correlation analyses.

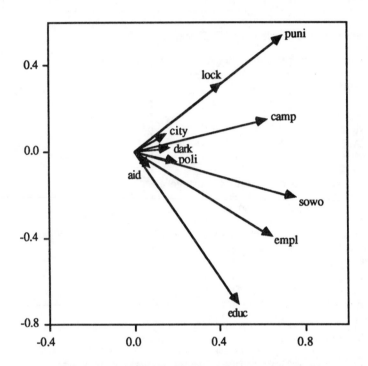

Figure 13.5o Crime and fear. CCA after HOMALS over CRIME and HOMALS over FEAR.

13.5.6 Rank orders for the categories of the background variables

In Tables 13.5c to 13.5f rank orders of the categories, obtained from the different analyses, are given. For multiple join HOMALS solutions rank orders on the first dimension have been taken; the same applies to PRINCMULT, which means PRINCALS results with background variables multiple nominal (PRINCSING: the latter variables single nominal). HOMACT means optimal HOMALS quantifications and HOMPAS non-optimal quantifications.

Table 13.5c Rank orders for VOTE categories

	PROT1	PROT2	CATHP	CONSE	LABOR	RADIC
HOMPAS	1	4	3	2	5	6
HOMACT	1	2	3	4	5	6
PRINCMULT	1	2	3	4	5	6
PRINCSING	2	1	4	3	5	6
CANALS	1	2	4	3	5	6

Table 13.5d Rank orders for RELI categories

	CALV	PROT	OTHER	CATHO	NOREL
HOMPAS	2	1	5	4	3
HOMACT	2	1	5	3	4
PRINCMULT	2	1	5	3	4
PRINCSING	1	3	2	4	5
CANALS	3	4	1	2	5

Table 13.5e Rank orders for AGE categories

	65–70	50–64	35–49	25–34	18–24	16–17
HOMPAS	2	1	3	4	5	6
HOMACT	2	1	3	4	5	6
PRINCMULT	2	1	3	4	5	6
PRINCSING	2	1	3	4	5	6
CANALS	2	1	3	4	5	6

Table 13.5f Rank orders for OCCU categories

	HIGHM	MIDDM	BUSIM	LOWM	SKILM	UNSKM	NOPRM
HOMPAS	14	11	1	7	9	2	10
HOMACT	13	12	1	6	10	2	11
PRINCMULT	13	12	1	6	9	2	10
PRINCSING	8	7	1	5	11	3	10
CANALS	12	14	13	8	9	11	10

	HIGHF	MIDDF	BUSIF	LOWF	SKILF	UNSKF	NOPRF
HOMPAS	13	12	5	6	4	8	3
HOMACT	9	14	4	8	5	7	3
PRINCMULT	11	14	4	8	5	7	3
PRINCSING	2	9	6	12	13	14	4
CANALS	7	6	2	3	1	4	5

Underlined rank orders indicate that these categories should be thought of as lying in the direction of the background vectors plotted in Figures 13.5g to 13.5p.

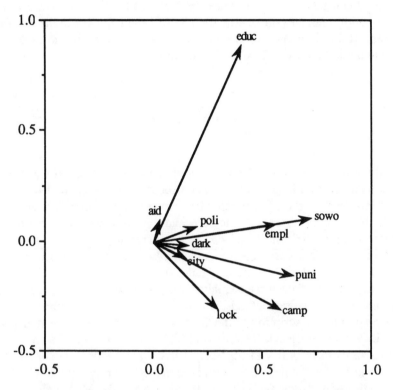

Figure 13.5p Crime and fear. CANALS over CRIME and FEAR single ordinal.

In Table 13.5c PRINCMULT gives the same rank orders as HOMACT; HOMPAS changes rank numbers for PROT2 and CONSE. Results for LABOR, RADIC, CATHP and PROT1 are rather stable.

The rank orders for the RELI categories (Table 13.5d) CALV and PROT are very similar. CANALS gives rather deviating rank numbers, especially for PROT, but also for OTHER and CATHO. Rank orders for AGE are stable over all analyses (Table 13.5e).

In Table 13.5f we see striking differences for TRADM, UNSKM, MIDDF, LOWF, SKILF, and UNSKF if we compare HOMPAS, HOMACT, and PRINCMULT (multiple quantifications) with CANALS (single quantifications). In relation with solutions where OCCU is connected with the CRIME variables, there is not much difference between males and females in general. Between males and females with the same occupational status the largest difference is between the 'no profession' categories and between the 'unskilled' categories. Within males TRADM and UNSKM have a deviant opinion, within females this is true for TRADF and NOPRF. In the CANALS solution OCCU is related to DARK: males are not afraid, and neither are higher and middle ranked females, but all other women are.

13.6 EPILOGUE

Special considerations that arise in the application of homogeneity analysis to multiple choice data in general are also discussed in Nishisato (1980, ch. 5). Further analyses of the controversial issue data can be found in Kiers (1989, ch. 10), who applies three-way methods and reports a cross-validation study. Van der Burg (1988, ch. 7 and ch. 8) has performed various OVERALS analyses on the 'As years go by' data, including a post-hoc random permutation study and a comparison of the bootstrap and the jackknife. The Parliament survey data were reanalysed by Meulman (1986) with various multidimensional scaling methods, which yield better distance representations. A more complete description of the CANALS bootstrap on these data can be found in Van der Burg and De Leeuw (1983), and the results of a redundancy analysis are described in Van der Burg (1988, ch. 3). A data set similar to the 'Crime and fear' example is the victimization study analysed in Israëls (1987, ch. 8) with redundancy analysis and rotation methods. More generally, the book by Israëls (1987) contains many careful applications in the area of survey research and social statistics. Greenacre (1984, ch. 9) gives a wide variety of applications of homogeneity analysis, covering the fields of genetics, social psychology, clinical research, education, criminology, food science, linguistics, ecology, palaeontology, and meteorology.

APPENDIX A
NOTATION

Bold upper case letters refer to matrices; bold lower case letters refer to vectors. Scalars and elements of matrices are set in italics.

➺	Refers to the epilogue of the chapter in which the subject is expanded upon.
≡	'is defined as' (in contrast to = which means 'is equal to').
∝	'is proportional to'.
⊗	Tensor or Kronecker product.
I	Identity matrix: matrix with all diagonal elements equal to 1 and all off-diagonal elements equal to 0.
u	Column vector with all elements equal to 1.
A'	Transpose of matrix **A**.
A^{-1}	Inverse of matrix **A**.
A^{+}	Moore–Penrose inverse of matrix **A**.
det **A**	Determinant of matrix **A**.
n	Number of row objects.
i, k	Running index for objects: $i = 1,...,n$.
Σ_i	Abbrevation of $\sum\limits_{i=1}^{n}$.
m	Number of variables.
m^{-1}	$1/m$: division by the number of variables.
j, l	Running index for variables: $j = 1,...,m$.
k_j	Number of categories of variable j.

r Running index for categories: $r = 1,...,k_j$.

$\sum_j$ Abbrevation of $\sum_{j=1}^{m}$.

K Number of sets.

k, l Running index for sets: $k = 1,...,K$.

$\sum_k$ Abbrevation of $\sum_{k=1}^{K}$.

J_k Index set that enumerates variables within sets: $\{J_1,...,J_k,...,J_K\}$ forms a partitioning of $\{1,...,m\}$.

p Number of dimensions.

r Number of successive solutions.

s, t Running index for dimensions: $s = 1,...,p$.

$\sum_s$ Abbrevation of $\sum_{s=1}^{p}$.

p_{max} Upper bound for dimensionality, or the maximum number of possible solutions in HOMALS; $p_{max} = (\sum_j k_j) - m$.

h Arbitrary variable. The jth variable is denoted by h_j.

$\underline{h}$ Arbitrary stochastic variable.

H $n \times m$ data matrix, consisting of columns h_j .

h_j Column j of data matrix H, containing values for variables h_j.

G_j Indicator matrix of variable h_j : $g_{(j)ir} = 1$ if the ith object is mapped in the rth category of h_j and $g_{(j)ir} = 0$ if the ith object is not mapped in the rth category of h_j.

G Partitioned matrix consisting of all indicator matrices G_j, concatenated rowwise.

$\mathbb{G}$ Partitioned matrix of the same shape as G, but not necessarily consisting of indicator matrices.

G^k Partitioned matrix for set k, consisting of G_j with $j \in J_k$.

G_k Interactive indicator matrix for set k.

D_j Diagonal matrix containing the column sums of G_j : $D_j = G_j'G_j$. These are the univariate marginals of h_j.

$\mathbf{C}_{jl}$ Cross table: $\mathbf{C}_{jl} = \mathbf{G}'_j\mathbf{G}_l$, containing the bivariate marginals for variables h_j and h_l.

$\mathbf{M}_j$ Diagonal matrix of row sums of $\mathbf{G}_j$. If $\mathbf{G}_j$ is a complete indicator matrix $\mathbf{M}_j = \mathbf{I}$.

$\mathbf{M}_*$ Sum of all $\mathbf{M}_j$. If $\mathbf{G}$ is complete $\mathbf{M}_* = m\mathbf{I}$.

$\mathbf{q}_j$ Vector of length n with quantifications of objects with respect to variable h_j: $\mathbf{q}_j = \mathbf{G}_j\mathbf{y}_j$.

$\mathbf{x}$ Vector of length n with object scores: $\mathbf{x} = m^{-1}\sum_j \mathbf{q}_j$.

$\mathbf{X}$ Matrix with object scores, of order $n \times p$.

$\mathbf{y}_j$ Vector of length k_j with category quantifications for variable h_j.

$\mathbf{Y}_j$ Matrix with category quantifications, of order $k_j \times p$.

$\mathbf{Y}$ Partitioned matrix of category quantifications consisting of $\mathbf{Y}_j$ concatenated columnwise.

$\mathbf{Y}^k$ Partitioned matrix of category quantifications for set k.

σ Loss function.

σ_J Join loss function.

σ_M Meet loss function.

λ^2 Eigenvalue.

$\tilde{\mathbf{X}}$ Auxiliary $\mathbf{X}$.

$\mathbf{X}^*$ Optimal $\mathbf{X}$.

$\mathbf{X}^\#$ Suboptimal $\mathbf{X}$.

$\mathbf{a}$ Vector of m weights.

η_j^2 Discrimination measures for variable h_j.

C_j Convex cone.

$\mathbf{Z}_k$ Canonical variables, i.e. linear composites of set k: $\mathbf{Z}_k = \mathbf{H}_k\mathbf{A}_k$.

$\mathbf{z}_{(k)s}$ Linear composite scores of set k in dimension s.

$\mathbf{R}_{kl}$ Matrix with correlations between the linear composite scores of set k and set l.

$\mathbf{R}_{kk}$ Matrix with correlations between the linear composite scores of set k in all dimensions.

$\mathbf{R(Z)}$ Matrix with correlations between the linear composite scores of all sets in all dimensions.

$\mathbf{D(Z)}$ Block diagonal matrix with submatrices $\mathbf{R}_{kk} = \mathbf{Z}'_k\mathbf{Z}_k$.

$\mathbf{F}$ Matrix with nonnegative numbers (e.g., contingency table, correspondence table).

$\mathbf{D}_r$ Diagonal matrix of row totals of $\mathbf{F}$.

$\mathbf{D}_c$ Diagonal matrix of column totals of $\mathbf{F}$.

n_r Number of row objects of $\mathbf{F}$.

n_c Number of column objects of $\mathbf{F}$.

N Total number of observations, total frequency, grand total of $\mathbf{F}$.

$\delta_{(j)ik}$ Dissimilarity between row i and row k with respect to column j.

$w_{(j)ik}$ Weight of residual associated with $\delta_{(j)ik}$.

$d_{ik}(\mathbf{X})$ Euclidean distance between row i and row k of $\mathbf{X}$.

SSQ( )   If $\mathbf{X}$ is a matrix, then SSQ($\mathbf{X}$) is the sum of squares of the elements of $\mathbf{X}$.

AVE( )   If $\mathbf{X}$ is a matrix, then AVE($\mathbf{X}$) is the vector of column averages of the elements of $\mathbf{X}$. If $\underline{x}$ is a random vector, then AVE($\underline{x}$) is the expected value of $\underline{x}$.

VAR( )   if $\mathbf{X}$ is a matrix, then VAR($\mathbf{X}$) is the vector of column variances of the elements of $\mathbf{X}$. If $\underline{x}$ is a random vector, then VAR($\underline{x}$) is the vector of variances.

GRAM( )   If $\mathbf{X}$ is a matrix, then GRAM($\mathbf{X}$) is the orthogonal matrix obtained by applying Gram–Schmidt orthogonalization to $\mathbf{X}$.

APPENDIX B
MATRIX ALGEBRA

B.1 IMAGES

Let x be an n–tuple of n real valued numbers. Let $\Re^n$ be the field of all such n-tuples. An interpretation of x as a vector in space becomes possible by defining in $\Re^n$ a set of coordinate vectors. Usually one takes for these coordinate vectors the elementary vectors e_i ($i = 1,...,n$), corresponding to the column vectors of the n x n identity matrix I. Such elementary vectors satisfy $e_i'e_i = 1$ (for all i) and $e_i'e_j = 0$ (for all $i{\neq}j$), so that they form an orthogonal coordinate system, with unit-length coordinates, also called a *basis* of $\Re^n$.

The vector x is defined in space as the weighted sum

$$x = x_1e_1 + x_2e_2 + ... + x_ie_i + ... + x_ne_n.$$

An example for $n = 3$ is

$$x = \begin{pmatrix} 2 \\ 3 \\ 4 \end{pmatrix} = (2) \begin{pmatrix} 1 \\ 0 \\ 0 \end{pmatrix} + (3) \begin{pmatrix} 0 \\ 1 \\ 0 \end{pmatrix} + (4) \begin{pmatrix} 0 \\ 0 \\ 1 \end{pmatrix}.$$

Let A be an n x m matrix. $A'x$ is said to be an *image* of x. The reason for this name becomes immediately obvious by writing

$$A'x = x_1a_1 + x_2a_2 + ... + x_ia_i + ... + x_na_n,$$

which shows that $A'x$ is constructed in the same way as a weighted sum of the columns a_i of A', as x was constructed from the columns of I. The image $A'x$ of x is an m-vector, and 'pictures' the n-vector x as its image in $\Re^m$. In the same way, let y be an m-vector in $\Re^m$; then its image Ay is an n-vector in $\Re^n$.

As an example, let $n = 3$, $m = 2$, and

$$A = \begin{pmatrix} 0.5 & 0.3 \\ 0.3 & 0.1 \\ 0.4 & 0.5 \end{pmatrix}.$$

Let $\mathbf{y}$ be some vector in $\Re^2$. For simplicity we take $\mathbf{y}$ in such a way that $\mathbf{y'y} = 1$, with

$$\mathbf{y} = \begin{pmatrix} 0.200 \\ 0.980 \end{pmatrix}.$$

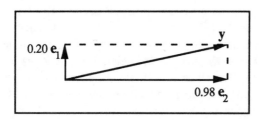

Figure B.1

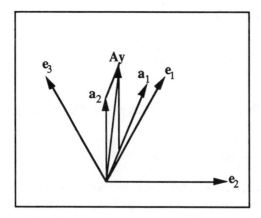

Figure B.2

Figure B.1 shows how $\mathbf{y}$ is constructed as the weighted sum $\mathbf{y} = (0.200)\mathbf{e_1} + (0.980)\mathbf{e_2}$. Figure B.2 represents $\Re^3$, with elementary vectors $\mathbf{e_1}$, $\mathbf{e_2}$, $\mathbf{e_3}$. In the figure, the two column vectors $\mathbf{a}_1$ and $\mathbf{a}_2$ of $\mathbf{A}$ have been drawn. According to the definition, $\mathbf{a}_1$ is the image of $\mathbf{e}_1$ (since $\mathbf{Ae}_1 = \mathbf{a}_1$) and $\mathbf{a}_2$ is the image of $\mathbf{e}_2$. Also

$$\mathbf{Ay} = \begin{pmatrix} 0.394 \\ 0.158 \\ 0.570 \end{pmatrix}$$

becomes the image of $\mathbf{y}$, and in Figure B.2 is constructed in the same way from $\mathbf{a}_1$ and $\mathbf{a}_2$ as $\mathbf{y}$ was constructed in Figure B.1 from $\mathbf{e}_1$ and $\mathbf{e}_2$.

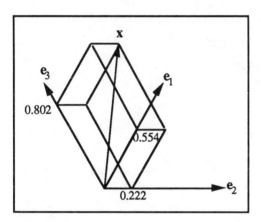

Figure B.3

We now extend $\mathbf{Ay}$ to a vector of unit length and call this extended vector $\mathbf{x}$. Since $\mathbf{y'A'Ay} = 0.505$, $\mathbf{x} = \mathbf{Ay}(0.505)^{-1/2} = \mathbf{Ay}/(0.711)$. In Figure B.3 $\mathbf{x}$ is drawn as a vector in $\Re^3$. Note that Figures B.2 and B.3 can be superimposed upon each other; both figures represent the same $\Re^3$ in the same way. In Figure B.3 $\mathbf{x}$ has the same direction as $\mathbf{Ay}$ in Figure B.2. Since

$$\mathbf{x} = \mathbf{Ay} / (0.711) = \begin{pmatrix} 0.554 \\ 0.222 \\ 0.802 \end{pmatrix},$$

the vector $\mathbf{x}$ can be constructed as the sum $0.554\mathbf{e}_1 + 0.222\mathbf{e}_2 + 0.802\mathbf{e}_3$, as shown in Figure B.3, where $\mathbf{x}$ appears as the body diagonal of a rectangular parallelepiped.

Figure B.4 shows the image $A'x = (0.664\ 0.589)'$ of x in $\mathfrak{R}^2$. Figure B.4 also shows the images $A'I$ of the elementary vectors e_1, e_2, e_3 of $\mathfrak{R}^3$. Their coordinates are given in the rows of A. In Figure B.4 it remains true that $A'x = 0.554A'e_1 + .222A'e_2 + .802A'e_3$. This construction is shown in the figure, and results in a flat image of the three-dimensional rectangular parallelepiped of Figure B.3.

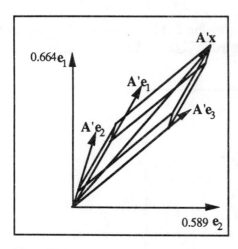

Figure B.4

Figures B.1 and B.4 can be superimposed; they portray the same $\mathfrak{R}^2$. If we do that, it is seen that $A'x$ in Figure B.4 does not have the same direction as y in Figure B.1. However, $A'x$ has the same direction as $A'Ay = (0.771)A'x$. It follows that where Ay is the image of y (going from $\mathfrak{R}^2$ to $\mathfrak{R}^3$) and $A'Ay$ the image of Ay (going backwards from $\mathfrak{R}^3$ to $\mathfrak{R}^2$), y and $A'Ay$ have not the same direction.

B.2 HYPERELLIPSOIDS

Continuing the example of Section B.1, let y be an arbitrary vector in $\mathfrak{R}^2$ with unit length ($y'y=1$). The class of all vectors of this type describes in $\mathfrak{R}^2$ a circle with unit radius. Such a circle is shown in Figure B.5 (this figure could be superimposed upon Figures B.1 or B.4). The image of this unit circle is described in $\mathfrak{R}^3$ by the set of vectors of the form Ay. It describes an ellipse, shown in Figure B.6. Note that the vectors $\mathbf{a}_1$ and $\mathbf{a}_2$ are 'spokes' (or *pseudo-radii*) of this ellipse: in Figure B.5, $\mathbf{e}_1$ and $\mathbf{e}_2$ are radii of the unit circle, and in Figure B.6, their images $\mathbf{a}_1$ and $\mathbf{a}_2$ are pseudo-radii of the ellipse. The ellipse in Figure B.6 is located in the plane spanned by $\mathbf{a}_1$ and $\mathbf{a}_2$ (a two-dimensional subspace of $\mathfrak{R}^3$). The proof that the set of vectors of the form $\mathbf{Ay}$ describes an ellipse would require showing that the procedure outlined above is equivalent to the definition of an ellipse in terms of a familiar construction rule. We omit that proof.

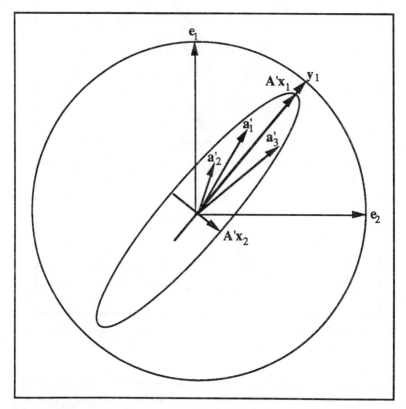

Figure B.5

Conversely, in $\mathfrak{R}^3$ the set of vectors $\mathbf{x}$, with familiar $\mathbf{x'x} = 1$, describes a sphere with unit radius. In Figure B.6 the outer contour of this sphere is drawn as a circle. The vectors $\mathbf{e}_1, \mathbf{e}_2, \mathbf{e}_3$ are located on the surface of the sphere, but appear in the figure as vectors interior to the outer contour (in the same way as cities, located on the surface of the earth, appear in a two-dimensional world map as points interior to a circle). The image in $\Re^2$ of the sphere is described by the set of vectors $\mathbf{A'x}$. It is shown in Figure B.5 as an ellipse, with the vectors $\mathbf{a}_1'$, $\mathbf{a}_2'$, and $\mathbf{a}_3'$ (the images of $\mathbf{e}_1$, $\mathbf{e}_2$, $\mathbf{e}_3$) interior to the ellipse. In fact, the ellipse is a flat map of the unit sphere of $\Re^3$.

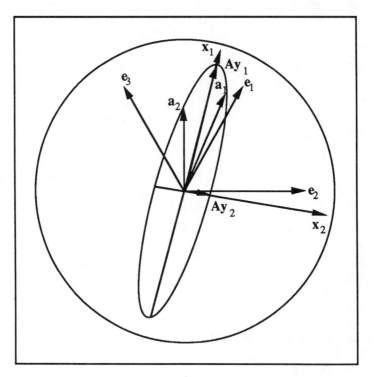

Figure B.6

Generalizing to the case where $\mathbf{A}$ is an $n \times m$ matrix, with $\mathbf{x}$ a vector of unit length in $\Re^n$ and $\mathbf{y}$ a vector of unit length in $\Re^m$, and assuming that $\mathbf{A}$ has full column rank m, with $m < n$, we will have the following results. The vector $\mathbf{x}$ describes in $\Re^n$ a *hypersphere* with unit radius. Its image in $\Re^m$ becomes a *hyperellipsoid* with dimensionality reduced to m. As a consequence vectors on the surface of the hypersphere in $\Re^n$ may be mapped as interior vectors of the hyperellipsoid in $\Re^m$. Conversely, $\mathbf{y}$ describes a hypersphere in $\Re^m$. The image $\mathbf{Ay}$ of $\mathbf{y}$ describes in $\Re^n$ a hyperellipsoid, confined to an m-dimensional subspace of $\Re^n$, and with $\mathbf{Ay}$ located on the surface of the hyperellipsoid.

Suppose now that $\mathbf{A}$ has column rank $p < m$. Its column vectors then span a p-dimensional subspace of $\mathfrak{R}^m$. It follows that $\mathbf{A}'\mathbf{x}$ will describe a hyper-ellipsoid also confined to the p-dimensional subspace of $\mathfrak{R}^m$. Conversely, the hypersphere described by $\mathbf{y}$ in $\mathfrak{R}^m$ becomes mapped as a p-dimensional hyperellipsoid described by $\mathbf{Ay}$ in $\mathfrak{R}^n$.

B.3 INVARIANT DIRECTIONS

In the example of Section B.1 it was shown that, if we take $\mathbf{x}$ proportional to the image $\mathbf{Ay}$ of $\mathbf{y}$, then the image $\mathbf{A}'\mathbf{x}$ of $\mathbf{x}$ (which must be proportional to $\mathbf{A}'\mathbf{Ay}$) is not necessarily proportional to $\mathbf{y}$ itself. However, there are special solutions $\mathbf{x}_i$ and $\mathbf{y}_i$ for which proportionality is retained, with the identical proportionality coefficient ψ_i. This implies the equations

$$\mathbf{Ay}_i = \mathbf{x}_i \psi_i , \tag{B.1a}$$

$$\mathbf{A}'\mathbf{x}_i = \mathbf{y}_i \psi_i , \tag{B.1b}$$

so that

$$\mathbf{A}'\mathbf{Ay}_i = \mathbf{y}_i \psi_i^2 , \tag{B.2a}$$

$$\mathbf{AA}'\mathbf{x}_i = \mathbf{x}_i \psi_i^2 . \tag{B.2b}$$

Such vectors $\mathbf{x}_i$ and $\mathbf{y}_i$ are said to constitute *invariant directions*. It can be shown that they correspond to the *principal axes* of the hyperellipsoids described by $\mathbf{Ay}$ and $\mathbf{A}'\mathbf{x}$.

(Why are the proportionality constants in equations B.1 identical? Suppose they were not, so that

$$\mathbf{Ay}_i = \mathbf{x}_i \lambda_i,$$

$$\mathbf{A}'\mathbf{x}_i = \mathbf{y}_i \gamma_i.$$

It then follows that

$$\mathbf{y}_i' \mathbf{A}'\mathbf{Ay}_i = \mathbf{y}_i' \mathbf{A}'\mathbf{x}_i \lambda_i = \mathbf{y}_i' \mathbf{y}_i \gamma_i \lambda_i = \gamma_i \lambda_i,$$

but also

$$y_i'A'Ay_i = x_i'x_i\lambda_i^2 = \lambda_i^2,$$

so that $\gamma_i\lambda_i = \lambda_i^2$. By a similar argument $\lambda_i\gamma_i = \gamma_i^2$. It follows that $\gamma_i = \lambda_i$.)

For the example of Sections B.1 and B.2 (Figures B.5 and B.6) let $y_1 = (0.773 \; 0.635)'$. Ay_1 is shown as a spoke of the ellipse in Figure B.6, with $Ay_1 = (0.577 \; 0.296 \; 0.626)'$. Define x_1 as Ay_1 extended to unit length (in Figure B.6, x_1 remains interior to the outer contour of the unit sphere). Since $y_1'A'Ay_1 = 0.812$, we have $x_1 = Ay_1/(0.901) = (0.640 \; 0.328 \; 0.695)'$. Its image is $A'x_1 = (0.696 \; 0.572)'$, and turns out to be proportional to y_1, with $A'x_1 = (0.901)y_1$. We therefore have a solution for invariant directions, with $\psi_1 = 0.901$. Figure B.5 illustrates that $A'x_1$ is the longest principal axis of the ellipse.

Conversely, in Figure B.6, Ay_1 is the longest principal axis of the ellipse described by Ay (but this ellipse is in the plane of a_1 and a_2, which is not the plane of drawing, so that the principal axis of the 'real' ellipse is not that of the drawn ellipse). Figure B.6 also shows the short principal axis, for $y_2 = (0.635 \; .773)'$, with $A'x_2 = (0.123 \; -.150)'$ and $\psi_2 = 0.194$.

Note that in Figure B.4 the image $A'x$ is much closer to the first principal axis than y in Figure B.1 (this could be seen by superimposing Figures B.1, B.4, and B.5). It can be shown that this result is generally valid. This at once suggests a computational algorithm for identifying the solution y_1: start with arbitrary y, find its image Ay, define x as Ay extended to unit length, find the image $A'x$, and redefine y as $A'x$ extended to unit length. Repeat this cycle, and y will become as close to y_1 as one wants. Such algorithms are described in detail in Section 3.3. They are called 'alternating' algorithms because the basic cycle has a step where x in $\Re^n$ is mapped 'backwards' in $\Re^m$.

B.4 SINGULAR VECTORS AND SINGULAR VALUES

Vectors y_i and x_i that satisfy equations (B.1) are called *singular vectors* of A. They can be collected in matrices Y and X that must satisfy

$$AY = X\Psi ,\tag{B.3a}$$

$$A'X = Y\Psi ,\tag{B.3b}$$

with $X'X = I$, $Y'Y = I$, and with Ψ a diagonal matrix with non-negative diagonal elements ψ_i, called *singular values* of A. The equations above imply

$$X'AY = \Psi, \tag{B.4}$$

where Ψ is called the *canonical form* of A. The equations (B.3) also imply

$$A = X\Psi Y', \tag{B.5}$$

which is called the *singular value decomposition* (SVD) of A.

Suppose the $n \times m$ matrix A $(n > m)$ has rank k $(k < m)$. The latter implies that there are $m-k$ independent solutions for y_i so that $Ay_i = 0$. They satisfy equation (B.1a) with $y_i = 0$. At the same time, there will be $n-k$ independent solutions $A'x_i = 0$; they satisfy equation (B.1b). On the other hand, equation (B.5) can be written as

$$A = \sum_i x_i \psi_i y_i', \tag{B.6}$$

where x_i is a column of X and y_i a column of Y. It follows that singular vectors with corresponding $\psi_i = 0$ can easily be omitted from equation (B.6), since their contribution is multiplied by the zero singular value. This in turn implies that for an $n \times m$ matrix of rank k, with $n \geq m$ and $m \geq k$, the SVD $A = X\Psi Y'$ can be redefined with X an $n \times k$ matrix, satisfying $X'X = I$, Y an $m \times k$ matrix, satisfying $Y'Y = I$, and Ψ a $k \times k$ diagonal matrix with positive diagonal elements $\psi_i > 0$.

B.5 EIGENVECTORS AND EIGENVALUES

Premultiplying the equations (B.3) with A' and A, respectively, we obtain

$$A'AY = A'X\Psi = Y\Psi^2, \tag{B.7a}$$

$$AA'X = AX\Psi = X\Psi^2. \tag{B.7b}$$

Column vectors of Y are called *eigenvectors* of the (square and symmetric) matrix $A'A$, and the diagonal elements of Ψ^2 are called their corresponding *eigenvalues*. As with singular vectors, we maintain the normalization convention $Y'Y = I$ (similarly for X). When there are multiple equal eigenvalues, the eigenvector structure is no longer uniquely defined — but even then it remains feasible to take a solution that satisfies $Y'Y = I$.

The algorithm suggested in Section B.3 remains applicable for eigenvectors. Let y be some arbitrary vector. Its image with respect to $A'A$ is $A'Ay$.

Extend $\mathbf{A'Ay}$ to unit length; call this vector $\mathbf{x}$. The image of $\mathbf{x}$ becomes $(\mathbf{A'A})'\mathbf{x} = \mathbf{A'Ax}$. However, this shows that the 'alternating' steps become in fact indistinguishable.

Obviously, once $\mathbf{Y}$ and Ψ^2 are identified for $\mathbf{A'A}$ on the basis of an eigenvector eigenvalue algorithm, $\mathbf{X}$ can be calculated afterwards using (B.3a): $\mathbf{X} = \mathbf{AY\Psi^{-1}}$, assuming that in Ψ only the p non-zero eigenvalues are retained (and in $\mathbf{Y}$ the p corresponding eigenvectors).

Given that $\mathbf{A}$ is a real-valued matrix, its singular values ψ_i also must be real-valued, and it follows that the eigenvalues ψ_i^2 of $\mathbf{A'A}$ never can be negative. $\mathbf{A'A}$ then is said to be a *positive semi-definite* (or Gramian) matrix. It may happen that a square and symmetric matrix $\mathbf{B}$ has negative eigenvalues. An example is

$$\mathbf{By} = \begin{pmatrix} 1 & 3 \\ 3 & 1 \end{pmatrix} \begin{pmatrix} 1 \\ -1 \end{pmatrix} = \begin{pmatrix} -2 \\ 2 \end{pmatrix} = \begin{pmatrix} 1 \\ -1 \end{pmatrix} (-2).$$

It follows that in such cases $\mathbf{B}$ never can be the matrix of squares and cross products of a real valued matrix $\mathbf{A}$, with $\mathbf{A'A} = \mathbf{B}$.

A matrix $\mathbf{B}$ with both positive and negative eigenvalues is called *indefinite*. Indefinite matrices may occur in applied data analysis (e.g. a correlation matrix based on a data matrix with many missing values could become indefinite when for the calculation of each pairwise correlation a different number of observations is used).

B.6 ALGEBRAIC APPLICATIONS OF EIGENVECTORS AND SINGULAR VECTORS

Let $\mathbf{B}$ (square and symmetric) have eigenvectors $\mathbf{Y}$ and eigenvalues Ψ^2. For any c, the powered matrix $\mathbf{B}^c$ will be equal to

$$\mathbf{B}^c = \mathbf{Y\Psi^{2c}Y'}. \tag{B.8}$$

Examples are

(a) $\mathbf{B}^2 = \mathbf{Y\Psi^2Y'Y\Psi^2Y'} = \mathbf{Y\Psi^4Y'}$.
(b) $\mathbf{B}^{-1} = \mathbf{Y\Psi^{-2}Y'}$, since $\mathbf{Y\Psi^2Y'Y\Psi^{-2}Y'} = \mathbf{YY'} = \mathbf{I}$ (if the inverse exists, $\mathbf{B}$ has no zero eigenvalues and $\mathbf{Y}$ is a square orthonormal matrix).
(c) $\mathbf{B}^{1/2} = \mathbf{Y\Psi Y'}$, since $\mathbf{Y\Psi Y'Y\Psi Y'} = \mathbf{Y\Psi^2Y'} = \mathbf{B}$.
(d) $\mathbf{B}^{-1/2} = \mathbf{Y\Psi^{-1}Y'}$ (defined if $\mathbf{B}$ has an inverse, so that $\mathbf{YY'} = \mathbf{I}$).

Let $\mathbf{X}$ be any $n \times m$ matrix $(n \geq m)$ of rank $k \leq m$. Its *generalized inverse* is defined as a matrix $\mathbf{X}^-$ that satisfies

$$\mathbf{X}\mathbf{X}^-\mathbf{X} = \mathbf{X}. \tag{B.9}$$

When $\mathbf{X}$ is square and of full rank $(n = m = k)$, $\mathbf{X}^-$ becomes identical to the proper inverse $\mathbf{X}^{-1}$. In all other cases, $\mathbf{X}^-$ is not unique. However, there is one solution for $\mathbf{X}^-$, for which the notation $\mathbf{X}^+$ is used, which is unique, and which satisfies in addition

$$\mathbf{X}^+\mathbf{X}\mathbf{X}^+ = \mathbf{X}^+, \tag{B.10a}$$

$$\mathbf{X}\mathbf{X}^+ = (\mathbf{X}\mathbf{X}^+)', \tag{B.10b}$$

$$\mathbf{X}^+\mathbf{X} = (\mathbf{X}^+\mathbf{X})'. \tag{B.10c}$$

$\mathbf{X}^+$ is often called the *Moore–Penrose* inverse. In terms of the singular value decomposition $\mathbf{X} = \mathbf{P}\Lambda\mathbf{Q}'$ the solution for $\mathbf{X}^+$ becomes $\mathbf{X}^+ = \mathbf{Q}\Lambda^{-1}\mathbf{P}'$. That this matrix satisfies the four requirements above is easily verified.

The advantage of $\mathbf{X}^+$ is that many algebraic expressions in which a proper inverse occurs remain valid for cases with deficient rank, provided that the proper inverse is replaced by the generalized inverse. Example: for a linear regression of $\mathbf{y}$ on $\mathbf{Z}$, the regression vector is $\mathbf{Z}\mathbf{b}$, with $\mathbf{b} = (\mathbf{Z}'\mathbf{Z})^{-1}\mathbf{Z}'\mathbf{y}$. Suppose $\mathbf{Z}$ has deficient column rank, so that $(\mathbf{Z}'\mathbf{Z})^{-1}$ does not exist. A valid solution remains $\mathbf{b} = (\mathbf{Z}'\mathbf{Z})^+\mathbf{Z}'\mathbf{y}$ (which in this particular example simplifies further to $\mathbf{b} = \mathbf{Z}^+\mathbf{y}$).

B.7 OPTIMIZATION PROPERTIES OF THE SVD

Section B.3 demonstrated that under the condition $\mathbf{y}'\mathbf{y} = 1$ the image $\mathbf{A}\mathbf{y}$ describes a hyperellipsoid, and that invariant directions correspond to the principal axes of such a hyperellipsoid. It follows immediately that the principal axes are related to stationary values for the sum of squares $\mathbf{y}'\mathbf{A}'\mathbf{A}\mathbf{y}$. This sum of squares is absolutely maximized by taking the stationary point $\mathbf{y}_1$ so that

$$\mathbf{y}_1'\mathbf{A}'\mathbf{A}\mathbf{y}_1 = \mathbf{y}_1'\mathbf{y}_1\psi_1^2 = \psi_1^2. \tag{B.11}$$

An unconditional minimum of function $y'A'Ay$ is represented by the shortest principal axis of the hyperellipsoid, and corresponds to

$$y'_m A'A y_m = y'_m y_m \psi^2_m = \psi^2_m. \tag{B.12}$$

Intermediate solutions y_j are conditional maxima in the sense that $y'_j A'A y_j$ is a maximum under the condition $y'y_i = 0$ ($i = 1, ..., j-1$) and is a minimum under the condition $y'y_k = 0$ ($k = j+1, ..., m$).

Section B.4 showed that we can write $A = \sum_i x_i \psi_i y'_i$. It can be proved that

$$A_{(k)} = \sum_{i=1}^{k} x_i \psi_i y'_i \tag{B.13}$$

is the best least squares rank k approximation to A, with the loss function equal to

$$tr(A'A - A'_{(k)}A_{(k)}) = \sum_{i=1}^{m} \psi^2_i - \sum_{i=1}^{k} \psi^2_i = \sum_{i=k+1}^{m} \psi^2_i. \tag{B.14}$$

This theorem is usually called the *Eckart–Young theorem* (Eckart and Young, 1936), although there are earlier formulations (e.g. Schmidt, 1906).

B.8 GENERALIZED EIGENVECTOR PROBLEM

Crucial in linear MVA is the solution of the *generalized eigenvector problem*

$$Cx = Dx\phi/m, \tag{B.15}$$

where C and D are square and symmetric matrices of dimension m x m. In Section B.9 examples will be given. For the present section we take the general and formal view that C and D are arbitrary square and symmetric matrices.

The generalized eigenvector problem can be reduced to a classical eigenvector problem by use of the following change of variables. Define $y=D^{1/2}x$, so that $x=D^{-1/2}y$. For the definition of $D^{1/2}$ see Section B.6; if D has no proper inverse, we may use the generalized inverse $(D^{1/2})^+$. Equation (B.15) now can be rewritten as

$$CD^{-1/2}y = D^{1/2}y\phi/m,$$

which can be brought into the symmetric form

$$D^{-1/2}CD^{-1/2}y = y\phi/m, \tag{B.16}$$

which is a classical eigenvector problem. Suppose (B.16) is solved by choosing y_i and ϕ_i; then x_i is identified from $x_i = D^{-1/2}y_i$. Since for the eigenvector y_i normalization $y_i'y_i = 1$ has been agreed upon, the resulting normalization for x_i becomes $x_i'Dx_i = 1$. As a consequence, the solution just described gives stationary values for the ratio

$$\frac{x'Cx/m}{x'Dx} = \phi, \tag{B.17}$$

which is called the *Rayleigh quotient*.

B.9 APPLICATIONS IN LINEAR MVA

Essentially all problems of MVA, viewed as techniques of dimension reduction, can be brought into one and the same general format: that of the generalized eigenvector equation (B.15). We shall illustrate this for a number of common MVA applications.

Principal components analysis starts from an $n \times m$ data matrix H, assumed to be in deviations from column means. In PCA the first step is usually to give columns of H equal normalization, say unity. Let $C = H'H$ and define D as the diagonal matrix of C. Then $HD^{-1/2}$ solves the first step. The matrix of sums of squares and cross products of $HD^{-1/2}$ becomes

$$R = D^{-1/2}H'HD^{-1/2}.$$

Actually, R is the correlation matrix. PCA is usually described in terms of the eigenvector decomposition $R = Y\Psi^2Y'$, from which we derive the factor matrix $F = Y\Psi$ so that $R = FF'$. But PCA could as well be formulated in terms of the SVD of $HD^{-1/2} = P\Psi Y' = PF'$, where P is the matrix of component scores. Since $P'P = I$, it follows that $D^{-1/2}H'P = Y\Psi = F$ can be interpreted as a matrix of correlations between observed variables and component scores.

Obviously, PCA fits into the format of Section B.8. The eigenvector problem $RY = Y\Psi^2$ is equivalent to $D^{-1/2}CD^{-1/2}Y = Y\Psi^2$. Substitute $X = D^{1/2}Y$, and $\Psi^2 = \Phi/m$, and PCA becomes equivalent to solving equation (B.15).

In *canonical correlation analysis* (CCA) we have the following situation. Given are two data matrices H_1 of order $n \times m_1$ and H_2 of order $n \times m_2$, both in deviations from column means. We want to find a solution for x_1 and x_2 in such a way that the correlation between $H_1 x_1$ and $H_2 x_2$ is stationary.

The problem becomes much simplified by selecting normalization conditions $x_1' H_1' H_1 x_1 = 1$ and $x_2' H_2' H_2 x_2 = 1$. The expression for the correlation between $H_1 x_1$ and $H_2 x_2$ then becomes

$$r = x_1' H_1' H_2 x_2.$$

Still further simplification is obtained by introducing the change of variables $y_1 = (H_1' H_1)^{1/2} x_1$ and $y_2 = (H_2' H_2)^{1/2} x_2$. This implies normalization $y_1' y_1 = 1$ and $y_2' y_2 = 1$. The expression for the correlation function becomes

$$r = y_1' \{ (H_1' H_1)^{-1/2} (H_1' H_2)(H_2' H_2)^{-1/2} \} y_2 = y_1' A y_2,$$

where A is introduced as a symbol for the matrix expression between brackets. It then becomes immediately clear that y_1 and y_2 should be identified with the singular vectors of $A = Y_1 \Lambda Y_2'$, since in that case the correlation becomes equal to a singular value λ_i. Maximum correlation is obtained by taking the first singular vectors y_{11} and y_{12}, with correlation λ_1.

In order to bring the CCA problem into the format of Section B.8, we first write $H = (H_1, H_2)$ so that H becomes the combined $n \times m$ data matrix. Define

$$C = H'H = \begin{pmatrix} H_1' H_1 & H_1' H_2 \\ H_2' H_1 & H_2' H_2 \end{pmatrix},$$

and let D be the partitioned block diagonal matrix of C, so that we may write

$$D^{-1/2} = \begin{pmatrix} (H_1' H_1)^{-1/2} & \\ & (H_2' H_2)^{-1/2} \end{pmatrix}.$$

Thus we obtain

$$D^{-1/2} C D^{-1/2} = \begin{pmatrix} I & A \\ A' & I \end{pmatrix} = \begin{pmatrix} I & Y_1 \Lambda Y_2' \\ Y_2 \Lambda Y_1' & I \end{pmatrix}.$$

Now it can be verified immediately that an eigenvector equation for $D^{-1/2} C D^{-1/2}$ can be written as

$$\mathbf{D}^{-1/2}\mathbf{C}\mathbf{D}^{-1/2}\begin{pmatrix}\mathbf{Y}_1\\\mathbf{Y}_2\end{pmatrix}=\begin{pmatrix}\mathbf{Y}_1\\\mathbf{Y}_2\end{pmatrix}(\mathbf{I}+\Lambda)\,,$$

which is in the format of equation (B.16). An equivalent expression is

$$\mathbf{C}\begin{pmatrix}\mathbf{X}_1\\\mathbf{X}_2\end{pmatrix}=\mathbf{D}^{1/2}\begin{pmatrix}\mathbf{Y}_1\\\mathbf{Y}_2\end{pmatrix}(\mathbf{I}+\Lambda)\,,$$

which is the format of equation (B.15), with

$$\begin{pmatrix}\mathbf{X}_1\\\mathbf{X}_2\end{pmatrix}=\mathbf{D}^{-1/2}\begin{pmatrix}\mathbf{Y}_1\\\mathbf{Y}_2\end{pmatrix}.$$

(A small technical detail is that the SVD of $\mathbf{A}$ implies the normalization $\mathbf{Y}_1'\mathbf{Y}_1 = \mathbf{I}$, and $\mathbf{Y}_2'\mathbf{Y}_2 = \mathbf{I}$. It follows that in the solution above for the eigenvectors of $\mathbf{D}^{-1/2}\mathbf{C}\mathbf{D}^{-1/2}$ the eigenvectors are normalized at $2\mathbf{I}$.)

This exposition of CCA implicitly covers a number of other standard MVA techniques, such as *canonical discriminant analysis* (CADA) or *multivariate analysis of variance* (MANOVA). In CADA, $\mathbf{H}_1$ will be an indicator matrix that gives a coding for subgroups of individuals, whereas $\mathbf{H}_2$ contains observations on dependent variables. In MANOVA, $\mathbf{H}_1$ contains the same sort of coding, now for the subgroups that correspond to the conditions of the systematic experimental design ('dummy' variables).

Multiple regression is a special case of CCA, with $m_2 = 1$, so that $\mathbf{H}_2$ becomes a single vector of observations $\mathbf{h}_2$. For convenience, assume that all columns of $\mathbf{H} = (\mathbf{H}_1,\mathbf{h}_2)$ are normalized to unity, so that $\mathbf{H}'\mathbf{H}$ is the overall correlation matrix:

$$\mathbf{C}=\mathbf{H}'\mathbf{H}=\begin{pmatrix}\mathbf{R}&\mathbf{r}\\\mathbf{r}'&1\end{pmatrix}.$$

Define $\mathbf{D}$ again as the partitioned block diagonal matrix of $\mathbf{C}$, so that

$$\mathbf{D}^{-1/2}\mathbf{C}\mathbf{D}^{-1/2}=\begin{pmatrix}\mathbf{I}&\mathbf{R}^{-1/2}\mathbf{r}\\\mathbf{r}'\mathbf{R}^{-1/2}&1\end{pmatrix}.$$

It is easily verified that

$$\begin{pmatrix} \mathbf{y}_1 \\ \mathbf{y}_2 \end{pmatrix} = \begin{pmatrix} \mathbf{R}^{-1/2}\mathbf{r} \\ \rho \end{pmatrix}$$

is an eigenvector of $\mathbf{D}^{-1/2}\mathbf{C}\mathbf{D}^{-1/2}$, where ρ is defined as $(\mathbf{r}'\mathbf{R}^{-1}\mathbf{r})^{1/2} = \rho$, so that ρ is the multiple correlation between $\mathbf{H}_1$ and $\mathbf{h}_2$. It then also follows that we can write

$$\mathbf{C}\begin{pmatrix} \mathbf{x}_1 \\ \mathbf{x}_2 \end{pmatrix} = \mathbf{D}\begin{pmatrix} \mathbf{x}_1 \\ \mathbf{x}_2 \end{pmatrix}(1 + \rho)$$

with

$$\begin{pmatrix} \mathbf{x}_1 \\ \mathbf{x}_2 \end{pmatrix} = \mathbf{D}^{-1/2}\begin{pmatrix} \mathbf{x}_1 \\ \mathbf{y}_2 \end{pmatrix} = \begin{pmatrix} \mathbf{R}^{-1}\mathbf{r} \\ \rho \end{pmatrix} = \begin{pmatrix} \mathbf{b} \\ \rho \end{pmatrix},$$

where $\mathbf{b}$ is the vector of regression weights. Obviously, $\mathbf{D}^{-1/2}\mathbf{C}\mathbf{D}^{-1/2}$ has more than one solution for its eigenvectors. These other solutions are unrelated to the solution for multiple regression.

Again, it should be realized that the solution for multiple regression implicitly covers discriminant analysis with two subgroups of individuals (here $\mathbf{h}_2$ contains a binary code that separates between these two groups). It also covers ANOVA (where $\mathbf{H}_1$ contains a code for the subgroups prescribed by the experimental design and where $\mathbf{h}_2$ contains the single dependent variable).

A *generalized canonical correlation problem* (GENCAN) is indicated when the data matrix $\mathbf{H}$ is partitioned into K subsets $\mathbf{H} = (\mathbf{H}_1 \ \mathbf{H}_2 \ \dots \ \mathbf{H}_K)$, where again we assume that columns of $\mathbf{H}$ are in deviations from their means. Let $\mathbf{C} = \mathbf{H}'\mathbf{H}$, and define $\mathbf{D}$ as the partitioned diagonal block matrix of $\mathbf{C}$ with diagonal submatrices $\mathbf{D}_{kk} = \mathbf{C}_{kk}$ and off-diagonal submatrices $\mathbf{D}_{kl} = 0$. It follows that $\mathbf{D}^{-1/2}\mathbf{C}\mathbf{D}^{-1/2}$ will have diagonal submatrices equal to identity matrices. The GENCAN problem then involves finding a solution for $\mathbf{y}$ that satisfies

$$\mathbf{D}^{-1/2}\mathbf{C}\mathbf{D}^{-1/2}\mathbf{y} = \mathbf{y}\psi^2$$

or

$$\mathbf{C}\mathbf{x} = \mathbf{D}\mathbf{x}\psi^2,$$

with $\mathbf{x} = \mathbf{D}^{-1/2}\mathbf{y}$. All MVA problems discussed earlier are now special cases. The PCA problem has $K = m$ (with the implication that each submatrix $\mathbf{H}_k$ of $\mathbf{H}$ corresponds to a single vector $\mathbf{h}_j$). The CCA problem has $K = 2$. The multiple correlation problem has $K = 2$, with in addition $m_2 = 1$. GENCAN is closely related to HOMALS. One might say that HOMALS comes to the same as GENCAN applied to the partitioned indicator matrix $\mathbf{G}$.

In all examples above, the linear MVA problem can be phrased in terms of the SVD of $\mathbf{HD}^{-1/2}$. A particular MVA problem obtains its special characteristics only from the definition of $\mathbf{D}$. In PCA, $\mathbf{D}$ is the diagonal matrix of $\mathbf{H'H}$. In CCA, $\mathbf{D}$ is the partitioned diagonal block matrix corresponding to the two partitions of $\mathbf{H}$. In GENCAN $\mathbf{D}$ is a partitioned diagonal block matrix corresponding to the K partitions of $\mathbf{H}$.

B.10 JOINT PLOTS

A joint plot of rows and columns of a matrix $\mathbf{H}$ with SVD $\mathbf{H} = \mathbf{X}\Psi\mathbf{Y'}$ is obtained by plotting the rows of $\mathbf{H}$ as points with coordinates given in the rows of $\mathbf{X}\Psi^\alpha$ and the columns of $\mathbf{H}$ as points with coordinates given in the rows of $\mathbf{Y}\Psi^{1-\alpha}$. Such a plot is a representation of rows and columns of $\mathbf{H}$ in the following way. Draw a vector through the jth row point, and project the m column point on this vector. The vector or projections will be proportional to the jth row of $\mathbf{H}$. Conversely, draw a vector through the ith column point and project the n row points on this vector. This vector of projections will be proportional to the ith column of $\mathbf{H}$.

The property mentioned above is invariant under choice of α. Obvious choices for α are $\alpha = 0$, $\alpha = 1/2$, and $\alpha = 1$. Frequently α will be chosen to highlight further characteristics of the data analytic problem, i.e. on grounds extraneous to the SVD.

APPENDIX C
CONES AND PROJECTION ON CONES

A set C in $\Re^n$ is a *cone* if $\mathbf{x} \in C$ implies that $\alpha\mathbf{x} \in C$ for all $\alpha > 0$. In words, a set is a cone if it contains the *ray* through $\mathbf{x}$ whenever it contains $\mathbf{x}$. In Figure C.1 we give some examples of cones in $\Re^2$. The cone in Figure C.1b is the positive *orthant*, the cone Figure C.1c is a *half-space*, the cone in Figure C.1d is a one-dimensional *subspace*, and the cone in Figure C.1e consists of two subspaces that intersect only in the origin. The cones in Figures C.1a and

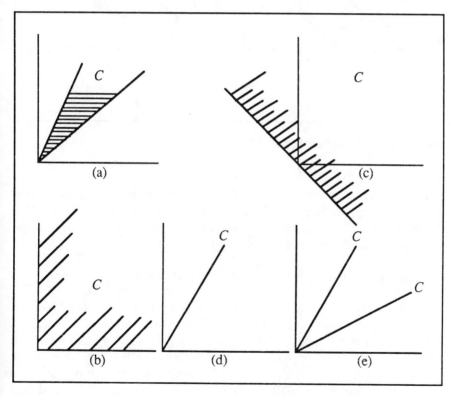

Figure C.1 Five cones.

C.1b are *pointed*, by which we mean algebraically that $x \in C$ implies that $-x \notin C$. The cones in Figure C.1, except the one in Figure C.1e, are *convex*, by which we mean that $x \in C$ and $y \in C$ imply that $x+y \in C$. Many writers use the word cone in the sense of convex cone. Some convenient references on cones are Goldman and Tucker (1956) and Berman (1973). Rockafellar (1970) also contains all of the relevant material (and much more).

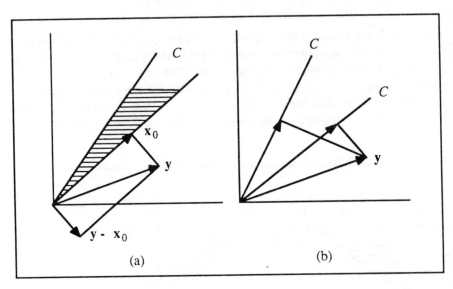

Figure C.2 Projection on a cone.

In our alternating least squares algorithms we often have to solve *cone projection problems*. Suppose C is a cone, y is a given vector in $\Re^n$, and W is a symmetric positive definite matrix. Then the *projection of y on C in the metric W* is that vector $x_0 \in C$ which satisfies

$$(y - x_0)'W(y - x_0) = \inf \{(y - x)'W(y - x) \mid x \in C\}.$$

We have not proved yet that such an x_0 acually exists. In fact it need not: if C is the interior of the positive orthant, $W = I$, then the infimum is not attained. In this case it is equal to $SSQ(y - x_0)$ with $x_0 = \max(0, y)$, but $x_0 \notin C$ if $y \notin C$. Consequently, we assume that the cones we are dealing with are *closed*, which implies by the way that they contain the origin. It is possible to prove that the projection on a closed cone in finite-dimensional space always exists, although it may not be unique. If the cone is both closed and convex, then the projection is unique.

It is possible to derive some simple properties of the projection by straightforward calculation. If x_0 is the projection, then

$$(y - x_0)'W(y - x_0) \leq (y - \alpha x_0)'W(y - \alpha x_0)$$

for all $\alpha \geq 0$. If we simplify this, we find the result that $y'Wx_0 = x_0'Wx_0$, or $(y - x_0)'Wx_0 = 0$, or $y - x_0$ is W-perpendicular to x_0. This is illustrated, for $W = I$, in Figure C.2a and b for, respectively, a convex cone and a cone consisting of two rays. In this last nonconvex case there are two solutions to $(y - x_0)'Wx_0 = 0$, in fact: if we have a y on the bisectrice of the acute angle formed by the two rays, then y is equally far from both rays, and the projection is not unique.

More precise statements are possible if the cone is convex. We then must have

$$[(y - x_0) - z]'W[(y - x_0) - z] \geq (y - x_0)'W(y - x_0)$$

for all $z \in C$. If we simplify this we find that $(y - x_0)'Wz \leq 0$ for all $z \in C$, which shows that $y - x_0$ makes a W-obtuse angle with all $z \in C$. In fact, for convex C this condition, together with $(y - x_0)'Wx_0 = 0$, characterizes a unique vector x_0 in C, which is consequently the projection we are looking for. In the nonconvex case the situation is considerably less simple.

The characterization of x_0 in the convex case implies the decomposition

$$y'Wy = x_0Wx_0 + (y - x_0)'W(y - x_0),$$

which is true for all y in $\Re^n$. Thus the squared W-length of a vector y is equal to the squared W-length of its projection on a convex cone C and the squared W-length of the residual. Conversely, y, x_0, and $y - x_0$ are a W-rectangular triangle. Another way to state this is by introducing the polar cone C^0, which is the set of all x such that $x'Wz \leq 0$ for all $z \in C$. The polar cone is illustrated in Figure C.3; it is convex and closed if C is convex and closed. If x_0 is the projection of y on C, then $y - x_0$ is the projection of y on C^0, and the two projections are W-orthogonal and their squared W-lengths add up to $y'Wy$. These results generalize more familiar results for subspaces. Observe that a subspace is a (closed, convex) cone for which $x \in C$ and $y \in C$ imply that $\alpha x + \beta y \in C$ for all real α, β. This means that x_0 is the projection of y on the subspace C in the metric W if and only if $(y - x_0)'Wz = 0$ for all $z \in C$, which means that $y - x_0$ must be in the orthogonal complement of C, which is the same thing as the polar cone of C. This is illustrated in Figure C.4.

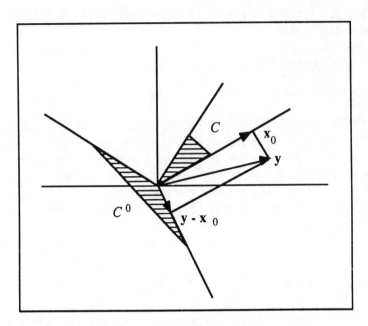

Figure C.3 Projection on a cone and its polar.

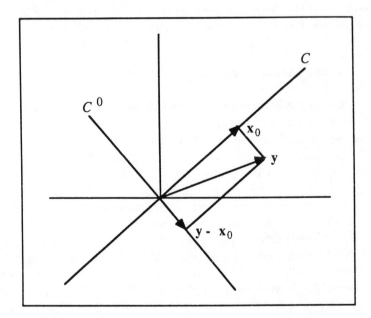

Figure C.4 Projection on a subspace and its complement.

An important case of cone projection in this book is *monotone regression*. We discuss this briefly in a comparatively simple case. Suppose

$$C = \{x \mid x_1 \leq \ldots \leq x_n\},$$

and we want to minimize $(x - y)'W(x - y)$ over $x \in C$, where we suppose now that W is diagonal (with positive elements on the diagonal). There are two results that make the monotone regression problem comparatively easy. We discuss them briefly. Suppose the (unique) solution to the problem is $\hat{x}$.

Result *A* says that $y_i > y_{i+1}$ implies that $\hat{x}_i = \hat{x}_{i+1}$. The proof is by contradiction. Suppose $\hat{x}_i < \hat{x}_{i+1}$. Define a new vector x^* by $x_k^* = \hat{x}_k$ if $k \neq i$ and $k \neq i +1$ and by $x_i^* = x_{i+1}^* = (w_i\hat{x}_i + w_{i+1}\hat{x}_{i+1}) / (w_i + w_{i+1})$, where the w_i are the diagonal elements of W. Clearly the new vector x^* is also in C, and some algebra gives

$$(y - x^*)'W(y - x^*) < (y - \hat{x})'W(y - \hat{x}),$$

which shows that $\hat{x}$ cannot be the projection.

Result *B* says that if we add the constraint $x_i = x_{i+1}$ to the definition of C, then the monotone regression problem can be reduced to a problem in $\Re^{n-1}$. The proof is by partitioning the sum of squares in the usual analysis of variance way. Details are in De Leeuw (1977), for example.

These two results suggest an algorithm. If y_i and y_{i+1} are in the wrong order, then we know that $\hat{x}_i$ and $\hat{x}_{i+1}$ must be equal from result *A*; we then use result *B* to reduce the problem to one in $\Re^{n-1}$. Again we look for violations, and use them to reduce the problem even further. If there are no violations left, then we are done; ultimately we could end up with a problem in $\Re^1$, for which there are by definition no violations.

In the book we sometimes use normalized cone regression. This can be either one of two things. In the first place minimization of

$$\frac{(y - x)'W(y - x)}{x'Wx}$$

over x in a cone C. This is called *implicit normalization*. This name suggests that there is also someting like *explicit normalization*. This is minimization of $(y - x)'W(y - x)$ over all $x \in C$ that satisfy in addition $x'Wx = 1$. It is a basic result of Kruskal and Carroll (1969) that in this simple case implicit normalization, explicit normalization, and no normalization all give essentially the same solution. All solutions are proportional to the projection of y on the cone, with only the proportionality constant different for the different problems. This result does not rely on convexity. There is one exception which should be noted. If y is in C^0, i.e. $y'Wz \le 0$ for all $z \in C$, then the origin is the projection of y on C. In the normalized problems the solution of x is one of the *extreme rays* of C, suitably normalized. An extreme ray is any ray in the cone that cannot be written as a nonnegative linear combination of two other rays in the cone. If C is a subspace and y is in the W-orthogonal complement, then the infimum in the implicitly normalized problem is equal to one – not attained, but approached by letting $x \to \infty$. The infimum in the explicitly normalized problem is attained for any $x \in C$ with $x'Wz = 1$; it is equal to $1 + y'Wy$. If C is the cone used in monotone regression, then the extreme rays are the vectors with the first k elements equal to zero and the last $n - k$ elements equal to one ($k = 1,\ldots, n-1$) together with the vectors u and $-u$ which span the intersection of C and C^0. Thus the cone is not pointed. The normalized regression problem must be solved by testing all these rays and by keeping the best one.

REFERENCES

Abelson, R.P. (1962). Scales derived by consideration of variance components in multiway tables. In H. Gulliksen and S. Messick (Eds.), *Psychological Scaling: Theory and Application.* New York: Wiley.

Abelson, R.P. (1964). Mathematical models of the distribution of attitudes under controversy. In N. Frederiksen and H. Gulliksen (Eds.), *Contributions to Mathematical Psychology.* New York: Rinehart and Winston.

Anderson, T.W. (1958). *An Introduction to Multivariate Statistical Analysis.* New York: Wiley.

Anderson, T.W. (1960). Some stochastic process models for intelligence test scores. In K.J. Arrow *et al.* (Eds.), *Mathematical Methods in the Social Sciences,* 1959. Stanford, Calif.: Stanford University Press.

Anderson, T.W. (1963). Asymptotic theory for principal components analysis. *Annals of Mathematical Statistics, 34,* 122–48.

Anderson, T.W. (1984). *An Introduction to Multivariate Statistical Analysis,* 2nd edition. New York: Wiley.

Andrich, D. (1988a). *Rasch Models for Measurement.* Beverly Hills, Calif.: Sage Publications.

Andrich, D. (1988b). The application of an unfolding model of the PIRT type to the measurement of attitude. *Applied Psychological Measurement, 12,* 33–51.

Anscombe, F.J. (1967). Topics in the investigation of linear relations fitted by the method of least squares. *Journal of the Royal Statistical Society, Series B (Methodological), 29,* 1–52.

Appell, P., and Kampé de Fériet, J. (1926). *Fonctions Hypergéometriques et Hyperspheriques. Polynomes de Hermite.* Paris: Gauthier-Villars.

Barcikowski, R.S., and Stevens, J.P. (1975). A Monte Carlo study of the stability of canonical correlations, canonical weights, and canonical variate–variable correlations. *Multivariate Behavioral Research, 10,* 353–64.

Barnett, V., and Lewis, T. (1984). *Outliers in Statistical Data.* New York: Wiley.

Barrett, J.F., and Lampard, D.G. (1955). An expansion for some second-order probability distributions. *Proceedings IEEE (Information Theory), 1,* 10–15.

Bartholomew, D.J. (1980). Factor analysis for categorical data. *Journal of the Royal Statistical Society, Series B (Methodological)*, **42**, 293–321.

Bartlett, M.S. (1948). Internal and external factor analysis. *British Journal of Statistical Psychology*, **1**, 73–81.

Bechtel, G.G. (1967). The analysis of variance and pairwise scaling. *Psychometrika*, **32**, 47–65.

Bechtel, G.G. (1971). A dual scaling analysis for paired comparisons. *Psychometrika*, **36**, 135–54.

Bechtel, G.G. (1976). *Multidimensional Preference Scaling*. The Hague: Mouton.

Bechtel, G.G., Tucker, L.R., and Chang, W.C. (1971). Scalar product model for the multidimensional scaling of choice. *Psychometrika*, **36**, 369–88.

Beckenbach, E.F., and Bellman, R. (1965). *Inequalities*. Berlin: Springer Verlag.

Bekker, P., and De Leeuw, J. (1988). Relations between variants of nonlinear principal component analysis. In J.L.A. Van Rijckevorsel and J. De Leeuw (Eds.), *Component and Correspondence Analysis*, pp. 1–31. New York: Wiley.

Bell, E.H., and Sirjamaki, J. (1967). *Social Foundations of Human Behaviour*. New York: Harper.

Beltrami, E. (1873). Sulle funzioni bilineari. *Giorn. Math. Battaglin*, **11**, 98–106.

Bennett, J.F. (1956). Determination of the number of independent parameters of a score matrix from the examination of rank orders. *Psychometrika*, **21**, 383–93.

Bentler, P.M., and Weeks, D.G. (1978). Restricted multidimensional scaling models. *Journal of Mathematical Psychology*, **17**, 138–51.

Benzécri, J.P. (1965). *Sur l'Analyse des Préférences*. Paris: Institute de Statistique de l'Université de Paris.

Benzécri, J.P. (1967). *Lois de Probabilité sur un Ensemble Produit. Les Diverses Notions d'Independence et le Critère d'Entropie Maximale* (Mimeo). Paris: Institute de Statistique de l'Université de Paris.

Benzécri, J.P. (1973). *Analyse des Données* (2 vols.). Paris: Dunod.

Berman, A. (1973). *Cones, Matrices, and Mathematical Programming*. Berlin: Springer Verlag.

Besse, P. (1988). Spline functions and optimal metric in linear principal component analysis. In J.L.A. Van Rijckevorsel and J. De Leeuw (Eds.), *Component and Correspondence Analysis*, pp. 81–101. New York: Wiley.

Bhattacharya, R.N., and Ghosh, J.K. (1978). On the validity of the formal Edgeworth expansion. *Annals of Statistics*, **6**, 434–51.

Bickel, P.J. (1976). Another look at robustness: a review of reviews and some new developments. *Scand. J. Statist.*, **3**, 145–68.

Billingsley, P. (1968). *Convergence of Probability Measures*. New York: Wiley.

Birkhoff, G. (1967). *Lattice Theory*. Providence: American Mathematical Society.

Bishop, Y.M.M., Fienberg, S.E., and Holland, P.W. (1975). *Discrete Multivariate Analysis: Theory and Practice*. Cambridge, Mass.: MIT Press.

Black, M. (1949). The definition of scientific method. In R.C. Staufler (Ed.), *Science and Civilization*. Madison: University Wisconsin Press.

Blalock, H.M.Jr. (1964). *Causal Inference in Nonexperimental Research*. Chapel Hill: University of North Carolina Press.

Bock, R.D. (1960). *Methods and Applications of Optimal Scaling*, Report 25. North Carolina: L. L. Thurstone Lab, University of North Carolina.

Bock, R.D., and Lieberman, M. (1970). Fitting a response model for n dichotomously scored items. *Psychometrika*, **35**, 179–97.

Boudon, R. (1967). *L'Analyse Mathématique des Faits Sociaux*. Paris: Plan.

Bourouche, J.M., and Saporta, G. (1980). *L'Analyse des Données*. Paris: Presses Universitaires de France.

Box, G.E.P. (1979). Some problems of statistics and everyday life. *Journal of the American Statistical Association*, **74**, 1-4.

Box, G.E.P., and Cox, D.R. (1964). An analysis of transformation. *Journal of the Royal Statistical Society, Series B (Methodological)*, **26**, 211–52.

Boyd, D.W. (1974). The power method for l^p norms. *Linear Algebra and Its Applications*, **9**, 95–102.

Breiman, L., and Friedman, J.H. (1985). Estimating optimal transformations for multiple regression and correlation. *Journal of the American Statistical Association*, **80**, 580–619.

Bunch, J.R., Nielsen, C.P., and Sorensen, D.C. (1978). Rank-one modification of the symmetric eigenproblem. *Num. Math.*, **31**, 31–48.

Burt, C. (1948a). A comparison of factor analysis and analysis of variance. *British Journal of Statistical Psychology*, **1**, 3–27.

Burt, C. (1948b). Factor analysis and canonical correlations. *British Journal of Statistical Psychology*, **1**, 95–106.

Burt, C. (1950a). The influence of differential weighting. *British Journal of Statistical Psychology*, **3**, 105–28.

Burt, C. (1950b). The factorial analysis of qualitative data. *British Journal of Statistical Psychology*, **3**, 166–85.

Burt, C. (1951). Test construction and the scaling of items. *British Journal of Statistical Psychology*, **4**, 95–129.

Cailliez, F., and Pagès, J.P. (1976). *Introduction à l'Analyse des Données*. Paris: SMASH.

Cambanis, S., and Liu, B. (1971). On the expansion of a bivariate distribution and its relationship to the output of a nonlinearity. *Proceedings IEEE (Information Theory)*, **17**, 17–25.

Campbell, N.A. (1982). Robust procedures in multivariate analysis II. Robust canonical variate analysis. *Applied Statistics*, **31**, 1–8.

Carroll, J.D. (1968). A generalization of canonical correlation analysis to three or more sets of variables. *Proceedings of the 76th Annual Convention of the American Psychological Association*, pp. 227–8.

Carroll, J.D. (1972). Individual differences and multidimensional scaling. In R.N. Shepard, A.K. Romney, and S.B. Nerlove (Eds.), *Multidimensional Scaling: Theory and Applications in the Behavioral Sciences*, Vol. I, pp. 105–55. New York: Seminar Press.

Carroll, J.D., and Arabie, P. (1980). Multidimensional scaling. *Annual Review of Psychology*, **31**, 607–49.

Carroll, J.D., and Chang, J.J. (1964). *Nonmetric Multidimensional Analysis of Paired Comparison Data*. Psychometric Society Meeting.

Chang, J.C., and Bargmann, R.E. (1974). *Internal Multidimensional Scaling of Categorical Variables*, Technical Report. Department of Statistics and Computer Science, University of Georgia.

Chatelin, F. (1979). *Approximation Spectrale d'Operateurs Lineaires avec Applications au Calcul des Elements Propres d'Operateurs Differentielles et Integraux*, Report RR 167. Grenoble: Dep. Math. Appl., Université de Grenoble.

Chesson, P.L. (1976). The canonical decomposition of bivariate distributions. *Journal of Multivariate Analysis*, **6**, 526–37.

Christofferson, A. (1975). Factor analysis of dichotomized variables. *Psychometrika*, **40**, 5–32.

Christofferson, A. (1977). Two-step weighted least squares factor analysis of dichomotized variables. *Psychometrika*, **42**, 433–8.

Cliff, N. (1966). Orthogonal rotation to congruence. *Psychometrika*, **31**, 33–42.

Cochran, W.G. (1972). Observational studies. In T.A. Bancroft (Ed.), *Statistical Papers in Honour of George Snedecor*. Ames: Iowa State University Press.

Collaris, J.W.M., and Kropman, J.A. (1978). *Van Jaar tot Jaar, Tweede Fase*. Den Haag: Staatsuitgeverij.

Coolen, H., and De Leeuw, J. (1987). Least squares path analysis with optimal scaling. In E. Diday *et al.* (Eds.), *Data Analysis and Informatics*, Vol. V. Amsterdam: North-Holland.

Coolen, H., Van Rijckevorsel, J., and De Leeuw, J. (1982). An algorithm for nonlinear principal component analysis with B-splines by means of alternating least squares. In H. Caussinus *et al.* (Eds.), *COMPSTAT 1982*, Part II. Vienna: Physika Verlag.

Cooley, W.W., and Lohnes, P.R. (1962). *Multivariate Procedures for the Behavioural Sciences.* New York: Wiley.

Cooley, W.W., and Lohnes, P.R. (1971). *Multivariate Data Analysis.* New York: Wiley.

Coombs, C.H. (1964). *A Theory of Data.* New York: Wiley.

Coombs, C.H., and Kao, R.C. (1960). On a connection between factor analysis and multidimensional unfolding. *Psychometrika,* **25**, 219–31.

Cooper, R.D., Hoare, M.R., and Rahman, M. (1977). Stochastic processes and special functions: on the probabilistic origin of some positive kernels associated with classical orthogonal polynomials. *Journal of Mathematical Analysis and Applications,* **61**, 262–91.

Coppi, R., and Bolasco, S. (Eds.) (1989). *Analysis of Multiway Data Matrices.* Amsterdam: North-Holland.

Cox, D.R. (1957). Note on grouping. *Journal of the American Statistical Association,* **52**, 543–7.

Cozijn, C., and Van Dijk, J.J.M. (1976). *Onrustgevoelens in Nederland.* Den Haag: WODC.

Curry, H.B., and Schoenberg, I.J. (1966). On Polya frequency functions IV; the fundamental spline functions and their limits. *J. d'Analyse Math.,* **17**, 71–107.

Daalder, H., and Rusk, J.G. (1972). Perceptions of party in the Dutch Parliament. In S.C. Patterson and J.C. Wahlke (Eds.), *Comparative Legislative Behaviour: Frontiers of Research.* New York: Wiley.

Daalder, H., and Van de Geer, J.P. (1977). Partij-afstanden in de Tweede Kamer. *Acta Politica,* **12**, 289–345.

Dagnelie, P. (1975). *Analyse Statistique à Plusieurs Variables.* Gembloux: Presses Agronomiques.

Dahmen, W. (1979). Multivariate B-splines: recurrence relations and linear combinations of truncated powers. In W. Schempp and K. Zeller (Eds.), *Multivariate Approximation Theory,* ISNM 51. Basel: Birkhäuser.

Dahmen, W. (1980a). On multivariate B-splines. *Journal of the Society for Industrial and Applied Mathematics (Numerical Analysis),* **17**, 179–91.

Dahmen, W. (1980b). Konstruktion mehr dimensionaler B-splines und ihre Anwendung auf Approximations Probleme. *Numerische Methoden der Approximations Theorie,* Vol. V, ISNM 52. Basel: Birkhäuser.

Darlington, R.B., Weinberg, S.L. and Walberg, H.L. (1973). Canonical variate analysis and related techniques. *Rev. Educ. Research,* **43**, 433–54.

Darroch, J.N. (1974). Multiplicative and additive interaction in contingency tables. *Biometrika,* **61**, 207–14.

Daudin, J.J. (1980). Régression qualitative: Choix d'espace prédicteurs. In E. Diday *et al.* (Eds.), *Data Analysis and Informatics*. Amsterdam: North Holland.

Daugavet, V.A. (1968). Variant of the stepped exponential method of finding some of the first characteristic values of a symmetric matrix. *USSR Comp. Math. Physics,* **8** (1), 212–23.

Dauxois, J., and Pousse, A. (1976). *Les Analyses Factorielles en Calcul des Probabilités et en Statistique: Essai d'Etude Synthétique.* Toulouse: Université Paul Sabatier (dissertation).

Davies, P.T., and Tso, M.K.-S. (1982). Procedures for reduced-rank regression. *Applied Statistics,* **31**, 244–55.

Davis, A.W. (1977). Asymptotic theory for principal component analysis: nonnormal case. *Australian Journal of Statistics,* **19**, 206–12.

Davis, C., and Kahan, W.M. (1970). The rotation of eigenvectors by a perturbation. *Journal of the Society for Industrial and Applied Mathematics (Numerical Analysis),* **7**, 1–46.

De Boor, C (1978). *A Practical Guide to Splines.* Berlin: Springer Verlag.

De Leeuw, J. (1968). *Canonical Analysis of Categorical Data.* Leiden: Department of Data Theory, University of Leiden.

De Leeuw, J. (1969). *Some Contribution to the Analysis of Categorical Data,* Report RN 004-69. Leiden: Department of Data Theory, University of Leiden.

De Leeuw, J. (1973). *Canonical Analysis of Categorical Data.* Leiden: University of Leiden (dissertation).

De Leeuw, J. (1977a). Correctness of Kruskal's algorithms for monotone regression with ties. *Psychometrika,* **42**, 141–4.

De Leeuw, J. (1977b). *Normalized Cone Regression* (mimeo). Leiden: Department of Data Theory, University of Leiden.

De Leeuw, J. (1977c). Applications of convex analysis to multidimensional scaling. In J.R. Barra, F. Brodeau, G. Romier, and B. Van Cutsem (Eds.), *Recent Developments in Statistics,* pp. 133–45. Amsterdam: North-Holland.

De Leeuw, J. (1984). The Gifi-system of non-linear multivariate analysis. In E. Diday, M. Jambu, L. Lebart, J. Pages, and R. Tomassone (Eds.), *Data Analysis and Informatics,* Vol. III. Amsterdam: North-Holland.

De Leeuw, J. (1985). *Jackknife and Bootstrap in Multinominal Situations,* Internal Report RR-85-16. Leiden: Department of Data Theory, University of Leiden.

De Leeuw, J. (1986). Regression with optimal scaling of the dependent variable. In O. Burke (Ed.), *Proceedings of the 7th International Summer School on Problems of Model Choice and Parameter Estimation in Regression Analysis,* Report 84, pp. 99–111. Berlin, GDR: Department of Mathematics, Humboldt University.

De Leeuw, J. (1987). Path analysis with optimal scaling. In P. Legendre and L. Legendre (Eds.), *Developments in Numerical Ecology*, pp. 381–404. NATO ASI Series, Vol. G 14. Berlin: Springer Verlag.

De Leeuw, J. (1988). Multivariate analysis with linearizable regressions. *Psychometrika*, 53, 437–54.

De Leeuw, J. (1989). Multivariate analysis with optimal scaling. In S. Das Gupta (Ed.), *Progress in Multivariate Analysis*. Calcutta: Indian Statistical Institute.

De Leeuw, J., and Heiser, W.J. (1980). Multidimensional scaling with restrictions on the configuration. In P.R. Krishnaiah (Ed.), *Multivariate Analysis, Vol. V*, pp. 501–522. Amsterdam: North-Holland.

De Leeuw, J., and Heiser, W.J. (1982). Theory of multidimensional scaling. In P.R. Krishnaiah and L. Kanal (Eds.), *Handbook of Statistics, Vol. 2*. Amsterdam: North-Holland.

De Leeuw, J., and Meulman, J.J. (1986). A special jackknife for multidimensional scaling. *Journal of Classification*, 3, 97–112.

De Leeuw, J., and Stoop, I. (1979). Sekundaire analyse 'Van Jaar tot Jaar' met behulp van niet-lineaire multivariate technieken. In J. Peschar (Ed.), *Van Achteren naar Voren*. Den Haag: Staatsuitgeverij.

De Leeuw, J., and Van der Burg, E. (1986). The permutational limit distribution of generalized canonical correlations. In E. Diday *et al.* (Eds.), *Data Analysis and Informatics*, Vol. IV, pp. 509–21. Amsterdam: North-Holland.

De Leeuw, J., and Van Rijckevorsel, J. (1980). HOMALS and PRIN-CALS: some generalizations of principal components analysis. In E. Diday *et al.* (Eds.), *Data Analysis and Informatics*. Amsterdam: North-Holland.

De Leeuw, J., and Van Rijckevorsel, J.L.A. (1988). Beyond homogeneity analysis. In J.L.A. Van Rijckevorsel and J. De Leeuw (Eds.), *Component and Correspondence Analysis*, pp. 55–80. New York: Wiley.

De Leeuw, J., Van Rijckevorsel, J., and Van der Wouden, H. (1981). Nonlinear principal component analysis with B-splines. *Methods of Operations Research*, 33, 379–93.

De Leeuw, J., and Verhelst, N. (1986). Maximum likelihood estimation in generalized Rasch models. *Journal of Educational Statistics*, 11, 183–96.

De Leeuw, J., Young, F.W., and Takane, Y. (1976). Additive structure in qualitative data: An alternating least squares method with optimal scaling features. *Psychometrika*, 41, 471–503.

Demjanov, V.F., and Malozemov, V.N. (1974). *Introduction to Minimax*. New York: Wiley.

Dempster, A.P. (1969). *Elements of Continuous Multivariate Analysis*. Reading: Addison-Wesley.

Dempster, A.P. (1971). An overview of multivariate data analysis. *Journal of Multivariate Analysis*, 1, 316–46.

Deville, J.C., and Saporta, G. (1980). Analyse harmonique qualitative. In E. Diday *et al.* (Eds.), *Data Analysis and Informatics*. Amsterdam: North-Holland.

DiCiccio, T.J., and Romano, J.P. (1988). A review of bootstrap confidence intervals (with discussion). *Journal of the Royal Statistical Society, Series B (Methodological)*, 50, 338–70.

Dillon, W.R., and Goldstein, M. (1984). *Multivariate Analysis, Methods and Applications*. New York: Wiley.

Divgi, D.R. (1979). Calculation of the tetrachoric correlation coefficient. *Psychometrika*, 44, 169–172.

Dronkers, J. (1978). *Manipuleerbare Variabelen in de Schoolloopbaan*. Uppsala: 9th World Congress of Sociology.

Dronkers, J., and Jungbluth, J.J.M. (1979). Schoolloopbaan en geslacht. In J. Peschar (Ed.), *Van Achteren naar Voren*. Den Haag: Staatsuitgeverij.

Drouet d'Aubigny, G. (1975). *Description Statistique des Données Ordinales: Analyse Multidimensionelle*. Grenoble: Thex.

Eagleson, G.K. (1964). Polynomial expansions of bivariate distributions. *Annals of Mathematical Statistics*, 35, 1208–15.

Eagleson, G.K. (1969a). Canonical expansions of birth and death processes. *Theory Prob. Appl.*, 14, 209–18.

Eagleson, G.K. (1969b). A characterization theorem for positive definite sequences on the Krawtchouk polynomials. *Australian Journal of Statistics*, 11, 29–38.

Eagleson, G.K., and Lancaster, H.O. (1967). The regression system of sums with random elements in common. *Australian Journal of Statistics*, 9, 119–25.

Eckart, C., and Young, G. (1936). The approximation of one matrix by another of lower rank. *Psychometrika*, 1, 211–8.

Edgerton, H.A., and Kolbe, L.E. (1936). The method of minimum variation for the combination of criteria. *Psychometrika*, 1, 183–7.

Edgeworth, F.Y. (1888). The statistics of examinations. *Journal of the Royal Statistical Society, Series A (General)*, 51, 599–635.

Edwards, A.L. (1957). *Techniques of Attitude Scale Construction*. New York: Appleton-Century-Crofts.

Efron, B. (1979). Bootstrap methods: another look at the jackknife. *Annals of Statistics*, 7, 1–26.

Elias, P. (1970). Bounds on performance of optimum quantizers. *Proceedings IEEE (Information Theory)*, 16, 172–84.

Erdelyi, A. (1953). *Higher Transcendental Functions*. New York: McGraw-Hill.

Escofier, B. (1983). Analyse de la différence entre deux mesures sur le produit de deux mêmes ensembles. *Cahiers de l'Analyse des Données*, **8**, 325–9.

Escofier, B. (1984). Analyse factorielle en référence à un modèle: Application à l'analyse de tableaux d'échanges. *Revue de Statistique Appliquée*, **32**, 25–36.

Escofier, B., and Le Roux, B. (1972). Etude de trois problèmes de stabilité en analyse factorielle. *Publ. ISUP*, **21**, 2–48.

Everitt, B.S., and Dunn, G. (1983). *Advanced Methods of Data Exploration and Modelling*. London: Heinemann.

Fan, Ky (1951). Maximum properties and inequalities for the eigenvalues of completely continuous operators. *Proceedings National Academy of Sciences (Washington)*, **37**, 760–6.

Ferguson, G.A. (1941). The factorial interpretation of test difficulty. *Psychometrika*, **6**, 323–9.

Fienberg, S.E. (1977). *The Analysis of Cross-classified Categorical Data*. Cambridge, Mass.: MIT.

Fischer, G.H. (1974). *Einführung in die Theorie Psychologischer Tests*. Bern: Huber.

Fisher, R.A. (1925). *Statistical Methods for Research Workers*. London: Oliver and Boyd.

Fisher, R.A. (1940). The precision of discriminant functions. *Annals of Eugenics*, **10**, 422–9.

Forsythe, A., and Hartigan, J.A. (1970). Efficiency of confidence intervals generated by repeated subsample calculations. *Biometrika*, **57**, 629–40.

Fortier, J.J. (1966a). Simultaneous linear prediction. *Psychometrika*, **31**, 369–381.

Fortier, J.J (1966b). Simultaneous nonlinear prediction. *Psychometrika*, **31**, 447–55.

Gabriel, K.R. (1971). The biplot-graphic display of matrices with application to principal component analysis. *Biometrika*, **58**, 453–67.

Gabriel, K.R. (1981). Biplot display of multivariate matrices for inspection of data and diagnosis. In V. Barnett (Ed.), *Interpreting Multivariate Data*, pp. 147-74. Chichester: Wiley.

Galton, F. (1888). Co-relations and their measurement, chiefly from anthropometric. *Proceedings of the Royal Society (London)*, **45**, 135–45.

Gantmacher, F.R., and Krein, M.G. (1937). Sur les matrices oscillatoires et complètement non-négatives. *Composition Math.*, **4**, 445–76.

Gantmacher, F.R., and Krein, M.G. (1960).*Oszillationsmatrizen, Oszillationskerne und Kleine Schwingungen Mechanischer Systeme*. Berlin: Akademie Verlag.

Gebelein, H. (1941). Das statistische Problem der Korrelation als Variations- und Eigenwertproblem und sein Zusammenhang mit der Ausgleichsrechnung. *Zeitschrift für Angewandte Mathematik und Mechanik*, **21**, 364–79.

Gersho, A. (1979). Asymptotically optimal block quantization. *Proceedings IEEE (Information Theory)*, **25**, 373–80.

Gifi, A. (1980). *Niet-lineaire Multivariate Analyse*. Leiden: Department of Data Theory, University of Leiden.

Gifi, A. (1985). *PRINCALS*, Report UG-85-03. Leiden: Department of Data Theory, University of Leiden.

Gilbert, E.S. (1968). On discrimination using qualitative variables. *Journal of the American Statistical Association*, **63**, 1399–412.

Giri, N.C. (1977). *Multivariate Statistical Inference*. New York: Academic Press.

Girshick, M.A. (1936). Principal components. *Journal of the American Statistical Association*, **31**, 519–28.

Gish, H., and Pierce, J.N. (1968). Asymptotically optimal quantizing. *Proceedings IEEE (Information Theory)*, **14**, 676–83.

Gittins, R. (1985). *Canonical Analysis: A Review with Applications in Ecology*. Berlin: Springer Verlag.

Glass, D.V. (Ed.) (1954). *Social Mobility in Britain*. Glencoe: Free Press.

Gnanadesikan, R. (1977). *Methods for Statistical Data Analysis of Multivariate Observations*. New York: Wiley.

Gokhale, D.V., and Kullback, S. (1978). *The Information in Contingency Tables*. New York: Dekker.

Goldberg, S. (1958). *Introduction to Difference Equations: With Illustrative Examples from Economics, Psychology and Sociology*. New York: Wiley.

Goldman, A.J., and Tucker, A.W. (1956). Polyhedral convex cones. In H.W. Kuhn and A.W. Tucker (Eds.), *Linear Inequalities and Related Systems*. Princeton: Princeton University Press.

Good, I.J. (1963). Maximum entropy for hypotheses formulation, especially for multidimensional contingency tables. *Annals of Mathematical Statistics*, **34**, 911–34.

Goodman, L.A. (1965). On the statistical analysis of mobility tables. *American Journal of Sociology*, **70**, 564–85.

Goodman, L.A. (1969). On the measurement of social mobility: an index of status persistence. *Amer. Sociol. Rev.*, **34**, 832–50.

Goodman, L.A. (1978). *Analyzing Qualitative/Categorical Data: Loglinear Models, and Latent Structure Analysis*. Reading: Addison-Wesley.

Goodman, L.A., and Kruskal, W.H. (1954). Measures of association for cross classification I. *Journal of the American Statistical Association*, **49**, 732–64.

Goodman, L.A., and Kruskal, W.H. (1959). Measures of association for cross classification II. *Journal of the American Statistical Association*, **54**, 123–63.

Goodman, L.A., and Kruskal, W.H. (1963). Measures of association for cross classification III. *Journal of the American Statistical Association*, **58**, 310–364.

Goodman, L.A., and Kruskal, W.H. (1972). Measures of association for cross classification IV. *Journal of the American Statistical Association*, **67**, 415–21.

Goodman, L.A., and Kruskal, W.H. (1979). *Measures of Association for Cross Classification*. Berlin: Springer Verlag.

Gordon, L. (1974a). Completely separating groups in subsampling. *Annals of Statistics*, **2**, 572–8.

Gordon, L. (1974b). Efficiency in subsampling. *Annals of Statistics*, **2**, 739–50.

Gower, J.C. (1984). Multivariate analysis: ordination, multidimensional scaling and allied topics. In E. Lloyd (Ed.), *Handbook of Applicable Mathematics*, Vol. IV, *Statistics, B*, pp. 727–81. New York: Wiley.

Green, P.E., with Carroll, J.D. (1976). *Mathematical Tools for Applied Multivariate Analysis*. New York: Wiley.

Green, P.E., with Carroll, J.D. (1978). *Analyzing Multivariate Data*. New York: Dryden Press.

Green, P.E., Halbert, M.H., and Robinson, P.J. (1966). Canonical analysis: an exposition and illustrative application. *Journal of Marketing Research*, **3**, 32–9.

Greenacre, M.J. (1984). *Theory and Applications of Correspondence Analysis*. London: Academic Press.

Griffiths, R.C. (1969). The canonical correlation coefficients of bivariate gamma distributions. *Annals of Mathematical Statistics*, **40**, 1401–8.

Griffiths, R.C. (1970). Positive definite sequences and canonical correlation coefficients. *Australian Journal of Statistics*, **12**, 162–5.

Groenen, P. and Gifi, A., (1989). *ANACOR*, Internal Report UG-89-01. Leiden: Department of Data Theory, University of Leiden.

Guilford, J.P. (1936). *Psychometric Methods*. New York: McGraw-Hill.

Guilford, J.P. (1941). The difficulty of a test and its factor composition. *Psychometrika*, **6**, 67–77.

Guilford, J.P. (1954). *Psychometric Methods*, 2nd edition. New York: McGraw-Hill.

Gulliksen, H. (1950). *Theory of Mental Tests*. New York: Wiley.

Guttman, L. (1941). The quantification of a class of attributes: a theory and method of scale construction. In P. Horst (Ed.), *The Prediction of Personal Adjustment*. New York: SSRC.

Guttman, L. (1944). A basis for scaling qualitative data. *Amer. Sociol. Rev.*, **9**, 139–50.

Guttman, L. (1946). An approach for quantifying paired comparisons and rank order. *Annals of Mathematical Statistics*, **17**, 144–63.

Guttman, L. (1950a). The basis for scalogram analysis. In S. A. Stouffer *et al.* (Eds.), *Measurement and Prediction*. Princeton: Princeton University Press.

Guttman, L. (1950b). The principal components of scale analysis. In S. A. Stouffer *et al.* (Eds.), *Measurement and Prediction*. Princeton: Princeton University Press.

Guttman, L. (1954). The principal components of scalable attitudes. In P.F. Lazarsfeld (Ed.), *Mathematical Thinking in the Social Sciences*. Glencoe: Free Press.

Guttman, L. (1959). Metricizing rank-ordered or unordered data for a linear factor analysis. *Sankhya, A*, **21**, 257–68.

Guttman, L. (1968). A general nonmetric technique for finding the smallest coordinate space for a configuration of points. *Psychometrika*, **33**, 469–506.

Haberman, S.J. (1974). *The Analysis of Frequency Data*. Chicago: University of Chicago Press.

Halmos, P.R. (1948). *Finite Dimensional Vector Spaces*. Princeton: Princeton University Press.

Hampel, F.R. (1973). Robust estimation: a condensed partial survey. *Z. Wahrscheinlichkeitstheorie verw. Geb.*, **27**, 87–104.

Hampel, F.R. (1974). The influence curve and its role in robust estimation. *Journal of the American Statistical Association*, **69**, 383–93.

Hampel, F.R., Ronchetti, E.M., Rousseeuw, P.J., and Stahel, W.A. (1986). *Robust Statistics: The Approach Based on Influence Functions*. New York: Wiley.

Hannan, E.J. (1961). The general theory of canonical correlation and its relation to functional analysis. *Journal of the Australian Mathematical Society*, **2**, 229–42.

Harris, R.J. (1975). *A Primer of Multivariate Statistics*. New York: Academic Press.

Hartigan, J.A. (1969). Using subsample values as typical values. *Journal of the American Statistical Association*, **64**, 1303–17.

Hartigan, J.A. (1971). Error analysis by replaced samples. *Journal of the Royal Statistical Society, Series B (Methodogical)*, **33**, 98–110.

Hartigan, J.A. (1975a). *Clustering Algorithms*. New York: Wiley.

Hartigan, J.A. (1975b). Necessary and sufficient conditions for asymptotic joint normality of a statistic and its subsample values. *Annals of Statistics*, **3**, 573–80.

Hartmann, W. (1979). *Geometrische Modelle zur Analyse Empirischer Daten*. Berlin: Akademie Verlag.

Heiser, W.J. (1981). *Unfolding Analysis of Proximity Data*. Unpublished Ph.D. Thesis, University of Leiden, Leiden.

Heiser, W.J. (1985). Undesired nonlinearities in nonlinear multivariate analysis. In E. Diday *et al.* (Eds.), *Data Analysis and Informatics*, Vol. IV, pp. 455–75. Amsterdam: North-Holland.

Heiser, W.J. (1987a). Joint ordination of species and sites: the unfolding technique. In P. Legendre and L. Legendre (Eds.), *Developments in Numerical Ecology*, pp. 189–221. New York: Springer.

Heiser, W.J. (1987b). Correspondence analysis with least absolute residuals. *Computational Statistics and Data Analysis*, 5, 337–56.

Heiser, W.J. (1989). Order invariant unfolding analysis under smoothness restrictions. In G. De Soete, H. Feger, and K.C. Klauer (Eds.), *New Developments in Psychological Choice Modeling*. Amsterdam: North-Holland.

Heiser, W.J., and De Leeuw, J. (1979a). Metric multidimensional unfolding. *MDN*, 4, 26–50.

Heiser, W.J., and De Leeuw, J. (1979b). *How to use SMACOF-3*. Leiden: Department of Data Theory, University of Leiden.

Heiser, W.J., and Meulman, J. (1983). Constrained multidimensional scaling, including confirmation. *Applied Psychological Measurement*, 7, 381–404.

Hill, M.O. (1974). Correspondence analysis: a neglected multivariate method. *Journal of the Royal Statistical Society, Series C (Applied)*, 23, 340–54.

Hinkley, D.V. (1988). Bootstrap methods (with discussion). *Journal of the Royal Statistical Society, Series B (Methodological)*, 50, 321–37.

Hiriart-Urruty, J.B. (1978). Gradients généralises de functions marginals. *Journal of the Society for Industrial and Applied Mathematics (Control and Optimization)*, 16, 301–16.

Hirschfeld, H.O. (1935). A connection between correlation and contingency. *Proceedings Cambridge Philosophical Society*, 31, 520–4.

Hirshi, T., and Selvin, H.C. (1973). *Principles of Survey Analysis*. Glencoe: Free Press.

Hogan, W.W. (1973a). Review of point-to-set maps in mathematical programming. *Journal of the Society for Industrial and Applied Mathematics*, 15, 591–603.

Hogan, W.W. (1973b). Directional derivation for extremal value functions, with applications to the completely convex case. *Operations Research*, 21, 188–209.

Horst, P. (1935). Measuring complex attitudes. *Journal of Social Psychology*, 6, 369–74.

Horst, P. (1936). Obtaining a composite measure from a number of different measures of the same attribute. *Psychometrika*, 1, 53–60.

Horst, P. (1961a). Relations among *m* sets of variables. *Psychometrika*, 26, 129–49.

Horst, P. (1961b). Generalized canonical correlations and their applications to experimental data. *Journal of Clinical Psychology*, 17, 331–47.

Hotelling, H. (1933). Analysis of a complex of statistical variables into principal components. *Journal of Educational Psychology*, 24, 417–41, 498–520.

Hotelling, H. (1935). The most predictable criterion. *Journal of Educational Psychology*, 26, 139–42.

Hotelling, H. (1936). Relations between two sets of variates. *Biometrika*, 28, 321–77.

Huber, P.J. (1972). Robust statistics: a review. *Annals of Mathematical Statistics*, 43, 1041–67.

Huber, P.J. (1981). *Robust Statistics*. New York: Wiley.

Hurt, J. (1976). Asymptotic expansions of functions of statistics. *Appl. Math.*, 21, 444–56.

Instituut voor Toegepaste Sociologie (1968). *Onderzoeksvoorstel 'Van Jaar tot Jaar'*. Nijmegen: Instituut voor Toegepaste Sociologie.

Israëls, A.Z. (1984). Redundancy analysis for qualitative variables. *Psychometrika*, 49, 331–46.

Israëls, A.Z. (1987a). *Eigenvalue Techniques for Qualitative Data*. Leiden: DSWO Press.

Israëls, A.Z. (1987b). Path analysis for mixed qualitative and quantitative variables. *Quality and Quantity*, 21, 91–102.

Izenman, A.J. (1980). Assessing dimensionality in multivariate regression. In P.R. Krishnaiah (Ed.), *Handbook of Statistics*, Vol. 1, *Analysis of Variance*, pp. 571–91. Amsterdam, North-Holland.

Jansen, P.G.W. (1983). *Rasch Analysis of Attitudinal Data*. Unpublished Ph.D. thesis, University of Nijmegen, Nijmegen, The Netherlands.

Jensen, D.R. (1971). A note on positive dependence and the structure of bivariate distributions. *Journal of the Society for Industrial and Applied Mathematics (Applied Mathematics)*, 20, 749–53.

Johnson, P.O. (1950). The quantification of qualitative data in discriminant analysis. *Journal of the American Statistical Association*, 45, 65–76.

Joliffe, I.T. (1986). *Principal Component Analysis*. New York: Springer.

Jordan, C. (1874). Mémoire sur les formes bilinéaires. *J. Math. Pures. Appl.*, 19, 35–54.

Jöreskog, K.G., and Sörbom, D. (1979). *Advances in Factor Analysis and Structural Equation Models*. Cambridge, Mass.: Abt Books.

Karlin, S. (1964). Oscillation properties of eigenvectors of strictly totally positive matrices. *J. Anal. Math. Jerusalem*, 9, 247–66.

Karlin, S. (1968). *Total Positivity*. Stanford: Stanford University Press.

Kato, T. (1976). *Perturbation Theory for Linear Operators*. Berlin: Springer Verlag.

Kendall, M.G. (1957). *A Course in Multivariate Analysis*. London: Griffin, 1957/1975.

Kendall, M.G. (1962). *Rank Correlation Methods*. London: Griffin.

Kendall, M.G. (1972). The history and future of statistics. In T.A. Bancroft (Ed.), *Statistical Papers in Honour of George Snedecor*. Ames: Iowa State University Press.

Kendall, M.G. (1975). *Multivariate Analysis*. London: Griffin.

Kettenring, J.R. (1971). Canonical analysis of several sets of variables. *Biometrika*, 58, 433–60.

Kiefer, J. (1964). Review of M. G. Kendall and A. Stuart: the advanced theory of statistics, Vol. 2. *Annals of Mathematical Statistics*, 35, 1371–80.

Kiers, H.A.L. (1989). *Three-way Methods for the Analysis of Qualitative and Quantitative Two-way Data*. Leiden: DSWO Press.

Kolata, W.G. (1978). Approximation in variationally posed eigenvalue problems. *Numer. Math.*, 29, 159–71.

Kroonenberg, P.M. (1983). *Three-mode Principal Component Analysis*. Leiden: DSWO Press.

Kropman, J.A., and Collaris, J.W.M. (1974). *Van Jaar tot Jaar: Eerste Fase*. Nijmegen: Instituut voor Toegepaste Sociologie.

Kruskal, J.B. (1964a). Multidimensional scaling by optimizing goodness of fit to a nonmetric hypothesis. *Psychometrika*, 29, 1–27.

Kruskal, J.B. (1964b). Nonmetric multidimensional scaling: a numerical method. *Psychometrika*, 29, 115–29.

Kruskal, J.B. (1965). Analysis of factorial experiments by estimating monotone transformations of the data. *Journal of the Royal Statistical Society, Series B (Methodological)*, 27, 251–63.

Kruskal, J.B. (1977). Multidimensional scaling and other methods for discovering structure. In K. Enslein, A. Ralston, and H.S. Wilf (Eds.), *Statistical Methods for Digital Computers*, Vol. III. New York: Wiley.

Kruskal, J.B., and Carroll, J.D. (1969). Geometric models and badness-of-fit functions. In P.R. Krishnaiah (Ed.), *Multivariate Analysis*, Vol. II. New York: Academic Press.

Kruskal, J.B., and Shepard, R.N. (1974). A nonmetric variety of linear factor analysis. *Psychometrika*, 39, 123–57.

Kshirsagar, A.N. (1978). *Multivariate Analysis*. New York: Marcel Dekker Inc.

Kuhfeld, W.F., Young, F.W., and Kent, D.P. (1987). New developments in psychometric and market research procedures. In *SUGI Proceedings*, Vol. 12, pp. 1101-6. Cary, N.C.: SAS Institute.

Kullback, S. (1959). *Information Theory and Statistics*. New York: Wiley.

Lafaye de Michaux, D. (1978). *Approximation d'Analyses Canoniques Non-linéaires de Variables Aléatoires*. Nice: Université de Nice (dissertation).

Lammers, C.J. (1969). Is de universiteit een politieke leerschool? *Universiteit en Hogeschool*, **15**, 1-43.

Lancaster, H.O. (1957). Some properties of the bivariate normal distribution considered in the form of a contingency table. *Biometrika*, **44**, 289-92.

Lancaster, H.O. (1958). The structure of bivariate distributions. *Annals of Mathematical Statistics*, **29**, 719-36.

Lancaster, H. O. (1959). Zero correlation and independence. *Australian Journal of Statistics*, **1**, 53-6.

Lancaster, H.O. (1960a). On tests of independence in several dimensions. *Journal of the Australian Mathematical Society*, **1**, 241-54.

Lancaster, H.O. (1960b). On statistical independence and zero correlation in several dimensions. *Journal of the Australian Mathematical Society*, **1**, 492-6.

Lancaster, H.O. (1969). *The Chi-squared Distribution*. New York: Wiley.

Lancaster, H.O. (1971). The multiplicative definition of interaction. *Australian Journal of Statistics*, **13**, 36-44.

Lancaster, H.O. (1975a). The multiplicative definition of interaction: an addendum. *Australian Journal of Statistics*, **17**, 34-5.

Lancaster, H.O. (1975b). Joint probability distributions in the Meixner classes. *Journal of the Royal Statistical Society, Series B (Methodological)*, **37**, 434-43.

Lancaster, H.O., and Hamdan, M.A. (1964). Estimation of the correlation coefficient in contingency tables with possibly nonmetrical characteristics. *Psychometrika*, **29**, 383-91.

Law, H.G., Snijder, C.W.Jr., Hattie, J.A., and McDonald, R.P. (1984). *Research Methods for Multimode Data Analysis*. New York: Praeger.

Lawley, D.N. (1944). The factorial analysis of multiple item tests. *Proceedings of the Royal Society (Edinburgh)*, **62**, 74-82.

Lazarsfeld, P.F. (1950). The logical and mathematical foundations of latent structure analysis. In S.S. Stouffer (Ed.), *Measurement and Prediction*. Princeton: Princeton University Press.

Lazarsfeld, P.F., and Henry, N.W. (1968). *Latent Structure Analysis*. New York: Houghton Mifflin.

Lebart, L. (1976). The significance of eigenvalues issued from correspondence analysis of contingency tables. In *Proceedings COMPSTAT 1976*. Wien: Physika Verlag.

Lebart, L., Morineau, A., and Warwick, K.M. (1984). *Multivariate Descriptive Statistical Analysis*. New York: Wiley.

Lee, P.A. (1971). A diagonal expansion for the 2-variate Dirichlet probability density function. *Journal of the Society for Industrial and Applied Mathematics (Applied Mathematics)*, **21**, 155–65.

Levine, M.V. (1970). Transformations that render curves parallel. *Journal of Mathematical Psychology*, **7**, 410–43.

Levine, M.V. (1972). Transforming curves with the same shape. *Journal of Mathematical Psychology*, **9**, 1–16.

Levine, M.V. (1975). Additive measurement with short segments of curves. *Journal of Mathematical Psychology*, **12**, 212–24.

Lingoes, J.C. (1968). The multivariate analysis of qualitative data. *Multivariate Behavioral Research*, **3**, 61–94.

Loevinger, J. (1947). A systematic approach to the construction and evaluation of tests of ability. *Psychol. Monograph*, **61** (4).

Loevinger, J. (1948). The technique of homogeneous tests compared with some aspects of 'scale analysis' and factor analysis. *Psychological Bulletin*, **45**, 507–30.

Lord, F.M. (1958). Some relations between Guttman's principal components of scale analysis and other psychometric theory. *Psychometrika*, **23**, 291–6.

Lorenz, G. (1953). *Bernstein Polynomials*. Toronto: University of Toronto Press.

Lubin, A. (1950). Linear and nonlinear discriminant functions. *British Journal of Statistical Psychology*, **3**, 90–104.

MacDonell, W.R. (1901). On criminal anthropology and the identification of criminals. *Biometrika*, **1**, 177–227.

MacKenzie, D. (1978). Statistical theory and social interests: a case study. *Social Studies of Science*, **8**, 35–83.

Mallet, J.L. (1982). Propositions for fuzzy characteristic functions in data analysis. In H. Caussinus *et al.* (Eds.), *COMPSTAT 82, Proceedings in Computational Statistics*. Vienna: Physika Verlag.

Marchetti, G.M. (1978). *Three-way Analysis of Two-mode Matrices of Qualitative Data*, Research Report. Florence: Department of Statistics, University of Florence, Italy.

Mardia, K.V., Kent, J.T., and Bibby, J.M. (1979). *Multivariate Analysis*. New York: Academic Press.

Masson, M. (1974). Analyse non-linéaire des données. *Comptes Rendues de l'Academie des Sciences (Paris)*, **287**, 803–6.

Maung, K. (1941a). Measurement of association in a contingency table with special reference to the pigmentation of hair and eye colours of Scottish school children. *Annals of Eugenics*, **11**, 189–223.

Maung, K. (1941b). Discriminant analysis of Tocher's eye colour data. *Annals of Eugenics*, **11**, 64–76.

Max, J. (1960). Quantizing for minimum distortion. *Proceedings IEEE (Information Teory)*, **6**, 7–12.

McDonald, R.P. (1967). Nonlinear factor analysis. *Psychometric Monograph*, **15**.

McFadden, J.A. (1966). A diagonal expansion in Gegenbauer polymonials for a class of second-order probability densities. *Journal of the Society for Industrial and Applied Mathematics (Applied Mathematics)*, **14**, 1433–6.

McGraw, D.K., and Wagner, J.T. (1968). Elliptically symmetric distributions. *Proceedings IEEE (Information Theory)*, **14**, 110–20.

McKinley, S.M. (1975). The design and analysis of the observational study. *Journal of the American Statistical Association*, **70**, 503–23.

Mehler, F.G. (1866). Über die Entwicklung einer Funktion von beliebig vielen Variablen nach Laplaceschen Funktionen höherer Ordnung. *J. Reine Angew. Math.*, **66**, 161–76.

Meulman, J. (1982). *Homogeneity Analysis of Incomplete Data*. Leiden: DSWO Press.

Meulman, J.J. (1986). *A Distance Approach to Nonlinear Multivariate Analysis*. Leiden: DSWO Press.

Meulman, J.J. (1988a). OVERALS: nonlinear generalized canonical analysis. In A. Di Ciaccio and G. Bove (Eds.), *Multiway '88. Software Guide*, pp. 59–66. Rome: Dipartimento di Statistica, probabilità e Statistiche Applicate, Università La Sapienza.

Meulman, J.J. (1988b). Nonlinear redundancy analysis via distances. In H.H. Bock (Ed.), *Classification and Related Methods of Data Analysis*, pp. 515–22. Amsterdam: North-Holland.

Meulman, J.J. (1989). Reduced space analysis. In R. Coppi and S. Bolasco (Eds.), *The Analysis of Multiway Data Matrices*, pp. 233–44. Amsterdam: North-Holland.

Meulman, J.J., and Heiser, W.J. (1988). Second order regression and distance analysis. In W. Gaul and M. Schader (Eds.), *Data, Expert Knowledge and Decisions*, pp. 368–80. Berlin: Springer Verlag.

Miller, R.G. (1974). The jackknife: a review. *Biometrika*, **61**, 1–15.

Mokken, R.J. (1970). *A Theory and Procedure of Scale Analysis*. The Hague: Mouton.

Mokken, R.J. (1971). *A Theory and Procedure of Scale Analysis with Applications in Political Research*. The Hague: Mouton.

Molenaar, I.W. (1983). Some improved diagnostics for failure of the Rasch model. *Psychometrika*, **48**, 49–72.

Morrison, D.F. (1967). *Multivariate Statistical Methods*. New York: McGraw Hill, 1967/1976.

Mosteller, F. (1949). *A Theory of Scalogram Analysis using Noncumulative Types of Items: A New Approach to Thurstone's Method of Scaling Attitudes*, Report 9. Harvard: Lab. of Social Relations, Harvard University.

Muirhead, R.J. (1982). *Aspects of Multivariate Statistical Theory*. New York: Wiley.

Muthèn, B. (1978). Contributions to factor analysis of dichotomous variables. *Psychometrika*, **43**, 551–60.

Naouri, J.C. (1970). Analyse factorielle des correspondences continues. *Institute de Statistique de l'Université de Paris*, **19**, 1–100.

Nevels, K. (1974). *Generalized Canonical Variates and Correlations*, Report HB-158-EX. Groningen: Psychological Department, University of Groningen.

Niemöller, B. (1976). *Schaalanalyse volgens Mokken*. Amsterdam: Technisch Centrum, University of Amsterdam.

Nishisato, S. (1978). Optimal scaling of paired comparison and rank order data: an alternative to Guttman's formulation. *Psychometrika*, **43**, 263–71.

Nishisato, S. (1980). *Analysis of Categorical Data: Dual Scaling and Its Applications*. Toronto: University of Toronto Press.

Nishisato, S. (1984). Forced classification: a simple application of a quantification method. *Psychometrika*, **49**, 25–36.

Nishisato, S. (1988). Forced classification procedure of dual scaling: its mathematical properties. In H.H. Bock (Ed.), *Classification and Related Methods of Data Analysis*, pp. 523–32. Amsterdam: North-Holland.

Norton, B.J. (1976). Biology and philosophy: the methodological foundations of biometry. *Journal of the History of Biology*, **8**, 85–93.

Norton, B.J. (1978). Karl Pearson and statistics: the social origin of scientific innovation. *Social Studies of Science*, **8**, 3–34

Okamoto, M. (1968). Optimality of principal components. In P.R. Krishnaiah (Ed.), *Multivariate Analysis*, Vol. II. New York: Academic Press.

Olsson, U. (1979). Maximum likelihood estimation of the polychoric correlation coefficient. *Psychometrika*, **44**, 443–60.

O'Neill, M.E. (1978a). Asymptotic distribution of the canonical correlation coefficient from contingency tables. *Australian Journal of Statistics*, **20**, 75–82.

O'Neill, M.E. (1978b). Distributional expansion for canonical correlations from contingency tables. *Journal of the Royal Statistical Society, Series B (Methodological)*, **40**, 303–12.

O'Neill, M.E. (1980). The distribution of higher-order interactions in contingency tables. *Journal of the Royal Statistical Society, Series B (Methodological)*, **42**, 357–65.

Parzen, E. (1979). Nonparametric statistical data modelling. *Journal of the American Statistical Association*, **74**, 105–31.

Pearson, K. (1892). *The Grammar of Science*. London: Scott, 1892/1900/1910.

Pearson, K. (1901). On lines and planes of closest fit to points in space. *Phil. Magazine*, **2**, 559–72.

Pearson, K. (1904). On the theory of contingency and its relation to association and normal correlation. *Drapers Company Research Memoirs (Biometric Series)*, no. 1.

Pearson, K., and Heron, D. (1913). On theories of association. *Biometrika*, **9**, 159–315.

Peay, E.R. (1988). Multidimensional rotation and scaling of configurations to optimal agreement. *Psychometrika*, **53**, 199–208.

Popper, K.R. (1963). *Conjectures and Refutations*. London: Routledge and Kegan Paul.

Press, S.J. (1982). *Applied Multivariate Analysis: Using Bayesian and Frequentist Methods of Inference*. Malabar, Florida: Krieger Publications.

Puri, M.L., and Sen, P.K. (1971). *Nonparametric Methods in Multivariate Analysis*. New York: Wiley.

Ramsay, J.O. (1978). Confidence regions for multidimensional scaling analysis. *Psychometrika*, **43**, 145–60.

Ramsay, J.O. (1988). Monotone regression splines in action. *Statistical Science* (in press).

Ramsay, J.O. (1989). The analysis of replicated spatial functions of time. In R. Coppi and S. Bolasco (Eds.), *Analysis of Multiway Data Matrices*, pp. 363–73. Amsterdam: North-Holland.

Rao, B.R. (1969). Partial canonical correlations. *Trabajos de Estadistica*, **20**, 211–19.

Rao, C.R. (1960). Multivariate analysis: an indispensable statistical aid in applied research. *Sankhya*, **22**, 317–38.

Rao, C.R. (1964). The use and interpretation of principal component analysis in applied research. *Sankhya*, **26**, 329–58.

Rao, C.R. (1965). *Linear Statistical Inference and Its Applications*. New York: Wiley.

Rao, C.R. (1980). Matrix approximation and reduction of dimensionality in multivariate statistical analysis. In P.R. Krishnaiah (Ed.), *Multivariate Analysis*, Vol. V. Amsterdam: North Holland.

Rao, C.R., and Slater, P. (1949). Multivariate analysis applied to differences between neurotic groups. *British Journal of Statistical Psychology*, **2**, 17–29.

Rasch, G. (1960). *Probabilistic Models for Some Intelligence and Attainment Tests*. Copenhagen: Danish Institute for Educational Research.

Rasch, G. (1961). On general laws and meaning of measurement in psychology. In *Proceedings IV Berkeley Symposium on Mathematical Statistics and Probability*. Berkeley: University of California Press.

Rasch, G. (1966). An item analysis which takes individual differences into account. *British Journal of Statistical Psychology*, **19**, 49–57.

Reissman, L. (1953). Levels of aspiration and social class. *Amer. Sociol. Rev.*, **18**, 233–42.

Rényi, A. (1959). On measures of dependence. *Acta Math. Acad. Ac. Sc. Hungar.*, **10**, 441–51.

Richardson, M. with Kuder, G.F. (1933). Making a rating scale that measures. *Personnel Journal*, **12**, 36–40.

Robert, F. (1967). Calcul du rapport maximal de deux normes sur $\mathbb{R}^n$. *RIRO*, **1**, 97–118.

Rockafellar, R.T. (1970). *Convex Analysis*. Princeton: Princeton University Press.

Roe, G.M. (1964). Quantizing for minimum distortion. *Proceedings IEEE (Information Theory)*, **10**, 384–5.

Roskam, E.E.Ch.I. (1968). *Metric Analysis of Ordinal Data in Psychology*. Voorschoten: Drukkerij/Uitgeverij VAM.

Roskam, E.E. (1977). A survey of the Michigan–Israel–Netherlands-integrated series. In J. C. Lingoes (Ed.), *Geometric Representations of Relational Data*. Ann Arbor: Mathesis Press.

Ross, J., and Cliff, N. (1964). A generalization of the interpoint distance model. *Psychometrika*, **29**, 167–76.

Rousseeuw, P.J., and Leroy, A.M. (1987). *Robust Regression and Outlier Detection*. New York: Wiley.

Roux, B., and Rouanet, H. (1979). L'analyse statistique des protocolles multidimensionelles: Analyse en composantes principales. *Publ. Institute de Statistique de l'Université de Paris*, **24**, 47–74.

Rowney, D.K., and Graham, J.Q. (Eds.) (1969). *Quantitative History*. Homewood.

Roy, S.N. (1957). *Some Aspects of Multivariate Analysis*. New York: Wiley.

Rozeboom, W.W. (1979). Sensitivity of a linear composite of predictor items to differential item weighting. *Psychometrika*, **44**, 289–96.

Russett, B.M. (1964). Inequality and instability. *World Politics*, **21**, 442–54.

Saporta, G. (1975). *Liaisons entre Plusieurs Ensembles de Variables et Codage de Données Qualitatives*. Paris: Université Paris VI (thèse).

Saporta, G. (1985). Data analysis for numerical and categorical individual time series. *Applied Stochastic Models and Data Analysis*, **1**, 109–19.

Saporta, G., and Hatabian, G. (1986). Régions de confiance en analyse factorielle. In E. Diday et al. (Eds.), *Data Analysis and Informatics*, Vol. IV, pp. 499–508. Amsterdam: North-Holland.

Sarmanov, O.V. (1963). Investigation of stationary Markov processes by the method of eigenfunction expansions. *Selected Translations in Math. Statist. and Probability*, **4**, 245–69.

Sarmanov, O.V., and Bratoeva, Z.N. (1967). Probabilistic properties of bilinear expansions of Hermite polynomials. *Theory Probability and Applications*, **12**, 470–81.

Sarmanov, O.V., and Zacharov, V.K. (1960). Maximum coefficients of multiple correlation. *Doklady Akademie Nauk*, **130**, 269–71.

Scheffé, H. (1952). An analysis of variance for paired comparisons. *Journal of the American Statistical Association*, **47**, 381–400.

Schmidt, E. (1906). Zur Theorie der linearen und nichtlinearen Integralgleichungen. *Mathematische Annalen*, **63**, 433–76.

Schriever, B.F. (1985). *Order Dependence*. Centrum voor Wiskunde en Informatica, Amsterdam (unpublished Ph.D. thesis).

Seber, G.A.F. (1984). *Multivariate Observations*. New York: Wiley.

Segijn, R. (1985). *Projectverslag Locale Minima in Princals*, Internal Report. Leiden: Department of Data Theory, University of Leiden.

Sharma, D.K. (1978). Design of absolutely optimal quantizers for a wide class of distortion measures. *Proceedings IEEE (Information Theory)*, **24**, 693–702.

Shepard, R.N. (1962). The analysis of proximities: multidimensional scaling with an unknown distance function. *Psychometrika*, **27**, 125–40 and 219–45.

Shepard, R.N. (1966). Metric structures in ordinal data. *Journal of Mathematical Psychology*, **3**, 287–315.

Sheppard, W.F. (1898). On the calculation of the most probable values of frequency constants for data arranged according to equidistant divisions of a scale. *Proceedings London Mathematical Society*, **29**, 353–80.

Shye, S. (1985). *Multiple Scaling*. Amsterdam: North-Holland.

Sibson, R. (1972). Order invariant methods of data analysis. *Journal of the Royal Statistical Society, Series B (Methodological)*, **34**, 311–49.

Slater, P. (1960). The analysis of personal preferences. *British Journal of Statistical Psychology*, **13**, 119–35.

Slooff, N., and Van der Kloot, W.A. (1985). *The Perceived Structure of 281 Personality Trait Adjectives*, Internal Report PRM-85-05. Leiden: Department of Psychometrics, University of Leiden.

Spearman, C. (1913). Correlation of sums and differences. *BJP*, **5**, 417–23.

Spearman, C. (1927). *The Abilities of Man*. New York: Macmillan.

Springarn, J.E. (1980). Fixed and variable constraints in sensitivity analysis. *Journal of the Society for Industrial and Applied Mathematics (Control and Optimizai ᴐᴉ),* **18**, 297–310.

Steel, R.G.D. (1951). Minimum generalized variance for a set of linear functions. *Annals of Mathematical Statistics,* **22**, 456–60.

Stevenson, C.L. (1938). Persuasive definitions. *Mind,* **47**, 331–53.

Stewart, D., and Love, W. (1968). A general canonical correlation index. *Psychological Bulletin,* **70**, 160–3.

Stewart, G.W. (1973a). Review of error and perturbation bounds for subspace associated with certain eigenvalue problems. *Journal of the Society for Industrial and Applied Mathematics,* **15**, 727–64.

Stewart, G.W. (1973b). *Introduction to Matrix Computation.* New York: Academic Press.

Stewart, G.W. (1975). Gershgorin theory for the generalized eigenvalue problem $Ax = \lambda Bx$. *Mathematical Computation,* **29**, 600–6.

Stewart, G.W. (1978). Perturbation theory for the generalized eigenvalue problem. In C. De Boor and G.H. Golub (Eds.), *Recent Advantages in Numerical Analysis.* New York: Academic Press.

Stewart, G.W. (1979). Perturbation bounds for the definite generalized eigenvalue problem. *Linear Algebra and Its Applications,* **23**, 69–85.

Stigler, S.M. (1977). Do robust estimators work with real data? *Annals of Statistics,* **5**, 1055–98.

Stoop, I. (1980). *Sekundaire Analyse van de 'Van Jaar tot Jaar' Data met Behulp van Niet-lineaire Multivariate Technieken: Verschillen in de Schoolloopbaan van Meisjes en Jongens,* Report RB 001-80. Leiden: Department of Data Theory, University of Leiden.

Styan, G.P. (1973). Hadamard products and multivariate statistical analysis. *Linear Algebra and Its Applications,* **6**, 217–40.

Sugiyama, M. (1975). *Religious Behaviour of the Japanese: Execution of a Partial Order Scalogram Analysis Based on Quantification Theory.* La Jolla: US–Japan Seminar of Multidimensional Scaling and Related Techniques.

Sylvester, J.J. (1889). Sur la réduction biorthogonal d'une forme linéo-linéaire à sa forme canonique. *Comptes Rendues de l'Academie des Sciences (Paris),* **108**, 651–3.

Symm, H.J., and Wilkinson, J.H. (1980). Realistic error bounds for a simple eigenvalue and its associated eigenvector. *Numerische Math.,* **35**, 113–26.

Takane, Y. (1987). Analysis of contingency tables by ideal point discriminant analysis. *Psychometrika,* **52**, 493–513.

Takane, Y., Young, F.W., and De Leeuw, J. (1979). Nonmetric common factor analysis. An alternating least squares method with optimal scaling features. *Behaviormetrika,* **6**, 45–56.

Takane, Y., Young, F.W., and De Leeuw, J. (1980). An individual differences additive model. An alternating least squares method with optimal scaling features. *Psychometrika*, **45**, 183–209.

Tatsuoka, M.M. (1971). *Multivariate Analysis: Techniques for Educational and Psychological Research*. New York: Wiley.

Ten Berge, J.M.F (1977). *Optimizing Factorial Invariance*. Unpublished doctoral dissertation, University of Groningen, Groningen, The Netherlands.

Ten Berge, J.M.F. (1985). On the relationship between Fortier's simultaneous linear prediction and Van den Wollenberg's redundancy analysis. *Psychometrika*, **50**, 121–2.

Ten Berge, J.M.F. (1988). Generalized approaches to the MAXBET problem and the MAXDIFF problem, with applications to canonical correlations. *Psychometrika*, **53**, 487–94.

Ten Berge, J.M.F., and Knol, D.L. (1984). Orthogonal rotations to maximal agreement for two or more matrices of different column orders. *Psychometrika*, **49**, 49–55.

Tenenhaus, M. (1977). Analyse en composantes principales d'un ensemble de variables nominales et numériques. *Revue de Statist. Appl.*, **25**, 39–56.

Tenenhaus, M. (1988). Canonical analysis of two convex polyhedral cones and applications. *Psychometrika*, **53**, 503–24.

Tenenhaus, M., and Young, F.W. (1985). An analysis and synthesis of multiple correspondence analysis, optimal scaling, dual scaling, homogeneity analysis, and other methods for quantifying categorical multivariate data. *Psychometrika*, **50**, 91–119.

Ter Braak, C.J.F. (1985). Correspondence analysis of incidence and abundance data: properties in terms of a unimodel response model. *Biometrics*, **41**, 859–73.

Ter Braak, C.J.F. (1986). Canonical correspondence analysis: a new eigenvector technique for multivariate direct gradient analysis. *Ecology*, **67**, 1167–79.

Ter Braak, C.J.F. (1987). *Unimodel Models to Relate Species to Environment*, Report. Agricultural Group, Wageningen, The Netherlands.

Ter Braak, C.J.F., and Barendregt, L.G. (1986). Weighted averaging of species indication values: its efficiency in environmental calibration. *Mathematical Biosciences*, **78**, 57–72.

Thompson, R.C. (1975). Singular value inequalities for matrix sums and minors. *Linear Algebra and Its Applications*, **11**, 251–69.

Thompson, R.C. (1976). The behaviour of eigenvalues and singular values under perturbations of restricted rank. *Linear Algebra and Its Applications*, **13**, 69–78.

Thompson, R.C., and Freede, L.J. (1970). On the eigenvalues of sums of Hermitean matrices. *Aequationes Math.*, **5**, 103–15.

Thompson, R.C., and Freede, L.J. (1971). On the eigenvalues of sums of Hermitean matrices II. *Linear Algebra and Its Applications*, 4, 369–76.

Thompson, R.C., and Therianos, S. (1972a). Inequalities connecting the eigenvalue of Hermitean matrices with the eigenvalues of complementary principal submatrices. *Bull. Aust. Math. Soc.*, 6, 117–32.

Thompson, R.C., and Therianos, S. (1972b). The eigenvalues of complementary principal submatrices of a positive definite matrix. *Canadian J. Math.*, 24, 658–67.

Thomson, G.H. (1934). Hotelling's method modified to give Spearman's g. *J. Educational Psychol.*, 25, 366–74.

Thomson, G.H. (1947). The maximum correlation of two weighted batteries. *British Journal of Statistical Psychology*, 1, 27–34.

Thorndike, R.M. (1977). Canonical analysis and predictor selection. *Multivariate Behavioral Research*, 12, 75–87.

Thorndike, R.M. (1978). *Correlational Procedures for Research*. New York: Gardner.

Thorndike, R.M., and Weiss, D.J. (1973). A study of the stability of canonical correlations and canonical components. *Educational and Psychological Measurement*, 33, 123–34.

Thurstone, L.L. (1947). *Multiple Factor Analysis*. Chicago: University of Chicago Press.

Tijssen, R., and De Leeuw, J. (1989). Multi-set nonlinear canonical correlation analysis via the Burt-matrix. In R. Coppi and S. Bolasco (Eds.), *Multiway Data Analysis*, pp. 257–67. Amsterdam: North-Holland.

Timm, N.H., and Carlson, J.E. (1976). Part and bipartial canonical correlation analysis. *Psychometrika*, 41, 159–76.

Tintner, G. (1946). Some applications of multivariate analysis to economic data. *Journal of the American Statistical Association*, 41, 472–500.

Topsøe, F. (1967). Preservation of weak convergence under mappings. *Annals of Mathematical Statistics*, 38, 1661–5.

Torgerson, W.S. (1958). *Theory and Methods of Scaling*. New York: Wiley.

Tricomi, F.G. (1955). *Vorlesungen über Orthogonalreihen*. Berlin: Springer Verlag.

Tucker, L.R. (1960). Intra-individual and inter-individual multidimensionality. In H. Guliksen and S. Messick (Eds.), *Psychological Scaling: Theory and Applications*, pp. 155–67. New York: Wiley.

Tukey, J.W. (1962). The future of data analysis. *Annals of Mathematical Statistics*, 33, 1–67.

Tukey, J.W. (1977). *Exploratory Data Analysis*. Reading: Addison-Wesley.

Tukey, J. W. (1980). We need both exploratory and confirmatory. *American Statistician*, 34, 23–5.

Tyan, S., and Thomas, J.B. (1975). Characterization of a class of bivariate distribution functions. *Journal of Multivariate Analysis,* **5**, 227–35.

Vainikko, G. (1976). *Funktionalanalysis der Diskretisierungsmethoden.* Leipzig: Teubner.

Van Blokland-Vogelesang, A.W. (1987). Unidimensional unfolding complemented by Feigin and Cohen's error model. In W.E. Saris and I.N. Gallhofer (Eds.), *Sociometric Research,* Vol. 2, *Data Analysis,* pp. 123–36. London: Macmillan.

Van Buuren, S., and Heiser, W.J. (1989). Clustering n objects into k groups under optimal scaling of variables. *Psychometrika,* **54**, 699–706.

Van de Geer, J.P. (1967). *Inleiding in de Multivariate Analyse.* Arnhem: Van Loghem Slaterus.

Van de Geer, J.P. (1968). *Matching K Sets of Configurations,* Report RN 005-68. Leiden: Department of Data Theory, University of Leiden.

Van de Geer, J.P. (1971). *Introduction to Multivariate Analysis for the Social Sciences.* San Francisco: Freeman.

Van de Geer, J.P. (1980). *Introduction to Multivariate Linear Analysis.* Part V: *Relations between K Sets of Data.* Leiden: Department of Data Theory, University of Leiden.

Van de Geer, J.P. (1984). Linear relations among *k* sets of variables. *Psychometrika,* **49**, 79–94.

Van de Geer, J.P. (1985). *HOMALS,* Report UG-85-02. Department of Data Theory, University of Leiden.

Van de Geer, J.P. (1986a). Relations among *k* sets of variables, with geometrical representation, and applications to categorical variables. In J. De Leeuw *et al.* (Eds.), *Multidimensional Data Analysis,* pp. 67–79. Leiden: DSWO Press.

Van de Geer, J.P. (1986b). *Introduction to Linear Multivariate Data Analysis* (2 vols.). Leiden: DSWO Press.

Van de Geer, J.P. (1987a). *Analysis of Linear Relations among Categorical Variables,* Research Report RR-87-10. Leiden: Department of Data Theory, University of Leiden.

Van de Geer, J.P. (1987b). *Algebra and Geometry of OVERALS,* Research Report RR-87-13. Leiden: Department of Data Theory, University of Leiden.

Van de Geer, J.P. (1988). *Lower Bounds for Canonical Correlations or Reduncdancies Derived from an OVERALS Solution,* Research Report RR-88-05. Leiden: Department of Data Theory, University of Leiden.

Van den Wollenberg, A.L. (1977). Redundancy analysis: an alternative for canonical correlation. *Psychometrika,* **42**, 207–19.

Van den Wollenberg, A.L. (1982). Two new test statistics for the Rasch model. *Psychometrika,* **47**, 123–40.

Van der Burg, E. (1985). *CANALS*, Internal Report UG-85-05. Leiden: Department of Data Theory, University of Leiden.

Van der Burg, E. (1988). *Nonlinear Canonical Correlation and Some Related Techniques*. Leiden: DSWO Press.

Van der Burg, E., and De Leeuw, J. (1983). Non-linear canonical correlation. *British Journal of Mathematical and Statistical Psychology*, 36, 54–80.

Van der Burg, E., De Leeuw, J., and Verdegaal, R. (1988). Homogeneity analysis with k sets of variables: an alternating least squares method with optimal scaling features. *Psychometrika*, 53, 177–97.

Van der Heijden, P.G.M., and De Leeuw, J. (1985). Correspondence analysis used complementary to loglinear analysis. *Psychometrika*, 50, 429–47.

Van der Heijden, P.G.M., De Falguerolles, A., and De Leeuw, J. (1989). A combined approach to contingency table analysis with correspondence analysis and log-linear analysis (with discussion). *Applied Statistics*, 38, 249–92.

Van der Lans, I.A., and Heiser, W.J. (1988). *Nonlinear Multiple Regression Analysis with Common Scale Transformation across Predictor Variables*, Research Report RR-88-10. Leiden: Department of Data Theory, University of Leiden.

Van Rijckevorsel, J.L.A. (1987). *The Application of Fuzzy Coding and Horseshoes in Multiple Correspondence Analysis*. Leiden: DSWO Press.

Van Rijckevorsel, J.L.A, Bettonvil, B., and De Leeuw, J. (1980). *Recovery and Stability in Nonlinear Principal Component Analysis*. Groningen: European Meeting of the Psychometric Society, 1980.

Van Rijckevorsel, J.L.A., and De Leeuw, J. (Eds.) (1988). *Component and Correspondence Analysis: Dimension Reduction by Functional Approximation*. New York: Wiley.

Veenhoven, R., and Hentenaar, F. (1975). *Nederlanders over Abortus*. Den Haag: Vereniging Stimezo Nederland.

Venter, J.H. (1966). Probability measures on product spaces. *South African Statistical Journal*, 1, 3–20.

Verdegaal, R. (1985). *Meer-sets Analyse voor Kwalitatieve Gegevens*, Research Report RR-85-14. Leiden: Department of Data Theory, University of Leiden.

Verdegaal, R. (1986). *OVERALS*, Internal Report UG-86-01. Leiden: Department of Data Theory, University of Leiden.

Wainer, H. (1976). Estimating coefficients in linear models: it don't make no nevermind. *Psychological Bulletin*, 83, 213–17.

Waternaux, C.M. (1976). Asymptotic distribution of the sample roots for a nonnormal population. *Biometrika*, 63, 639–45.

Weeks, D.G., and Bentler, P.M. (1979). A comparison of linear and monotone multidimensional scaling models. *Psychological Bulletin*, **86**, 349–54.

Weinberger, H.F. (1960). Error bounds in the Rayleigh–Ritz approximation of eigenvalues. *J. Research National Bureau of Standards*, **64B**, 217–25.

Weinberger, H.F. (1974). *Variational Methods for Eigenvalue Approximation*. Philadelphia: Society for Industrial and Applied Mathematics.

Weiss, D.J. (1972). Canonical correlation analysis in counseling psychology research. *J. Counseling Psychol.*, **19**, 241–52.

Werner, B. (1974). Optimale Schranken für Eigenelemente selbstadjungierter Operatoren in der Hilbertraumnorm. In *ISNM* **24**. Basel: Birkhauser.

Wilkinson, J.H. (1961). Rigorous error bounds for computed eigensystems. *Computer J.*, **4**, 230–41.

Wilks, S.S. (1938). Weighting systems for linear functions of correlated variables when there is no independent variable. *Psychometrika*, **3**, 23–40.

Williams, E.J. (1952). Use of scores for the analysis of association in contingency tables. *Biometrika*, **39**, 274–89.

Wilson, E.B. (1926). Empiricism and rationalism. *Science*, **64**, 47-57.

Wilson, E.B., and Worcester, J. (1939). Note on factor analysis. *Psychometrika*, **4**, 133–48.

Winsberg, S. (1988). Two techniques: monotone spline transformations for dimension reduction in PCA and easy-to-generate metrics for PCA of sampled functions. In J.L.A. Van Rijckevorsel and J. De Leeuw (Eds.), *Components and Correspondence Analysis*, pp. 115–35. New York: Wiley.

Winsberg, S., and Ramsay, J.O. (1980). Monotonic transformations to additivity using splines. *Biometrika*, **67**, 669–74.

Winsberg, S., and Ramsay, J.O. (1981). Analysis of pairwise preference data using integrated B-splines. *Psychometrika*, **46**, 171–86.

Winsberg, S., and Ramsay, J.O. (1982).Monotone splines: a family of tansformations useful for data analysis. In H. Caussinus *et al.* (Eds.), *COMPSTAT 82, Proceedings in Computational Statistics*, pp. 451–6. Vienna: Physika Verlag.

Winsberg, S., and Ramsay, J.O. (1983). Monotone spline transformations for dimension reduction. *Psychometrika*, **48**, 575–95.

Wolfowitz, J. (1969). Reflections on the future of mathematical statistics. In R.C. Bose *et al.* (Eds.), *Essays in Probability and Statistics*. Chapel Hill: University of North Carolina Press.

Wong, E., and Thomas, J.B. (1962). On polynomial expansions of second order distributions. *Journal of the Society for Industrial and Applied Mathematics (Applied Mathematics)*, **10**, 507–16.

Wood, D.A., and Erskine, J.A. (1976). Strategies in canonical correlation with applications to behavioural data. *Educational and Psychological Measurement*, **36**, 861–78.

Wood, R.C. (1969). On optimum quantization. *Proceedings IEEE (Information Theory)*, **15**, 248–52.

Woodward, J. A., and Overall, J.E. (1976). Factor analysis of rank-ordered data: An old approach revisited. *Psychological Bulletin*, **83**, 864–7.

Yamamoto, T. (1980). Error bounds for computed eigenvalues and eigenvectors. *Numerische Math.*, **34**, 189–99.

Yanai, H. (1986). Some generalizations of correspondence analysis in terms of projection operators. In E. Diday *et al.* (Eds.), *Data Analysis and Informatics*, Vol. IV, pp. 193–207. Amsterdam: North-Holland.

Yates, F. (1933). The analysis of replicated experiments when the field results are incomplete. *The Empire Journal of Experimental Agriculture*, **1**, 129–42.

Yates, F. (1948). The analysis of contingency tables with grouping based on quantitative characters. *Biometrika*, **35**, 176–81.

Young, F.W., De Leeuw, J., and Takane, Y. (1976). Regression with qualitative and quantitative variables: an alternating least squares method with optimal scaling features. *Psychometrika*, **41**, 505–29.

Young, F.W., Takane, Y., and De Leeuw, J. (1978). The principal components of mixed measurement level multivariate data: an alternating least squares method with optimal scaling features. *Psychometrika*, **43**, 279–81.

Zangwill, W.I. (1969). *Nonlinear Programming: A Unified Approach*. Englewood Cliffs, N.J.: Prentice Hall.

AUTHOR INDEX

A

Abelson, R.P. 244
Amiard, J.-C. 210
Anderson, T.W. 4, 5, 8, 10, 13–15, 16,
    39, 40, 62, 302, 412
Andrich, D. 328
Anscombe, F.J. 24
Appell, P. 276, 385
Arabie, Ph. 261

B

Barcikowski, R.S. 219
Barendregt, L.G. 328
Bargmann, R.E. 196, 197
Barnard, G.A. 23, 28, 33
Barnett, V. 63
Barrett, J.F. 276
Bartholomew, D.J. 328
Bartlett, M.S. 46, 218, 219
Bechtel, G.G. 177, 345
Beckenbach, E.F. 389
Bekker, P. 149
Bell, E.H. 179, 180, 181, 182
Bellman, R. 389
Beltrami, E. 152
Bennett, J.F. 154
Bentler, P.M. 176
Benzécri, J.P. 11, 21, 25–27, 31, 35,
    44, 49, 57, 177, 258, 266, 267, 272,
    333, 334, 376
Berman, A. 325, 526
Besse, P. 390
Bettonvil, B. 405
Bhattacharya, R.N. 412
Bibby, J.M. 62
Bickel, P.J. 27, 28
Billingsley, P. 412
Bineau, J. 370
Birkhoff, G. 352
Birnbaum, A. 84
Bishop, Y.M.M. 20, 44, 278
Black, M. 32

Blalock, H.M.Jr., 45
Bock, R.D. 272, 275, 328
Bolasco, S. 215
Boudon, R. 45
Bourbaki, N. 11
Bourouche, J.M. 209, 214
Box, G.E.P. 23, 24, 28–31, 33, 242
Boyd, D.W. 390
Bratoeva, Z.N. 276
Breiman, L. 254
Bunch, J.R. 317
Burt, C. 83, 84, 102, 108, 218, 318

C

Cahours, Ch. 370
Cailliez, F. 11, 12, 15, 16, 20, 31, 49,
    209, 210
Cambanis, S. 276
Campbell, N.A. 254
Carlson, J.E. 247
Carnap, R. 32
Carroll, J.D. 10, 11, 15, 49, 62, 81,
    177, 198, 199, 203, 257, 261, 323,
    333, 334, 529
Cartan, H. 11
Chang, J.C. 196, 197
Chang, J.J. 333, 334
Chang, W.C. 177
Chatelin, F. 406
Chesson, P.L. 276
Choquet, G. 11
Christofferson, A. 328
Cliff, N. 171, 198, 334
Cochran, W.G. 43, 45
Collaris, J.W.M. 447
Coolen, H. 254, 390
Cooley, W.W. 5, 6, 10, 14, 15
Coombs, C.H. 6, 35, 153–155, 157,
    187, 255, 256, 303, 319, 332
Cooper, R.D. 387
Coppi, R. 215
Cox, D.R. 242, 405

561

SUBJECT INDEX